ORNAMENTAL HORTICULTURE:
SCIENCE, OPERATIONS & MANAGEMENT

2nd edition

JACK E. INGELS

State University of New York
College of Agriculture and Technology

Cobleskill, New York

I T P™

Delmar Publishers Inc.

NOTICE TO THE READER

Cover photo courtesy of Jack Ingels
Cover Design: Judy Orozco and Wendy Troeger

For information, address Delmar Publishers Inc.
3 Columbia Circle, Box 15-015
Albany, New York 12212

Delmar Staff
Publisher: Tim O'Leary
Associate Editor: Cathy L. Carter
Production Editor: Wendy Troeger

Printed in the United States of America
published simultaneously in Canada
by Nelson Canada,
a division of The Thomson Corporation

2 3 4 5 6 7 8 9 10 XXX 00 99 98 97 96 95 94

Library of Congress Cataloging-in-Publication Data

Ingels, Jack E.
Ornamental horticulture : science, operations & management / Jack E. Ingels.
p.   cm.
Includes bibliographical references (p. 525) and index.
ISBN 0-8273-6364-8 (text)
1. Ornamental horticulture. I. Title.
SB404.9.I54 1994
635.9–dc20
93-28344
CIP

Contents

Key Tables and Figures

The following list may be used as a quick reference guide to key tables and figures in the text.

Preface

Ornamental horticulture is a multifaceted industry which offers challenging employment opportunities. Those who have a basic understanding of plant science, its applications to practical growing situations, the crafts of horticulture, and business practices as applied in horticulture will find themselves better prepared to accept the challenges of the industry.

ORNAMENTAL HORTICULTURE: SCIENCE, OPERATIONS & MANAGEMENT was written because the author saw an educational need not being served by texts previously available to students. The texts either presented the science of horticulture at too high a level and in too much depth, or provided merely an overview of the industry techniques without giving the necessary scientific basis of those techniques. As a result, these texts fail to provide the necessary blend of theory and practical application that makes the subject more meaningful to the beginning student. ORNAMENTAL HORTICULTURE: SCIENCE, OPERATIONS & MANAGEMENT fulfills this need for the beginning student who has no prior academic training or work experience in the industry. It is the author's intent to write a truly comprehensive text that would inspire the beginning ornamental horticulturist with an initial enthusiasm for the subject and sustain it throughout the program.

SCOPE OF THE TEXT

Unlike other texts that emphasize one aspect of the subject more than others, this text offers a balanced study of ornamental horticulture as an applied science, a craft, a profession, and a business. Section I consists of six chapters devoted to the science of ornamental horticulture. Plant structure and the mechanisms by which plants survive (photosynthesis, respiration, and transpiration) are described. This information is essential to an understanding of how the manipulation of the environment will affect plant growth and response. The role of soil (both natural and that made by horticulturists) in plant growth and nutrition is thoroughly explained. The basic binomial plant classification system is described. Methods of controlling plant growth through the use of natural and synthetic growth regulators are also explained. The very important area of plant reproduction, both sexual and asexual, is described in detail. The discussion includes the most common methods of propagating plants and their application in the professions of ornamental horticulture. The plant science section concludes with a discussion of common plant pests/diseases, including common symptoms of injured plants, principles of pest control, types of pesticides and their safe use, and the concept of biological controls. A color section is also provided in the text and gives full color examples of common plant pests/diseases.

Once the student has a basic understanding of plant physiology and how plants interact with their environment, the author discusses the application of plants in the various segments of ornamental horticulture. In Section II, the author presents the principles of floral design and describes the experiences and skills needed by a floral designer; discusses the status of the interior foliage plant industry, the problems unique to the interior use of plants, and the cultural requirements; presents the principles of landscape design with checklists and numerous examples; describes the procedures involved in the installation of trees, shrubs, groundcovers, vines, bedding plants, and bulbs and their cultural requirements; covers the maintenance of landscape plantings, including watering, fertilizing, mulching, pruning, winterizing and weed, insect, and disease control; presents basic principles of using vines in the landscape, espalier pruning, topiary pruning, and bonsai; discusses turfgrasses to meet the needs of differing environmental conditions and uses, presents common methods of turf installation, and outlines a maintenance program for professional lawn care; and concludes with a discussion of propagation techniques, including tissue and organ culture.

Section III discusses the major professions of ornamental horticulture: the floriculture industry, the nursery industry, and the landscape industry, as well as specialized, nontraditional career opportunities. Each chapter describes the unique characteristics of the industry segment, including any specialized educational requirements for workers. The author also points out that interest, training, and skills in ornamental horticulture can be adapted to an assortment of career fields.

Section IV concentrates on the business and production techniques of ornamental horticulture. The chapter on growing structures covers the types of structures used, systems utilized to grow plants, and the physical layout of plant production areas. This is followed by a discussion of specific growing techniques used for greenhouse production. The author then distinguishes between the different types of nurseries and provides typical production techniques for each. The final chapters of the section concentrate on business management techniques. As a business, ornamental horticulture is governed by the same forces to which every business must respond if it is to be successful. Too many other texts overlook or ignore the fact that ornamental horticulture is a consumer-driven industry requiring skillful business management methods. This text is the first to offer equal coverage of this aspect of the profession.

STRUCTURE OF THE TEXT

The structure of ORNAMENTAL HORTICULTURE: SCIENCE, OPERATIONS & MANAGEMENT is designed to support and enhance the learning process. Each chapter opens with clearly stated learning objectives. These objectives are accomplished by means of the well-organized text material and the most extensive collection of photographs and drawings of any textbook written on this subject. Student mastery of the learning objectives is measured at the end of each chapter in the achievement review. The answers to the achievement reviews are provided in an Instructor's Guide. Wherever possible, the information has been summarized in tables for ready reference. All terms are defined upon first use in the text and again in a comprehensive glossary at the end of the text. Two appendices are included, following the glossary, for the student's reference. One appendix is a list of professional and trade organizations for horticulture and related areas, while the other is a list of selected readings for the study of ornamental horticulture.

ABOUT THE AUTHOR

Jack E. Ingels holds a Board of Trustees' Distinguished Teaching Professorship at the State University of New York's College of Agriculture and Technology at Cobleskill. His undergraduate schooling was at Purdue University and his graduate schooling was at Rutgers University. His fields of specialization include ornamental horticulture, landscape design, plant physiology, and plant pathology. He is an experienced university teacher with a diverse practical background in industry as well. He is also the author of *Landscaping: Principles and Practices,* one of the most successful texts in Delmar's agriculture series.

Acknowledgments

The following individuals devoted their time and considerable professional experience to reviewing the manuscript. Their critical insights were a valuable component in ensuring the integrity of this text.

Richard Austin
Olympia Technical Community College
Olympia, WA 98502

Charlie P. Giedeman
Belleville Area College
Belleville, IL 62221

John J. Ball, Ph.D.
University of Minnesota Technical College
Waseca, MN 56093

John M. Centko
Valencia Community College
Orlando, FL 32802

James W. Boodley, Ph.D.
Development Horticulturist
Smithers-Oasis Company
Kent, OH 44240

Cathy Haas
Hartnell College
Salinas, CA 93901

The author wishes to thank the following individuals, organizations, and companies for providing illustrations or technical information for this textbook.

Sharlotte Albert
James and Linda Angell, Floral Designers
Arnold Arboretum of Harvard University
Mark Barry
Beverly Becher
Beckman Industrial Corporation, Cedar Grove Division
A. T. Bianco Landscaping and Nursery
Big John Tree Transplanter Manufacturing
Brooklyn Botanic Garden

J. A. Buck, General Electric Lamp Marketing Department
Carefree Garden Products
ChemLawn Corporation
Chevron Chemical Company
Council of Linnean Society of London
Design Imaging Group
Stephen DiCerbo, Illustrator
Economy Label Sales Company
John Farfaglia

Frank Ferraro, Jr.
Terry Forsyth
James Glavin, Landscape Architect
Albert Glowacki Landscape, Inc.
Cari Goetcheus, Illustrator
Carol Holliday
Michael Horaz, Bonsaist
Rodney Jackson, Photographer
Thomas Kenly, Landscape Designer
Lake County Nursery Exchange
Lord and Burnham, Division of Burnham
   Corporation
J. J. Mauget Company
Monrovia Nursery Company
Michael Montario
Emily Morgan
Sheryl Morzella
National Garden Bureau, Inc.
Jean Oppenheim
Stanley Pendrak

Greg Perez
Kelly Pottenburgh
Scott Raas, Landscape Designer
Olga Ressler
Mark Scelza
Schaefer Nursery
Peter Snopsky
Solar Sunstill, Inc.
Speedling Florist
Linda Stacey
Jere Tatich, Landscape Architect
3M Agrichemicals Corporation
Timothy Toland, Illustrator
Kathy Tripp
United States Department of Agriculture
Richard Vedder
Weather-Matic Corporation, Division Telsco
   Industries
Ray Wyatt
Robert Yates, Landscape Designer

Foreword:
A Brief History of
Ornamental Horticulture

It is easier to define and distinguish the crops and crafts of ornamental horticulture from the rest of horticulture than to separate the historical antecedents of ornamental horticulture from those of pomology, forestry, and vegetable production. Several factors account for the difficulty of assigning specific dates to the horticultural time line for ornamentals. Foremost among these factors is the changing way people have used and regarded many of the plants of horticulture over time. Today's flowering specimen trees were commonly more prized for their fruit in centuries past. Even their importance as food suffered inconsistencies. For example, peaches were once as important for hog feed as for peach brandy. Today, certain cultivated varieties of peaches may be used for the landscape value of their flowers, with their fruit regarded as little more than a maintenance nuisance. Olive trees have undergone a similar change of purpose in landscape use.

The development of ornamental horticulture has accompanied the evolution of a worldwide system of agriculture, but it has not paralleled that development. Agriculture reaches back to primitive cultures and to the cultivation of edible plants, which began when the reliable availability of wild game waned.

The biblical Garden of Eden is not possible to date, but its influence in ornamental horticulture has been great. Several Western cultures have set the Garden of Eden as an ideal standard to strive toward in the development of their gardens. Eastern cultures have similar romanticized ideals of the garden as a spiritual paradise, and those ideals have frequently influenced the design of their earthly landscapes.

Aside from the religion-based garden influences, there are documented records of gardens as far back as the ancient Egyptians and Sumerians, over 5,000 years ago. Irrigation made arid lands productive, and the selective production of preferred plants began. Although food provision was undoubtedly the major purpose of these ancient gardens, the Egyptians were among the earliest civilizations to cultivate plants for their aesthetic values. Their interest in plants as sources of spices, fragrant oils, and fibers eventually progressed to the development of formal gardens around the homes of the affluent. For centuries the Egyptians refined their techniques of horticulture production. Their pictorial and written documentation of the plants important to them throughout their thirty-five centuries of cultural dominance comprise a great legacy that will keep scientists and historians busy for years.

As human society evolved and advances were made

in agricultural technology, those civilizations that were attentive to the stewardship of their land tended to last longer than those such as ancient Greece, which flourished brilliantly but briefly on the calendar of Mankind. After the Egyptians it was not until the era of the ancient Romans, about 2,500 years ago, that ornamental horticulture was again recognized as having value in the lives of people. The Romans were the photocopy society of history, borrowing extensively from the agricultural and horticultural knowledge of Egypt and Greece. However, to their credit, they refined and improved the techniques and proved themselves to be much better stewards of their land than the Greeks, whose culture they supplanted. It was the Romans who introduced grafting and budding as common propagative techniques.

Ancient Rome exemplified what many would regard as a regrettable accompaniment to the appreciation and advancement of ornamental horticulture. As a result of its total cultural dominance of the known world, its unchallenged military might, its enslavement of weaker societies, and its supreme self-assurance, Rome's wealthy citizens were free to practice civility at the highest levels. They built gracious villas at the center of large farms. Directly adjoining the villas were large gardens that extended the spaciousness of the indoors to the outside. Those villa gardens were usually walled to permit greater control of the growing environment and the development of the owners' fantasies. Formal patterns, water features, sculpture, flower plantings, pruned shrubbery, and carefully planned paving patterns created outdoor rooms whose primary purposes were visual enjoyment and leisure time pleasure.

Had Rome not fallen, it is interesting to wonder how the history of ornamental horticulture would be written today; but decline and fall it did. With the demise of Rome, Western civilization plunged into the Dark Ages. Surviving only fragmentally in the small protected gardens of monasteries scattered across Europe, ornamental horticulture did not return to favor until the Renaissance, which began in Italy in the sixteenth century.

The gardens of the Italians of that century were joyout rediscoveries of their rich Roman ancestry, whose ruins were literally strewn at their feet. Once again confident of their power to exercise control over their environment, Italians of the sixteenth century built large formal gardens that were as remarkable for their engineering as for their aesthetics. The technology of the time allowed the Italians to transform their steep hillsides into broad terraces for pleasurable human activity. Lavish water displays resulted from the redirection of rivers and streams into and out of the gardens. With the sculpture of heathen Rome still abundant, former marbleized deities assumed new roles as novel ornaments in the gardens of Christian Italy.

In the seventeenth century, France continued the tradition of lavish formal garden development on a scale even larger and grander than that of the Italians. As the lead player on the stage of Western civilization, France, during the reigns of the Louises, affirmed again that gardens and ornamental horticulture paralleled the growth of military prowess, power, and wealth more than the improvement of agricultural technology. While many less fortunate citizens of France were barely subsisting, the great formal garden master André LeNotre was creating living works of art at Versailles and other sites throughout France. His apprentices spread throughout Europe, attempting to bring the gardens of every other country into compliance with the formal, Baroque tradition of France. While the French gardens were seemingly simple—symmetrical balance, walk intersections marked by a fountain or piece of sculpture, plants sheared into sculptural shapes (topiary), and a design most intricate and complex near the building, becoming less intense as it advanced toward the surrounding countryside—few were able to match the genius of LeNotre. Also, not every garden site could duplicate the vast flatness of France, where the gardens could cover hundreds of square miles and create the desired impression of limitless luxury and extravagance.

England, in the eighteenth century, brazenly rejected the belief that landscapes must be formalized and display the heavy hand of Man in order to qualify as a garden. With a court life less formal than that of the French and a more free-willed aristocracy, England was receptive to the influences of men like William Kent, Lancelot "Capability" Brown, and Humphrey Repton, who believed formality to be anti-Nature. The English Naturalism style gained such popularity throughout the British Isles that centuries-old formal gardens were swept away, to be replaced with grassy landscapes whose groupings of trees, carefully shaped lakes, serpentine streams, distant views of grazing sheep, fabricated grottoes, and manufactured ruins atop distant hillsides fit their designers' ideas of how a romantic Eden-like paradise landscape should appear. Nowhere was a straight, formal line, sheared plant, or splashing fountain to be seen. Yet these naturalistic gardens were just as contrived as their formal Continental predecessors. Observing these gardens today, many wonder what was so special about them. They are so parklike and so evocative of a drive through the open countryside, it is easy to forget that at the time of their development they were as extraordinary as was the horseless carriage at the start of the Industrial Revolution.

The other great contribution of the English in the history of ornamental horticulture was their role as plant collectors. Their fondness for plants is almost genetic. As they moved back and forth across the civilized and not-so-civilized world, English explorers took and/or collected plants with them. In addition, they wrote about the plants they collected and even preserved pressed samples of them, much to the benefit of later botanists and taxonomists. They established great botanic gardens for the propagation, study, and public display of their plant bounty collected from the far reaches of the empire. Also, though the English did not invent greenhouses, the orangeries of the aristocracy planted the benefits of indoor plant production into our historic consciousness.

Concurrent with the evolution of ornamental horticulture in Europe, but separated both geographically and attitudinally, were the gardens of Asia. First the Chinese, then the Koreans, and later the Japanese used plants within and surrounding their homes in ways that most Westerners can never fully comprehend. Closely tied to their Buddhist and Shinto faiths, oriental Asians perceived themselves as being a part of the natural world, not separate from and dominant over it. In their cities, crowded even centuries ago as they are today, gardens were small by comparison to Europe's. The gardens were walled and within their confines the Orientals developed gardens that ranged from lush and green (the Buddhist paradise gardens) to stark and minimal (the Zen

gardens). Plants were pruned and trained to represent their larger counterparts outside the walls. In more abstract uses, plants represented mountains, clouds, islands, and other nonplant elements of the natural world. The intent was to represent the larger, not-so-perfect, natural world in scaled-down, perfect form. Shinto gardeners frequently believed that certain unusual plants possessed spirits or the souls of departed friends or relatives or important persons. As such those plants were regarded with special reverence. Unlike the formal gardens of Europe, which were intended as a grand stage for the display and glorification of the human players, with plants subjugated to a minor role, the gardens of the Orient were designed to feature plants and other elements of the natural world in a way that provoked an intellectual thoughtfulness by the visitors. In many cases, a walk through an oriental garden was a psychodrama intended to remind even the most aristocratic Asians of their natural place in the world.

If any one thing stands out as a landmark along the historic progression of ornamental horticulture, it is the discovery of the New World. The Americas were a rich repository of plants, already being cultivated for both food and ornamental purposes by Indian cultures, long before the invasion by European explorers began. Once underway, the exploration and settlement of the New World promoted large-scale transplantation of foreign plant species and plant products between Europe and the Americas.

As the various nations of Europe sought to stake their territorial claims in the New World, numerous colonies were established in the lands that years later would become the United States. The seventeenth century saw the Spanish, the Dutch, and the English heavily committed to the colonization of the coastal areas of North America. While indigo, tobacco, timber, rice, and other economic crops were of far greater importance to the settlers, they still found room in their lives to value and grow ornamentals too. The New England Puritans established cottage gardens to produce flowering plants that scented their houses, spiced their cooking, and decorated their celebrations. The Dutch discovered and sent back to Holland the bulbs that would later establish their international preeminence as bulb purveyors to the world.

For much of the next century, even as the early

colonists were replaced by sons and daughters who sought and eventually seized their independence from their European homelands, American gardens and uses of ornamental plants were copies of Old World ideas. Twentieth-century tourists visiting historic Williamsburg, Virginia, see eighteenth-century formal gardens everywhere. Savannah, Georgia, is a picturesque city built to a European ideal of public squares at the center of residential clusters.

The great estates of George Washington's Mount Vernon and Thomas Jefferson's Monticello were profoundly influential in America's perception of ornamental horticulture, due largely to the godlike esteem that both men enjoyed from their contemporary countrymen and the admiring millions who have followed. Both estates were a combination of farm and English naturalistic garden. Spectacularly sited, Mount Vernon and Monticello represented their owners' responsible stewardship of their lands and their willingness to try new agricultural techniques. While neither estate made lavish uses of flowers, their shade tree plantings and expansive lawn areas were reminiscent of the designs of Kent and Brown in England.

As the post-Revolution nation grew stronger economically, the favorable climate and tobacco/cotton economy of the Southeast fostered a lifestyle and garden culture that were aristocratic and formalized. Bowling greens, Elizabethan flower gardens, and serpentine walks lined with graceful shade trees typified the gardens of early America in the South. It is tempting to seek similarities between the self-confident, slave labor centered, plantation lifestyle of the pre–Civil War southern United States and that of the Ancient Romans, wherein both societies found time to cultivate their appreciation of ornamental plants under similar circumstances.

As the nation pressed its boundaries westward through the visionary efforts of Jefferson, Lewis and Clark, and others, the vegetation of the prairies and the fertile soil of the American Midwest were added to the American treasure chest of resources. Far away, at the western edge of the continent, Spanish settlers were building missions throughout the land that would become California. Mission plantings were generally typified by orchard groves, vine-

yards, kitchen gardens, and small flower plantings. As the Spanish Mexican influence took root in southwestern North America, it laid the foundation for a garden design legacy that reaches back to the fourteenth century, yet is still viable today. The Moorish gardens, which were prevalent in southern Spain in that early time, were typically walled and paved and closely related to the buildings they adjoined. They often used water features for their psychological cooling effect, and the term *Spanish Patio* came to define a distinctive type of outdoor area development. Six centuries later, that garden style is still popular, particularly around the adobe and mission-like architecture of California, New Mexico, and Arizona.

The history of ornamental horticulture in America is as difficult to chart on a time line as is its history worldwide. Perceptions of the role of plants change over time. Exploration and discovery of new species continue. Nevertheless, certain individuals, places, and events are worthy of note, even at the risk of omitting others of comparable merit.

- Dutch settlers brought bowling greens to America. These set the precedent for our village greens, city parks, and athletic fields.
- Botanic gardens, featuring all types of horticultural plants, were first established in Colonial America in the eighteenth century and facilitated the exchange of plant species between the country and the city and between the Old World and the New. One of the earliest and most famous botanic gardens was that of John Bertram of Pennsylvania, established in 1728.
- Greenhouses were also being built in America, originally as orangeries for citrus fruits, prior to the Revolution. In appearance and construction materials used, they were far cruder and darker structures than those of today.
- Nurseries, although not unknown elsewhere in the country, were most important in western New York, around Rochester. Both fruit trees and ornamentals were grown and shipped to all parts of the nation and the world. For at least half of the nineteenth century, Rochester reigned as the center of nursery production in America.

- Commercial seed houses and the mail order seed businesses began in the mid-1800s, further diversifying the regional floras of America.
- Jacob Bigelow, a Boston physician and botanist, took responsibility for the development of the Mount Auburn Cemetery in Massachusetts about 1817. It was done in the naturalistic style of eighteenth-century England and set a new national standard for burial grounds. So widespread was the enthusiasm for naturalistic cemeteries and their inclusion of a wide variety of trees and shrubs that the best ones became comparable to botanic gardens. These cemetery grounds also whetted the American appetite for more numerous and more naturalistic city parks.
- Frederick Law Olmstead, regarded as the Father of Landscape Architecture in America, drew upon the designs of the English naturalists to win the competition for the design of New York City's Central Park, the first great city park in America. He and his partner, Calvert Vaux, built city parks throughout the eastern United States and firmly established the democratic concept of public landscaping for all citizens, not just the wealthy and powerful.
- Late in the nineteenth century, the lawnmower was technically improved to the point that lawns could be kept trimmed in a manner similar to today. Prior to that, scythes and/or grazing animals were the means of controlling turf height.
- Pierre du Pont, of the Pennsylvania du Ponts, established what is arguably America's premier horticultural display garden over a forty-five year period, beginning in 1906. Now open to the public in Kennett Square, Pennsylvania, Longwood Gardens is an Americanized eclectic interpretation of sixteenth-century Italian Renaissance, seventeenth-century French Baroque, and eighteenth-century English Naturalism.

Ornamental horticulture in the United States is not as much a state of being as it is a state of becoming. As America's ethnic heritage expands from a majority of Euro-Americans to include more Asian-Americans, African-Americans, Hispanic-Americans, and other immigrant groups, it is predictable that the uses and appreciation of ornamental plants will continue to evolve. That is probably as it should be. One can only wonder how this Foreword will be written when the third edition is published.

SECTION I

THE SCIENCE OF ORNAMENTAL HORTICULTURE

1
The Green Plant

Objectives

Upon completion of this chapter, you will be able to

- list the important roles played by green plants in our lives and the earth's ecosystem.
- describe the natural classification system for the plant kingdom.
- list the parts of a typical higher green plant and describe their functions.
- describe how plants grow.
- define the processes of photosynthesis, respiration, and transpiration.
- describe the environmental factors that affect plant growth.

THE VALUE OF PLANTS IN OUR LIVES

Twentieth-century men and women live within a culture of superlatives—the tallest building, fastest automobile, most populous city, costliest motion picture, and so on. Always seeking an event or superhero to top all others, we look farther and farther into space, probe deeper into the oceans, and contrive an endless array of mechanical gadgetry. Yet within sight of us all at nearly all times are marvels of nature whose simple existence allows our survival and whose wonders are taken for granted although only partially understood.

Green plants produce and recycle the oxygen upon which animal life depends. They capture the energy of the sun and convert it into forms usable by humans and other animals. Plants are the only living organisms capable of manufacturing their own food. Surely if only one plant existed on the entire earth, it would rank as the foremost wonder of the world—an oxygen and food factory.

Fortunately for us, the plant kingdom dominates the biological world and gives generously to us while asking only for a pollution-free environment and suitable habitat in return. In addition to their principle services of energy conversion, food manufacture, and oxygen production (through the process known as photosynthesis), plants offer us food products, medicines, fermenting agents, lumber, dyes, clothing, lubricants, fuel, flowers, fragrances, shade, glorious autumn colors, vines to swing on, and challenges to climb. Plants are synonymous with natural landscapes. They nurture us physically as well as psychologically. They inspire us, refresh us, and regenerate our lives.

CLASSIFYING PLANTS

The plants of the world include a rich array that may be either macroscopic or microscopic. Simply stated, the difference is whether we can see them with the unaided eye or whether we need a microscopic or other magnifying agent. To organize the world's plants into understandable categories has been the objective of both scientists and tradespeople for many years. However, their reasons for categorizing plants are as varied as the groups themselves, resulting in a number of different classification systems. Some are based on natural factors, others on physical appearance.

A classification system is an attempt to arrange current knowledge into a tidy system to which new knowledge can later be added. The first attempt to classify plants was made by a Greek named Theophrastus about 2,300 years ago. That first system dealt with the known macroscopic plants of the time.

The current systems for the scientific classification of plants are *natural systems* based upon the genetic and evolutionary relationship among plants, Table 1-1. As science advances, botanists and taxonomists (scientists who specialize in the classification of plants) learn more about familiar plants and frequently find reasons to change their placement within the classification. Over time, the classification system itself has changed. Even organisms such as the bacteria and fungi, which were once considered to be plants, have been reclassified as *plantlike* members of the natural world. Because the knowledge of plants is dynamic, not static, it is proper that the classification system be equally flexible and dynamic.

As Table 1-1 indicates, the higher plants, which include the ornamentals, are classified within the Division Tracheophyta and the Classes Gymnospermopsida and Angiospermopsida. The larger classes of plants are further subdivided into subclasses, orders, families, genera, species, forma, varieties, and cultivars. Each plant fits into the classification system in a unique way, based upon its specific combination of characteristics.

This classification system represents a hierarchy of similarities. From any point in the classification, the groups of plants share increasingly more genetic and anatomical similarities as their placement progresses down the ranking system, moving from Division toward species. For example, the botanical classification of a thornless honeylocust tree is:

Kingdom: Plantae
Division: Tracheophyta
Class: Angiospermopsida
Order: Rosales
Family: Leguminosae
Genus: Gleditsia
Species: triacanthos
Forma: (none for this particular plant)
Variety: inermis
Cultivar: "Sunburst"

All plants are given their scientific names by using the genus and species in which they are classified. Known as *binomial nomenclature,* this method of designating specific plants is described more fully in Chapter 3. It is worth noting here that some species can be further described but do not vary so greatly that they constitute truly different species. A *variety* of a species is such a plant. The species may differ naturally from plant to plant by some small botanical variation such as flower color or leaf variegation. The different forms of the plant might represent natural varieties of a single species. Some varieties exist, not naturally, but through the intentional intervention of horticulturists. Sterile plants with showy flowers that produce little or no fruit exemplify such varieties. Known as *cultivars* (meaning cultivated variety), they are not truly a part of the natural classification system. However, in ornamental horticulture, they represent a highly important category of plants.

Forma is a lesser-used subcategory. It includes a group of plants within a species that are not distinctive enough even to comprise a variety, yet stand apart in some way. Frequently, they are of greater interest to botanists and taxonomists than to horticultural practitioners.

Importantly, scientific classification systems are not only dynamic, as mentioned, but they also are not universally accepted by all botanists. Scientific disagreement is healthy and justifiable because it en-

TABLE 1-1.

An Abbreviated Natural Classification of the Plant Kingdom

Category	Scientific Name	Common Name
Kingdom	Plantae	Plant kingdom
Subkingdom	Thallobionta	Thallophytes
Division	Rhodophyta	Red algae
Division	Cryptophyta	Cryptomonads
Division	Dinophyta	Dinoflagellates
Division	Xanthophyta	Chloromonads and yellow-green algae
Division	Chrysophyta	Golden brown algae and coccolithophorids
Division	Bacillariophyta	Diatoms
Division	Phaeophyta	Brown algae
Division	Chlorophyta	Green algae
Division	Prasinophyta	Prasinophytes
Division	Charophyta	Stone-worts
Division	Euglenophyta	Euglenoids
Subkingdom	Embryobionta	Embryophytes and cormophytes
Division	Bryophyta	Bryophytes
Division	Tracheophyta	Vascular plants (the Higher Plants)
Subdivision	Psilophytina	Whisk ferns
Subdivision	Lycophytina	Club-mosses
Subdivision	Sphenophytina	Horse-tails
Subdivision	Filicophytina	Ferns
Class	Cycadopsida	Cycads
Class	Ginkgopsida	Ginkgo
Class	Coniferopsida	Conifers
Class	Angiospermopsida	Flowering plants
Subclass	Dicotyledonidae	Dicotyledons
Subclass	Monocotyledonidae	Monocotyledons

courages taxonomists worldwide to converse with one another through conferences and research publications and to check and recheck one another's work.

Since this is not a botany text, the temptation will be great to skip all of the foreign words and press on toward a more interesting topic, particularly after discovering that the ornamental plants are contained almost entirely within the last three classes, Ginkgopsida, Coniferopsida, and Angiospermopsida. However, few divisions are without some members that affect the plants and people of the ornamental horticulture industry. The more primitive plants and plantlike organisms, such as the lichens, mosses, and liverworts, may be attractive elements of a natural landscape, or they may be destructive pathogens of disease, like certain bacteria and fungi. They may also create problems merely by growing where they are not wanted, as with Spanish moss and pond weeds. An appreciation of the diversity and complex relationships of the plant kingdom and the plantlike kingdoms is essential to a successful career in ornamental horticulture.

PARTS OF A PLANT

The most logical starting point for a scientific study of ornamental plants is their macroscopic anatomy (what we can see with the unaided eye).

Members of the higher plants are made up of roots, stems, leaves, and flowers or cones, Figure 1-1. Flowers are the reproductive structures of the Angiospermopsida as cones are for the Coniferopsida. Hereafter, these two Classes will be referred to simply as the flowering plants and the conifers.

Roots

Roots are the below-ground portion of the plant. They may be *fibrous,* with a network of roots reaching out horizontally and vertically through the soil; or they may be *tap roots* in which one central root grows larger and is more dominant than the others. In both types of root system the larger roots are supplemented by many smaller *root hairs,* Figure

1-2. The principal function of the roots is to absorb water and mineral nutrients from the soil. Much of the absorption occurs through the root hairs. In addition, the roots serve to anchor the plant against toppling by wind and rain. Roots also store food materials produced in the leaves. Specialized roots called *adventitious roots* develop from stems in some plants such as philodendron and from leaves and cut stems of various plants being propagated vegetatively, Figure 1-3.

Stems

Stems are the central axis of plants. They are usually above ground and may be long or short, single or multiple, and herbaceous or woody (these terms are defined later in the chapter). The principal functions of a stem are to conduct water and minerals absorbed by the roots to the leaves and other above-ground plant parts and to conduct food materials produced in the leaves to the roots and other plant parts.

Leaves

Leaves may be thin and flat, thick and fleshy, broad or needle-like. They are appendages of the

FIGURE 1-1.
The principal parts of a flowering plant

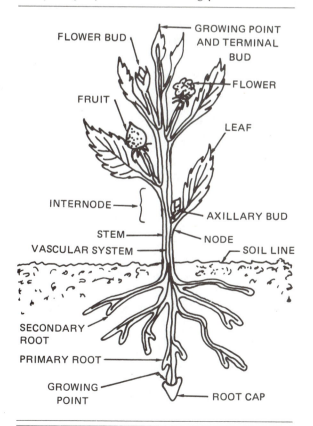

FIGURE 1-2.
Typical root systems of plants

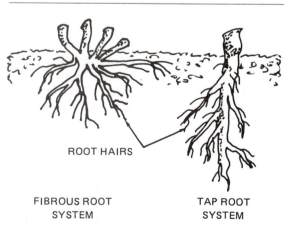

FIGURE 1-3.
Adventitious roots are seen on this philodendron. They originate on the lower stem.

stem and are the major food manufacturers of the plant. Leaves contain a green pigment, *chlorophyll,* which allows them to use the energy in light to convert carbon dioxide and water into food and oxygen. This process is called *photosynthesis.*

Cones or Flowers

Cones are the reproductive structures of the conifers, such as pines, spruce, and firs. Cones contain naked, unenclosed seeds on the upper surface of each cone scale. The conifers are slightly more primitive in evolutionary development than the flowering plants.

Flowers epitomize the peak of evolutionary development in plants and make the flowering plants dominant in the plant kingdom. They are reproductive structures that produce seeds enclosed in fruit. Because the varying anatomical features of their flowers are one means of identifying plants, a knowledge of flower parts and types of flowers is necessary for horticulturists.

A *complete* flower is one that possesses all the floral organs (sepals, petals, stamens, and pistils), Figure 1-4. An *incomplete* flower lacks one or more of these organs. In addition to being complete or incomplete, flowers may be perfect or imperfect. A *perfect* flower has both stamens (male reproductive organs) and pistils (female reproductive organs). An *imperfect* flower lacks one or the other. Imperfect flowers may be termed *pistillate* or *staminate* depending upon which of the two essential organs they possess. If both pistillate and staminate flowers occur on the same plant, it is said to be *monoecious.* If the two imperfect flowers occur on separate plants, the plants are termed *dioecious.* Willows and hollies are examples of dioecious ornamental plants.

Flowers may be produced as *single* blossoms

FIGURE 1-4.
Longitudinal section of a complete flower (From J. Boodley, *The Commercial Greenhouse,* © 1981 by Delmar Publishers Inc.)

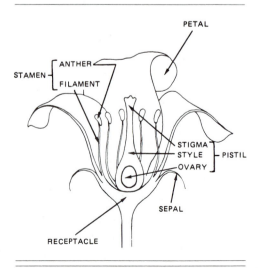

FIGURE 1-5.
A single flower

FIGURE 1-6.
A cluster (inflorescence)

such as roses, Figure 1-5, in clusters (*inflorescences*) such as gladioli or snapdragons, Figure 1-6, or as *composite* flowers such as chrysanthemums, Figure 1-7. Composite flowers give the appearance of a single blossom but are actually a grouping of many tiny flowers.

FIGURE 1-7.
A composite flower (Courtesy National Garden Bureau, Inc.)

THE STRUCTURE OF PLANT PARTS

All living organisms except viruses have the *cell* as their basic structural unit. A single plant part, such as a leaf, may be composed of millions of cells. The cells of plants may range in size from 1/25,000 of an inch to nearly 4/10 of an inch. Plant cells can best be visualized as three-dimensional chambers which, when joined together, are responsible for the shape, size, appearance, and function of all the earth's plants. Plants grow from seeds to maturity by the enlargement of existing cells and the production of new ones.

All plant cells are basically alike, Figure 1-8. Nevertheless, leaves do not look like stems, and roots and flowers are equally different. Some parts of the plant photosynthesize, and other parts do not. Some parts are rigid, like stems, and other parts are pliable, like leaves. As cells group together, they become differentiated in their functions. Large groups of similar cells carrying on the same function are termed *tissues.* Groups of tissues make up the *organs* of plants, Figures 1-9, 1-10, and 1-11. From the diagrams, several facts should be noted.

FIGURE 1-8.
A typical plant cell with major parts identified and labeled (Courtesy Stephen DiCerbo)

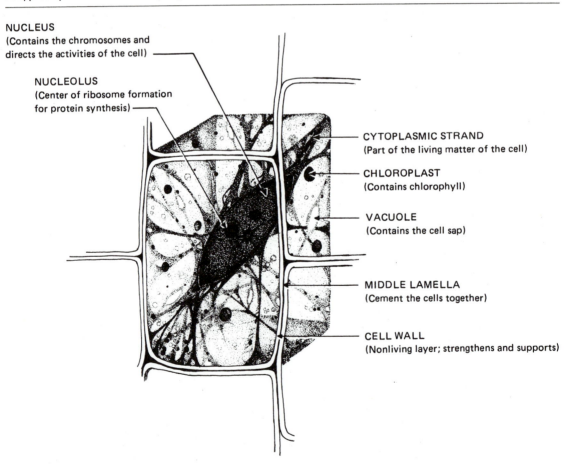

NUCLEUS
(Contains the chromosomes and
directs the activities of the cell)

NUCLEOLUS
(Center of ribosome formation
for protein synthesis)

CYTOPLASMIC STRAND
(Part of the living matter of the cell)

CHLOROPLAST
(Contains chlorophyll)

VACUOLE
(Contains the cell sap)

MIDDLE LAMELLA
(Cement the cells together)

CELL WALL
(Nonliving layer; strengthens and supports)

FIGURE 1-9.
Cross section of a leaf (Courtesy Stephen DiCerbo)

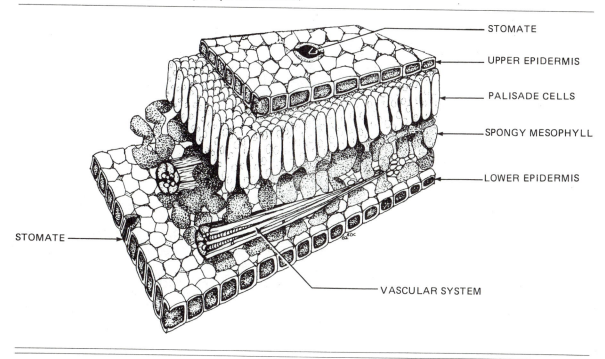

STOMATE

UPPER EPIDERMIS

PALISADE CELLS

SPONGY MESOPHYLL

LOWER EPIDERMIS

STOMATE

VASCULAR SYSTEM

FIGURE 1-10.
Cross sections of different types of plant stems

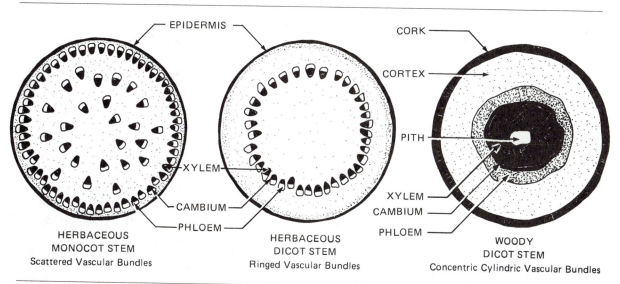

EPIDERMIS

CORK

CORTEX

PITH

XYLEM

CAMBIUM

PHLOEM

XYLEM

CAMBIUM

PHLOEM

**HERBACEOUS
MONOCOT STEM**
Scattered Vascular Bundles

**HERBACEOUS
DICOT STEM**
Ringed Vascular Bundles

**WOODY
DICOT STEM**
Concentric Cylindric Vascular Bundles

FIGURE 1-11.

Cross section of a root tip (Courtesy Stephen DiCerbo)

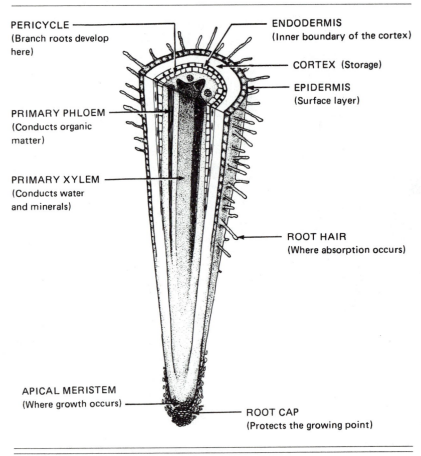

PERICYCLE
(Branch roots develop
here)

ENDODERMIS
(Inner boundary of the cortex)

CORTEX (Storage)

EPIDERMIS
(Surface layer)

PRIMARY PHLOEM
(Conducts organic
matter)

PRIMARY XYLEM
(Conducts water
and minerals)

ROOT HAIR
(Where absorption occurs)

APICAL MERISTEM
(Where growth occurs)

ROOT CAP
(Protects the growing point)

The Cell

Tiny though they are, plant cells are not hollow like empty boxes. Each cell is comprised of a variety of components, and each component has a vital role to play in the growth, reproduction, and differentiation of the cell.

- *The cell wall* is the foremost distinction between plant and animal cells, since animal cells lack walls. Usually multilayered, the cell wall has a *primary wall* on the outside, a *secondary wall* on the inside, and a cementing agent between them termed the *middle lamella*. The wall is composed of a matrix of carbohydrates reinforced by cellulose molecules arranged in long, rodlike structures.

As cells age they acquire deposits of *lignin* (complex polymers) within the carbohydrate matrix. The result is hardening of the cells, which explains why plants become woody or stiffer as they grow and age.

- *The protoplast* is the living matter of the cell.
- *The nucleus* is the vessel in the cell that contains the chromosomes, the nucleolus, and the nucleoplasm. The chromosomes carry the genes that direct heredity; the nucleolus is the site of ribosome production; and the nucleoplasm supports it all.

- *Chloroplasts* contain the chlorophyll pigment, which is vital to photosynthesis.
- *Cytoplasm* is all of the living material in the cell other than the nucleus. (Cytoplasm + nucleus = the protoplast)
- *The vacuole* is a cavity within the cytoplasm. It is lined with a membrane and filled with salts, various pigments, and organic materials that are collectively termed the *cell sap.*
- *The plasma membrane* surrounds the protoplast like a thin plastic bag, separating it from the cell wall. The membrane is semipermeable and controls what substances pass into and out of the cell.
- *Cytoplasmic strands* connect the protoplasts of adjacent cells, making the living material of the cells continuous within the plant.

Although most plant cells contain the same components, they do not all serve the same function. Some remain simple (embryonic) and primarily divide and create new cells, permitting the plant to grow. *Meristematic cells* are such cells. They are concentrated at the tips of shoots and roots in most plants, which explains why those plants grow from their extreme ends (their meristems), not their bases. An obvious exception is the grasses, which do grow at the base, not at the tips of the their blades. Grassy monocots have their meristems located at their base, a convenient property that allows mowing without cutting off the growth cells.

Other cells, although originating from the meristematic cells, assume other roles in the plant as a result of their differentiation. *Parenchyma cells* are specialized cells comprising the cortex and pith tissues in stems and the spongy mesophyll tissue in leaves. Most present in leaves, flowers, and fruits, these cells allow the plant to heal its wounds, secrete and excrete materials, and store food. They are also the site of photosynthesis within the plant, where they contain chloroplasts. *Collenchyma cells* are also specialized to provide plants the structural strength they need for support. Their cell walls are unevenly thick, permitting flexibility of the plants' stems. *Sclerenchyma cells* have the thickest walls and are also involved in structural support of the plant when the cells assume the shape of long slender *fibers.* Sclerenchyma cells can also assume assorted other globular shapes, termed *sclerids.* Enmassed, sclerid cells form the hardest plant features, such as the pits in cherries and peaches.

Stems

The transport of water, nutrients, and food materials between leaves and roots occurs in the *vascular bundles.* Water and minerals are carried upward in the *xylem,* while food materials move downward in the *phloem.*

Stems increase in diameter due to the activity of the *cambium* tissue, which produces the xylem and phloem.

Leaves

Where the vascular tissues extend into leaves, they form a network of veins. The pattern of the veins (venation) is important in the classification and identification of many plants.

Gaseous exchange between the air outside the plant and the intercellular spaces inside the plant occurs through pores termed *stomata* or *stomates* (singular *stomate*). The plant loses water vapor through the stomata when they are open. That water loss is known as *transpiration.* When the stomata are closed, transpiration is reduced.

The upper and lower epidermis are covered by a waxy *cuticle* (not shown), which keeps the leaf basically impermeable to water and helps some plants to retain internal moisture.

Leaf Color

Plant color in general, and leaf color in particular, results from the presence of pigments within the cells. In the majority of higher plants, including the ornamentals, chlorophyll is the pigment in greatest abundance. That is why most living, nondormant plants appear green. Other pigments are present in plants, however. *Xanthophyll* (bright yellow), *carotene* (orange), and *anthocyanins* (red) are pigments that occasionally dominate in plants such as the coleus or in the blossoms of most plants, but they usually await the aging of the leaf and the loss of the chlorophyll pigment to reveal their presence.

Roots

The root grows in the region of the *apical meristem*. That is the region where the cells are most actively dividing. The apical meristem is protected by the *root cap* as the root presses through the soil.

JUVENILITY AND MATURITY IN PLANTS

Although *maturation* is most commonly associated with animal aging or human emotional development, it also finds its place in plant science. Most commonly observed as a morphological (structural) change in the plant, it can also be manifested as a change in flowering or fruiting habit or in its mode of growth.

Juvenility is a state of vegetative growth during which a plant cannot flower. In annual plants, this state is very short (perhaps a month or two), whereas in longer-lived species, a healthy, actively growing plant may exist in the juvenile stage for years. *Maturity* is the state of growth during which the plant becomes capable of flowering.

While the unseen changes that must necessarily occur in the plant to initiate its change of state from juvenile to adult are still not entirely understood, the physical evidence of change is often striking. English ivy (*Hedera helix*) in its juvenile state is a trailing vine with deeply lobed leaves. Many people do not recognize it at maturity because the leaves become unlobed (entire) and the plant easily supports itself for upright growth. *Philodendron* leaves have an opposite appearance, that is, unlobed when juvenile and lobed when mature. Oak trees (*Quercus* species) retain their dead leaves through much of the winter as long as the tree is in its juvenile state. Old oaks drop their leaves much earlier.

MAJOR PLANT PROCESSES

A green plant is comparable to a machine that operates nonstop. However, unlike a machine that would perform a single function or programmed series of functions, the plant "machine" performs different functions simultaneously. At different times and under varying environmental conditions, certain plant processes increase and others decrease.

The physiological (functional) processes of the higher plants are numerous, but three stand out: photosynthesis, respiration, and transpiration.

Photosynthesis

Photosynthesis has been referred to earlier. It is a process unique to green plants, in which food (sugar) is manufactured from water and carbon dioxide in the presence of *chlorophyll*. Light energy, from the sun or other sources, drives the chemical reaction, and oxygen is released in the process. Water, used in the process, is also produced. The chemical equation for photosynthesis is believed to be:

$$6CO_2 + 12H_2O \xrightarrow[\text{light energy}]{\text{chloroplast}} C_6H_{12}O_6 + 6O_2 + 6H_2O$$

The rate of the process varies with the light intensity, temperature, and concentration of carbon dioxide in the plant's atmosphere. Also, excessive accumulation of the end product, sugar, can slow the reaction.

Respiration

Respiration is the process that permits living cells to obtain energy from organic material (usually glucose sugar). It is a breaking-down process, unlike photosynthesis, which is a manufacturing process. Respiration uses oxygen and enzyme catalysts to oxidize the sugar to carbon dioxide and water. In the process, energy is produced. The reaction can be stated as follows:

$$\text{glucose} + \text{oxygen} \xrightarrow{\text{enzymes}} \text{carbon dioxide} + \text{water} + \text{energy}$$

This process converts the chemical energy captured through photosynthesis into a form of energy available to the plant for growth, reproduction, and cell maintenance. The process does not create energy but simply changes its form.

Transpiration

Transpiration is the loss of water in vapor form from a plant. Water enters the plant through the roots and saturates the intercellular spaces throughout the plant. Then, because the amount of water vapor in the atmosphere is nearly always less than the amount of water vapor inside the plant tissue, the vapor leaves the plant. Most transpiration occurs through the stomata, 90 percent of which are located on the lower surface of the leaves. Smaller amounts of water vapor are lost through the cuticle. When water uptake exceeds the rate of transpiration, water passes out of the plant in liquid form through leaf openings called *hydathodes.* This slow exudation of liquid water is called *guttation.*

Transpiration rates are accelerated by increased temperatures or light. As the humidity in the air around the plant increases or decreases, the rate of transpiration decreases or increases in response.

When soil dries out, transpiration may cause the plant to lose water vapor faster than its roots can absorb replacement water. If the condition is prolonged, the plant wilts. As long as there is water in the soil the wilted plant will recover as soon as the rate of transpiration slows. This is called *temporary wilting.* If there is so little water in the soil that the plant cannot replace the water vapor it has lost even after transpiration slows, *permanent wilting* occurs. Kept too long in the permanent wilting state, a plant may become incapable of recovering even after water is added to the soil.

WHAT PLANTS NEED FOR GROWTH

Plants can and do exist under incredibly adverse environmental conditions. They survive in the arctic as well as the desert. They live atop mountains and in the depth of canyons. They survive on land and in both fresh and salt water.

The production of ornamental plants requires more than helping plants survive, however. It requires maximizing their rate of growth, their quality, or both. The growth potential of a plant is determined by two things: its genetic heritage and the environment in which it develops. Its *genetic heritage* is predetermined; there is nothing a grower can do to change it. Its *environment,* on the other hand, is under the grower's control to some extent. Environmental factors play a significant role in how quickly, if ever, the plant reaches its maximum size, when it blooms, how large its leaves become,

whether it sets fruit, and many other qualitative factors.

A plant's environment exists both above and below ground. It consists of the soil and the air and those natural elements common to them both—water and gases.

Soil

The soil is the environment of the root zone. It provides the mineral elements needed by plants for use during photosynthesis. It also collects and supplies water for uptake by the plants. To function most effectively, the soil must be porous, with sufficient space between the soil particles to store air and water. It must be loose enough to allow roots to grow through it, yet supportive enough to stabilize the plant.

Not all the mineral elements in soil are essential for plant growth. Likewise, not all the elements essential for plant growth are present in necessary amounts in all soils. Altogether, sixteen elements are essential at some time during each plant's life for optimum growth. These elements are basically the same for all plant species. The elements are categorized as *macronutrients* and *micronutrients,* depending upon the quantities needed by the plant. Designation as a macro- or micronutrient is not a measure of importance, since all are essential for growth. (Mineral nutrients are discussed in more detail in Chapter 2.)

When soil is less than ideal for optimum plant growth, it is usually the result of:

1. a deficiency of essential elements (especially nitrogen, phosphorus, and potassium, used in greatest amounts by plants)
2. poor *aeration* (too little air space between soil particles)
3. improper drainage (either too little or too much)

Changing the soil environment is one way to improve plant growth. For example, fertilizer can add missing nutrients, and sand/or peat moss can improve aeration or drainage.

Atmosphere

The atmosphere surrounds the above-ground portion of the plant. It supplies the carbon dioxide needed for photosynthesis and the oxygen needed for respiration. In closed environments such as greenhouses, the addition of carbon dioxide to the air has had the effect of increasing the rate of photosynthesis in certain ornamental plants such as roses and carnations. The addition of gaseous pollutants to the atmosphere can have a harmful effect on plants. It may harm their physical appearance and may reduce their rate of photosynthesis.

Water

Water is essential to all life. For plants to grow, they need water for photosynthesis. Water also carries various essential elements within the plant.

Water in liquid form enters the plant through the roots. Water as air vapor also plays a role since the humidity of the air around the plant influences the rate of transpiration. The greater the humidity, the more slowly will the plant transpire.

The presence or absence of water and the natural high or low humidity of a region influence not only the rate but the nature of plant growth, Figures

FIGURE 1-12.
Abundant rainfall and constant high humidity result in lush plant growth, as in this rain forest. (Courtesy United States Department of Agriculture)

1-12 and 1-13. The leafy, lush growth of a rain forest contrasts sharply with the sparse, stark plants of a desert. Epidermal thickness, type of root system, population numbers—all are influenced by the presence or absence of water.

Light

Light is the energy that permits all life on earth to exist. The sun radiates its light energy toward the planet's surface, where it would go unused if not for green plants. They capture and transform the energy, using it first for their own growth and later providing food for the growth of animals.

Plants respond to light whether it is naturally or artificially produced. Without light, the green color of a plant quickly fades and the plant dies. The direction of plant growth is affected by the source of light. It is not uncommon to see house plants growing toward a nearby window. Such growth movement in response to light is called *phototropism*. Phototropism results from the buildup of a natural growth regulator called an *auxin* on the side of a plant away from the light source. Since auxins promote cell elongation, the side of a stem with a higher concentration of auxins will grow faster than the side with a lesser concentration. The result is a plant that appears to be growing toward the light, Figure 1-14.

Another influence of light is *photoperiodism*, the effect of varying periods or durations of light exposure on plant growth and development. Most dramatic is the ability of light to initiate or delay blooming. Certain plants known as *short-day plants* will flower only after exposure to day lengths less than a critical amount. For example, chrysanthemums will bloom only if the day length is less than fourteen hours per day; poinsettias will flower only with less than twelve hours of light per day. *Long-day plants* such as the Rose-of-Sharon (*Hibiscus syriacus*) will

FIGURE 1-13.
Typical plant growth of the American Southwest, a region of limited rainfall and high temperature (Courtesy Ray Wyatt, Tempe, AZ)

FIGURE 1-14.

An explanation of phototropism

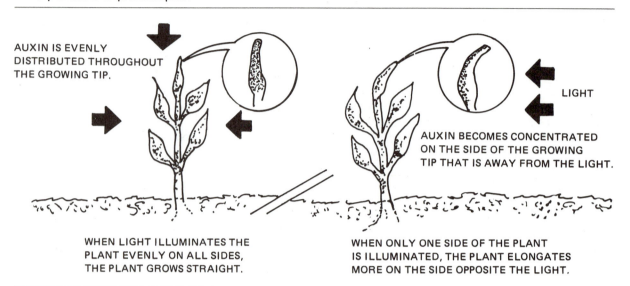

AUXIN IS EVENLY DISTRIBUTED THROUGHOUT THE GROWING TIP.

LIGHT

AUXIN BECOMES CONCENTRATED ON THE SIDE OF THE GROWING TIP THAT IS AWAY FROM THE LIGHT.

WHEN LIGHT ILLUMINATES THE PLANT EVENLY ON ALL SIDES, THE PLANT GROWS STRAIGHT.

WHEN ONLY ONE SIDE OF THE PLANT IS ILLUMINATED, THE PLANT ELONGATES MORE ON THE SIDE OPPOSITE THE LIGHT.

flower only when day length exposure exceeds a critical amount. Still other plants are termed *day neutral* and bloom freely within a wide range of photoperiods. Proper manipulation of the day length is an important step in the production of certain commercially valuable horticulture crops.

Temperature

Temperature can control the rate of chemical reactions occurring within the plant. Processes such as germination, transpiration, respiration, and flowering are accelerated by increased temperatures and slowed by reduced temperatures. Normally plants will grow satisfactorily within a range of temperatures. Exceeding these temperature limits may result in injury or death to the plant.

Certain plants, both ornamentals and nonornamentals, grow best in a cool temperature range (from 60° to 75°F); others do best in a warm temperature range (from 75° to 90°F). For example, in greenhouse production, carnations do best at a day temperature of about 65°F and a night temperature of 50° to 55°F. In contrast, greenhouse chrysanthemums prefer a warmer growing environment both day and night. Similarly, one group of turfgrasses does best in cooler, temperate regions, and another group does best in warmer, subtropic and tropical regions. *Cool season crops* and *warm season crops* are the general terms applied to these plants. The subtleties of temperature tolerance and temperature control include growing environments that provide different temperatures for the roots and above-ground portions of the same plant.

The chief danger of an introductory chapter such as this one is oversimplification. A discussion of the botanical sciences in so few pages may create the impression that the green plant is like a Tinker Toy®, simplistic in form and easy to understand. Nothing could be more incorrect. The green plant is a marvel of biological and evolutionary engineering, about which much is still unknown. Any student of ornamental horticulture should first be a student of botany.

SUMMARY

The survival of the human race and all other animal life depends upon the oxygen provided and recycled by the green plants of the earth. Green plants also capture the sun's energy and convert it into forms usable by members of the animal kingdom. Plants are the only organisms capable of manufacturing their own food. Our dependence upon them is total.

The classification of plants is a dynamic process, changing as new knowledge is acquired and new members are identified. The scientific systems of classification are natural ones based upon the genetic and evolutionary relationships among plants.

Members of the higher plants (the Division Tracheophyta) have roots, stems, and leaves, and may have cones (Class Coniferopsida) or flowers (Class Angiospermopsida). Flowers and cones are the reproductive structures. Roots serve to absorb water and mineral nutrients from the soil, anchor the plant, and store food materials produced in the leaves. Stems are the central axis of plants. Their function is to conduct water and minerals from the roots to the leaves and food materials from the leaves to the roots and other plant parts. Leaves are the major sites of food manufacturing in the plant.

The formation of roots, stems, leaves, cones, and flowers in the higher plants is attributable to the ability of plant cells to differentiate and assume assorted roles for the growth and development of the plant. Plant cells may be meristematic, parenchyma, collenchyma, or sclerenchyma in type. All contain similar materials but differ in the comparative amounts of those materials. They also appear in differing numbers, depending upon the tissue or organ they comprise.

It is in the leaves where the important processes of photosynthesis, respiration, and transpiration occur most actively. Photosynthesis is the process by which food in the form of sugar is manufactured from water and carbon dioxide in the presence of the green pigment, chlorophyll, and light. Respiration is the process that permits living cells to transform organic material into energy. Oxygen and enzyme catalysts oxidize sugar to carbon dioxide and water, releasing energy simultaneously. Transpiration is the process by which a plant loses water vapor through the stomata in the leaves.

Color in plants is caused by the presence of pigments, with the dominant pigment, usually chlorophyll in a healthy, growing plant, giving the plant its tone. Lacking the dominance of chlorophyll, one or more colors may appear due to the presence of other pigments such as Xanthophyll, carotene, or anthocyanins in the tissue.

The growth potential of a plant is determined by its genetic heritage and its environment. The genetic heritage is predetermined, but a grower can exercise some control over environmental factors: the soil, atmosphere, water, light, and temperature. The specific response of the plant to environmental change is sometimes related to whether the plant is in its juvenile or its mature stage.

Achievement Review

A. ESSAY

What relationship exists between the plant kingdom and the animal kingdom? Which is most dependent upon the other? What is the basis for the dependency? What responsibility do human beings have for the well-being of the plant kingdom? Write a short essay on the value of green plants that incorporates answers to these questions.

B. MULTIPLE CHOICE

From the choices given, select the answer that best completes each of the following statements.

1. Macroscopic plants are _____ to the unaided eye.
   a. invisible   c. green
   b. visible   d. nongreen

2. Microscopic plants require _____ to be seen.
   a. a microscope
   b. a magnifying glass
   c. an unaided eye
   d. an electron microscope

3. The current systems that use genetic relaships as the basis for classifying plants are _____ systems.
   a. artificial
   b. Latinized
   c. horticultural
   d. natural

4. Flowering plants are in the division _____ .
   a. Phaeophyta
   b. Chlorophyta
   c. Bryophyta
   d. Tracheophyta

5. The basic structural unit of plants is the _____.
   a. nucleus
   b. cell
   c. cambium
   d. tissue

6. The major difference between plant cells and animal cells is the presence of a _____ in plants.
   a. cell membrane
   b. vacuole
   c. cell wall
   d. nucleus

7. The process of converting energy from solar to chemical form through the manufacture of sugar is _____.
   a. respiration
   b. transpiration
   c. guttation
   d. photosynthesis

8. The green pigment plants need to capture the sun's energy is _____.
   a. chlorophyll
   b. protoplasm
   c. photosynthesis
   d. lignin

9. Water, nutrients, and food materials travel up and down the stem in the _____.
   a. phloem
   b. intercellular spaces
   c. xylem
   d. vascular bundles

10. Water and minerals are transported in the _____.
   a. phloem
   b. intercellular spaces
   c. xylem
   d. vascular bundles

11. Food materials are transported down the stem in the _____.
   a. phloem
   b. intercellular spaces
   c. xylem
   d. vascular bundles

12. Stems increase in diameter due to the production of vascular tissue by the _____.
   a. cell walls
   b. cambium
   c. middle lamella
   d. epidermis

13. The many pore-like openings in a leaf through which transpiration releases water vapor and through which gaseous exchange occurs are the _____.
   a. hydathodes
   b. stomata
   c. guard cells
   d. vascular bundles

14. Roots grow in the region called the _____.
   a. root hairs
   b. root cap
   c. endodermis
   d. apical meristem

15. Photosynthesis uses the raw materials of water and _____ to produce simple sugar, water, and _____.
   a. carbon dioxide/oxygen
   b. oxygen/carbon dioxide
   c. glucose/oxygen
   d. sunlight/carbon dioxide

16. Respiration uses the raw materials of glucose and _____ in the presence of enzymes to produce carbon dioxide, water, and _____.
   a. carbon dioxide/oxygen
   b. water/oxygen
   c. sunlight/simple sugar
   d. oxygen/energy

17. A cultivar is a variety that is sustained by _____ .
   a. nature
   b. propagators and growers
   c. volunteers
   d. cell differentiation

18. The simplest embryonic cells are the _____ cells.
    a. parenchyma   c. meristematic
    b. collenchyma   d. scleroid
19. Complex _____ can result when cells differen-
    tiate, then group together to carry on the same
    function.
    a. cell walls   c. tissues
    b. pigments   d. photosynthesis

C. SHORT ANSWER

Answer each of the following questions as briefly
as possible.

1. What are the four major parts of a higher
   green plant?
2. What are the reproductive structures of
   Angiospermopsida? Coniferopsida?
3. What are the two forms of root systems?
4. What is the name of the specialized root
   type formed during vegetative reproduction?
5. Why do trees and shrubs change colors in
   the autumn in temperate regions of the
   country?
6. Label the parts in this cross-sectional diagram of a root at right.

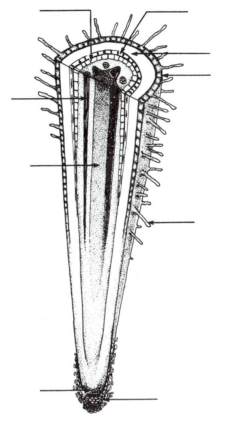

7. Name the three types of stems and label
   each of the parts in the cross-sectional
   diagrams.

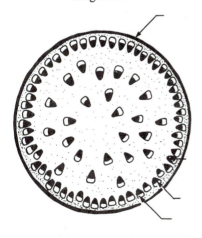
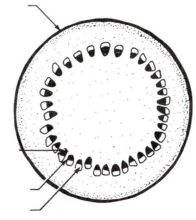
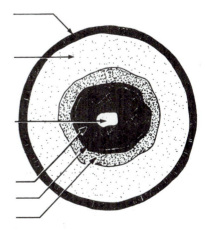

8. Label the parts in this cross-sectional diagram of a flower.

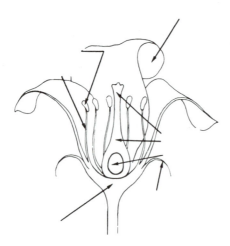

9. What is the term used for a flower that:
   a. has all of the floral organs
   b. lacks one or more of the floral organs
   c. has both pistils and stamens
   d. lacks either pistils or stamens
10. What is the term used for a plant that carries:
    a. both pistillate and staminate flowers on the same plant
    b. pistillate and staminate flowers on separate plants

D. TRUE/FALSE

Indicate if the following statements are true or false.

1. The growth potential of a plant is determined solely by its genetic heritage.
2. Plant size can be modified by a grower.
3. Not all mineral elements in the soil are essential for plant growth.
4. Macronutrients are more important to plant growth than micronutrients.
5. Nitrogen, phosphorus, and potassium are used in the greatest quantities by green plants.
6. Air pollution can affect plant growth.
7. A plant growing in an environment of low humidity can be expected to transpire more than one growing in a highly humid area.
8. Phototropism is the influence of varying durations of light on plant growth and development.
9. Photoperiodism is plant movement in response to light.
10. Auxins promote cell elongation.
11. Long-day plants require shorter night periods than short-day plants to initiate flowers.
12. The terms cool season and warm season crops refer to the temperature conditions most suitable for the plants' growth.

2
The Soil

Objectives

Upon completion of this chapter, you will be able to

- state how and why soils differ.
- list the components of soil, major soil separates, and the soil textures they create.
- list the sixteen elements essential to plant growth and their functions and symptoms of their deficiency in plants.
- define good soil structure and list the factors that promote it.
- define soil acidity and alkalinity in terms of pH.
- compare the qualities of fertilizers.
- describe how essential elements in the soil become available for plant use.

WHAT IS SOIL?

If asked to define *soil*, most people would probably describe where it is rather than what it is. They might also describe what it does, but not how it does it. Like so much of the natural world, soil is taken for granted—praised when the backyard garden is bountiful but disparaged when it is tracked in on the new carpet. Perhaps the greatest evidence of knowing little about the soil is to label it *dirt*.

Soil is the underground environment of plants, and that part of the definition (*where* it is) is generally understood by all. What it does and how it originates are less widely understood. Most people are aware that the plant life of the continent changes greatly from region to region. It should not be too surprising then to learn that the soil also changes

considerably from place to place. Therefore any attempt to define and describe soil must be approached in general terms, applicable from the red clay regions of Georgia through the loam fields of Iowa to the deserts of Utah.

Soil is the thin outer layer of the earth's crust, made up of weathered minerals, living and nonliving organisms, water, and air. To understand fully that definition is to understand much of modern soil science.

Weathered Minerals

Imagine a cross-sectional slice made down into the earth's crust, Figure 2-1. This is called a *soil profile.* The mineral content of soil results from the weathering of solid bedrock over long periods

FIGURE 2-1.
A soil profile

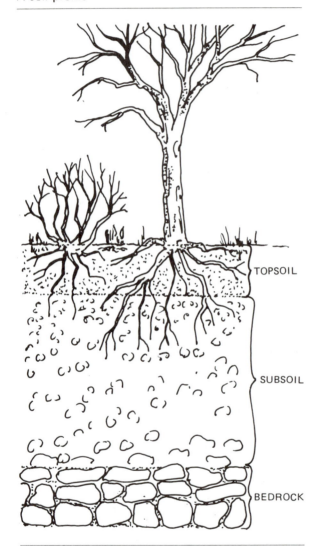

TOPSOIL

SUBSOIL

BEDROCK

of bedrock takes place over eons of geological time, it is nonetheless a simple breakdown of large pieces of rock into smaller particles.

Living and Nonliving Organisms

Over time, distinctive layers develop in undisturbed soils. Above the bedrock layer is the *subsoil*. It is finely weathered, like the layer above it, the topsoil, but it lacks organic matter in the quantity found in the topsoil layer. The roots of green plants flourish in the topsoil, richest in organic matter and shallowest in depth, and rely on it for nutrients, support, water, and air.

The organic matter in soil comes from the decomposition of plant and animal tissue. When green plants are plowed into the soil they are immediately acted upon by soil organisms, which rapidly break the plant tissue down into a form usable for their own growth. Organic compounds that do not decompose quickly eventually succumb to enzymatic action, forming a complex mixture called *humus*. Humus as well as *green manure* (plowed-under green plants) are important to the soil's structure. This organic matter increases both the water-holding and the mineral-holding capacity of the soil.

Water and Air

Water and air exist around and between the soil particles. As much as 50 percent of the topsoil may be air and water in liquid or vapor form. The ratio of air to water depends on the texture of the soil and how wet it is. A wet soil leaves less space for the air to occupy than a dry soil.

WHY SOILS DIFFER

Soils vary in many ways. They vary in color and weight. Some drain easily while others stay wet and bog-like. Some are rocky, breaking tools and backs, while others are easy to dig. Even though the original parent stone may be the same or similar,

of time. The solid rock is acted upon by an assortment of natural forces including temperature alternations that crack the rock, water that freezes within the cracks, and plant roots that further pry the cracks open. Then follows abrasive grinding by wind, water, and sometimes ice. Lichens and microbes may produce organic acids that react with the rock to further weaken it. While the weathering

differences in the subsoil and topsoil may result from variations in:

- weathering elements
- soil movement
- topography
- climate
- amount of organic matter

Soils that weather from bedrock and remain in place are termed *sedentary,* in contrast to *transported* soils. Transported soils have been moved by the forces of nature.

1. *Colluvial soils* have moved in response to gravity, as after a landslide or mudslide.
2. *Alluvial soils* are carried in water such as rivers. They are eventually deposited on flood plains and at deltas.
3. *Aeolian soils* are transported and deposited by winds.
4. *Glacial till* is soil deposited by glaciers.

The best agricultural soils are usually alluvial and glacial till. Colluvial soils are characterized by coarse textures and other undesirable chemical and physical qualities. Aeolian soils are finely textured but vary greatly in their productivity. Sedentary soils may be useful for agriculture if they have not lost their nutrient elements.

SOIL SEPARATES AND SOIL TEXTURE

When bedrock weathers, it forms particles of differing sizes. Based upon their diameter, these particles are classified into groups called *soil separates.* In decreasing order of size, the separates are:

1. gravel (coarse and fine)—2.0 mm or more in diameter
2. sand (very coarse [2.00–1.00 mm], coarse, [1.00–.50 mm], medium [.50–.25 mm], fine [.25–.10 mm] and very fine [.10–.05 mm])
3. silt (.05–.002 mm in diameter)
4. clay—less than 0.002 mm in diameter

The relative proportions of these separates of different sizes in any one soil create the *soil texture.* The proportions can only be determined precisely in a soils laboratory. There, a series of sieves are used to separate out the sand (and gravel, if present), while a suspension and settling technique is used to separate and measure the percentages of silt and clay, Figure 2-2.

Most soils in nature contain sand, silt, and clay in some proportions. The textural names given to soils are ways of describing these proportions. For example, if a soil contains about 50 percent sand and 50 percent of the finer clay and silt separates, it is termed a *loam.* Loams are generally favored for horticultural field production because they have enough sand to provide good drainage and aeration, yet enough of the finer particles to retain moisture and provide necessary plant nutrients (as will be discussed later). When one or two separates dominate the mixture, the textural name given the soil reflects that domination, as in *sandy loam, silt loam, silty clay loam, sandy clay, silty clay,* and so on.

In order to appreciate soil textural names, the physical and chemical properties of the separates should be considered.

Sand

Sand particles have assorted shapes and sizes depending upon how they were weathered. They range from smooth and round to sharp and angular. The spaces between sand particles are large compared to the spaces between silt and clay particles. Water passes through sand quickly because of this large pore space, and air is present in greatest quantity in sand. Sand is low in mineral nutrients and is generally inactive chemically.

Silt

Silt particles are irregularly shaped and much smaller than most sand particles. Given identical volumes of sand and silt, there would be many more silt than sand particles; hence silt has a greater surface area than sand. Since water clings to particle

FIGURE 2-2.
To use soil sieves, a premeasured amount of dried soil is passed through the stack. Each sieve's mesh is finer than the one before. Sands are separated by particle size. An additional suspension and setting technique is required to measure the percentage of silt and clay.

surfaces, the result is that silt holds water in the soil far better than sand. It does not provide as much space for air, however. Like sand, silt has a low nutrient level and is not very active chemically.

Clay

Clay has very small, plate-like particles. It possesses the greatest surface area of all the separates. Water is held tightly to the clay particles and passes very slowly through the soil. Predictably, air is often in short supply in a heavy clay soil, especially when it is wet. Clay has an adhesive quality when moistened and squeezed. This is what gives cohesiveness to soil, sometimes too much, creating sticky, hard-to-plow fields. Clay is active chemically. Many of the particles have surface charges, which attract water and ions. The term for such a chemical state

is *colloidal*. It is the colloidal quality of clay that makes it important for chemical activity and nutrient exchange in the soil.

The names of soil textures should now assume descriptive meaning. For example, a sandy loam would possess the following characteristics:

- sand, silt, and clay separates present, with sand dominant
- drainage good and perhaps slightly excessive
- nutrient content good since clay is present
- aeration good due to the sand, assuming that it is not too fine
- water-holding capability fair to good, depending on the amount of organic material and clay

The United States Department of Agriculture (U.S.D.A.) soil texture triangle illustrates how soils are named based upon a laboratory determination of their composition, Figure 2-3.

FIGURE 2-3.

The soil texture triangle

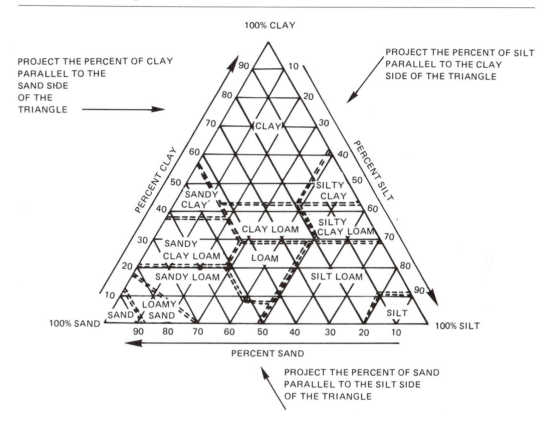

HOW TO USE THE TRIANGLE:

I.   Needed information: the mechanical analysis data for the soils to be named.

	Examples:	Soil #1	Soil #2	Soil #3
		60 percent sand	22 percent sand	28 percent sand
		30 percent silt	60 percent silt	36 percent silt
		10 percent clay	18 percent clay	36 percent clay

II.   Using any two of the percentages for each soil, project them into the triangle following the direction indicated by the arrows and the grid lines. Within the compartment where the two lines intersect, the textural name of the soil is read.

Soil #1	Soil #2	Soil #3
Sandy loam	Silt loam	Clay loam

SOIL NUTRIENTS

As described in Chapter 1, green plants require nutrients, which they obtain from the soil. A rich loam soil will provide a balanced supply of the elements essential to the growth of most plants. A soil that is predominantly sand or silt will be nutritionally poor.

The need by green plants for at least sixteen separate chemical elements has been proven repeatedly through tests that demonstrate growth abnormalities when any one of these *essential elements* is lacking. As previously mentioned, the *amount* of the chemical element required by a plant is not a measure of the element's essentiality. Whether required in large amounts (macronutrients) or very small amounts (micronutrients), the element is essential if the plant cannot grow and develop normally without it.

The elements presently known to be vital to the survival of green plants are shown, along with their chemical symbols, in Table 2-1. One method of remembering the essential elements is to associate them with a catchy phrase. For example:

See MG men mob Cousin Hopkins clean cafe

C Mg Mn MoB CuZn HOPKNS Cl CaFe

Of the essential elements, the plant obtains only carbon, hydrogen, and oxygen from sources other than the soil. The remainder are absorbed as minerals from the soil around the plant's roots.

Knowing that the elements are necessary still does not address the question of what each element does for the plant that makes it so essential. The roles of several have not yet been clearly defined; they are believed to allow certain enzyme systems

to function normally in the plant. At least some of the functions of other essential elements are known, as well as the symptoms shown by the plant when the element is lacking or in short supply. A brief summary of the functions of twelve essential elements and the symptoms of their deficiency is given in Table 2-2. Both the functions and symptoms are discussed further in later chapters.

Plants may exhibit symptoms of nutrient deficiency for several reasons:

- The element may be lacking totally or not be present in sufficient quantity.
- The element may be bound in a chemical form unavailable or too slowly available to the plant.
- There may be an overall imbalance of nutrients in the soil

While micronutrients can be and often are deficient in soils, macronutrients are most often deficient. Nitrogen is foremost among the elements regularly lacking in sufficient quantities to produce strong, healthy plants. When nitrogen, in the nitrate form, is not adsorbed by the colloidal particles of the soil, it passes quickly through the root region of the soil in an action called *leaching*.

SOIL STRUCTURE AND ORGANIC MATERIAL

In good loam soils, small soil particles adhere together to form larger particles or *aggregates*. This arrangement of soil particles into aggregates is termed the *soil structure*. Structure resulting from small porous aggregates is highly desirable since it blends the desirable qualities of looseness, drainage, and aeration with water and mineral retention. In the bare hand, good loam soil feels like short pastry dough. This may explain why such a structure is often referred to as a *crumb structure* or a *granular structure*.

Since the granular structure of soil is a desirable attribute, resulting in high-quality horticultural and field crops, it is important to understand how such soil structure develops and can be encouraged.

TABLE 2-1.
The Essential Elements

Macronutrients		Micronutrients	
Calcium	(Ca)	Boron	(B)
Carbon	(C)	Chlorine	(Cl)
Hydrogen	(H)	Copper	(Cu)
Magnesium	(Mg)	Iron	(Fe)
Nitrogen	(N)	Manganese	(Mn)
Oxygen	(O)	Molybdenum	(Mo)
Phosphorus	(P)	Zinc	(Zn)
Potassium	(K)		
Sulfur	(S)		

TABLE 2-2.
Essential Elements with Their Functions and Symptoms of Deficiency

Element	Function in the Plant	Symptoms of Deficiency
Boron	• Role not clearly defined except in translocation of sugar	• Dead shoot tips • Leaves thicken, curl, and become brittle • Flowers fail to form • Stunted roots
Calcium	• A component of the cell wall • Needed for cell division	• Stems, leaves, and roots die at tips, where growth is normally most active • Chlorosis of young leaves, then necrosis along margins
Copper	• A component of several important enzymes • Needed for photosynthesis	• Necrosis in young leaf tips and margins
Iron	• Needed for chlorophyll synthesis	• Chlorosis of younger leaves only, usually in an interveinal pattern
Magnesium	• Essential for photosynthesis as a component of chlorophyll molecule • An important enzyme activator	• Interveinal chlorosis, appearing first in older leaves, followed by red or purple color and necrotic spots
Manganese	• An enzyme activator in respiration • Makes nitrogen available for plant use	• Interveinal chlorosis and necrosis
Molybdenum	• Makes nitrogen available for plant use	• Interveinal chlorosis in lower leaves • Marginal necrosis • Flowers fail to form
Nitrogen	• Important to synthesis and structure of protein molecules • Encourages vegetative growth • Promotes rich green color	• Chlorosis, first noticeable in older leaves • Stunting of growth
Phosphorus	• Essential to energy transfer • Stimulates cell division • Needed for flowering • Promotes maturation • Promotes disease resistance • Promotes root development	• Red or purple discoloration of older leaves • Premature leaf drop • Stunting of growth

TABLE 2-2.
Continued

Potassium	• Role not clearly defined • Believed to activate important plant systems and enzymes	• Mottled chlorosis first noticeable on lower leaves • Necrosis at tips and margins of leaves
Sulfur	• Important to structure of protein molecules • Needed for enzyme activity to occur	• Leaf chlorosis first noticeable in younger leaves
Zinc	• Important to the synthesis of plant auxins • An important enzyme activator • Needed for protein synthesis	• Interveinal chlorosis on lower leaves • *White* necrotic spots • Leaf dwarfing • Distortion of leaves

The most significant factor in the development of a granular structure is organic matter, including green manure, soil organisms, decomposing plant roots, and especially humus. The organic material and colloidal clay bind the small mineral particles together as crumb-like aggregates. Humus is highly significant as a binding agent.

The characteristics of organic matter have already been described briefly. When a field of grasses, weeds, and other herbaceous plants is turned under, the green manure represented by their plant parts is rapidly acted upon by organisms of the soil. Macrobial life, such as insects and earthworms, feeds upon the plant parts to obtain the chemical energy bound within them. Microbial life—principally algae, fungi, bacteria, and actinomycetes—may directly attack the decomposing plant parts or contribute indirectly through digesting the excretions of the macrobes and the dead macrobes themselves.

The breakdown of dead plant tissue and other organic material by the microorganisms of the soil is accomplished with digestive enzymes in a manner similar to the way an animal's stomach digests food. Some chemical compounds within the plant break down more quickly than others. The simple proteins, sugars, and starches decompose quickly; the more complex organic compounds, like lignin, a component of the cell wall, decompose more slowly. Eventually, though, all organic matter decomposes into either humus, energy, or a number of other end products, including carbon dioxide, water, nitrates, phosphates, sulfates, and calcium compounds. The energy release explains why the temperature rises inside a compost pile.

SOIL ACIDITY AND ALKALINITY

The soil's water, held between the particles and granules of the soil, contains dissolved mineral salts. This liquid is known as the *soil solution*. The way the soil solution reacts determines the acidity, alkalinity, or neutrality of the soil. Many farmers and home gardeners still refer to the "sweetness" or "sourness" of the soil, harkening back to a time when differences in the soil's reaction could be observed and dealt with even if the causes were not fully understood. Today it is understood that some soils contain more hydrogen ions (H^+) than hydroxyl ions (OH^-). This makes them acidic. Other soils contain more hydroxyl ions than hydrogen ions. They are termed alkaline. When a soil contains equal concentrations of hydrogen and hydroxyl ions, it is termed neutral. The exact relationship

FIGURE 2-4.
Diagrammatic representation of pH

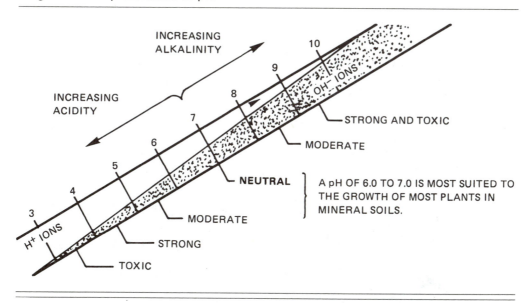

between the hydrogen and hydroxyl ions is expressed as a pH number, Figure 2-4.

- A pH of 7.0 is neutral.
- A pH of less than 7.0 is acidic.
- A pH of more than 7.0 is alkaline.

The pH of a soil can only be measured precisely using an instrument known as a pH meter, Figure 2-5. Commercial growers and homeowners can either send soil samples to their state agricultural college for a pH test at a nominal charge or purchase their own portable pH meter for immediate results. There are also pH test kits on the market, but their results are imprecise compared to those obtained with a pH meter. In situations where the crop is very sensitive to soil pH, the pH meter test should be used.

Additions to the soil that increase the number of H^+ ions will lower the pH of the soil; conversely, soil additions that increase the number of OH^- ions will raise the soil pH. Many of the materials used to improve the structure and texture of the soil will also modify its pH. For example, peat moss is highly

acidic, and its addition to the soil as a source of organic material will have a direct impact on the acidity of the soil solution. Limestone has the opposite effect, contributing alkalinity to the solution. These changes in the pH may or may not be desirable. Thus, additives should be used with caution and with a knowledge of their total impact. Obviously, it is easier to adjust the pH of a greenhouse bench or pot crop where the soil mixture is totally within the control of the grower than it is to change the pH of a 100-acre nursery field, especially since many soils have a strong *buffer resistance* to pH change. *Buffering* occurs when hydrogen ions that are held in adsorbed form dissociate from the clay particles and enter into the soil solution to replace those hydrogen ions neutralized by the addition of lime. No pH change will result until enough lime is added to deplete the supply of hydrogen ions that constitute a reserve of acidity in the soil. If the reserve acidity is strong and the field is large, a significant change in pH may be impossible. Where strong buffer resistance is not a factor, the pH of nursery fields can be altered to improve crop production.

FIGURE 2-5.
A pH meter provides the most accurate measure of a soil's acidity or alkalinity.

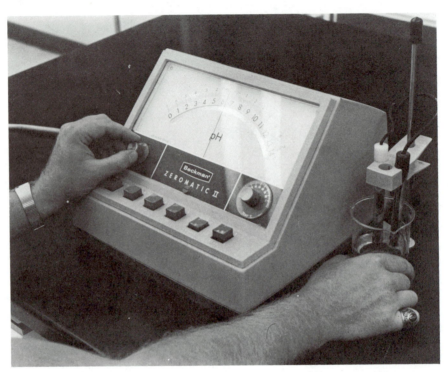

Within the pH range of 4.0 to 9.0, the availability of many mineral nutrients is determined by the acidity or alkalinity of the soil. For example, many plants will exhibit a distinctively patterned yellowing, or *chlorosis,* when grown in soil having a high pH. The cause of the chlorosis is lack of iron in the plant tissue, and it results because iron compounds, needed by the plant, are precipitated out of the soil solution and rendered unavailable to the plant. At a lower pH, the iron will remain in the soil solution and be available for plant uptake.

CATION EXCHANGE

To understand how colloidal clay particles and humus contribute to the chemical reactions of the soil, how the soil's pH can be modified, and how the application of chemical fertilizer can increase the nutrients in soil requires an understanding of *cation exchange.* The term refers to the capacity of colloidal particles to attract positively charged ions (*cations*) and to exchange one ion for another. Without cation exchange, nutrients would be readily leached from the soil. With cation exchange, the hydrogen cations held by the colloidal particles can be replaced by cations furnished through the decomposition of organic material, the weathering of rocks, or the application of fertilizers. It follows that soils having a higher percentage of colloidal particles, such as clay soils and organic soils, have a higher capacity for cation exchange than sandy soils that are lower in colloidal particles.

Role in pH

It is the replacement of hydroxyl cations on the colloidal particles of the soil by hydrogen ions that makes a soil acidic. To make a soil more acidic (lower the pH), elemental sulfur is usually added. In the soil, bacteria convert the sulfur to sulfuric acid. To make the soil more alkaline (increase the pH), calcium or calcium-magnesium compounds are commonly used.

The influence of pH on cation exchange is centered around the availability of nutrients as described earlier and illustrated by iron chlorosis at high soil pH. There are other elements that become bound tightly within the soil and are unavailable to plants when the pH is high, just as there are elements made unavailable when the pH is too low. At extremely low or high pH ranges the mere excess of H^+ or OH^- ions can be toxic to the plants.

Role in Mineral Absorption

It seems logical to assume that minerals are absorbed into plant roots as water is absorbed. It is a logical assumption but an incorrect one. The uptake of water and the uptake of minerals are independent processes.

Minerals enter root cells through a permeable membrane when the concentration of the mineral salts in the soil solution is greater than in the root cell. Such a condition creates a *concentration gradient*. Since plants are continually using the mineral salts within the roots, the concentration gradient serves to explain how certain elements are absorbed. With others, absorption occurs even against a concentration gradient. The explanation is thought to reside with ion exchange or with contact exchange.

In *ion exchange*, a positively charged ion may be absorbed by a root cell if another positively charged ion is released from the cell. Another form of ion exchange can occur when both a positively charged ion (cation) and a negatively charged ion (anion) are absorbed by the root cell together, thus maintaining the electrostatic equilibrium in the cell.

In *contact exchange*, the intimate association between the soil particles and the root hairs is the key. A direct exchange occurs between the ions adsorbed to the particles of soil and those of the root cells.

FERTILIZERS

Fertilizers are nutrient additives applied to the soil periodically to maintain optimum crop productivity. The need for fertilization may result from a deficiency of one or more mineral elements in the soil, their presence in a form unavailable to the plant, or the leaching of elements into the soil to a depth below the root zone.

Chemical Fertilizers

Since nitrogen, phosphorus, and potassium are the soil elements used in greatest quantity by the green plant, a fertilizer that provides all three elements is termed a *complete fertilizer*. The actual percentage by weight of each of the three major elements in a fertilizer determines its *analysis*. For example, 100 pounds of 10-6-4 analysis fertilizer contains 10 pounds of nitrogen (N), 6 pounds of phosphoric acid (P_2O_5), and 4 pounds of potash (K_2O). The analysis figures are always expressed in the same order and represent the same nutrients. Note that phosphorus and potassium are not present in elemental form, but as chemical compounds. Thus the amount of actual element provided for the plant is less than the analysis implies, Figure 2-6.

With simple arithmetic, fertilizers can be compared on the basis of their *nutrient ratio*. The ratio is a reduction of the analysis to the lowest common denominator. For example, a 5-10-10 analysis has a ratio of 1-2-2. (Each of the numbers has been reduced through dividing by a common factor of 5.) A fertilizer analysis of 10-20-20 also has a ratio of 1-2-2. Thus a 5-10-10 fertilizer supplies the three major nutrients in the same proportion as a 10-20-20 fertilizer, but twice as much of the actual product must be applied to obtain the same amount of nutrients, Table 2-3.

When less than 30 percent of a complete fertilizer's weight represents available nutrients, it is termed a *low-analysis fertilizer*. When the amount of available nutrients is 30 percent or more, the

FIGURE 2-6.

Interpreting fertilizer analysis figures: The nutrients are always shown in the same order. (From J. Ingels, *Landscaping: Principles and Practices,* 2nd edition, © 1983 by Delmar Publishers Inc.)

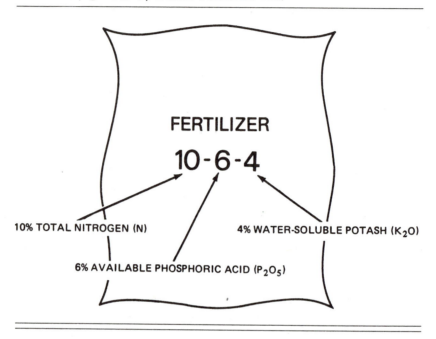

FERTILIZER

10 - 6 - 4

10% TOTAL NITROGEN (N)   4% WATER-SOLUBLE POTASH (K_2O)

6% AVAILABLE PHOSPHORIC ACID (P_2O_5)

product is a *high-analysis* fertilizer. The remaining material in a fertilizer is filler, either organic or chemical. The filler may provide some additional essential elements, and may even be important as a source of micronutrients, but essentially the filler is a carrier for the available macronutrients. It al-lows them to be applied evenly to the soil and crop. It also adds weight and bulk, both undesirable features. Table 2-4 compares high-analysis and low-analysis fertilizers.

The more common chemical fertilizers include: anhydrous ammonia, ammonium nitrate, urea, so-

TABLE 2-3.

Comparison of Fertilizers with Same Nutrient Ratio

50 Pounds of 5-10-10 Fertilizer Contain:	50 Pounds of 10-20-20 Fertilizer Contain:
2 1/2 pounds of N (nitrogen)	5 pounds of N
5 pounds of P_2O_5 (phosphoric acid)	10 pounds of P_2O_5
5 pounds of K_2O (potash)	10 pounds of K_2O

TABLE 2-4.

Comparison of High- and Low-Analysis Fertilizers

High-Analysis Fertilizers	Low-Analysis Fertilizers
• Contain more nutrients and less filler	• Contain fewer nutrients and more filler
• Cost less per pound of actual nutrient	• Cost more per pound of actual nutrient
• Weigh less; less labor required in handling	• Weigh more and are bulkier; more labor required in handling
• Require less storage space	• Require more storage space
• Require less material to provide a given amount of nutrients per square foot	• Require more material to provide a given amount of nutrients per square foot
• Require less time to apply a given amount of nutrients	• Require more time to apply a given amount of nutrients

dium nitrate, and ammonium sulfate as carriers of nitrogen; superphosphates, ammoniated phosphates, and ground rock phosphate as carriers of phosphorus; and potassium nitrate, potassium chloride, potassium sulfate, and potassium-magnesium sulfate as carriers of potassium.

Organic Fertilizers

Organic fertilizers tend to be low in nutrient content, especially nitrogen. Although they enjoy some popularity with organic gardeners and hobbyists, they have limited application to commercial ornamental horticulture. Not only are the nutrients limited in the organics, but they are often more slowly available to the plant than those in chemical fertilizers. Organic fertilizers include materials such as dried blood, cocoa meal, animal manures, dried sewage sludge, and bone meal. The latter two materials have some commercial usage: Sewage sludge is used as a top-dressing on golf greens and bone meal as a high-phosphorus fertilizer for flowering bulbs.

Ammonification and Nitrification

The presence of an element in the soil does not guarantee that the plant can make use of it. It may be an essential element but in a form that must undergo chemical change before being available to the plant. Nitrogen is made available to green plants as nitrate salts (NO_3^-) or ammonium salts (NH_4^+). In turn, members of the animal kingdom depend almost exclusively on green plants for their nitrogen, which they take from plants' proteins and amino acids.

Nitrogen salts are not found as minerals in the soil. They are formed by the decomposition of nonliving plants, plant parts, animals, or excretory products. Integral to this degradation are several separate groups of bacteria that are key factors in the twin conversions known as ammonification and nitrification.

Ammonification is the conversion of the nitrogen in organic compounds to ammonia. *Nitrification* is the conversion of ammonia to nitrite, then to nitrate as shown.

$$2NH_4^+ + 4O_2 \rightarrow 2NO_2^- + 4H_2O + \text{energy}$$
(ammonia)   (nitrite)

$$2NO_2^- + O_2 \rightarrow 2NO_3^- + \text{energy}$$
(nitrite)   (nitrate)

The energy produced by these reactions is used by the bacteria for their growth and development.

Collectively, the two processes can be diagrammed as follows:

$$\underbrace{\text{organic nitrogen} \rightarrow \text{ammonium salts (NH}_4^+\text{)}}_{\text{Ammonification}} \underbrace{\rightarrow \overset{\text{nitrite}}{\text{salts (NO}_2^-\text{)}} \rightarrow \overset{\text{nitrate}}{\text{salts (NO}_3^-\text{)}}}_{\text{Nitrification}}$$

Three additional points deserve mention. First, the nitrifying bacteria are sensitive to low temperatures, acidity, and poor drainage. In such conditions, the bacteria do not drive the biochemical changes as rapidly as they do under more favorable conditions. Second, the microorganisms of the soil need nitrogen for growth as much as green plants or animals do. As a result, the addition of large quantities of organic matter to the soil can result in a temporary decline in the amount of nitrogen available to crop plants as the soil microbial population increases. Finally, three genera of bacteria (*Azotobacter, Clostridium,* and *Rhizobium*) are able to convert atmospheric nitrogen gas (N_2) into organic compounds. After the bacteria die, the processes of ammonification and nitrification release the nitrogen for use by green plants. Especially important is the relationship between the *Rhizobium* bacteria and members of the bean family (the legumes or Leguminosae). The bacteria live in the legume roots where they obtain organic food materials. In return, the bacteria capture or fix nitrogen gas, making it available to the green plant.

Phosphorus

Phosphorus is present in very small amounts in mineral soils. An even smaller amount is present in a form usable by plants at any one time, since even simple phosphorus compounds are usually insoluble in the soil solution. Along with nitrogen, which is also present in small amounts, phosphorus is often a limiting factor in soil nutrition.

Much of the soil's phosphorus is bound within the organic matter of the soil and becomes available to plants as organic material decomposes. The phosphorus held within the mineral matter of the soil is much more slowly available to the plant. Only through a slow interaction between the insoluble phosphate, water, carbon dioxide, and various root exudates does the phosphate become water-soluble and hence available. Even then, the soluble phosphates can quickly revert to complex, insoluble forms, especially as the pH increases or decreases.

Potassium

Potassium is present in soil in much larger quantities than either nitrogen or phosphorus. Nearly all the soil's potassium is in inorganic forms, however, and inorganic potassium is not readily available to plants. It becomes available slowly through the reaction of water and carbonic acid in the soil with the feldspars, micas, and other sources of insoluble potassium.

The other essential elements require similar chemical reaction in the soil to become available for absorption by the plant. Soil chemistry is a complex field of study and the brief treatment given the subject here has only touched upon a few of the points necessary to an understanding of crop production.

SUMMARY

Soil is the thin outer layer of the earth's crust, made up of weathered minerals, living and nonliving organisms, water, and air. It provides the underground environment of plants. In profile, three distinctive layers can be seen: bedrock, subsoil, and topsoil. Both subsoil and topsoil are finely weathered from the bedrock, but topsoil contains more organic matter than subsoil. As a result, it is more supportive of plant growth.

Organic matter is composed of plant and animal compounds that soil microbes can rapidly break down and of humus. Humus is a complex colloidal mixture that originates with organic compounds that do not decompose easily. Humus and other organic matter are important in the development of good soil structure, causing small soil particles to bind and form larger particles or aggregates.

Soils differ in many ways: color, weight, drainage, rockiness, and texture. The differences result from weathering elements, soil movement, topography, climate, and variations in the amount of organic matter.

When bedrock weathers, particles of differing sizes are formed. The four groups of weathered particles or separates are gravel, sand, silt, and clay. The relative proportion of these separates in any one soil creates the soil texture. The textural name of a soil roughly describes the particular mixture of separates that it contains.

Because of their colloidal properties, clay and humus are instrumental in making nutrients available to plants. Sixteen nutrients have been found essential to the growth of green plants (C, H, O, P, K, N, S, Ca, Mg, Fe, Mo, B, Cu, Mn, Zn, Cl). Of these essential elements, only carbon, hydrogen, and oxygen are obtained by the plant from sources other than the soil. The remainder are absorbed as minerals from the soil around plant roots. When one or more element is lacking totally or in part, is bound in a form unavailable to the plant, or is part of an overall nutrient imbalance in the soil, the plant will exhibit symptoms of nutrient deficiency.

The soil solution is a liquid composed of water held within the soil and mineral salts dissolved in the water. How the soil solution reacts chemically determines its acidity or alkalinity and is expressed as a pH number. Soils that contain more hydrogen ions (H^+) than hydroxyl ions (OH^-) are acidic and have a pH of less than 7.0. Soils containing more hydroxyl ions than hydrogen ions are alkaline and have a pH greater than 7.0. A soil with equal concentrations of hydrogen and hydroxyl ions is neutral and has a pH of 7.0. Soil additives can raise or lower the pH of the soil depending upon whether they increase the number of H^+ ions or OH^- ions.

Modification of pH and nutrient uptake by plants depend upon cation exchange: the capacity of colloidal particles to attract positively charged ions (cations) and exchange one ion for another. Those soils with a higher percentage of colloidal particles, such as clays and organic soils, have the highest cation exchange capacity. If soil pH becomes too high or too low, certain elements become bound tightly within the soil and unavailable to the plants.

Fertilizers are nutrient additives applied to the soil periodically to maintain optimum crop productivity. They may be complete or incomplete, high-analysis or low-analysis, inorganic (chemical) or organic. Usually, the elements within the fertilizer must undergo chemical changes before becoming available to the plant. Ammonification and nitrification exemplify such changes.

Achievement Review

A. SHORT ANSWER

Answer each of the following questions as briefly as possible.

1. Define the following terms.
   a. soil
   b. topsoil
   c. subsoil
   d. humus
   e. soil texture
   f. essential element
   g. macronutrients
   h. micronutrients
   i. leaching
   j. soil structure
   k. cation
   l. cation exchange
   m. complete fertilizer
   n. fertilizer analysis
   o. nutrient ratio
   p. high-analysis fertilizer
   q. low-analysis fertilizer

2. List five ways in which the same bedrock could become quite different subsoil or topsoil.

3. Match the type of transported soil with the correct agent of transport.

   a. aeolian soil   1. water
   b. colluvial soil   2. sedentary
   c. glacial till   3. gravity
   d. alluvial soil   4. ice
              5. wind

4. List the four soil separates in order of increasing particle size.

5. From the soil texture triangle, identify soils having the following analysis:

   a. 60 percent sand   c. 25 percent sand
       30 percent silt   30 percent silt
       10 percent clay   45 percent clay
   b. 25 percent sand   d. 80 percent sand
       55 percent silt   10 percent silt
       20 percent clay   10 percent clay

6. Based upon the textural name, arrange the following soils in decreasing order of water retention: sandy loam, silty clay, clay, clay loam, loam, sand.

7. Indicate whether the following statements apply most appropriately to clay, silt, or sand.

   a. It is the most chemically active particle in the soil.
   b. The size of the air spaces between the particles is greatest.
   c. Water passes most quickly through this separate.
   d. This separate has small particle size and low nutrient value.
   e. The particles are small and plate-like.
   f. It is the separate most important to soil nutrition.

8. List the sixteen essential elements and their chemical symbols.

9. Compare high and low analysis fertilizers by placing an X in the appropriate column after each of the characteristics listed below.

Fertilizer Characteristic	Low-Analysis Fertilizer	High-Analysis Fertilizer	All Complete Fertilizers
Contains N, P, and K			
Less than 30 percent of its weight is in available nutrients			
Most bulky to store and handle			
Contains 30 percent or more of its weight in available nutrients			
Costs more per pound of nutrients			
Requires less time to apply a given amount of nutrient			
Contains filler material			
Requires less material to apply a given amount of nutrients per square foot			

B. ESSAY

1. Describe the natural decomposition of organic matter in the soil and its relationship to soil structure.
2. Explain how nutrients move from the soil into the plant.

C. TRUE/FALSE

Indicate if the following statements are true or false.

1. Excess hydrogen ions in the soil make it acidic.
2. Excess hydroxyl ions in the soil make it alkaline.
3. An equal concentration of hydrogen and hydroxyl ions results in a pH of 6.0.
4. A pH of 7.0 represents a state of chemical neutrality in the soil.
5. A large nursery field with strong buffer capacity could have its pH changed easily.
6. The soil pH of potted greenhouse crops can usually be changed easily.
7. A field test kit is the most precise way to measure soil pH.
8. Organic matter can lower soil pH.
9. Sulfur can lower soil pH.
10. Limestone can raise soil pH.

3
Describing and Identifying Plants

Objectives

Upon completion of this chapter, you will be able to

- use scientific plant names correctly.
- define common taxonomic and professional terms.
- describe plants using taxonomic terms.
- use a plant key.

PLANT TAXONOMY

In Chapter 1, you were introduced to a system for classifying plants based on their genetic and evolutionary relationships. It is a natural system of classification and one that accommodates all plants. The systematic classification of plants is called *taxonomy* and those botanists who are especially interested in plant classification are called *taxonomists*.

To review briefly, every plant in nature is classified into the following categories and subcategories:

Kingdom
Division
Class
Order
Family
Genus
Species

Every plant derives its *scientific name* from its generic and specific classification, the smallest two categories to which a plant belongs. Since there are two parts to each scientific name, it is termed a *binomial.* Scientific names are derived from the classical Latin and Greek languages, predominantly Latin. The use of languages that are no longer spoken assures that the language will not change and thus affect the scientific terminology.

A plant's scientific name gives it international recognition, since a specific scientific name can only be assigned to one plant, unique in some way from all others in the world. Although plants have common names by which they are known within a particular country or region, the common names have little acceptability to the scientific community because they are often misleading and are inconsistent from place to place.

To illustrate all of this, consider the trees we call the maples. All are in the Genus *Acer* because of certain genetic and structural similarities. Examples are the silver maple, amur maple, hedge maple, red maple, and Norway maple. To the residents of a particular region, the common names are an adequate identification for these plants, known in the Latin as *Acer saccharinum, Acer ginnala, Acer campestre, Acer rubrum,* and *Acer platanoides* respectively. However, there is also *Acer saccharum.* It is known in some areas of North America as sugar maple and in other areas as hard rock maple, yet it is the same plant. Also, if desiring a maple with red, not green leaves, throughout the summer season, should a red maple, a scarlet maple, or a crimson king maple be selected? All imply a red color, but with two of the choices, the red color occurs only in the autumn. Only one, the crimson king maple, has red foliage all summer long. Also consider the plant known as the box elder. The name does not suggest a maple, yet it too is a maple, *Acer negundo.* The common names given plants are often based upon physical similarities to more familiar plants. For example, the grape holly (*Mahonia aquifolium*) is neither a grape nor a holly, but it has leaves resembling an American holly and purple-clustered fruit suggestive of grapes. Common names may also be based upon legend, sentiment, or a desire to offer tribute, as, for example, the Judastree, the Tree of Heaven, or the Jacqueline Kennedy Tea Rose.

Professional horticulturists need to know both the scientific and common names of the plants they grow or work with. The layman will normally use the common names in dialogue with the horticulturist; but professionals will use the scientific names in business matters to assure understanding.

Scientific names are always underlined or printed in italics. The genus is capitalized, but the species epithet is not. To be totally correct, the genus and species are often followed by an abbreviated name or initial in reference to the person who first described the species. For example, in *Nerium oleander* L., the *L* after the species name refers to the Swede Karl von Linne, who described the oleander

in 1753. More often called Linnaeus, after the Latin form, he was the first to establish the binomial system of nomenclature for classifying plants and animals. During his seventy-one years (1707-78), Linnaeus described over thirteen hundred different plants from all corners of the globe. He truly earned his title as the father of taxonomy, Figure 3-1.

In the everyday use of scientific names, the reference to the pioneer taxonomist may be omitted; however, still another subcategory beyond the species can occur and is often needed to fully distinguish a certain plant. Within the cultivated species that are the economic backbone of ornamental horticulture are certain groups of plants sufficiently different in appearance from others of the same species

FIGURE 3-1.
Karl von Linne (Linnaeus), the father of taxonomy, was the first to establish a binomial system of nomenclature for classifying plants and animals. (Courtesy Council of Linnean Society of London)

to warrant special designation. Taxonomists term such a group a *variety*. When it is an intentionally cultivated variety, it is termed a *cultivar*. Consider the Norway maple. It has the scientific name of *Acer platanoides* and has distinctive morphological (structural) features, including deep green foliage. Within the species, one member shares all of the morphological characteristics except that its foliage has a reddish tinge. Its common name is the Schwedler maple or Schwedler Norway maple. Its scientific name is *Acer platanoides* 'Schwedleri.' The cultivar portion of the name is capitalized, not italicized, and set off by single quotation marks. Another way to indicate the cultivar is as *Acer platanoides* cv Schwedleri. Either method is considered correct.

HORTICULTURAL DESCRIPTIONS OF PLANTS

In addition to their botanical classification and possession of scientific and common names, most plants can be grouped into an assortment of other categories based upon physical appearance rather than genetic relationship. Such groupings are artificial yet serve many needs of professional plant workers, especially the ornamental horticulturists. An understanding of the terms and descriptions to follow is basic to the many career fields in ornamental horticulture.

Woody plants are those having a corky outer surface of bark covering their older stems. The woody plants usually survive the winter, and the woody stems normally increase in diameter each year.

Herbaceous plants are more succulent plants. They lack bark covering, and their twigs usually do not increase much in diameter. They are often unable to survive the winter in cold climates above ground.

Evergreens are plants that retain their leaves all year. While individual leaves drop and are replaced periodically, the overall appearance of the plant remains green.

Deciduous plants are those that drop their leaves and enter a period of dormancy once a year.

Semievergreens may retain their leaves during the winter months, but the leaves discolor and often winter-burn. In southern climates such plants usually are evergreen.

Trees are woody plants, either evergreen or deciduous, that produce a canopy of leaves atop a single stem.

Shrubs are basically the same as trees except that they seldom get as tall and have multiple stems instead of one.

Vines may be woody or herbaceous. Their stems are unable to support the weight of the plants. If the plants are to grow upright, they must be supported by a trellis, fence, or wall.

Groundcovers can be woody or herbaceous, flowering or nonflowering, trailing or compact. At maturity, their height is 18 inches or less.

Annuals are plants that complete their life cycle from seed to fruit to death within one growing season.

Perennials are plants that live several years and, where necessary, can survive cold winter months in a dormant state. They do not die after flowering.

Biennials live two years. The first year is spent in a vegetative stage of development. After a period of winter temperatures, necessary to initiate flower development, the plant flowers during the second year, then dies.

Hardy plants are those that will survive the winter temperatures of a locale. Ornamental plants, especially the woody ones, are assigned *hardiness zone* ratings. These correspond to ten zones on the North American continent with different average annual minimum winter temperatures. Zone 1 is the coldest zone and zone 10 is the warmest zone, Figure 3-2. For example, a plant rated as a zone 5 plant will survive the winters in zone 5 or warmer. It will not survive in zone 4 or colder.

The hardiness zones were established by the United States Department of Agriculture (U.S.D.A.). Other attempts to rate plant hardiness within regions have complicated this essentially simple rating system. It is important to know if the U.S.D.A. rating system or some other one is being referred to when using hardiness ratings for plants.

FIGURE 3-2.

Hardiness zones (Compiled by The Arnold Arboretum, Harvard University, May 1, 1967)

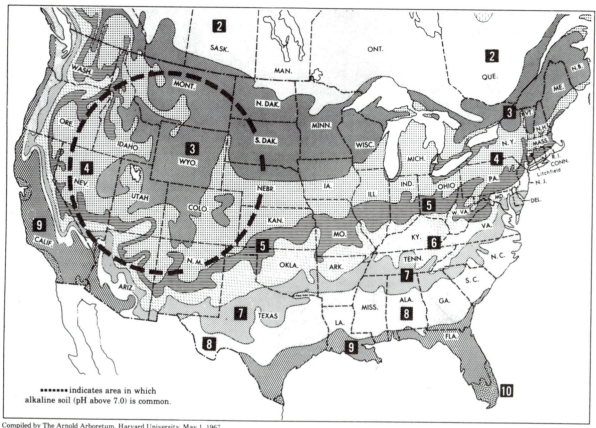

•••••• indicates area in which
alkaline soil (pH above 7.0) is common.

Compiled by The Arnold Arboretum, Harvard University, May 1, 1967

Tender plants are those that will not survive temperatures below freezing.

Nursery plants are those produced in nursery fields, greenhouses, and container operations for use in residential, commercial, public, and institutional landscapes.

Greenhouse crops are usually herbaceous and often flowering plants grown in greenhouses for sale to retail flower shops and other outlets.

Bedding plants are used to create flower beds and flower borders. They may be annual or perennial but are nearly always herbaceous. They are often grown and sold as multiples in strips of peat moss pots or in shallow plastic trays. Others are sold as singles in 4-inch pots.

Foliage plants are prized more for their leaves and habit of growth than for their flowers or fruit. In temperate regions they are used primarily as indoor plants. In tropical and semitropical regions they are used outdoors.

Native plants are those that evolved in a given area.

Exotic plants are those brought into an area to which they are not native. Very often their survival is dependent upon care and tending by humans.

Naturalized plants are those brought into an area as exotics that adapt well enough to cultivation and grow as successfully as native plants.

Aquatic plants are rooted under water. Those of importance to ornamental horticulture, such as water lilies, send their leaves and flowers to the surface.

Specimen plants are visually distinctive due to some feature such as growth habit, flower, bark, or fruit color. The term is applied most often in the landscape profession.

Accent plants, another landscape term, refers to plants different from others with which they may be grouped. The difference is not as marked as with specimen plants, however. It may be only a difference of height, shape, or texture.

Many more horticultural plant categories exist. They are a part of the technical language of the many specialized branches of ornamental horticulture.

HOW PLANTS ARE IDENTIFIED

The groupings just listed are especially helpful for narrowing the choices when plants are being selected to serve a function, play a role, or solve a problem. There still remains the need to identify plants specifically; for example, to recognize the difference between red oaks and live oaks or between zinnias and dahlias or to recognize deciduous plants by their winter twigs after leaves, flowers, fruit, or other helpful features are gone.

There are several kinds of identification. The simplest and at the same time the most complex is identification through recognition, often built upon a lifetime of association and familiarity. If someone asked how you identify your best friends, it might be difficult to describe the mental process behind the recognition. The combination of body build, facial features, hair and eye color, manner of walking, voice pattern, and more, all identify a certain individual. Having a mental picture of that friend then allows us to take only one feature, such as the voice on the telephone, and reconstruct the total person in our mind's eye. Similar identification becomes possible with plants after repeated exposure to the same species. How the plants were first learned is forgotten and the ability to recognize those plants in nearly all situations becomes a part of our permanent learning.

Another, less precise, kind of identification occurs when a plant species is recognized as belonging to a group but the exact species name is not known. For example, you may know it's a spruce, but not know *which* spruce. Such a level of identification serves many professional horticulturists who have no particular need for precision. The exact basis for the ability to recognize at least the genus of the plant may be as lost in memory as that described earlier.

Finally, there is the kind where the horticulturist has absolutely no idea what the plant is and must begin a methodical process of identification through searching and gradual elimination. Knowing that in all probability the plant has been identified, classified, named, and entered into the scientific literature offers some comfort when faced with the identification of an unknown. Even then it seems like a monumental task trying to identify one plant from among the thousands and thousands of species on the earth.

Sometimes there are shortcuts. For instance, if the plant is known to be a broadleaved evergreen, or a flowering woody shrub, or a cactus, you can go to specialized texts in which a description or photo may be found. Perhaps someone from the state university or the local county agent can identify the plant. Such questions can be directed to them with clear conscience, but it is still no substitute for knowing how to find out yourself.

The tracing of unknown plants, especially the economically important ornamentals, requires the use of a *dichotomous analytical key*. There are hundreds of such keys, some dealing with a limited group of plants, others more encompassing. All present the searcher with a series of couplets and pursue identification through a process of elimination. Each couplet consists of two contrasting statements. The searcher chooses the statement that best describes the unknown plant. Beneath that statement will be another couplet, again requiring the selection of one and elimination of the other. The process continues until all species have been rejected except the one to which the plant belongs.

These keys are based in large part upon physical features of the plant rather than physiological or evolutionary relationships. Very complex and inclusive keys such as the one published in 1949 by Liberty Hyde Bailey, *Manual of Cultivated Plants Most Commonly Grown in the Continental United States and Canada* (Revised edition, Macmillan, New York) often require information unknown to the searcher. For example, the couplets may require the choice between flowers with five petals or multiples of five and flowers with six petals or multiples of six. If the unknown plant has no flowers attached, the tracing can get off track very easily.

For the beginner it is best to begin with a simpler key. Consider the following simplistic example to understand how keys work. Only *genera* (the plural of genus) are being identified.

1. Plant evergreen
    2. Leaves needle-like
        3. Needles separate
            4. Needles stiff and sharp, four sided . . . Spruce (*Picea*)
            4. Needles flat with two white lines on underside . . . . . . Hemlock (*Tsuga*)
        3. Needles grouped in fascicles. . . . . Pine (*Pinus*)
    2. Leaves broad and flat
        5. Leaves oval in shape, not spiny
            6. Flowers large, compound, and showy. . . . (*Rhododendron*)
            6. Flowers small, fascicled, and bell-like . . . . Andromeda (*Pieris*)
        5. Leaves holly-like and spiny . . . . . . Oregon-grape (*Mahonia*)
1. Plant deciduous

Keys are seldom this simple but at least the concept of dichotomous couplets can be appreciated.

VISUAL DESCRIPTIONS OF PLANTS

Since keys rely heavily upon physical features of plants as the basis for separation, a trained horticulturist must have a working knowledge of the terms used to describe plants. Such terminology is the stuff of which dichotomous keys are made.

Leaves

A leaf is described in terms of its *shape*, its *margin*, the presence or lack of *lobes*, its *base*, the pattern of its *veins*, its *apex* style, whether it is *simple* or *compound*, and whether it is *smooth*-surfaced or *pubescent* (covered with fine epidermal hairs), Figures 3-3 to 3-9.

FIGURE 3-3.

Parts of a leaf (From H. E. Reiley and C. Shry, Jr., *Introductory Horticulture,* 2nd edition, © 1983 by Delmar Publishers Inc.)

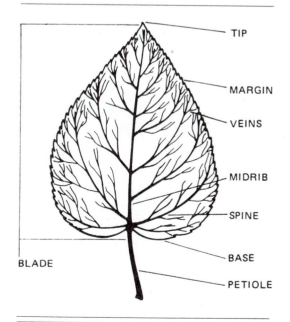

FIGURE 3-4.
Examples of a simple leaf and different types of compound leaves

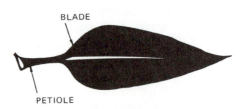

BLADE

PETIOLE

SIMPLE (ONE SOLID BLADE PER PETIOLE)

PETIOLE

LEAFLET

TRIFOLIATE COMPOUND
(THREE LEAFLETS PER PETIOLE)

PALMATE COMPOUND
(FIVE LEAFLETS JOINED AT ONE
POINT ON A SINGLE PETIOLE)

PINNATE COMPOUND
(MULTIPLE LEAFLETS ALIGNED
ALONG A SINGLE PETIOLE)

BIPINNATE COMPOUND
(MULTIPLE LEAFLETS ATTACHED TO
SHORT STALKS [RACHIS] AND ALIGNED
ALONG A SINGLE PETIOLE)

FIGURE 3-5.
Examples of typical leaf shapes

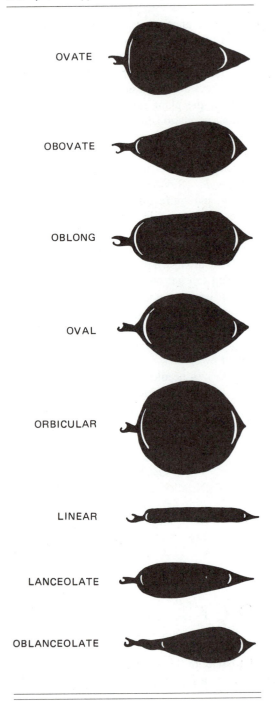

OVATE

OBOVATE

OBLONG

OVAL

ORBICULAR

LINEAR

LANCEOLATE

OBLANCEOLATE

FIGURE 3-6.
Examples of leaf apexes

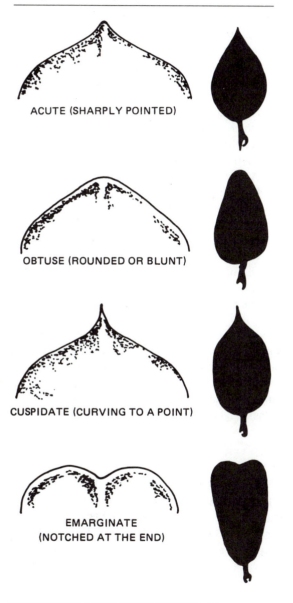

ACUTE (SHARPLY POINTED)

OBTUSE (ROUNDED OR BLUNT)

CUSPIDATE (CURVING TO A POINT)

EMARGINATE
(NOTCHED AT THE END)

FIGURE 3-7.
Examples of leaf bases

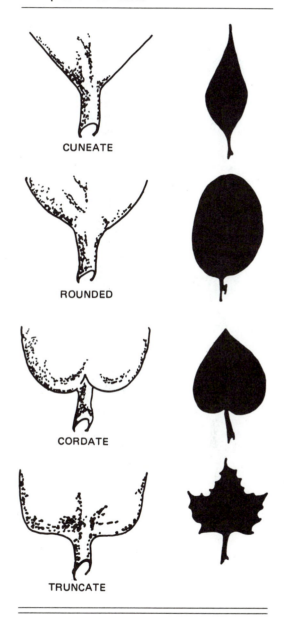

CUNEATE

ROUNDED

CORDATE

TRUNCATE

FIGURE 3-8.
Examples of leaf margins

ENTIRE   DENTATE

SERRATE   DOUBLY SERRATE

CRENATE   CRENATE-SERRATE

Twigs

A branch or twig is described on the basis of the *type of buds* it possesses, the nature of its *terminal* bud, the *arrangement* of its buds and leaves, the shape of its leaf *petiole scars,* prominence of its *lenticels* and *stipule scars,* and the type of *pith* it produces, Figures 3-10 to 3-13.

Flowers

The parts of a flower were illustrated and discussed in Chapter 1. In addition to the flower parts already described, plant keys may identify plants by the type of flower formed. Flowers may be described by the *position of the ovary* in relation to the flower parts, Figure 3-14.

- A *hypogynous flower* has a *superior* ovary, which means that the ovary is attached to the stem above the place where the other flower parts are attached.
- A *perigynous flower* also has a superior ovary, but the petals and sepals are fused to form a tube-like structure around but separate from the ovary.
- An *epigynous flower* has an *inferior* ovary, or one that is attached to the stem below where the other flower parts are attached. There is a floral tube, but it is united with the wall of the ovary.

Flowers may also be described as *singular* (one flower per stem as with a tulip), *composite* (multiple small ray and disc flowers as with the sunflower), or an *inflorescence* (clusters of small flowers arranged on an axis as with snapdragons or viburnums), Figure 3-15.

Fruits

A fruit is the ripened ovary of the flower. Generally, as a fruit develops the ovary enlarges beyond the size seen in a fresh flower before pollination. Since there are different types of flowers, it follows that there are also varying kinds of fruits.

Fruits are categorized into four major groups:

1. Simple fruit: develops from a single ovary
2. Aggregate fruit: develops from a single flower having a group of ovaries
3. Multiple fruit: develops from multiple ovaries of multiple flowers borne on a single stalk
4. Accessory fruit: develops from one or more ovaries and includes the calyx and/or receptacle

FIGURE 3-9.
Examples of lobing in simple leaves

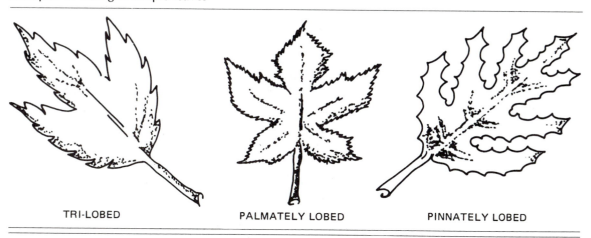

TRI-LOBED   PALMATELY LOBED   PINNATELY LOBED

FIGURE 3-10.
The parts of a woody twig important in identifying a species

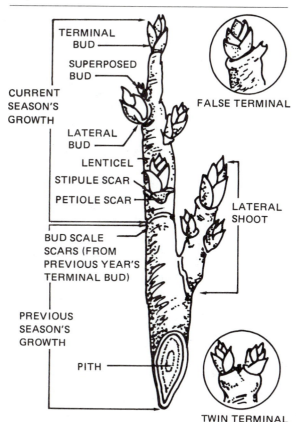

TERMINAL BUD

SUPERPOSED BUD

CURRENT SEASON'S GROWTH

FALSE TERMINAL

LATERAL BUD

LENTICEL

STIPULE SCAR

PETIOLE SCAR

LATERAL SHOOT

BUD SCALE SCARS (FROM PREVIOUS YEAR'S TERMINAL BUD)

PREVIOUS SEASON'S GROWTH

PITH

TWIN TERMINAL

Some of the fruit types are more characteristic of ornamental plants than other types; however, a complete understanding of fruit types is preferable to a partial one. The simple fruits are described by the mature appearance of the ovary wall, the *pericarp*. Depending upon the particular fruit type, the pericarp may be further divided into three separate layers:

1. exocarp (outer layer)
2. mesocarp
3. endocarp (inner layer)

The three layers are easily recognized in the drupes, or "pit" fruits, exemplified by plums and peaches. The thin outer skin is the exocarp, the fleshy part of the fruit is the mesocarp, and the hard pit is the endocarp. So distinctive are the many fruits formed by plants that keys exist which depend solely on the fruits as the basis for the separation, Figure 3-16.

THE ASSIMILATION OF TERMINOLOGY

By now the accumulation of terminology may be weighing heavily upon you. Rather than being put off by all of the new words, regard them as a new

FIGURE 3-11.
Types of buds

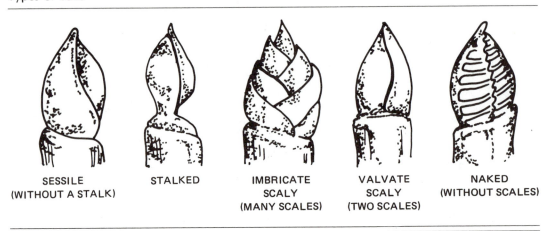

SESSILE (WITHOUT A STALK)   STALKED   IMBRICATE SCALY (MANY SCALES)   VALVATE SCALY (TWO SCALES)   NAKED (WITHOUT SCALES)

FIGURE 3-12.
Typical leaf and bud arrangements

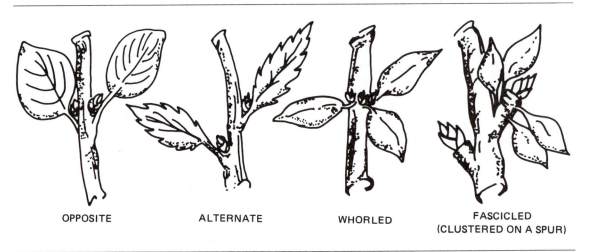

OPPOSITE   ALTERNATE   WHORLED   FASCICLED (CLUSTERED ON A SPUR)

language, as an expansion of your existing knowledge. Like any new words, they are best learned and understood through repeated use. Obtaining a good plant key and a supply of unknown plants, then identifying one from the other is the best means of learning the many terms. What begins as memorization can become permanent knowledge if approached properly.

SUMMARY

The systematic classification of plants is termed taxonomy. Taxonomy provides each plant in nature with a binomial scientific name composed of the genus and species to which it belongs. While many plants have common names as well, the scientific

FIGURE 3-13.
Different types of pith

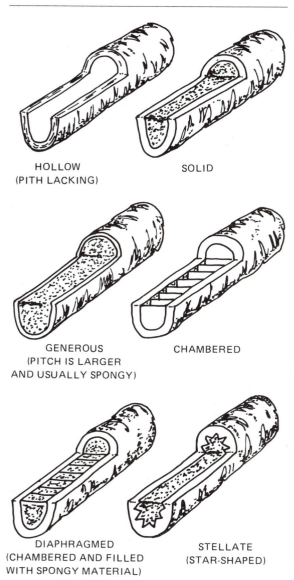

HOLLOW
(PITH LACKING)

SOLID

GENEROUS
(PITCH IS LARGER
AND USUALLY SPONGY)

CHAMBERED

DIAPHRAGMED
(CHAMBERED AND FILLED
WITH SPONGY MATERIAL)

STELLATE
(STAR-SHAPED)

as varieties. Intentionally cultivated varieties are termed cultivars.

In addition to their botanical classification, plants are commonly grouped into assorted other categories based upon their physical appearance rather than their genetic relationships. For example, plants may be categorized as:

- woody or herbaceous
- evergreen, deciduous, or semi-evergreen
- trees, shrubs, vines, or groundcovers
- annuals, perennials, or biennials
- hardy or tender
- nursery or greenhouse crops
- bedding or foliage plants
- native, exotic, or naturalized
- aquatic
- specimen or accent

Unknown plants can be identified by using a dichotomous analytical key. Hundreds of keys exist, some dealing with a limited group of plants, others more encompassing. All require the searcher to choose the plant's identifying features from among a series of couplets. Use of the keys requires that the searcher know the terminology used to describe leaves, twigs, flowers, and fruits.

Leaves are described in terms of their shapes, margins, lobing, bases, vein patterns, apex styles, surface pubescence, and whether they are simple or compound. Twigs are described on the basis of their bud types, terminal buds, bud and leaf arrangements, petiole scar shapes, prominence of their lenticels and stipule scars, and type of pith. Flowers are described by the position of the ovary in relation to the flower parts, or by the arrangement of flowers on the stem. Fruit are described as simple, aggregate, multiple, or accessory depending upon the number of flowers and ovaries involved in their formation.

binomial identifies the plant worldwide. Within a particular species, certain groups of plants are sufficiently distinctive to warrant further designation

FIGURE 3-14.
Types of flowers based upon the position of the ovary

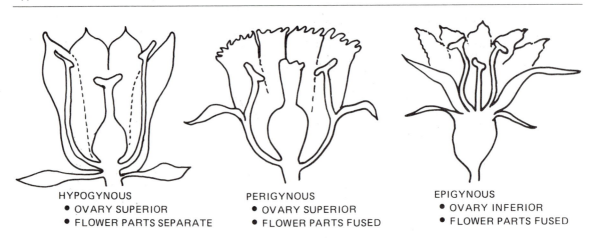

HYPOGYNOUS
• OVARY SUPERIOR
• FLOWER PARTS SEPARATE

PERIGYNOUS
• OVARY SUPERIOR
• FLOWER PARTS FUSED

EPIGYNOUS
• OVARY INFERIOR
• FLOWER PARTS FUSED

Achievement Review

A. SHORT ANSWER

Answer each of the following questions as briefly as possible.

1. For each of the following statements, list the word that would complete it correctly.
   a. The systematic classification of plants is called _____.
   b. Botanists specializing in plant classification and relationships are called _____.
   c. The two-part scientific name of a plant is called a _____.
   d. Every plant derives its scientific name from its genus and _____.
   e. In the scientific name, only the _____ is capitalized.
   f. Every plant has only _____ scientific name.
   g. Scientific names are either italicized or _____.
   h. The father of taxonomy is _____.
   i. Within a single species of plants, differences may result in a subcategory known as a _____.
   j. When differences among members of a species are intentionally encouraged by horticulturists, the plants are termed _____.

2. Explain the differences between the following types of plants:
   a. woody and herbaceous plants
   b. evergreen and deciduous plants
   c. trees and shrubs

FIGURE 3-15.
Examples of singular and clustered inflorescences

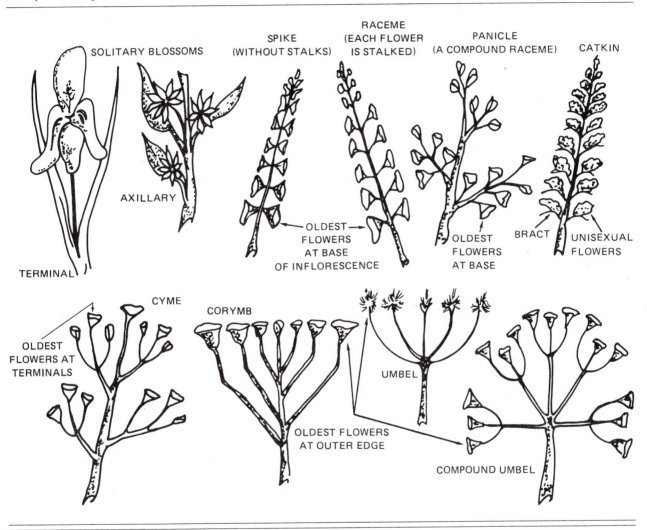

d. vines and groundcovers
e. annuals and perennials
f. biennials and perennials
g. hardy and tender plants

h. bedding plants and foliage plants
i. native and exotic plants
j. naturalized and exotic plants
k. specimen and accent plants

FIGURE 3-16.
Typical fruit types

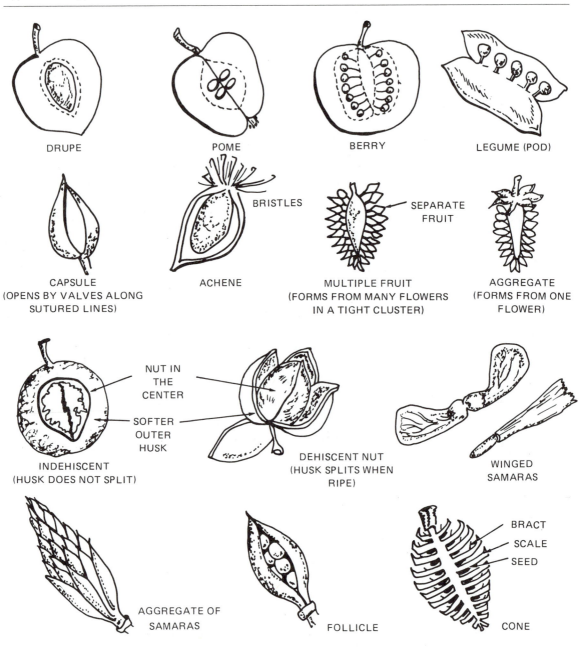

DRUPE

POME

BERRY

LEGUME (POD)

CAPSULE
(OPENS BY VALVES ALONG
SUTURED LINES)

BRISTLES

ACHENE

SEPARATE
FRUIT

MULTIPLE FRUIT
(FORMS FROM MANY FLOWERS
IN A TIGHT CLUSTER)

AGGREGATE
(FORMS FROM ONE
FLOWER)

NUT IN
THE
CENTER

SOFTER
OUTER
HUSK

INDEHISCENT
(HUSK DOES NOT SPLIT)

DEHISCENT NUT
(HUSK SPLITS WHEN
RIPE)

WINGED
SAMARAS

AGGREGATE OF
SAMARAS

FOLLICLE

BRACT

SCALE

SEED

CONE

3. Label these three types of leaves and the types of apex, base, and margin of each.

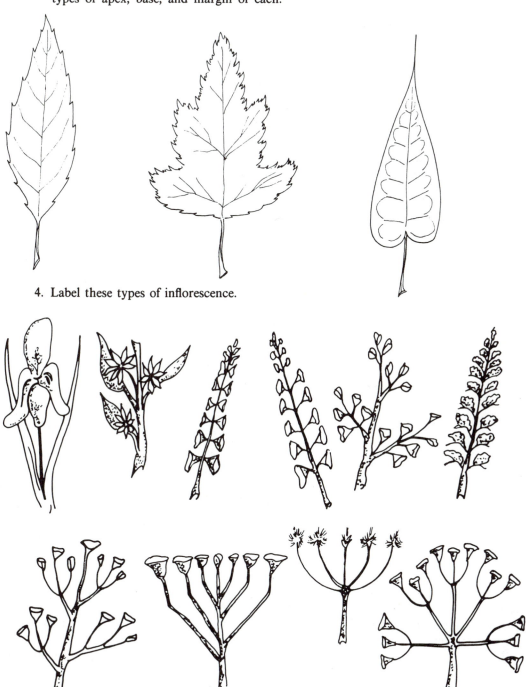

4. Label these types of inflorescence.

5. Label these types of fruits.

6. From the definitions that follow, list the correct terms.

a. a flower that has an inferior ovary with a floral tube united with the wall of the ovary

b. a flower that has a superior ovary and a floral tube separated from the ovary

c. a flower that has a superior ovary and no floral tube

d. a flower having only one blossom per stem

e. a flower head composed of many small ray and disc flowers

f. a fruit that develops from a single ovary

g. a fruit that develops from a single flower having a group of ovaries

h. a fruit that develops from one or more ovaries and includes the calyx and/or receptacle

i. a fruit that develops from multiple ovaries of multiple flowers borne on a single stalk

j. the innermost part of the pericarp

k. the outermost part of the pericarp

l. the central part of the pericarp

4
Plant Growth Regulators

Objectives

Upon completion of this chapter, you will be able to

- list at least three different types of naturally occurring growth regulators and their effects on plants.
- list several commercial products that regulate plant growth.

GROWTH REGULATORS DEFINED

Green plants contain a number of *organic compounds* that control plant growth. These compounds are present only in minute amounts, yet they stimulate, inhibit, and otherwise modify the rate, the direction, and the nature of plant development. The compounds that occur naturally are called *hormones*. They may be produced in one part of the plant yet exert their effect in another part, far removed from the source. Still other modifiers of plant development are synthesized by chemists and made available to commercial horticulturists for the production of more desirable crops. All of the products, either naturally occurring or commercially synthesized, are given the name of *growth regulators*.

TYPES OF GROWTH REGULATORS

Most plant scientists would agree that our knowledge of plant growth regulators is still incomplete.

The regulators have probably not all been isolated; their effects at different concentrations vary; their effects in different species of plants vary; and even the sites of their production and activity within plants vary. At present, we have a fair knowledge of auxins, gibberellins, and cytokinins. We have a limited knowledge of several others.

Auxins

Auxins are plant hormones that both promote and inhibit plant growth. *Indoleacetic acid* (IAA) is the most commonly occurring natural auxin, Figure 4-1. Auxins have been found responsible for phototropism, geotropism, and apical dominance.

Our knowledge of auxins began with the curiosity of scientists who wondered why stems grow up and roots grow down, why plant stems seem to bend toward the light, and why the removal of an apical (or terminal) bud stimulates the growth of lateral

FIGURE 4-1.

Chemical structure of IAA (indole-3-acetic acid)

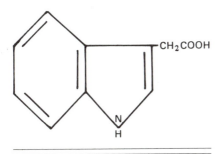

FIGURE 4-2.

Using grass seedlings, Darwin demonstrated that seedlings bend naturally toward a unilateral light source (A); if lightproof caps are placed over the tips of the seedlings, they do not bend (B).

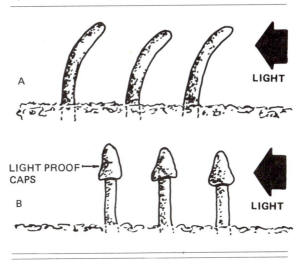

buds beneath it. Studies with the sheathlike covering (coleoptile) of the young stem of oat seedlings (*Avena*) have provided the evidence upon which our understanding of auxins is based.

Seedlings exposed to light from only one side bend toward the light source. Charles Darwin first recorded the observation that seedlings bend toward the light only if the tips are exposed. When the seedling tips are covered with caps that do not allow the passage of light, the seedling tips do not bend, Figure 4-2. The conclusion drawn from Darwin's studies was that a substance produced in the tips can exert an influence on the growth of tissue some distance from the tip. Such a conclusion is compatible with the definition of a hormone.

Researchers after Darwin continued to experiment. In one experiment, the tips of seedlings were removed, a bit of gelatin was added to each cut stump, and the tips were replaced. The seedlings still bent toward a unilateral light source. When the tips were not replaced, the seedlings did not bend, Figure 4-3. The conclusion was that whatever accelerated the growth on the unlighted side of the stem was able to diffuse through the gelatin.

A test called the *Avena* coleoptile test has shown that a freshly cut tip replaced on one side of the coleoptile stump will promote more growth on that side than on the other side, Figure 4-4. The curvature will occur even without the unilateral light source, leading to the conclusion that the coleoptile tips are the source of the hormone, and that tissue growth stimulation is its effect.

In an actively growing stem, the concentration of auxin in the tip (or apex) is so great that it can inhibit the growth of buds and shoots beneath it. This influence is termed *apical dominance*. When the apex bud or shoot is removed, thus diminishing the supply of auxin, the lateral buds and shoots are allowed to grow. The horticultural practice of pruning to promote denser growth involves overcoming apical dominance by removing the tips of branches.

Auxins are present in plant roots as well as in shoots. The research of K. V. Thimann provided evidence that roots are much more sensitive than shoots to auxin. The same amount of auxin that promotes cell elongation in shoots can inhibit growth in roots, Figure 4-5.

The ability of auxin to both promote and inhibit plant growth explains why shoots grow upward and roots grow downward, Figure 4-6. When a coleoptile or stem is placed in a horizontal position, auxin accumulates in higher concentrations on the lower

FIGURE 4-3.

A seedling that normally bends toward a unilateral light source (A) will not do so if its tip is removed (B). If the tip is reattached with gelatin (C-1), bending occurs (C-2).

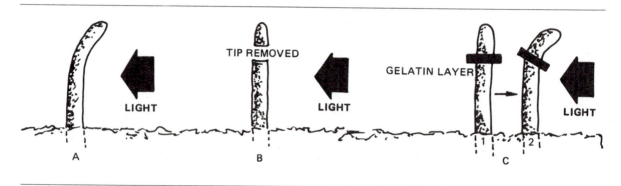

FIGURE 4-4.

The *Avena* (oat) coleoptile test: Coleoptiles kept in the dark do not bend (A). If their tips are removed and then replaced on one side (B), coleoptile curvature will result, and it will be away from the side holding the tip (C).

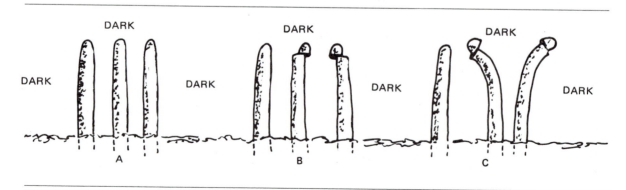

FIGURE 4-5.

Auxin concentrations needed to promote growth vary in different parts of the plant. If the concentrations are altered, growth can be inhibited.

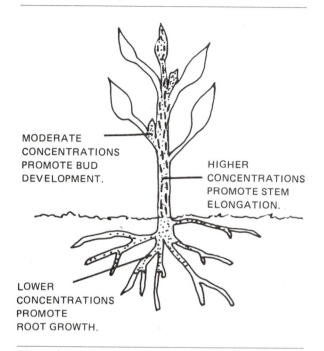

MODERATE CONCENTRATIONS PROMOTE BUD DEVELOPMENT.

HIGHER CONCENTRATIONS PROMOTE STEM ELONGATION.

LOWER CONCENTRATIONS PROMOTE ROOT GROWTH.

FIGURE 4-6.

The response of auxin to gravity explains why shoots grow upward and roots downward.

AUXIN PROMOTES CELL ELONGATION IN SHOOTS.....

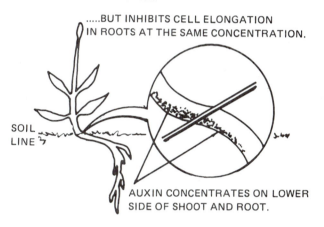

.....BUT INHIBITS CELL ELONGATION IN ROOTS AT THE SAME CONCENTRATION.

SOIL LINE

AUXIN CONCENTRATES ON LOWER SIDE OF SHOOT AND ROOT.

side of the plant. The auxin causes the cells on the underside of the shoot to elongate faster than the cells on the upper side, so the shoot grows upward. The same concentration of auxin inhibits the growth of the cells on the lower side of the roots allowing the cells on the upper surface to grow faster and direct the root downward. The response of auxin to gravity and the resulting growth reaction in plants is termed *geotropism.*

The mechanism of auxin action is not fully understood, although several theories exist. One line of research has correlated the presence of auxin with the production of ethylene in plants. Since some of the responses associated with auxins can be duplicated by exposing the plants to ethylene, the theory is that the growth responses attributed to auxins actually result from ethylene produced through the auxin presence. Another theory is that auxin increases the plasticity of the cell walls, permitting the cells to expand to larger sizes. Still another theory is that auxin increases the energy supply in the tissue where it is present, and this increased metabolic activity stimulates plant growth.

Gibberellins

Gibberellins are a group of thirty or more closely related plant hormones that promote cell enlargement, often causing dramatic increases in plant height. One of the common forms is gibberellic acid, Figure 4-7. They will not cause curvature in *Avena* coleoptiles but will cause dwarf corn plants to elongate and grow to normal size. Since auxins will not affect dwarf corn plants, the two growth hormones can be clearly distinguished by scientists.

The number of plant processes known to be affected by gibberellins is not as great as those known to be influenced by auxins, but their influence is nonetheless impressive. Rosetted plants can be made to grow tall following the application of gibberellic

FIGURE 4-7.

Chemical structure of GA₃ (gibberellic acid)

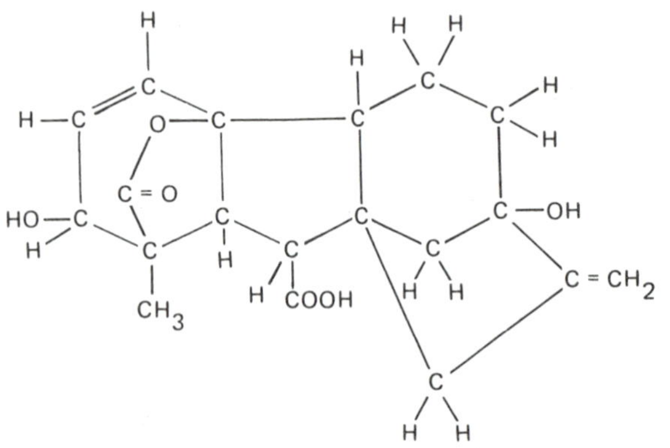

acid. Dormant seeds that normally require a cold treatment (stratification) for germination can be made to germinate without the treatment when gibberellic acid is applied instead. Also, various other plant processes normally initiated in response to temperature and/or photoperiod stimulations can be prompted instead by the application of gibberellins.

Rosetted plants are those having good leaf development but retarded internodal growth. Under natural growing conditions such plants will respond to long-day photoperiods or a period of cold treatment with a dramatic elongation of the internodes (bolting) and the production of flowers atop a stem five or six times the original height of the plant. When kept under short-day conditions or in warm temperatures, the plants remain rosetted. Since the treatment of plants with gibberellin under conditions normally promotive of rosetting can induce bolting and flowering, it is suggested that the stage of growth of the plant is controlled by the amount of natural gibberellin present within the plants. Both cell division and cell enlargement occur in the plants treated with the gibberellin.

Both gibberellins and auxins, when applied to the pistils of certain flowers, can cause them to set fruit without being pollinated. The process is called *parthenocarpy,* and the resulting seedless fruit are called *parthenocarpic fruit.* Examples most readers would recognize are the naval orange, the banana, and the seedless grape. In ordinary pollination, growth hormones are introduced to the pistil from pollen produced on the stamen or synthesized in the ovary following a stimulus from pollen.

Through observation and study of these and other plant processes, scientists are able to theorize about the role of gibberellins in plants. Gibberellins have been shown to stimulate cell division and cell enlargement in young embryos of germinating grains. Evidence suggests that the gibberellin activates genes that are normally repressed. As the genes become active, new ribonucleic acid (RNA, the carrier of genetic information) and new enzymes result. The new enzymes digest and release materials stored in the endosperm. These materials then become available to the developing embryo, thereby promoting cellular activity that might otherwise be inhibited. It is doubtful whether such an explanation covers all the effects of gibberellins, however.

Cytokinins

Cytokinins are naturally occurring hormones, of which the best known is *kinetin,* Figure 4-8. The principal role of the cytokinins is in the division of cells, but an expanding body of evidence suggests

FIGURE 4-8.

Chemical structure of kinetin

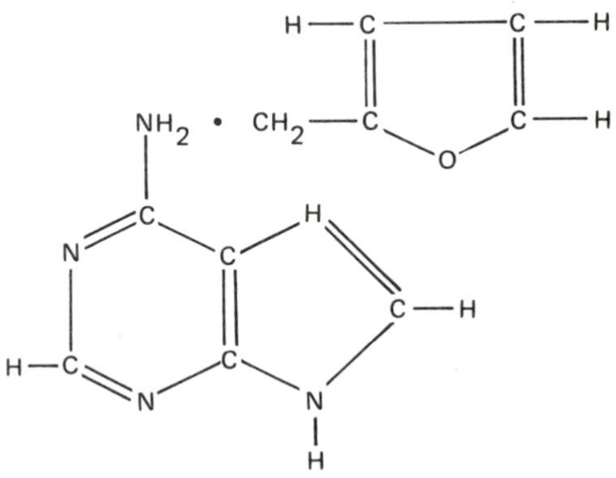

their involvement in many other plant growth processes. That involvement seems to be in association with gibberellins, auxins, and other growth regulators.

In their interaction with auxins, cytokinins have been found to promote either cell enlargement or cell division. The initiation and growth of roots and shoots as well as the overcoming of apical dominance have all been linked to the presence of cytokinins.

Cytokinins can be distinguished from auxins and gibberellins with a diagnostic test that involves the culture of tobacco parenchymal cells in a nutrient solution. When kinetin and auxin are added to the solution, cell division is accelerated in the tobacco tissue. No effect upon dwarf corn plants or *Avena* coleoptile curvature results when cytokinins are applied, so the three tests are important biological detectors for the growth regulating hormones.

Other Growth Regulators

In addition to the three hormones already described, other plant hormones have been either isolated or theorized to exist. Among them are:

- ethylene—thought to influence geotropism, apical dominance, and the ripening of fruit
- florigen—not yet isolated, but believed to initiate flowering in plants
- phytochrome—a participant in the flowering of short-day plants
- anti-gibberellins—a group of synthetic compounds that have a dwarfing effect upon plant growth

It is likely that further research will find that many plant growth responses stem from the synergistic action of two or more growth regulators. Since they can coexist within the same tissue and often promote similar responses in the plant, their combined action seems predictable.

COMMERCIAL GROWTH REGULATORS

Even without fully understanding the way growth regulators work, scientists have been able to synthesize products that influence the growth of plants in a similar manner. Many products of commercial importance are growth retardants and are used in the floriculture trade to produce shorter potted crops. Several are used by greenhouse and nursery propagators to promote rooting of cuttings, Figure 4-9. More recently, products have been developed that reduce the rate of growth of turfgrasses.

FIGURE 4-9.

The effect of growth regulators is illustrated dramatically in these results of a day length/regulator study. The control plant on the left was given short days to delay flowering. It received no growth regulators. The center plant was given long day exposures and no growth regulator. It grew tall and flowered. The plant on the right received the same long day exposures, which promoted flowering. However, it also received sprays with a growth retardant which has kept it shorter and fuller. (Courtesy United States Department of Agriculture)

FIGURE 4-10.
Examples of growth retardants that are available for commercial use

TABLE 4-1.
Growth Retardants for Florist Crops

Product Name	Crop	Effects
A-Rest®	Chrysanthemums Poinsettias Easter lilies	Shortens stem length
B-Nine®	Chrysanthemums Poinsettias Hydrangeas	Shortens stem length
	Azaleas	Promotes heavier flower set in azaleas
Cycocel®	Poinsettias	Shortens stem length
	Azaleas	Promotes heavier flower set in azaleas
Florel®	Poinsettias	Shortens stem length
	Carnations	Promotes branching in carnations
	Roses	Promotes new growth from base of roses
Phosfon®	Chrysanthemums Easter lilies	Shortens stem length

Under regular growing conditions, some greenhouse plants grow too tall to make desirable potted plants. Poinsettias, Easter lilies, chrysanthemums, and some hydrangeas can all get "leggy" if started too early or subjected to overly warm temperatures during the period of most rapid vegetative growth. The application of growth retardants to the plants will help to regulate their size, Figure 4-10. With potted azaleas, more flower buds can be induced and stem growth limited with the application of growth regulating chemicals.

Table 4-1 lists the commercial growth retardants most commonly used by greenhouse growers and describes their effects.

In the nursery and florist trades, where vegetative cuttings are rooted, commercially available growth regulators are usually used to promote and accelerate root formation. The most popular root-promoting hormones are the auxins indoleacetic acid (IAA), indolebutyric acid (IBA), and napthalene-acetic acid (NAA). Applied as dusts or solutions to the ends of cuttings, the auxins can speed rooting time considerably, assuming other requirements for growth are not limiting, Figure 4-11.

Where there is a large expanse of turfgrass there is also a need to keep it mowed. In temperate zones, lawn areas may require mowing thirty or more times during a normal growing season. Often the terrain is sloping and difficult or dangerous to mow. With the application of growth retardants, turf growth can be restrained with a significant saving in the cost of fuel and manpower normally required for mowing. Use of the retardants is wisely restricted to low-quality, minimum use, difficult-to-mow areas, since they often cause discoloration, thinning, and reduced ability to recover from winter injury. The chemical regulators are often applied in combination with broad-leaved weed killers, since the retardants are often ineffective against many weeds. Table 4-2 describes the commercial growth retardants most commonly used by turf specialists.

Most of the turf growth regulators are applied twice each season, with six weeks or more between applications. While the chemical products are not a substitute for mowing, they do save money and

FIGURE 4-11.
A stem cutting being dipped in auxin to promote rooting.

time as part of an overall maintenance program, Figure 4-12.

SUMMARY

Plant growth regulators are organic compounds that control the rate, direction, and nature of plant growth. Knowledge of plant growth regulators is incomplete; however, a great deal is known about auxins, gibberellins, and cytokinins.

Auxins are hormones (naturally occurring growth regulators) that both promote and inhibit plant growth. Indoleacetic acid (IAA) is the most commonly occurring natural auxin. Auxins occur in both shoots and roots. In the shoots, they are responsible for apical dominance. They are also responsible for various tropisms in plants.

TABLE 4-2.
Growth Retardants for Turfgrasses

Generic Name	Product Name	Effects
Chlorflurenol	Maintain CF-125®	Absorbed by foliage; carried upward and downward; inhibits or slows cell division
Maleic hydrazide	Chemform® De-Cut® Retard® Siesta® Slo-Gro® Super Sprout Stop®	Absorbed by foliage; carried to areas of active growth; inhibits cell division in shoots, buds, and roots
Maleic hydrazide plus chlorflurenol	Posan®	Absorbed by foliage; carried to areas of active growth; suppresses seedhead formation
Mefluidide	Embark®	Absorbed by foliage; carried upward more than downward; suppresses growth and seedhead formation

FIGURE 4-12.
The growth retardant Embark 2-S® is seen to dramatically reduce the growth of turfgrass in this comparative trial. (Courtesy 3M Agrichemicals)

UNTREATED

TREATED

Gibberellins are hormones that promote cell division and enlargement in plants, often causing dramatic increases in plant height. Thirty or more gibberellins are known. They can overcome rosetting in plants, cause dormant seeds to germinate, and cause flowers to set fruit without pollination.

Cytokinins are also hormones. The best known is kinetin. Evidence indicates that the principle role of cytokinins is to accelerate cell division. In association with auxins, they may also promote cell enlargement. The initiation of roots and shoots and the overcoming of apical dominance have all been linked to the presence of cytokinins.

Each of the plant growth regulators can be identified by a distinctive diagnostic test. Even without fully understanding how these hormones work, scientists have been able to synthesize products that influence the growth of plants in a similar manner. The synthetic products are most frequently used either to retard the growth of florist crops that tend to get too tall or to promote branching or heavier flower set. They are also used to restrict the growth of turfgrass in low-quality, minimum-use, or difficult-to-mow areas.

Achievement Review

A. SHORT ANSWER

Answer each of the following questions as briefly as possible.

1. Which of the following are characteristic of naturally occurring growth regulators?
   a. organic compounds
   b. inorganic compounds
   c. present in plants in large amounts
   d. present in plants in minute amounts
   e. produced and active in the same plant part
   f. produced in one part and active in another

2. List the diagnostic test that distinguishes each of the following growth regulators from all others.
   a. auxins
   b. gibberellins
   c. cytokinins

3. Which one or more of the three major growth regulators causes each of the following characteristic plant responses?
   a. apical dominance
   b. increased cell elongation
   c. increased cell division
   d. geotropism
   e. parthenocarpy
   f. phototropism
   g. bolting

4. If you were a commercial horticulturist, which commercially produced growth regulator(s) could you choose to solve the following problems?
   a. Vegetative cuttings need to be rooted quickly.
   b. Potted poinsettias need to be kept short.
   c. Carnations need their branching increased.
   d. Easter lilies need to be kept short.

5

Plant Reproduction

Objectives

Upon completion of this chapter, you will be able to

- describe the characteristics of sexual and asexual reproduction in plants.
- describe the processes of mitosis and meiosis.
- list the most common methods of propagating plants.
- recognize the contributions of Mendel to the science of genetics.
- discuss methods of plant improvement.

PLANT REPRODUCTION DEFINED

When organisms duplicate themselves, the process is termed *reproduction*. While only partially understood, plant reproduction is recognized as a complicated process that results in the multiplication of cells and the organisms that they make up. When reproduction is deliberately controlled and manipulated it becomes *propagation*. Ornamental horticulturists who control the reproduction of plants are called *plant propagators*. In large operations, this can be a full-time occupation.

SEXUAL AND ASEXUAL REPRODUCTION

The reproduction of plants through the formation of *seeds* is called *sexual reproduction*. It requires the fusion of two sex cells or *gametes*, each having

one set of chromosomes (*haploidal*), to form a cell with two sets of chromosomes (*diploidal*) that is termed a *zygote*. The gametes are formed through a sequence of cell divisions that reduces the number of chromosomes in each cell by half. The process is termed *meiosis*, Figure 5-1.

The creation of a diploidal cell occurs at the time of fertilization. Through sexual reproduction, the genetic material of the cell recombines in a way that can produce new plants, differing in physical and physiological characteristics from their parent plants. Such differences as stronger stems, larger flowers, and greater winter tolerance may appear. In other instances, sexual reproduction results in new plants identical in nearly every way to their parent plants. The degree of difference is determined largely by the characteristics of the *homologous chromosomes* (those that associate in pairs during meiosis). When plants self-pollinate, the pairs of

FIGURE 5-1.

The stages of meiosis

1. DIPLOIDAL CELL AT THE BEGINNING OF PROPHASE

THE CELL IS IN A DIP-LOIDAL CONDITION AND TERMED A ZY-GOTE. THE CHROMO-SOMES IN THE NUCLEUS BECOME SHORT AND THICK.

2. PROPHASE I

HOMOLOGOUS CHRO-MOSOMES PAIR ALONG THEIR ENTIRE LENGTH. THEY DO NOT FUSE. THE PAIRING IS TERMED SYNAPSIS. EACH CHROMOSOME DOUBLES TO PRODUCE FOUR CHROMATIDS PER PAIR.

3. METAPHASE I

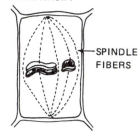

SPINDLE FIBERS

THE CHROMOSOME PAIRS MOVE TO THE HORIZONTAL CENTER OF THE CELL. SPINDLE FIBERS FORM BETWEEN THE ENDS OF THE CELL. SOME ATTACH TO THE CHROMATIDS.

4. ANAPHASE I

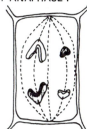

THE HOMOLOGOUS CHROMOSOMES SEPA-RATE. THE GROUPS OF CHROMOSOMES AT EACH END OF THE CELL ARE NOT THE SAME. THE CHROMA-TIDS HAVE NOT SEPARATED.

5. TELOPHASE I

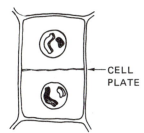

CELL PLATE

6. METAPHASE II

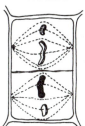

7. ANAPHASE II

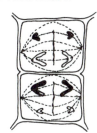

8. HAPLOIDAL CELLS

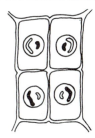

genes on the homologous chromosomes are basically alike (*homozygous*), and the plants that are produced are nearly exact duplicates. When plants cross-pollinate, the pairs of genes on the homologous chromosomes are often dissimilar. The plants that result are not exactly like either parent plant.

When an exact duplicate of a plant is desired, especially if it is not self-pollinating, asexual propagation methods must be used.

Asexual reproduction is a vegetative process that eliminates genetic variation. The result is the multiplication of plants possessing the same genetic complex. Asexual reproduction can perpetuate an individual plant essentially unchanged for generations over a period of many years.

The asexual reproduction of plants is made possible by a combination of two processes termed mitosis and cytokinesis. *Mitosis* is the normal division of a cell nucleus that occurs as a plant grows, enlarging from embryo to maturity, Figure 5-2. It is mitosis that occurs when new roots form on a cutting or new shoots break from a stem. The chromosomes in the cells do not reduce in number or recombine as they divide, thus allowing no opportunity for a new plant to be initiated. *Cytokinesis* completes the division of the non-nuclear remainder of the cell's contents and the formation of a new cell wall. Table 5-1 summarizes the major distinguishing features of mitosis and meiosis.

HOW PLANTS ARE PROPAGATED

In the natural world, plants reproduce both sexually and asexually. For example, new plants may arise vegetatively from the root system of a plant while the same plant is producing flowers, pollinating, and setting fruit and seed. Many plant species reproduce asexually in assorted ways, each a variation of the basic mitosis process. Although the result is the same—that is, continuation of the genetic complex in the new plants—some of the reproduc-

FIGURE 5-2.
The stages of mitosis

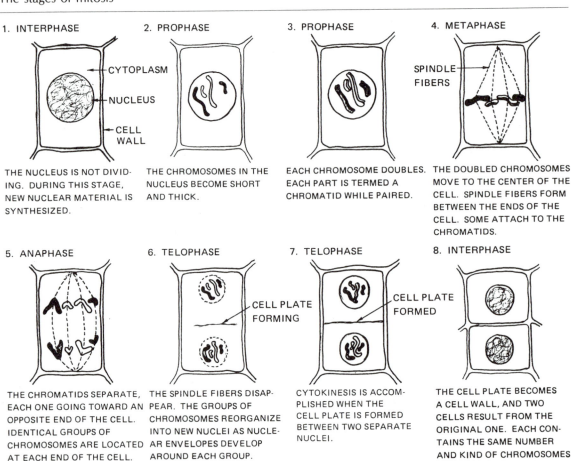

1. INTERPHASE

CYTOPLASM
NUCLEUS
CELL WALL

THE NUCLEUS IS NOT DIVIDING. DURING THIS STAGE, NEW NUCLEAR MATERIAL IS SYNTHESIZED.

2. PROPHASE

THE CHROMOSOMES IN THE NUCLEUS BECOME SHORT AND THICK.

3. PROPHASE

EACH CHROMOSOME DOUBLES. EACH PART IS TERMED A CHROMATID WHILE PAIRED.

4. METAPHASE

SPINDLE FIBERS

THE DOUBLED CHROMOSOMES MOVE TO THE CENTER OF THE CELL. SPINDLE FIBERS FORM BETWEEN THE ENDS OF THE CELL. SOME ATTACH TO THE CHROMATIDS.

5. ANAPHASE

THE CHROMATIDS SEPARATE, EACH ONE GOING TOWARD AN OPPOSITE END OF THE CELL. IDENTICAL GROUPS OF CHROMOSOMES ARE LOCATED AT EACH END OF THE CELL.

6. TELOPHASE

CELL PLATE FORMING

THE SPINDLE FIBERS DISAPPEAR. THE GROUPS OF CHROMOSOMES REORGANIZE INTO NEW NUCLEI AS NUCLEAR ENVELOPES DEVELOP AROUND EACH GROUP. FORMATION OF A CELL PLATE BEGINS.

7. TELOPHASE

CELL PLATE FORMED

CYTOKINESIS IS ACCOMPLISHED WHEN THE CELL PLATE IS FORMED BETWEEN TWO SEPARATE NUCLEI.

8. INTERPHASE

THE CELL PLATE BECOMES A CELL WALL, AND TWO CELLS RESULT FROM THE ORIGINAL ONE. EACH CONTAINS THE SAME NUMBER AND KIND OF CHROMOSOMES AS THE ORIGINAL CELL.

TABLE 5-1.

Characteristics of Mitosis and Meiosis in Plants

Characteristic	Mitosis	Meiosis
Occurs during sexual reproduction		X
Occurs during asexual reproduction	X	
Reduces the number of chromosomes by half		X
The basic process of vegetative growth	X	
Plants resulting are genetically identical to parents	X	
The nucleus divides twice		X
Homologous pairs of chromosomes separate once, while the cells divide twice		X
Creates haploidal cells, or gametes		X

tive methods are faster or in other ways preferable. The method of propagation selected by a grower may be based upon:

- ease of propagation
- number of plants needed
- rate of growth of the species
- characteristics desired in the new plant (when grafting of two species is involved)
- desire to avoid disease present in the parent plant
- desire to perpetuate a mutation developing in a parent plant
- cost

Sexual Propagation

Sexual propagation utilizes seeds. Some seeds germinate immediately after planting. Some require light to germinate, while others prefer darkness. Some seeds require pregermination treatment, such as scarification or stratification and in some cases both. *Scarification* is the breaking of a seed coat otherwise impervious to water to permit water uptake by the embryo. *Stratification* is the exposure of the seeds to low temperatures. The cold period is believed to initiate the formation of growth regulators or the destruction of growth inhibitors necessary before the seed can germinate. In nature, scarification may be accomplished by soil microorganisms or passage through the digestive tract of an animal. Stratification occurs naturally during the winter. The horticulturist may simulate and accelerate the activities of nature by treating the hard seed coats of certain plants with acid or by mechanically scratching their seed coats to promote water uptake. Bulk quantities of dormant seeds may be placed under refrigeration to hasten their stratification.

Asexual Propagation

Asexual propagation utilizes the vegetative parts of a plant to grow new plants. Stems or roots are more commonly used, but leaves can be used as well. Each vegetative cell has the inherent ability to reproduce an entire plant that is genetically identical to the one from which the cell originated.

When many plants are reproduced asexually from a single plant, the group of new plants is termed a *clone*. Each individual plant within the clone is referred to as a *ramet*. Plants of horticultural importance that are propagated almost totally by asexual means are termed *clonal varieties*. They are a type of cultivar. Some plants have been perpetuated by horticulturists for so many years that their original parental origins have become lost in time. They depend totally upon humans for survival in their present form.

Some of the techniques used by professional horticulturists to propagate plants asexually do little more than allow the plants to grow and reproduce normally. Other methods of propagation are totally dependent upon the efforts of the horticulturist for their success. Techniques such as grafting and budding may replace the root system of a plant with a hardier one or create a plant having multiple colors of flowers on a single stem. Such plants are artificial contrivances, but nevertheless important in the industry. Still other propagative techniques are basically natural in origin but subject to manipulation by horticulturists. The following list summarizes various asexual techniques.

Runners These are stems that grow along the ground and form new plants at one or more of their nodes. The strawberry plant is an example likely to be familiar to most readers. The spider plant (*Chlorophytum comosum variegatum*) is an example of an ornamental plant, Figure 5–3. Propagators need only separate the new rooted plants from the parent and transplant them.

Stolons Stolons are aerial shoots that take root after coming into contact with the soil. Numerous grasses spread in this fashion, as do several of the shrubby dogwoods (*Cornus* sp.). As with runners, the propagators simply separate the rooted shoot from the parent and transplant.

Sucker shoots Certain plants produce new shoots from adventitious buds that develop on the roots, Figure 5-4. The black locust (*Robinia pseudoacacia*) is a prolific sucker producer. Suckers can be cut from the parent root system and transplanted.

Bulbs and similar organs Vegetative structures termed either bulbs, corms, tubers, tuberous roots, or rhizomes are modified stem or root tissues that store food during dormant periods. Seasonal perennials like tulips, daffodils, crocuses, and gladioli exemplify these structures. The organs form as part of the root system and are easily collected and separated by propagators, Figure 5-5.

Layering In layering, roots develop on a stem that is still attached to the parent plant. Runners are characteristic of naturally layered plants. Some plants root better this way because food production and water uptake are not reduced or totally severed

FIGURE 5-3.
The spider plant gets its common name from the appearance created by its runners and new plants. (Courtesy Rodney Jackson, Photographer)

FIGURE 5-4.
Suckers of this tulip tree are originating from the root system.

FIGURE 5-5.
Examples of true bulbs and bulb-like reproductive structures

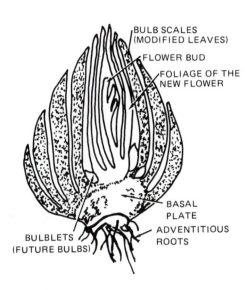

BULB SCALES (MODIFIED LEAVES)
FLOWER BUD
FOLIAGE OF THE NEW FLOWER
BASAL PLATE
ADVENTITIOUS ROOTS
BULBLETS (FUTURE BULBS)

CROSS-SECTIONAL VIEW OF A TRUE BULB. A TRUE BULB IS A SHORT, THICK STEM WITH MODIFIED LEAF SCALES AND ROOTS AT THE BASE.

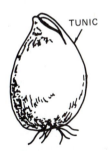

TUNIC

EXTERIOR VIEW OF A TUNICATE BULB WITH ITS PAPER-LIKE OUTER SCALE (TUNIC) INTACT. EXAMPLE: TULIP

SCALES

A NON-TUNICATE BULB WITH LOOSE SCALES BUT NO OUTER TUNIC. EXAMPLE: LILY

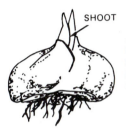

SHOOT

A CORM. IT IS ALL STEM MATERIAL. ROOTS GROW FROM THE BASE AND THE LEAVES EMERGE FROM THE TOP. EXAMPLE: GLADIOLUS

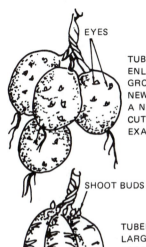

EYES

TUBERS. THESE ARE FLESHY ENLARGEMENTS ON UNDERGROUND STEMS. EACH EYE IS A NEW SHOOT AND WILL PRODUCE A NEW PLANT IF THE TUBER IS CUT APART. EXAMPLE: POTATO

SHOOT BUDS

TUBEROUS ROOTS ARE ENLARGED ROOTS THAT STORE FOOD. SHOOTS FORM NEAR THE STEM END ONLY. EXAMPLE: DAHLIA

CUT LEAVES

RHIZOMES ARE HORIZONTAL UNDERGROUND STEMS. THEY PRODUCE SHOOTS ON THEIR UPPER SURFACE AND ROOTS ON THE LOWER SURFACE. EXAMPLE: IRIS

as when a cutting is taken. The rubber plant (*Ficus elastica*) and many temperate woody ornamentals can be reproduced by layering. There are various ways of layering plants depending upon the species involved. In general, the stem is partially severed, a rooting hormone is applied, and the cut stem is buried in soil or wrapped in a cool, moist medium promotive of root formation, Figure 5-6.

Cuttings When segments of roots, leaves, or stems are cut off a plant and placed under appropriate environmental conditions, new shoots and roots form, and eventually a new plant is produced. Propagation by stem cuttings is the most common means of reproducing plants asexually. Chrysanthemums, geraniums, and woody shrubs are among the many

FIGURE 5-6.

A Jiffy-7® pot is used to air-layer a containerized rubber plant. (Courtesy Carefree Garden Products)

Air layering

important horticultural plants perpetuated by cutting propagation, Figure 5-7.

Grafting With compatible species, the upper portion of one plant can be joined with the lower portion of a different plant, and they will fuse to become a single new plant expressing characteristics of the two from which it originated. Roses are commonly grafted to produce a plant having showy flowers and a hardy root system. Many woody ornamentals are propagated by grafting, Figure 5-8.

Budding This is a type of grafting in which buds of one plant are implanted into the stem of another compatible species. Roses are the major ornamentals utilizing this propagative technique. Fruit and nut trees are the most common plants grafted in this manner, Figure 5-9.

Division of the crown As shrubs and herbaceous perennials mature, the *crown* (junction of the roots and shoots) enlarges and can be divided into root-shoot units that will then grow into separate plants. Even in plants whose shoots die back at the end of the growing season, the crown divisions will regenerate new plants as long as adventitious shoot buds exist, Figure 5-10.

Tissue culture Small sections of meristematic shoot tissue or callus tissue are surgically removed from the parent plant, placed into balanced nutrient media, and grown under carefully controlled environmental conditions, Figure 5-11. Through mitosis and tissue differentiation new plants result that are copies of the plant from which they originated. Orchids and foliage plants are commonly reproduced in this way. A major benefit of this method is to assure that the propagated plants are disease-free by culturing with active apical meristem (meristematic) tissue. Such tissue usually grows faster than any disease-causing agents that might be infecting a parent plant and that could be transported into the new plants through other propagative techniques. Tissue culture is only successful if done under aseptic conditions.

Apomictic embryos In certain cases, an embryo can be produced without meiosis and fertilization. The resulting seed is thus asexually produced. Several grass species propagate in this manner, but it

FIGURE 5-7.
Vigorous roots have formed on the stem cutting of this Pothos after two weeks in the propagation bench.

is of greater significance with fruit crops than with ornamentals.

PLANT IMPROVEMENT

No study of plant reproduction is complete without some discussion of how plants change and improve through the years. The essence of vegetative reproduction is that there is no change; that is, each new plant is exactly like the parent plant. While this may be desirable in the short term, it would be detrimental in the long run if species were unable to change to respond to new environmental conditions. Such changes can only occur through sexual reproduction or spontaneous mutation. When the sexual reproduction process is controlled by the deliberate actions of human beings, *plant breeding* results and a specialized science, *plant genetics,* takes effect.

Plant genetics is a young science tracing its origin to Gregor Mendel, an Austrian, whose work with the common garden pea in 1865 began the understanding of how and why plants change. Weighty texts and entire university courses are devoted to the subject of genetics; it is not the intent here to explain the intricasies of plant improvement. Rather, it is hoped that you will gain a basic understanding of a science and specialized technology integral to the long-term improvement of ornamentals and other crop plants.

The physiological activities of each cell, tissue, organ, and total plant are the result of biochemical reactions. These reactions are driven by *enzymes,* which are synthesized from coded information carried on the *chromosomes* in each cell's nucleus. Each enzymatic effect is controlled by an individual *gene,* the determiners of heredity located on the chromosomes. Genes are part of the deoxyribonucleic acid (DNA) portion of the chromosome. As the chromosome duplicates during meiosis, the genes and the DNA also duplicate. This duplication determines the transmission of hereditary features.

FIGURE 5-8.
The graft union of this beech tree is readily apparent. The scion is growing more vigorously than the stock, yet the tree is healthy. Not all graft unions are so obvious.

FIGURE 5-9.
New shoots break from the bud graft on a rose

cept, the fusion of gametes from a 6-foot-tall parent and a 2-foot-tall parent might produce a 4-foot-tall offspring, as though inheritance resembled mathematical averaging. Since such results were not the case, Mendel looked for other explanations.

He worked with the common garden pea and selected two strains that were notably different in height; one grew to a mature height of 6 or 7 feet and the other to a mature height of 1 or 2 feet. He declined to study other differences between the parents, isolating for his observations only the height variable. He cross-bred selected parents, carrying out the pollinations under carefully controlled conditions and keeping precise records of the offspring that resulted.

When Mendel allowed two tall parents to self-pollinate, the offspring were all tall. Similarly, self-pollinated short parents always produced short offspring. When a tall parent was pollinated with a short parent, the offspring were also all tall. These first generation offspring are termed F_1. When F_1 generation plants were then allowed to self-polli-

THE CONTRIBUTION OF GREGOR MENDEL

A look at the work of Gregor Mendel provides an appreciation of his contribution to learning along with a basic introduction to the study of plant improvement. Gregor Mendel was an Augustinian monk who lived from 1822 to 1884. Throughout his adult life, Mendel enjoyed plants and entertained a curiosity about how their hereditary features were transmitted from generation to generation. Prior to his studies, scholars accepted the idea that inheritance in plants was not predictable and that the characteristics of the parents just blended together in the offspring. According to that mistaken con-

FIGURE 5-10.
The crown of an African violet is being divided into three parts. Each will develop into a new plant.

FIGURE 5-11.
Tissue cultured implants of geranium and chrysanthemum

nate, the second generation (F$_2$) offspring were divided, with three-fourths of the plants tall and one-fourth of the plants short. In neither generation, F$_1$ nor F$_2$, were there any plants of intermediate or average height.

Explaining Mendel's observations with modern terminology, it is now recognized that factors such as height are controlled by genes and that these genes occur in pairs, as with short and tall, on homologous chromosomes. When the sexual gametes are formed, the paired genes separate. To illustrate, the genes for tall and short are usually represented as T and t respectively. The homologous chromosomes from the original tall plants are therefore represented as TT. Those from the original short plants are represented as tt. The chromosomes and genes separate during meiosis, and the resulting haploid gametes each contain a single gene for height. They recombine after pollination to create a diploidal Tt offspring, Figure 5-12.

Since all the F$_1$ generation mature as tall pea plants despite containing a gene for shortness, the

T gene is regarded as *dominant* and the t gene as *recessive.* Thus the *phenotype* (external appearance) of Mendel's plants was the same (tall) whether the *genotype* (genetic composition) was TT or Tt.

The self-fertilization of the F_1 generation plants produces a 3:1 ratio of tall to short plants, three tall to one short, Figure 5-13.

A simpler method of diagramming the crosses is shown in Table 5-2. The genotype of one parent is written across the top and the genotype of the other parent is written down the side. By multiplying the letter symbols down and across, the possible types of offspring and their predicted ratio result.

The example of Mendel's work just described refers to parents that differ in only one characteristic. A cross between such parents is termed a *monohybrid cross.* When two or more independently inherited characters distinguish the parents, a cross is termed *dihybrid.* Gregor Mendel selected a parent pea plant that always produced yellow, round seeds when self-pollinated. Another parent always produced green, wrinkled seeds when self-pollinated. When the two plants were cross-pollinated, the F_1 offspring all produced yellow, round seeds. He deduced that the yellow and round traits were dominant over the green and wrinkled traits.

Using Y and y to represent yellow and green and R and r to represent round and wrinkled, the two original parents can be represented as YYRR and yyrr. The F_1 plants would be YyRr, heterozygous for the two characteristics. Since the characteristics exist on two different chromosomes, the two pairs of genes separate independently of each other. The result is that F_1 parents produce four types of haploid gametes during meiosis: YR, Yr, yR, and yr. When F_1 plants are self-pollinated, the checkerboard method of predicting how the gametes will combine corresponds closely with what Mendel actually observed: nine yellow and round, three yellow and wrinkled, three green and round, and one green and wrinkled. Refer to Table 5–3.

As self-pollination continues, the proportion of homozygous offspring increases. That is why self-pollinating plants usually breed true. In animals, especially humans, self-pollination or inbreeding

FIGURE 5-12.

First generation of cross between TT and tt parents

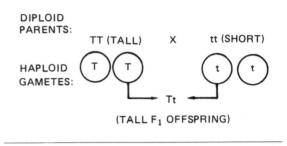

(TALL F_1 OFFSPRING)

FIGURE 5-13.

Second generation of cross between TT and tt parents

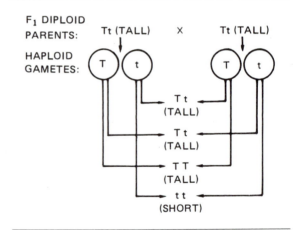

TABLE 5-2.

Ratio of Phenotypes 3:1

#2 Parent \ #1 Parent	T	t
T	TT (tall)	Tt (tall)
t	Tt (tall)	tt (short)

TABLE 5-3.
Ratio of Phenotypes 9:3:3:1

Parent #2 \ Parent #1	YR	Yr	yR	yr
YR	YYRR	YYRr	YyRR	YyRr
Yr	YYRr	YYrr	YyRr	Yyrr
yR	YyRR	YyRr	yyRR	yyRr
yr	YyRr	Yyrr	yyRr	yyrr

may be detrimental because a homozygous offspring manifesting recessive harmful genes is likely to result. With plants, the inbreeding technique can be used to produce homozygous offspring that exhibit desirable recessive genes.

The great danger inherent to a simplified discussion of a complex subject is that readers will assume the limited information is all they need to know. Therefore, it bears repeating that what is sought here is not substantive insight into the science of genetics or the technology of plant breeding, but rather a basic understanding of the kind of knowledge needed for plant improvement.

The work of Gregor Mendel illustrates many of the principles upon which the science of genetics is based, but the science continues to grow and knowledge to accumulate. A few more concepts are worth mentioning.

- Dominance does not exist between all gene pairs. F_1 may appear intermediate in a characteristic compared to the parents.
- Some plants are self-sterile, or unable to self-pollinate and produce seed. Cross-fertilization gains importance in such species and results in increasing the number of variations in the offspring.

- Certain genes are termed *lethal.* Their presence on the chromosome of a plant will kill the plant. For example, plants need chlorophyll to photosynthesize. If a lethal recessive gene for no chlorophyll appears as a homozygous individual, the plant will eventually die.
- When inbred plant species are crossed, the F_1 hybrid generation may have qualities superior to those of either parent. The phenomenon is known as *hybrid vigor.* It can be predicted and created by selective breeding, but its cause is not clearly understood.

SUMMARY

When organisms duplicate themselves, the process is termed reproduction. If deliberately controlled and manipulated, it is termed propagation. Plant reproduction may be either sexual or asexual.

Sexual reproduction is accomplished through the formation of seeds and requires the fusion of two sex cells (gametes). Each gamete has one set of chromosomes and is thus termed haploidal. Fusion results in a cell or zygote that has two sets of chromosomes and is termed diploidal. The gametes are formed in the process of meiosis, which reduces the number of chromosomes in each cell by half. Through sexual reproduction, the genetic material of the cell recombines, resulting in new plants that differ from their parent plants in physical or physiological characteristics.

Asexual reproduction is a vegetative process that eliminates genetic variation. It can perpetuate an individual plant unchanged for many years. Asexual reproduction is made possible by a combination of the processes of mitosis and cytokinesis. Mitosis is the normal division of a cell nucleus that occurs as a plant grows. Cytokinesis is the division of the remainder of the cell's contents and the formation of a new cell wall. There is no reduction and recombination of the chromosomes as the cells divide and no opportunity for a new plant to be initiated.

As reproduction is both sexual and asexual, so too is plant propagation. Sexual propagation utilizes seeds. Not all seeds require the same environment for germination. Some seeds need light, while others require darkness to germinate. Some seeds remain dormant due to an impervious seed coat and require scarification, or breaking of the seed coat, in order to germinate. Others require a period of low temperatures, or stratification, before germination can occur.

Asexual propagation utilizes the vegetative parts of a plant to grow new plants. Stems, roots, and leaves can be used since each vegetative cell has the ability to reproduce an entire plant that is genetically identical to the one from which the cell originated. Asexual propagation techniques include use of:

- runners
- stolons
- sucker shoots
- bulbs and similar organs
- layering
- cuttings
- grafting
- budding
- division of the crown
- tissue culture
- apomictic embryos

The essence of asexual reproduction is that there is no change in the new plants. While this may be desirable in the short term, over time plants must be able to change and respond to new environmental conditions. Such changes can only occur through sexual reproduction or spontaneous mutation.

Plant breeding is the deliberate control of the sexual reproduction process. The science of plant breeding is known as plant genetics and began in the late 1800s with Gregor Mendel's study of the common garden pea. His work disproved the previously held belief that inheritance in plants is not predictable. It is now commonly accepted that all physiological activities of plants as well as their physical appearance are controlled by individual genes, the determiners of heredity. The genes are located on chromosomes in the nucleus of each plant cell. Chromosomes and their genes are duplicated and distributed through meiosis. Depending upon whether plants are self-pollinated or cross-pollinated and how many independently inherited characteristics distinguish the parents, the offspring may closely resemble the parents or be very different. Survival and adaptability of plant species to new environments depend upon the variations that result from sexual reproduction.

Achievement Review

A. SHORT ANSWER

Answer each of the following questions as briefly as possible.

1. Briefly define the following terms:

a. reproduction
b. propagation
c. sexual propagation
d. asexual propagation
e. gamete
f. zygote
g. meiosis
h. mitosis
i. clone
j. clonal variety

2. Label each stage in this diagram of mitosis.

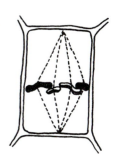

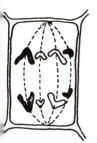

3. Label each stage in this diagram of meiosis.

  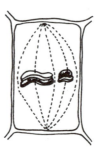 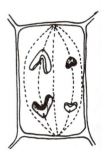

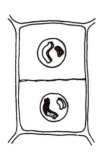 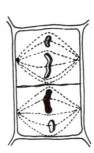 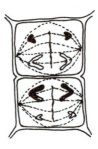 

4. Give two reasons why some seeds remain dormant and will not germinate.
5. Identify the type of asexual propagation described in the following statements:
   a. aerial shoots that take root after coming into contact with the soil
   b. roots that form on the stem at the point of a partial cut
   c. stems that grow along the ground and form new plants at the nodes
   d. asexual seed production
   e. the joining of two plant parts to create a single plant
   f. new shoots formed from adventitious buds on underground roots
   g. the most common method of reproducing plants asexually
   h. storage organs representing modified stem or root tissue and capable of growing into a new plant
   i. small sections of plant tissue grown in nutrient media and allowed to differentiate into new plants

B. MULTIPLE CHOICE

From the choices given, select the answer that best completes each of the following statements.

1. The science of plant breeding is termed _____.
   a. physiology
   b. reproduction
   c. botany
   d. genetics

2. A species responds naturally to environmental changes through _____.
   a. sexual reproduction
   b. asexual reproduction
   c. cuttings
   d. tissue culture

3. _____ is credited with the earliest important work in plant genetics.
   a. Luther Burbank
   b. Albert Schweitzer
   c. Gregor Mendel
   d. W. A. Burpee

4. Genes, the determiners of heredity, are located on the _____.
   a. nucleus
   b. chromosomes
   c. chromatids
   d. DNA

5. The external appearance of a plant is known as its _____.
   a. phenotype
   b. F_1 hybrid
   c. epidermis
   d. genotype

6. The genetic makeup of a plant is known as its _____.
   a. phenotype
   b. F_1 hybrid
   c. epidermis
   d. genotype

7. A cross between two parents that differ in only one characteristic is termed a (an) _____.
   a. F_1 hybrid
   b. F_2 hybrid
   c. dihybrid cross
   d. monohybrid cross

8. When the influence of a gene is masked by the presence of another gene, the latter is said to be _____.
   a. heterozygous
   b. homologous
   c. dominant
   d. recessive

9. The proportion of homozygous offspring is increased by continuing _____.
   a. dominance
   b. asexual propagation
   c. self-pollination
   d. cross-fertilization

10. The superior qualities of an F_1 generation of plants may be explained as _____.
    a. lethal genes
    b. hybrid vigor
    c. dominant genes
    d. recessive genes

6

Plant Pests and Their Control

Objectives

Upon completion of this chapter, you will be able to

- state the major causes of injury to plants.
- characterize insects and groups of pathogens as plant pests.
- list common symptoms of injured plants.
- list the principles of pest control.
- describe the types of pesticides and their safe use.
- explain the concept of integrated pest management.

PLANT INJURIES AND THEIR CAUSES

Anything that impairs the healthy growth and maturation of a plant may be regarded as an injurious agent. Some injurious agents cannot be transmitted. Others can be transmitted from one plant to another and are regarded as either *infectious* or *infestious.* An infected plant has the injurious agent active within it. An infested plant has the agent active on its surface. Some agents of injury are members of the plant or animal kingdom and are thus biological in character. Others are environmental or circumstantial and nonbiological in nature. Table 6-1 will help you understand these differences.

Injurious agents that are biological and infectious or infestious are also parasitic. A *parasite* is an organism incapable of manufacturing its own food. It derives its sustenance from the cells of other organisms, such as plants, which can manufacture food. Rodents, rabbits, deer, and other animals that injure plants are technically parasitic, but they are not involved with plants at the cellular level.

The parasitic insects, fungi, bacteria, viruses, and nematodes, as well as the weeds, are usually referred to as plant pests. Each has given rise to its own specialized branch of biological science.

- *Entomology* is the study of insects, their effects upon plants, and their control.

TABLE 6-1.
Injurious Agents

	Infectious	Infestious
Biological Agents of injury		
Insects	Occasionally	Yes
Fungi	Yes	Occasionally
Bacteria	Yes	No
Viruses	Yes	No
Nematodes	Yes	Yes
Rodents, rabbits, deer, and other animals	No	No
Weeds	No	No
Other Causes of Injury		
Snow, ice, wind, sun scald	No	No
Lawnmowers and other mechanical tools	No	No
Vandalism	No	No
Nutrient deficiency	No	No
Fertilizer burn, spray damage, and other chemical injury	No	No

- *Plant pathology* is the study of plant diseases, their causes and their control.
- *Bacteriology* is the study of bacteria.
- *Mycology* is the study of fungi.
- *Virology* is the study of viruses.
- *Nematology* is the study of nematodes.
- *Weed science* is the study of weeds and their control.

INSECTS

Regarded by nearly all biologists as the most interesting, economically significant, and certainly most prolific of all members of the animal kingdom, insects are a study in contrasts. For millions of years, insects have benefited and beleaguered human civilization. Essential as scavengers, pollinators, and reducers of organic matter to earth mold, insects also despoil food crops, invade our homes, attack our bodies and those of other animals, and lay waste our agricultural products.

Insects are primarily terrestrial animals that range from the arctic to the antarctic and from the depths of the earth to the mountain tops. They constitute over two-thirds of all the animal species on the planet and are surpassed only by microbes in sheer numbers of individuals.

The Classification of Insects

In the animal kingdom, the major divisions are known as *phyla* (singular *phylum*). The phylum Arthropoda contains the class Insecta, or the insects. Other classes of Arthropoda include the crayfish, spiders, centipedes, ticks, millipedes, mites, and scorpions. Certain of the noninsect Arthropoda are important as plant pests, especially the mites, but the plant pests that represent the greatest numbers and do the greatest amount of economic damage are the true insects.

Anatomy of Insects

Combining the characteristics of all members of the phylum Arthropoda with those unique to the class Insecta, the external anatomy of insects features:

- jointed appendages
- bilateral symmetry
- an exoskeleton
- three body divisions (head, thorax, and abdomen)

- six legs attached to the thorax
- wings (usually two pairs) attached to the thorax
- one pair of antennae attached to the head
- simple and compound eyes
- reproductive organs in the abdomen
- air tubes for respiration

The external anatomy of a typical insect is exemplified by the grasshopper, Figure 6-1. Several of the anatomical features warrant special mention.

FIGURE 6-1.
The external anatomy of a typical insect is exemplified by the grasshopper.

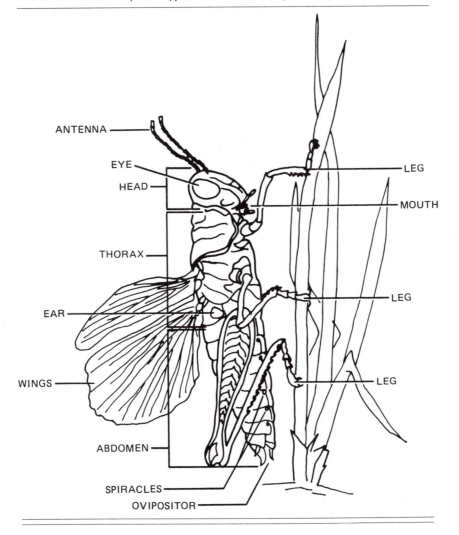

The Exoskeleton The exoskeleton is the hard protective covering of the insect's body. It functions like our more familiar endoskeleton, serving as the point of attachment for muscles. It also protects the softer tissues inside. The exoskeleton has helped assure the survival of insects through centuries of evolution and environmental stress. Excessive moisture, severe drying, and enemies of all types are thwarted by it. The exoskeleton is composed principally of *chitin,* a nitrogenous polysaccharide compound that resists water, alcohol, acids, and alkalis.

The Head The head bears the mouthparts, eyes, and antennae. Sensory perception is the function of the antennae, and they are one of the major distinguishing features in the identification of insects. Following are nine of the more common types of insect antennae, Figure 6-2.

- *pectinate:* comb-like
- *plumose:* feather-like
- *setaceous:* tapering to a point
- *serrate:* saw-like
- *lamellate:* expanded thin plates at the tips
- *aristate:* enlarged at the tip and bearing a bristle
- *clavate:* club-like
- *filiform:* thread-like; uniform the entire length
- *moniliform:* bead-like

The type of mouthpart is also highly important in identifying insects, and equally so in selecting a means of controlling them. The mouthparts of insects determine how they feed on plants, and the symptoms of injury that are manifested. These are the most common types of mouthparts, Figure 6-3.

- *chewing:* The mouth can tear, chew, and grind food. Examples: grasshopper and beetles.
- *siphoning:* The mouthpart is like a coiled tube. It is dipped into a liquid food, which is then drawn in. Examples: moths and butterflies.
- *sponging:* Two sponge-like structures collect the liquid food and move it to the insect's food canal. Example: housefly.
- *rasping-sucking:* The mouthpart rasps (breaks) the surface of the plant, then sucks the sap that exudes from the wound. Example: thrips.
- *piercing-sucking:* The mouth punctures the epidermis of plants, then sucks the sap. Examples: aphids, scale, leafhoppers.
- *chewing-lapping:* The insect is able to tear and chew the plant tissue as well as draw up released liquid through a temporary food tube. Example: bees.

Legs The legs project from the thorax. They and the wings propel insects in a variety of ways. They vary as widely as antennae and mouthparts, Figure 6-4.

Wings The wings also arise from the thorax. Unlike the legs, whose number (six) is part of the very definition of an insect, the number of wings can vary. Most adult insects have two pairs; some have only one pair; and still other insects have no wings. The venation pattern of the wings is a factor in the identification of insects.

Digestive System The digestive system is a tube that extends from the mouth of the insect to the anus. It is divided into three parts, termed the fore-intestine, mid-intestine, and hind-intestine. Food is digested in the mid-intestine.

Respiratory System The respiratory system combines expansion and contraction of the abdomen with diffusion, to exchange oxygen and carbon dioxide in the cells and tissues of the insect. Insects breathe through small openings along the side of the thorax and abdomen termed *spiracles.* The spiracles open into tracheae, tubes that carry essential oxygen directly to the cells and tissues. Carbon dioxide passes out of the tissues by a reversal of the process.

Nervous System The nervous system is made up of groups of specialized cells called *ganglia* (singular, *ganglion*) and nerve fibers that join the ganglia to other parts of the insect's body. The insect's brain is simply an enlarged ganglion located in the head.

Reproductive System The reproductive system is located in the abdomen. Reproduction normally requires mating between a male and female, although some insects are incapable of reproduction.

FIGURE 6-2.
Typical insect antennae (Courtesy Cari Goetcheus)

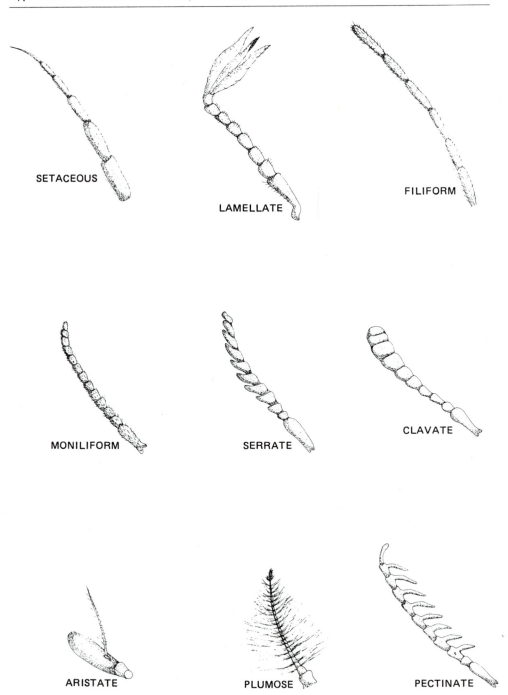

FIGURE 6-3.

All insect mouthparts are modifications of two main types: chewing (left) and sucking (right).

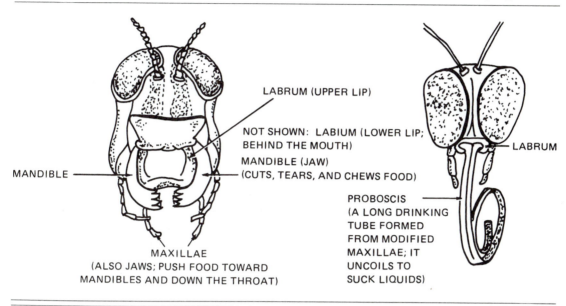

LABRUM (UPPER LIP)

NOT SHOWN: LABIUM (LOWER LIP;
BEHIND THE MOUTH)

MANDIBLE (JAW)
(CUTS, TEARS, AND CHEWS FOOD)

MANDIBLE

LABRUM

PROBOSCIS
(A LONG DRINKING
TUBE FORMED
FROM MODIFIED
MAXILLAE; IT
UNCOILS TO
SUCK LIQUIDS)

MAXILLAE
(ALSO JAWS; PUSH FOOD TOWARD
MANDIBLES AND DOWN THE THROAT)

Still others are able to reproduce through *parthenogenesis,* which allows eggs to develop even though they have not been fertilized.

In addition to the reproductive organs common to most animals, insects possess specialized organs for mating and depositing eggs. The male insect may have special structures for clasping the female. The female has an *ovipositor,* an organ for laying eggs.

Growth and Change of Insects

No members of the animal kingdom transform more dramatically during their development than the insects. The homely caterpillar becomes the graceful butterfly; the grub worm in last summer's lawn becomes the June bug buzzing around the porch light this year. The changes in insect form as they grow are termed *metamorphosis.* Some insects have no metamorphosis, going from the earliest stage to the adult form looking the same except for increasing size. Others have complete metamorphosis and include four stages of development: *egg, larva, pupa,* and *adult.* Still other insects have some form of metamorphosis that includes a *nymph* stage but not the larva or pupa stages; they are said to possess gradual or incomplete metamorphosis.

In those insects with a nymph stage, the young insect usually bears some resemblance to the adult into which it will develop. At first the resemblance may be slight, but as the insect grows it passes through a series of growth stages called *instars.* As it grows, the insect must shed its confining exoskeleton (*molt*) to permit continued growth into the next instar stage. Each molt and new instar brings the insect closer to the adult form.

An insect's life begins as an *egg.* The eggs may hatch inside the female (and are termed *ovoviviparous*) or they may be laid by the female in large or small clusters for later hatching (*oviparous*). The eggs are usually deposited near a food supply since the insects that hatch can move only short distances but possess insatiable appetites. The female insect may lay as few as one egg or as many as several

FIGURE 6-4.
Examples of insect legs modified for various purposes. The five parts of the leg are labeled in one example. (Courtesy Cari Goetcheus)

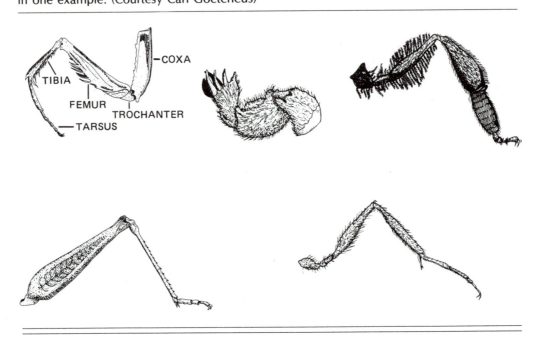

thousand daily depending upon the species. There is little evidence of a maternal instinct among insects; once the eggs are laid the female leaves them to hatch and fend for themselves.

Most insects undergo a complete metamorphosis, including the larva and pupa forms. The larval stage finds the insect in a worm-like form and possessing a chewing mouthpart, Figure 6-5. In this form the insect does little more than eat and grow. Much of the economic devastation wrought by insects against crops results from this larval stage. When larval maturity is reached, the larva ceases feeding and seeks a protected location for its next stage of growth, the pupal stage.

The pupa is totally helpless and physically inactive, Figure 6-6. However, while in this stage, the insect undergoes remarkable alterations in its physiology and morphology. It is during the pupal stage that the transformation from worm-like larva to adult occurs. Legless forms develop legs, wings develop where none were, mouthparts change from

chewing to another type. The miracle of metamorphosis is one of the true wonders of the natural world, Figure 6–7.

Whatever the type of metamorphosis, the adult stage is eventually reached. The most essential function of the adult insect is reproduction. Some insects die soon after mating or laying eggs. Other insects live for a long time in the adult stage. The majority of insects are of greatest economic significance in their nonadult stages, since that is when the greatest feeding and crop destruction occurs.

PLANT DISEASES

The effect of disease in plants is injury, which is manifested in *symptoms* of abnormal growth. The cause may be one or more continuous irritants termed *pathogens* that live off the *host plant* as parasites. Despite our best efforts at control, the United States loses an estimated 15 to 20 percent of crop productivity each year due to plant diseases. History

FIGURE 6-5.
Types of insect larvae (Courtesy Cari Goetcheus)

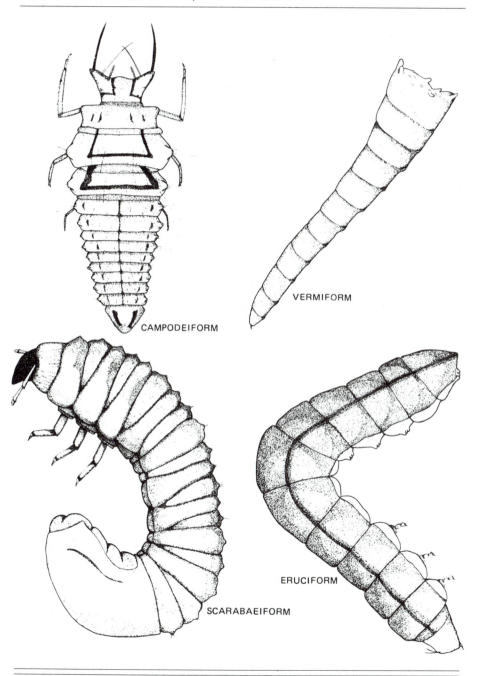

CAMPODEIFORM

VERMIFORM

SCARABAEIFORM

ERUCIFORM

FIGURE 6-6.
Three types of insect pupae (Courtesy Cari Goetcheus)

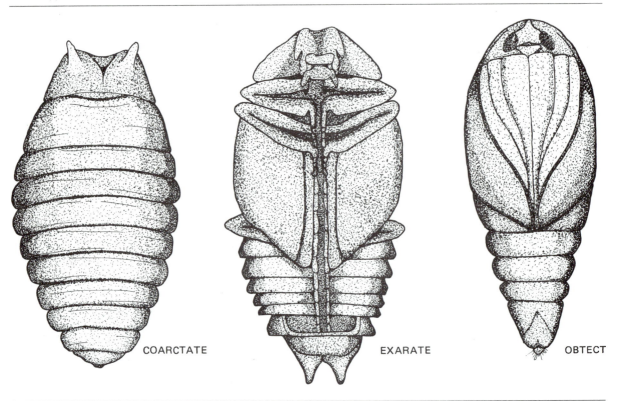

COARCTATE   EXARATE   OBTECT

is laden with accounts of starvation and pestilence resulting from crop failures that were themselves the result of widespread infection by plant pathogens. Plant diseases such as the potato blight in Ireland alter the destiny of an entire country. In our country, chestnut blight and Dutch elm disease have all but eliminated two important native North American tree species.

From the moment it begins as a seed, cutting, or other propagative structure, the plant is susceptible to injury and death from hundreds of pathogenic agents. Some affect the plant in its early stages of growth. Others are a threat to the maturing plant. Some pathogens affect the flowers while leaving the foliage unimpaired. Others bother only the fruit.

Still others injure any and all parts of the plant. To attain a basic understanding of plant diseases, you need to be familiar with both the causal agents (pathogens) and the diseases they create in their host plants.

Plant Pathogens

The most important causes of disease in plants are the bacteria, fungi, viruses, and nematodes. All are microscopic in size, reproduce prolifically, and have a remarkable ability to survive. Following is a summary of the important characteristics of the pathogen groups.

FIGURE 6-7.
The four types of insect metamorphosis

EGG
▼
SEVERAL INSTAR STAGES.
EACH RESEMBLES THE ADULT BUT IS LARGER
THAN THE PREVIOUS INSTARS.
▼
ADULT
● PRIMITIVE
● WINGLESS
● EXAMPLES:
SPRINGTAILS, SILVERFISH

NO METAMORPHOSIS

EGG
▼
SEVERAL NYMPH AND INSTAR STAGES,
WITH DISTINCTIVE CHANGES IN APPEARANCE
AND PHYSIOLOGY.
● BREATHE THROUGH TRACHED GILLS
● SPEND THESE STAGES IN WATER
▼
ADULT
● FULLY DEVELOPED WINGS
● BREATHE ATMOSPHERIC OXYGEN
● ASSORTED MORPHOLOGICAL
  MODIFICATIONS
● EXAMPLES:
MAYFLIES, STONEFLIES, DRAGONFLIES

INCOMPLETE METAMORPHOSIS

EGG
▼
SEVERAL NYMPH AND INSTAR STAGES.
EACH RESEMBLES THE ADULT BUT LACKS
WINGS AND GENITAL APPENDAGES. EACH
MOLT BRINGS THE RESEMBLANCE CLOSER
TO THE ADULT.

▼
ADULT
● FULLY DEVELOPED WINGS
● FULLY DEVELOPED GENITAL
  APPENDAGES
● ASSORTED MORPHOLOGICAL
  MODIFICATIONS
● EXAMPLES:
APHIDS, GRASSHOPPERS, SQUASH BUGS

GRADUAL METAMORPHOSIS

EGG
▼
LARVA
● ENLARGES VIA INSTARS AND MOLTS,
  BUT BEARS NO RESEMBLANCE TO THE
  ADULT
● WORM-LIKE OR GRUB-LIKE IN
  APPEARANCE
● CHEWING MOUTH PARTS
▼
PUPA
● DISTINCTIVE CHANGES IN APPEARANCE
  AND PHYSIOLOGY
● INACTIVE, DOES NOT FEED
▼
ADULT
● FULLY DEVELOPED MORPHOLOGICALLY
  AND PHYSIOLOGICALLY
● DIVERSE APPEARANCES; MOST COMMON
  FORM OF METAMORPHOSIS AND
  REPRESENT MOST INSECTS

COMPLETE METAMORPHOSIS

Bacteria

a. Members of the plant kingdom
b. Nonchlorophyllous
c. Single-celled
d. Three body forms: spherical (coccus), rod-shaped (bacillus), and spiral-shaped (spirillus), Figure 6-8. Only the bacillus form causes plant diseases.
e. Reproduction by simple cell division
f. Some forms survive unfavorable conditions as hard-coated spores termed *endospores.*
g. Some forms propel themselves by means of whip-like appendages termed *flagella.* Bacteria may have a single flagellum at one end (monotrichous), more than one flagellum at one end (lophotrichous), or flagella all over the cell (peritrichous), Figure 6-9.
h. All pathogenic species of bacteria are classified within six genera:
   • *Pseudomonas*
   • *Xanthomonas*
   • *Erwinia*
   • *Corynebacterium*
   • *Agrobacterium*
   • *Streptomyces*

FIGURE 6-8.
Body forms of bacteria

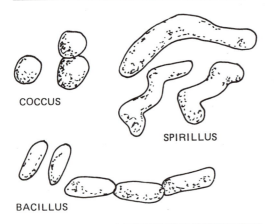

Fungi

a. Members of the plant kingdom
b. Nonchlorophyllous
c. Their vegetative body (called a thallus) ranges in size from a single cell to a thread-like multicellular structure termed a *mycelium.* The individual threadlike filaments making up the mycelium are *hyphae.* The functions of the hyphae are to absorb nutrients from the host plant, reproduce the fungus by formation of sexual and asexual fruiting structures, and survive adverse conditions, Figure 6-10.
d. Reproduction by formation of *spores*

FIGURE 6-9.
Flagellation forms of bacteria

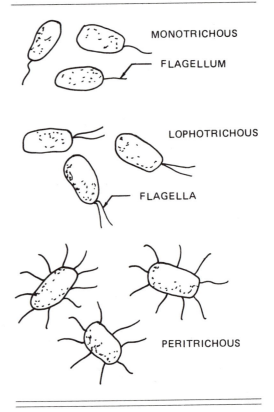

FIGURE 6-10.
Fruiting bodies common to many of the fungal pathogens

ASEXUAL FRUITING BODIES

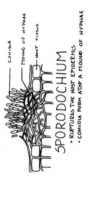

ACERVULUS
- SHALLOW AND SAUCER-SHAPED
- CONIDIA ARE PRODUCED ON THE SURFACE AFTER HOST CELLS RUPTURE

PYCNIDIUM
- HARD WALLED
- SAC-LIKE SHAPE
- CONIDIA ARE PRODUCED WITHIN

SPORODOCHIUM
- RUPTURES THE HOST EPIDERMIS
- CONIDIA FORM ATOP A MOUND OF HYPHAE

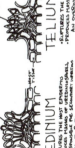

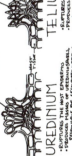

AECIUM
- BELL-SHAPED
- PRODUCES CHAINS OF DIKARYOTIC AECIOSPORES
- COMMON TO RUSTS

SEXUAL FRUITING BODIES

CLEISTOTHECIUM
- COMPLETELY ENCLOSED
- CONTAINS SAC-LIKE ASCI AND ASCOSPORES
- COMMON TO POWDERY MILDEWS

APOTHECIUM
- SAUCER SHAPED
- USUALLY GROWS ON HOST SURFACE
- ASCI AND ASCOSPORES LINE THE INNER SURFACE

PERITHECIUM
- HARD WALLED
- SAC-LIKE SHAPE
- OPEN AT THE TOP
- LINED WITH ASCI AND ASCOSPORES

PYCNIUM
- SAC-LIKE SHAPE
- CONTAINS HAPLOID PYCNIOSPORES
- CONTAINS RECEPTIVE HYPHAE FOR PLASMOGAMY
- COMMON TO RUSTS

UREDINIUM
- RUPTURES THE HOST EPIDERMIS
- PRODUCES MASSES OF UREDINIOSPORES
- RESPONSIBLE FOR SECONDARY SPREAD
- COMMON TO RUSTS

TELIUM
- RUPTURES THE HOST EPIDERMIS
- PRODUCES MASSES OF TELIOSPORES, AN OVERWINTERING STAGE
- COMMON TO SMUTS AND RUSTS

e. Classification is on the basis of life cycle, specifically how the vegetative and reproductive organs develop.

f. Pathogenic fungi are placed in the following six classes by most mycologists:
- Myxomycetes
- Plasmodiophoromycetes
- Phycomycetes
- Ascomycetes
- Basidiomycetes
- Deuteromycetes

A number of nuclear changes occur during the life cycle of the fungi, Figure 6-11. Fungi that reproduce sexually have haploid nuclei during the vegetative (gametophytic) stage. The sexual cells (gametes) form during the haploid phase. The fusion of two compatible gametes occurs during *plasmogamy*, marking the beginning of the diploid phase. When the two sexual nuclei actually fuse, the stage of *caryogamy*, is achieved. The result of the fertilization is a *zygote*. The zygote eventually becomes the organ in which reduction and division (meiosis) oc-

FIGURE 6-11.

A generalized life cycle of fungi

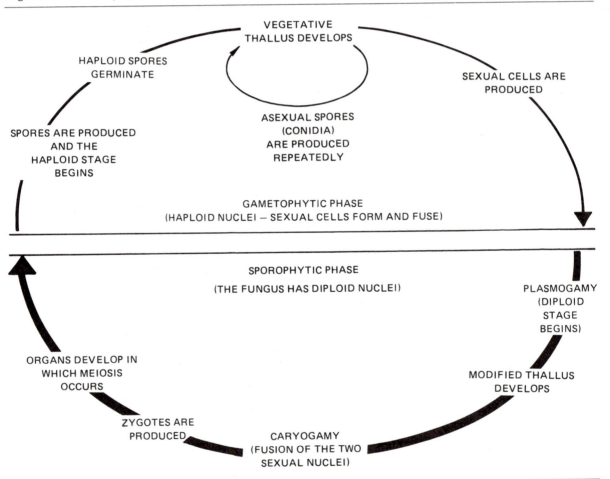

curs. Following meiosis, haploid spores are formed, which germinate to form the vegetative thallus. During the gametophytic stage, additional spores may be produced asexually. These are termed *conidia* (singular *conidiospore*). Each conidiospore is capable of germinating and forming a new thallus. With many pathogenic fungi, the number of asexual conidia produced far surpasses the number of sexual spores produced, making them the most significant pathogenic form.

Viruses

a. Their categorization as plants or animals is undetermined. Evidence suggests they are an entirely different life form.
b. Nonchlorophyllous
c. Composed of nucleic acids and proteins
d. Smallest of the pathogens, visible only with an electron microscope
e. Plant viruses appear most commonly in two morphological forms: a long, narrow, tube-like rod and an isometrical polyhedron (shape with many sides of equal length).
f. Reproduce only within a living host
g. No natural classification system is yet developed. Instead, viruses are grouped together on the basis of the symptoms they create in their hosts, their physical and chemical properties, the insects that transmit them, and other qualities.

Nematodes

a. Members of the animal kingdom
b. Feed on both plants and animals. Nearly a thousand species attack plant parts. Those that complete most of their life cycle within a host plant are termed *endoparasitic.* Those whose life cycle occurs outside the host plant are *ectoparasitic.*
c. Plant nematodes average about 1 millimeter in length.
d. Body wall usually transparent, with an outer covering termed a *cuticle*
e. Feeding apparatus consists of mouthparts equipped with a *buccal spear,* used to puncture the host cell to withdraw cellular fluids.
f. Favored by light, sandy soils with a high average temperature
g. Life cycle generally simple. The females lay eggs that hatch to form larvae. These larvae grow and pass through a series of four *molts,* finally reaching the adult stage, Figure 6-12.

Relationships Between Plants and Pathogens

Plant diseases do not arise spontaneously. There must be a causal agent (the pathogen) in the vicinity of a susceptible plant (the host). The pathogen must be in a form that will grow and develop when transferred to the host. The infectious form of the pathogen is termed *inoculum.* The inoculum is transferred to the host by an *agent of dissemination.* There it must arrive at an appropriate *site of infection* with environmental conditions favorable to the pathogen if a parasitic relationship between the pathogen and the host is to develop. Following are examples of these new terms.

FIGURE 6-12.
General life cycle of a plant nematode

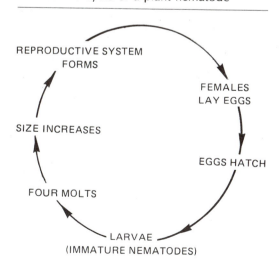

REPRODUCTIVE SYSTEM FORMS

FEMALES LAY EGGS

SIZE INCREASES

EGGS HATCH

FOUR MOLTS

LARVAE (IMMATURE NEMATODES)

Some *agents of dissemination* are:

- insects
- splashing water and rainfall
- wind
- animals, including humans
- equipment and tools

Some *forms of inoculum* are:

- spores of fungi
- bacterial ooze (a concentrated mass of bacteria)
- virus particles
- strands of hyphae

Some *sites of infection* are:

- stomates
- cuts
- wounds
- blossoms
- roots
- fruit

The precise nature of the relationship that develops between the host and the pathogen is determined by the type of parasitism.

- *Obligate parasites:* These pathogens can exist only on a living host.
- *Obligate saprophytes:* These irritants cannot survive on a living host, only on nonliving matter. An obligate saprophyte cannot be considered a pathogen.
- *Facultative saprophyte:* This pathogen is normally a parasite, growing on a living host. However, it can survive on nonliving matter and as such can be grown on nonliving agar for laboratory study.
- *Facultative parasite:* This pathogen is normally a saprophyte, growing on nonliving matter. However, it can grow as a parasite on plants. In so doing, it must kill the host cells with secreted enzymes before it is able to feed on them. Facultative parasites can survive in the soil for long periods of time feeding on nonliving organic matter.

As long as the inoculum is only *on* the host, the plant is termed *infested*. Once the pathogen pen-etrates the host's tissues, the plant is termed *infected*. Disease begins when the host responds to the injurious presence of the pathogen.

SYMPTOMS OF INJURED PLANTS

Responses to pathogenic irritants and insects are termed symptoms. Some symptoms are common to numerous insects and diseases. Other symptoms are almost unique to certain irritant-host relationships. The sum of all the symptoms expressed by a host from the time it is initially infected until it either recovers or dies is known as the *symptom complex*.

Infection of a plant by pests is not a static condition. The symptoms expressed early in the infection may be quite different than those expressed later. For any one pest, however, the symptom complex is usually specific. To diagnose a disease or insect problem correctly requires recognizing the specific changes each major pathogen or insect can create in a host.

Symptoms may be influenced by an assortment of factors including the species of the host, the environment, the quantity of inoculum or insects, and the stage of development of the pathogen or insect. Furthermore, symptoms can result from other causes such as a damaging environment, improperly applied chemicals, animal injury, and mechanical damage. For this reason, it is often necessary to isolate and identify the specific agent of plant injury or to consider other possible sources of irritation before the cause of plant symptoms can be established.

While specific symptoms cover a wide range often separated only by subtleties, collectively they can be grouped into major categories that permit description and comparison, Figure 6–13.

Wilting

Plants may wilt from lack of water. If such a symptom is environmental in origin, the plant will recover when watered. If insects attack a plant's root system or pathogens destroy the xylem tissue, the wilting may be permanent. When fungi invade

FIGURE 6-13.
Symptoms of plant injury

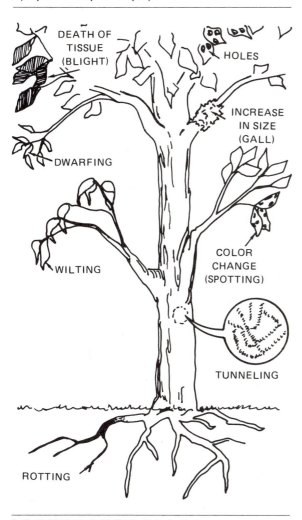

When the plant turns yellow but does not lack light, the symptom is termed *chlorosis*. Chlorosis is probably the most common of all color changes. It can be caused by insects, pathogens, or environmental problems.

Rotting

Rotting results from a destruction of the host cells that causes a release of the cellular fluids. It may be accompanied by a strong, often foul, odor. Rots may be dry or soft and occur in the roots, leaves, stems, buds, or fruits. They are usually the result of plant diseases or freezing.

Death of Tissue

When tissue becomes dessicated (dried out) and dies, it usually turns brown or black and is said to be *necrotic*. The dead tissue, or necrosis, may be localized on leaves as *spots* or centered in young buds. It may also be extensive, encompassing entire branches as in *blights*. Insects, pathogens, or environmental factors can cause necrosis. It ranks with chlorosis in the frequency of its presence. It is often the final symptom in the symptom complex.

Dwarfing

All or part of a plant can be reduced in size as a result of pathogens (especially viruses), insects, and nematodes. Dwarfing may result from the reduction of water uptake at a time when new tissue is expanding. Insects or nematodes in the root system of a plant may create such dwarfing, which results from a reduction in cell size (*hypotrophy*) or number (*hypoplasia*). Hypotrophic or hypoplastic reactions by the plant can result in local or overall dwarfing. Both conditions can occur simultaneously.

Increase in Size

Plant parts may become malformed in response to insect or pathogenic irritants. As cells increase

the tender stem tissue of a young seedling, *damping-off* develops, the plant wilts and drops over to die.

Color Changes

Resulting from the destruction of chlorophyll, color changes may be localized (*spots, rings, lesions*) or widespread through the plant. Lack of light turns a plant yellow in a condition termed *etiolation*.

in size (*hypertrophy*) or in number (*hyperplasia*), symptoms are expressed as *galls, witches brooms, swollen roots, abnormal shoot growth, scabs,* and *fasciations.*

Tunneling

Insects may bore into plant trunks, chew their way between the upper and lower epidermis of leaves, and tunnel up and down the stems. Borers and leaf miners commonly create such injury in plants.

Holes

Leaves filled with holes are symptomatic of both insect and pathogenic causes. Insects usually cause holes by feeding on the leaf tissue. When caused by pathogen activity, the holes are preceded by necrotic lesions and spots that represent dead cells. As the tissue dies in these localized areas, the necrotic areas drop out, creating the holes.

Hundreds of combinations and variations of symptoms can result from the many possible causal agents, reactions of individual host plants, and unpredictable modifications caused by the environment. If the quick and accurate diagnosis of what is troubling a particular host plant sounds difficult, be assured it is!

WEEDS

No discussion of the pests affecting ornamental plants is complete without considering the weeds. A weed may be defined as *a plant having no positive economic value and/or growing in a place where it is not desired.* Unlike insects and pathogens, weeds do not derive their sustenance directly from other plants as parasites. Instead, weeds *compete* with other plants for the materials both need to grow and thrive. In addition, weeds often serve as alternate hosts, providing sites for the overwintering of insects or pathogenic inoculum. Some fungi need weed species in which to produce one or more spore forms as part of a complex life cycle. In summation,

weeds affect ornamental plants adversely in the following ways:

- as competitors for space, nutrients, water, light, heat energy, and CO_2
- by shading crop plants, further inhibiting their growth
- as alternate hosts for insects and pathogens
- by crowding due to excessive vegetation and by the vining manner of some weeds
- through contamination of the soil with large quantities of seeds and root stock that permit proliferation of the weeds over many seasons
- by production of chemical exudates that inhibit germination and/or normal growth of crop plants

Classification of Weeds

Being plants, weeds share the binomial nomenclature of botanical classification. All are classified into families, genera, species, and varieties. Also like other plants, weeds are classified in other ways. Their place of growth may classify them as *aquatic* or *terrestrial.* Their leaf type may classify them as *grasses* or *broad leaved.* They may also be categorized as *herbaceous* or *woody* or as *annuals, biennials,* or perennials. These terms were defined in Chapter 3.

The Origin and Dissemination of Weeds

Even when the climate is hot and dry or the soil infertile, weeds seem to proliferate. They can appear in a raised greenhouse bench or beside a containerized street tree. They are as apt to be found in a manicured lawn as in a cultivated flower bed. Frustrated gardeners through the centuries have pondered how the weedy pests invade their crops.

Historically, the origin and distribution of weeds in an area have paralleled migration and colonization by early settlers. Many of our North American weeds came from Europe. No doubt there were weed seeds aboard the Mayflower. Certainly the forty-niners carried a good assortment of weeds with them aboard their wagons as they rolled westward.

Within a local area, weeds are distributed as seeds or other reproductive parts such as stolons, roots, bulbs, rhizomes, or tubers. Cultivated soil is likely to have a greater number and diversity of weeds than uncultivated land due to the dissemination of propagative parts by cultivating tools.

Many agents of dissemination account for the dispersal of weeds. They may be natural or artificial. Some *natural agents of dissemination* are:

- flowing water
- wind
- birds
- livestock and other animals
- forceful dispersal of seeds from pods or capsules

Some *artificial agents of dissemination* are:

- cultivating tools
- vehicles
- clothing
- transplanting desired plants from one area to another and moving weeds with them

Gardeners who work manure from the local stockyard into their soil may be planting weed seeds from fodder that originated far away. Landscapers who use inexpensive seed to develop a lawn may be planting more weed seeds than quality grasses. Greenhouse growers who permit the maturation of a few *Oxalis* under the greenhouse benches will soon find them growing above the benches as their pods split (dehisce) and forcefully eject seeds over a considerable distance. Homeowners who pull a wild morning glory or field bindweed from the shrub bed leave behind segments of the root system destined to reproduce the plant vegetatively a hundredfold.

Weeds have evolved to survive. They are prolific seed producers, with some species capable of producing more than a million seeds per parent plant. As with nonweed plants, some weed seeds germinate quickly and easily, while others remain dormant longer. Weed seeds are usually distributed throughout a soil's profile, so turning over the earth may

bring to the surface seeds that have been buried and dormant for several years.

Weed seeds and fruits usually benefit from a morphological development that assures their successful dissemination, Figure 6-14. Assorted appendages that catch the wind like sails or latch onto the clothing or fur of passing humans or animals provide transport over long distances. Hard seedcoats allow safe passage of the seeds through the digestive tract of birds and livestock to be deposited with the manure. Some weed seeds are even able to plant themselves. Their wedge-like appendages respond to alternating moisture levels in the soil and air by literally burrowing into the loose soil. Each change in the moisture state thrusts the seed deeper until it is completely buried.

THE CONTROL OF PLANT PESTS

In a text such as this, where insects, diseases, and weeds are treated in terms of their general effect on plants, a discussion of their control must be equally general. Specific products and formulations used in the control of specific pests can fill volumes. Fundamental to the training of professional horticulturists, however, is a solid understanding of *why* and *when* certain control measures can be effective against pests. The technology of pest control may change over time, but the principles that direct it will not.

The *purpose of pest control* is to reduce the damage that can result from an injurious agent. Damage can be either quantitative or qualitative. With *quantitative damage,* all or part of the host plant is destroyed by the antagonistic agent. With *qualitative damage,* the host suffers a loss of appearance and sale value. Damage can also be to land and equipment. Land left unusable by a long-lived, soil-borne pathogen can be considered damaged, as can equipment that depreciates in value without maximum use because of pest presence in a crop.

In seeking and comparing control measures, a grower must decide what degree of effectiveness is being sought. There are three *levels of pest control.*

FIGURE 6-14.

Devices produced by weed seeds and fruits that assure their dissemination

SPECIES *ERODIUM* (FILAREES)

MOISTURE-SENSITIVE APPENDAGES TWIST AND TURN AS THEY MOISTEN AND DRY, DRIVING THE SEED INTO THE SOIL.

SOLIVA

SHARP PRONGS ON THE SEED CLING TO PASSING ANIMALS AND CAN EASILY PIERCE THE SKIN OF PEOPLE.

BIDENS (SPANISH NEEDLE)

BARBS ON THE FRUIT CLING TO PASSING ANIMALS OR PEOPLE.

PHYSALIS (GROUNDCHERRY)

WING-LIKE MEMBRANE OF THE FRUIT CATCHES THE WIND.

STIPA (PORCUPINE GRASS)

TUFTED HAIRS ON THE FRUIT CLING TO PASSING ANIMALS OR PEOPLE.

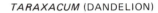

OXALIS

DRIED SEED PODS HAVE A TRIGGERING MECHANISM. WHEN TOUCHED, THEY PROPEL SEEDS FOR SEVERAL FEET IN ALL DIRECTIONS.

TARAXACUM (DANDELION)

TUFTS OF FINE HAIRS CATCH THE WIND TO PROPEL THE FRUIT GREAT DISTANCES.

MEDICAGO (BUR CLOVER)

HOOKED SPINES ON THE POD CLING TO PASSING ANIMALS OR PEOPLE.

XANTHIUM (COCKLEBUR)

HOOKS ON THE FRUIT CATCH THE FUR OF PASSING ANIMALS OR THE CLOTHING OF PEOPLE.

1. *Partial control* is the most common type. When a homeowner sprays shrubs with an all-purpose (broad-spectrum) pesticide, the shrubs may still exhibit some symptoms of insects or disease, but not as severely as if they had not been sprayed.
2. *Absolute control* is total control. All symptoms of pest injury are absent.
3. *Profitable control* is the level attained when monetary returns on the crop exceed the cost of the control measures.

While partial control is the most common type, profitable control is the type most sought by growers. Absolute control, even when possible, is usually too costly to be feasible. Absolute control may be practical on a noncommercial scale, with a single foliage plant in a home or a carefully tended flower bed in the backyard; but it seldom exists in commercial practice.

In determining the potential profitability of a control measure, three factors must be considered:

1. *The value of a single crop plant.* The value of individual plants can vary greatly. The cost of bringing one marigold plant to a saleable state is far less than the cost of one poinsettia or one Japanese maple. More valuable crop plants warrant more costly control measures.
2. *The ultimate value of the crop.* A grower may justify the expense of spraying a field of young sapling trees on the basis of their sale value five years hence.
3. *The average loss over a period of years.* Some insects or diseases may be troublesome only irregularly. However, rather than risk a devastating infestation, a grower may choose to invest in protective control every year to guard against a potential pest outbreak.

Principles of Control

The control of insects, pathogens, and weeds depends on how successfully the horticulturist applies one or more of the four basic principles of control.

Exclusion   Exclusion is the first principle of control. It includes all the measures designed to keep a pest from becoming established in an area. The area may be a single flat of soil, a greenhouse bench, a lawn, a nursery field, a geographic region, even an entire country. The measures vary depending upon the type of pest.

- *The use of pest-free propagative stock* is essential in the commercial production of horticultural crops. Growers must select their sources of supply carefully. Certain sources have excellent reputations as suppliers of disease- and insect-free cuttings and weed-free seed. *Certified seeds and plants* are produced under the direction of an agency that provides documentation of their varietal purity, viability, moisture content, and the absence of weed seed, insects, and pathogens.
- *The treatment of propagative material* can kill pathogens and insects. Immersing bulbs in a fungicidal dip, treating seeds with heat, disinfesting propagative tools, and disinfecting cuttings prior to placing them in a propagation bench are all ways to eliminate pests that may be in or on those materials.
- *Plant quarantines* are restrictions on the production, movement, or very existence of certain plants or plant products in an area. The restrictions are imposed by a legislative authority and enforced in an effort to prevent the introduction or spread of a pest in an area.

Eradication   Eradication is the principle that seeks to remove or eliminate pests that are already in, on, or near plants in infested areas. The measures of eradication attempt to reduce the quantity of pathogenic inoculum, insects and their eggs, weeds and their seeds.

- *Isolating and destroying individual infected plants* is time-consuming and expensive, yet it has its place in ornamental plant production. In agronomic production, it is usually too costly. When the removal of infected plants accompanies another job function, such as watering or disbudding, the cost of control is reduced. Nevertheless, isolating and destroying is more effective for the greenhouse pot crop grower than for the sod producer.

- *Hand-pulling and cultivation* are directed against weed pests in nursery fields and landscape plantings.
- *Destruction of alternate hosts,* such as weeds, can also aid pest control. The alternate hosts may allow completion of the life cycle of a fungus. They may also harbor insect vectors of virus pathogens.
- *Crop rotation and soil treatments* are both methods of eradicating pests that persist in the soil. After a nursery crop has been dug from a field, a season of planting with a cover crop such as barley or sudangrass will not only improve soil structure but may allow the cover crop to compete with and smother troublesome weeds. Soil treatment in the form of either heat or chemicals will eradicate pathogens, insects, and weeds. Greenhouse soils are commonly pasteurized by steam heat or hot water availa-

ble from the heating system, Figure 6-15. Nursery fields and landscape plantings can use chemical eradicants effectively.

- *Destroying host parts* that display evidence of insect or disease damage can reduce the amount of inoculum available for dissemination to nearby hosts. As growers tend their crops, they have the opportunity to eradicate potentially problematic host parts with little additional expense in time and money.
- *Removing infested refuse* eliminates a site for the overwintering of pathogens or insects. It makes a good case for the elimination of fallen leaves and grass clippings from landscapes.
- *Chemical sprays, dusts, and drenches* are generally the most expensive methods of eradication. The products, collectively termed *eradicants,* strive to: kill the pathogen before it can infect the host; kill the insect before it can do much damage or reproduce; or kill the weed before or shortly after it emerges. Chemical control

FIGURE 6-15.
Steam rising from a greenhouse bench during pasteurization

will be discussed in more detail later in the chapter.

Protection   Protection is the principle of control that sets up a barrier between the host plants and the pests to which they are susceptible. It is a shielding endeavor that can be accomplished either through manipulating the plants' growing environment or by applying chemicals.

- *Manipulating the environment* is an attempt to create conditions for growth more favorable to the ornamental host than to the pest. An obvious example of a shield offering protection is the growth of plants inside a screened area to protect the host from insects and the pathogens they may be carrying. Other examples of environmental alteration include: *modifications of the moisture level* to avoid creating a prolonged period of heavy moisture in which pathogens can thrive; *alteration of the soil pH* to attain a pH at which the host will grow better than the pest; *modification of the nutrient level of the soil* to produce healthy, actively growing plants that are less susceptible to injury by pests; maintenance of *cool temperatures* to slow the metabolic rate of both hosts and pests, thus reducing the extent of damage. When cut flowers are kept in refrigerated storage by a florist, the principle of protection is being applied.
- *Chemical sprays and dusts* can be applied to seeds, foliage, and wounds of plants to place a barrier between the host and the insect or pathogens. In this case, it is essential that the chemical be applied before the pests arrive. It then kills them after they arrive.

Resistance   Resistance is the fourth principle of control. It is an attempt to change the plant's morphology and/or genotype so that it will suffer less from diseases or insects. Resistance may originate naturally with a plant population or it may be developed through research by plant hybridizers.

Resistance in plants is not like immunity in animals. That is, plants do not produce antibodies as animals can. Instead, plant resistance is based upon either the physical structure of the plant or cell hypersensitivity. For example, physical features such as leaf pubescence or thick cuticle may discourage penetration by fungal spores or insects merely by holding them at a distance from potential sites of infection. *Hypersensitivity* is the extraordinary reaction of a plant cell to invasion, in which the cell dies so quickly that it fails to support further proliferation of the pest.

Resistance is seldom total. All plants have some degree of susceptibility, and that susceptibility often increases with time. The loss or lessening of resistance to a pest does not necessarily mean that the host plant has changed; it may mean that the pest has altered in a way that allows it to infect the plant. Considering the reproductive potential of most pests and the rapidity of their life cycles, it is not surprising that resistance, especially that developed by artificial hybridization, is often overcome by the natural mutation of insects and pathogens.

PESTICIDES

Understanding chemical pesticides requires knowing what they are and what they are not. What they are is *poisonous*. What they are not is *medicinal*. The belief that pesticides are medicines for ailing plants is misguided—it implies a curative quality that is lacking. An infected plant can seldom be cured. Necrotic tissue cannot regain its life, holes chewed in leaves will not restore themselves, galls will not diminish.

As noted already, injury to plants can be *quantitative* (reducing the number of marketable plants) and/or *qualitative* (rendering the plant unattractive to consumers). To be most effective, pesticides need to be on the plant before the pathogen or insect invader arrives and to kill it promptly upon its arrival. These pesticides will be regarded as *protectants*. If the pathogen is already at the site of infection or the insects are already feeding on the plant, then the pesticides must kill them immediately. Such pesticides are termed *eradicants*.

Thus far, the pesticides described are active against the pest only, not the host plant. Chemical control of weeds can become more complicated since the pest, like the host, is a higher plant. With the range of genetic differences narrowed, the chemicals must be formulated more precisely. The term used to describe all the chemicals that kill plants is *herbicides*. Those that kill all green plants are *nonselective herbicides*. Those that kill some kinds of plants and not others are *selective herbicides*. These assorted products are also characterized by whether they kill upon direct contact with the weed or after the chemical has been incorporated systemically into the weed. Herbicides may kill the weed before the crop emerges (preemergence), as the crop emerges (at emergence), or after the crop has emerged (postemergence).

Pesticide Product Formulations

Whether the pesticide is a fungicide, nematicide, insecticide, or herbicide, it is usually available in several different formulations. The choice of formulation is based upon:

- the size of the crop area being treated
- the amount of active ingredient being applied
- the other materials being applied along with the pesticide, such as other pesticides or fertilizers
- cost
- safety
- ease of application

Following is a summary of product formulations and their characteristics.

Solutions   The pesticide dissolves into its oil or water carrier as a homogeneous physical mixture. It does not precipitate out, so once dissolved the pesticide need not be agitated. The pesticide and carrier are in the molecular or ionic state and cannot be separated mechanically.

Emulsifiable Concentrates   Some pesticides are not soluble in water yet must be applied in a water carrier. A typical emulsifiable concentrate contains the pesticide, a suitable solvent, an emulsifier, and often a wetting agent, sticker, or antifoaming agent. The concentration of pesticide is usually high, so the cost per pound of active ingredient is rather low. The elements do not settle out, once mixed, so they do not need continuous agitation. Due to their high concentration, they can be dangerous to handle.

Wettable Powders   These pesticides are of limited solubility in water. They are combined with dilating agents and surfactants that prevent flocculation of the pesticide particles. Wettable powders require continuous agitation to assure uniform coverage. They are relatively low in cost, easily stored and handled, but are hazardous if inhaled or absorbed through the skin.

Granules and Pellets   These pesticides are formulated as coarse, solid particles for easy application. The carrier may be sand, clay, ground corn cobs, fertilizer granules, or vermiculite. No dilution is required, since the percentage of active ingredients is lower than in other formulations, usually between 4 and 10 percent.

Fumigants   These pesticides come in the form of poisonous gases that kill when absorbed or inhaled. They are most useful in greenhouses or other enclosed areas, where they penetrate into cracks and crevices and into the soil. They are often highly toxic.

Pesticide Safety

The safe use of chemical pesticides requires a respect for their toxicity. All manufacturers and distributors of pesticides are required by federal law to provide explicit information about their products' ingredients, formulations, toxicity, and proper rates of application and about the specific pests controlled and proper means of safe handling. This information is provided on the pesticide label, Figure 6-16. The signal words that indicate the product's level of toxicity should be committed to memory by everyone who must work with these important but dangerous tools of modern agriculture, Table 6-2.

Equally important is the *material safety data sheet* (MSDS) that is prepared for each chemical product and included with it at delivery, Figure 6-17. The MSDS provides a complete compilation of all the information a user should know about the product. Information includes the manufacturer; the product's chemical name, physical properties, and chemical reactivity; fire and explosion data; health hazard data including symptoms of exposure and carcinogenicity;

FIGURE 6-16.
A typical pesticide label (Courtesy Chevron Chemical Company)

RESTRICTED USE PESTICIDE
For retail sale to and use only by Certified Applicators or persons under their direct supervision and only for those uses covered by the Certified Applicator's certification.

Chevron
ORTHO

MONITOR®

4 Spray

(INSECTICIDE)

POISON

Active Ingredient	By Wt.
* Methamidophos	40.0%
Inert Ingredients	60.0%

* O,S-dimethyl phosphoramidothioate
Contains 4 pounds active per gallon at 68° F.

KEEP OUT OF REACH OF CHILDREN.
DANGER—PELIGRO
FOR AGRICULTURAL USE ONLY.
DO NOT STORE IN AREAS ACCESSIBLE TO CHILDREN.
READ PRECAUTIONARY STATEMENTS ON SIDE PANEL.
PRECAUCION AL USUARIO: Si usted no lee ingles, no use este producto hasta que la etiqueta le haya sido explicada ampliamente.

NET CONTENTS 1 GALLON

special protection and handling information; as well as an emergency telephone number to call at any time for information regarding treatment for exposure or spillage. MSDS sheets should be kept in a place where they can be easily referenced if needed. Copies should be posted where everyone working with the pesticides can read them.

To protect against pesticide misuse, the following safeguards should always be observed:

- Use only products tested and recommended by the state college of agriculture.
- Follow the manufacturer's directions for safe application of the product. *NOTE:* It is a violation of federal law to use pesticides in a manner inconsistent with the label directions.
- Use the lowest recommended concentration. Do not arbitrarily increase dosage or frequency of application.
- Do not apply on a windy day or before a rain. This precaution will prevent drifting of sprays and dusts in the wind and runoff of pesticides in rainwater.
- Mix only the amount of pesticide that can be applied at one time. If excess remains, try to use it up on tolerant crops rather than disposing of the chemical elsewhere.
- Avoid spraying near streams and lakes or on slopes where runoff flows directly into a body of water. Also do not spray near bee hives or at a time when bees are pollinating.

TABLE 6-2.
Levels of Pesticide Toxicity and Safety Equipment Needed

Pesticide Signal Word	Level of Toxicity	Label Symbol	Special Equipment Required
DANGER—POISON	High	Skull and crossbones	Rubber boots, gloves, rubber pants, hat, and raincoat; face shield and gas mask with its own air supply
WARNING	Moderate	None	*Same as for high toxicity*
CAUTION	Low	None	Rubber boots and gloves; respiratory equipment recommended for prolonged indoor use

FIGURE 6-17.
Example of a material safety data sheet (MSDS)

TELEPHONE: 518-234-5315

AJAX CHEMICAL & FERTILIZER CORPORATION

BOX 123
COBLESKILL, NEW YORK 12043 U.S.A.

MATERIAL SAFETY DATA SHEET

Conforms to U.S. Department of Labor Bureau of Labor Standards

SECTION I

MANUFACTURER'S NAME AJAX Chemical & Fertilizer Corp.	**EMERGENCY TELEPHONE NO.** 518-234-5315

ADDRESS *(Number, Street, City, State, and ZIP Code)* Box 123 Jayridge Road, Cobleskill, NY 12043

CHEMICAL NAME AND SYNONYMS Oxy-Doxyl	**TRADE NAME AND SYNONYMS** Oxy-D 23
CHEMICAL FAMILY Insecticide-Nematicide	**FORMULA** Chemical mixture

SECTION II HAZARDOUS INGREDIENTS

	%	TLV
A. ACTIVE INGREDIENTS Methyl N'N'-dimethyl-N-		
(Methylcarbamoyl)-Oxy) -2-Thioxamimidate CAS NO. 23135-22-0	10	NA
B. SOLVENTS	%	TLV

C. OTHER Inert	PURPOSE	% 90

SECTION III PHYSICAL DATA

BOILING POINT (°F.)	NA	**SPECIFIC GRAVITY (H$_2$O =1)**	NA
VAPOR PRESSURE (mm HG.) Negligible		**PERCENT VOLATILE BY VOLUME (%)**	NA
VAPOR DENSITY (AIR =1)	NA	**EVAPORATION RATE (_____ =1)**	NA
SOLUBILITY IN WATER 28G/100 ML (Oxamyl)		Bulk Density	27.5 lbs.
APPEARANCE AND ODOR Blue-green granular			cu ft.

SECTION IV FIRE AND EXPLOSION HAZARD DATA

		Lel	Uel
FLASH POINT (Method used) NA	**FLAMMABLE LIMITS** NA		

EXTINGUISHING MEDIA On small fires use dry chemical, carbon dioxide, foam or water spray.

SPECIAL FIRE FIGHTING PROCEDURES
If area is heavily exposed to fire and if conditions permit this extinguishing with water spray.If conditions permit,cool containers with water if exposed to fire.

UNUSUAL FIRE AND EXPLOSION HAZARDS
Wear self contained breathing apparatus. Protective inhalation equipment should be worn in the vicinity of the fire until the ashes are cold. All unprotected people should be removed from the area and upwind.   NA NOT Applicable or not available

FIGURE 1.
Arborvitae Tip Blight

FIGURE 2.
Arborvitae Bag Worms

FIGURE 3.
Web of Fall Webworms

FIGURE 4.
Birch Borer Tunnels

FIGURE 5.
Black Spot of Rose

FIGURE 6.
Camellia Flower Blight

Reprinted with permission of ChemLawn Corporation.

FIGURE 7.
Eastern Tent Caterpillar on Cotoneaster

FIGURE 8.
Dogwood – Leaf Spot

FIGURE 9.
Dogwood – Canker

FIGURE 10.
Gypsy Moth Larva

FIGURE 11.
Dutch Elm Disease

FIGURE 12.
English Ivy – Bacterial Leaf Spot

Reprinted with permission of ChemLawn Corporation.

FIGURE 13.
Euonymus Scale

FIGURE 14.
Euonymus – Crown Gall

FIGURE 15.
Pyracantha Scab

FIGURE 16.
Adult White Flies

FIGURE 17.
Hawthorn Leaf Blight

FIGURE 18.
Hemlock Needle Blight

FIGURE 19.
Hemlock Scale

FIGURE 20.
Holly Spine Spot

FIGURE 21.
Holly Leaf Miner

FIGURE 22.
Horsechestnut – Anthracnose or Blotch

FIGURE 23.
Juniper Phomopsis Twig Blight

FIGURE 24.
Juniper – Cedar Apple Rust Gall

Reprinted with permission of ChemLawn Corporation.

FIGURE 25.
Lilac – Powdery Mildew

FIGURE 26.
Honeylocust Podgall Midge

FIGURE 27.
Maple Tar Spot

FIGURE 28.
Red Maple Anthracnose

FIGURE 29.
Gypsy Moth Adult with Egg Mass

FIGURE 30.
Maple – Verticillium Wilt

Reprinted with permission of ChemLawn Corporation.

FIGURE 31.
Maple Bladder Gall Mite

FIGURE 32.
Mountain Ash – Septoria Leaf Spot

FIGURE 33.
Pin Oak Leaf Blister

FIGURE 34.
White Oak Anthracnose

FIGURE 35.
Pin Oak Sawfly

FIGURE 36.
Oak Leaf Gall

Reprinted with permission of ChemLawn Corporation.

FIGURE 37.
Oak Leaf Miner

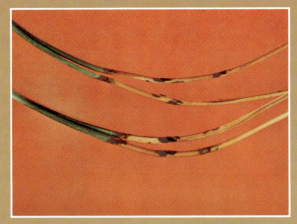

FIGURE 38.
Austrian Pine Needle Blight

FIGURE 39.
Red Headed Pine Sawfly

FIGURE 40.
Red Pine Needle Rust

FIGURE 41.
Austrian Pine – Diplodia Twig Blight

FIGURE 42.
Pine Needle Scale

FIGURE 43.
Poplar Canker

FIGURE 44.
Spruce – Cooley Gall Aphid

FIGURE 45.
Sycamore Anthracnose

FIGURE 46.
Black Weevil Feeding

FIGURE 47.
Birch Leaf Miner

FIGURE 48.
Weigela – Root Knot Nematode Gall

Reprinted with permission of ChemLawn Corporation.

FIGURE 6-17.

Example of a material safety data sheet (MSDS) (continued)

```
Oral LD50 for Oxamyl 10 G: 110 mg/kg
Dermal LD50 for Ocamyl 10 G: 2000 + mg/kg
```

SECTION V HEALTH HAZARD DATA

THRESHOLD LIMIT VALUE LD50 for Tech Oxamyl: 5.4 mg/kg acute oral-danger!

EFFECTS OF OVEREXPOSURE May be fatal if swallowed, or inhaled, or absorbed through skin

EMERGENCY AND FIRST AID PROCEDURES
Call a physician immediately. Contact local poison control center and or hospital.
Remove victim to fresh air. Remove contaminated clothing. Wash from skin with soap
and water. Flush eyes with running water for at least 15 minutes.

SECTION VI REACTIVITY DATA

STABILITY	UNSTABLE		CONDITIONS TO AVOID Excessive heat, moisture
	STABLE	X	

INCOMPATABILITY (Materials to avoid) NONE

HAZARDOUS DECOMPOSITION PRODUCTS NONE KNOWN

HAZARDOUS POLYMERIZATION	MAY OCCUR		CONDITIONS TO AVOID
	WILL NOT OCCUR	X	

SECTION VII SPILL OR LEAK PROCEDURES

STEPS TO BE TAKEN IN CASE MATERIAL IS RELEASED OR SPILLED
Do not breathe dust. Do not get in eyes, on skin, on clothing. Absorb material with
inert absorbant. Sweep up and place in a waste disposal container. Treat area with
household bleach. Flush area with water.

WASTE DISPOSAL METHOD
Dispose of containers and waste in accordance with local/state/federal regulations.

SECTION VIII SPECIAL PROTECTION INFORMATION

RESPIRATORY PROTECTION (Specify type) Self-contained respirator. (check list at bottom)

VENTILATION	LOCAL EXHAUST Yes		SPECIAL
	MECHANICAL (General) Yes		OTHER

PROTECTIVE GLOVES Impervious type	EYE PROTECTION Chemical goggles

OTHER PROTECTIVE EQUIPMENT
Wear clean clothes daily, wash thoroughly after use. No smoking or drinking while
applying

SECTION IX SPECIAL PRECAUTIONS

PRECAUTIONS TO BE TAKEN IN HANDLING AND STORING
Keep container closed. Not for use or storage around the home. Do not store near
feed or food stuffs.

OTHER PRECAUTIONS
Keep animals and persons from stored areas.
Personnel applying Oxy-Doxyl should wear all protective gear at all times.

Approved respirators: Mine Safety Appliances "Comfo Type H"; "Ultra Filter" Cartridges;
American Optical Sureguard Filter Cartridge No. R58

Notice: The information herein is given in good faith, but no warranty, express or
implied, is made.

- Keep pesticides in a locked area that is well ventilated and is clearly marked with the statement that it contains pesticides. Do not stockpile pesticides, as the danger of container corrosion and leakage increases with age.
- Wash all application equipment and clothing thoroughly after use: soap and water on the gloves before removal; three water flushings of spray tanks and hoses for equipment; all contaminated clothing hosed off and hung to dry. Store protective clothing away from the chemicals.
- For certain pesticides, such as nonselective herbicides, the sprayer should be marked and not used for any other purpose lest the residue harm an important crop.
- Careful records of pesticide use must be kept, including what was applied to what, in what amounts, when, and by whom.
- Empty pesticide containers should be disposed of carefully. They should *never* be set out for municipal collection or just hauled off to a local dump site. Many states have designated chemical dump sites where materials are carefully packaged and identified before disposal. With less toxic pesticides, empty containers should be buried or transported to a sanitary landfill.
- Telephone numbers of nearby doctors, hospitals, and the closest poison control center should be posted in a conspicuous location.

INTEGRATED PEST MANAGEMENT

For the latter half of the twentieth century the most accepted defense against plant pests has been the use of chemical pesticides. Pesticides have enabled, perhaps even encouraged, horticulturists to overlook nonchemical ways of applying the principles of control. Control measures that favored chemical use provided quick, expensive results and encouraged the development of a successful pesticide research and manufacturing industry in the United States. The natural controls of the biological world were passed over either because they were too slow or simply because they were not understood.

Currently the use of chemical pesticides is being reassessed, not to eliminate them from use, but to incorporate them into a more broad-based package of control measures. That package returns some of the balance of the natural world to horticultural production by allowing beneficial insect and microbial predators of destructive insects and pathogens to exist within the production range. By so doing, the predators keep the population of harmful insects and pathogens to a level that does not necessitate excessive use of chemical pesticides. When they are used, chemical pesticides are applied only when and where needed, rather than when their use is excessive, wasteful, and expensive. This multifaceted approach to pest control is termed *integrated pest management* (IPM). It applies the principles of control by carefully balancing the relationship existing among the crop hosts, the production environment, and the plant pests to which the crops are susceptible, Figure 6-18

The techniques of IPM vary with the circumstances of production. In greenhouses, for example, the pro-

FIGURE 6-18.
Integrated pest management balances the relationship between host, pest, and environment.

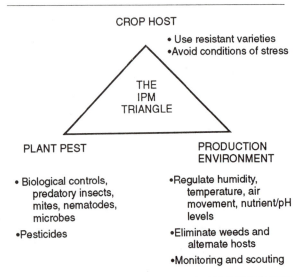

TABLE 6-3.

Greenhouse Predators

Predator	Pest controlled
Aphid midge	Aphids
Ladybugs	Aphids, mites, scales, mealybugs, other soft-bodied insects
Lacewings	Aphids and other soft-bodied insects
Predatory mites	Spider mites and thrips
Parasitic wasps	Whitefly and mealybugs
Entomogenous nematodes	Insects that spend all or part of their life cycle in the root zone

duction facility is closed and contained. This structure permits the screening of vents and thus the introduction of predatory insects into the crop with the knowledge that they will remain and justify their cost. Table 6-3 lists some of the predators currently available for insect and disease control in greenhouses.

Due to the openness of the production site, field and container nurseries are less able to use predatory insects in their IPM programs. Instead, they rely heavily on cultural practices such as proper spacing and pruning, balanced watering, fertilizer monitoring, elimination of weeds and other alternate hosts, and the use of resistant varieties to reduce pest populations. Other aids to control include *horticultural oils* that suffocate the insects, *insecticidal soaps* that kill insects upon contact, and *microbial insecticides* that are poisonous only to their targeted pest. As the American public look less and less favorably upon chemical pesticides, the federal government has responded by withdrawing approval for the use of many pesticides and by requiring certification of the applicators who use restricted pesticides. This trend can be expected to continue, resulting in a drastically reduced arsenal of chemical weapons to direct against the pests of ornamental crops. Further, pests continue to develop resistance to the approved chemicals still available, and pesticide rotation becomes increasingly difficult as the number of products slowly diminishes. Growers who rely solely on chemical pesticides for

control are clinging to an outdated concept that may soon leave them defenseless and untrained in the production methods of this final decade of the twentieth century.

SUMMARY

Biological pests that injure plants include insects, fungi, bacteria, viruses, nematodes, foraging animals, and weeds. Animals usually damage plants by feeding on their foliage or lower branches. Weeds compete with more desirable plants for space, sunlight, nutrients, and water. The remaining biological pests either infect or infest their host plants and live as parasites, deriving their sustenance from the cells of the host.

Insects are the most prolific and economically significant of all the infestious pests. They constitute over two-thirds of all the animal species on the earth. Grouped within the class Insecta, insects have a distinctive external anatomy that separates them from other Arthropoda. Many of these features are important in their identification.

Reproduction in insects is usually sexual and requires mating between a male and female, although several species can reproduce through parthenogenesis. As young insects develop from egg to adult, they may change dramatically in their appearance. The change in form is termed metamorphosis and may include as many as four stages of development (egg, larva, pupa, and adult). The majority of insects are of greatest economic significance in their non-adult stages.

Disease in plants is caused by pathogens. The most important plant pathogens are the bacteria, fungi, viruses, and nematodes. Each group of pathogens has distinctive characteristics that separate it from the others, although the symptoms they create are often similar.

For disease to develop, the pathogen must be transferred to a susceptible host plant by an appropriate agent of dissemination. The pathogen must be in an infectious form (termed inoculum) and must arrive at an appropriate site of infection on the host. Environmental conditions favoring the pathogen's development are also needed.

Following infection by a pathogen or infestation by insects the host may begin to exhibit symptoms of the injury. They can include:

- wilting
- color changes
- rotting
- death of tissue
- dwarfing
- increase in size
- tunneling
- holes

Weeds are plants that have no positive economic value and/or are growing where they are not desired. Their major damage is done as competitors with desired crop plants. In addition, weeds can serve as alternate hosts for overwintering insects or inoculum. Weeds are prolific seed producers. They include species that disseminate on every breeze and germinate quickly. Others remain dormant, distributed throughout a soil's profile awaiting their turn to be brought to the surface where they can germinate.

The purpose of pest control is to reduce the damage that can result from an injurious agent. Most growers seek a level of profitable control that allows the monetary return on a crop to exceed the cost of the control measures. To determine the potential profitability of a control measure, a grower must consider the value of a single crop plant, the ultimate value of the crop, and the average losses over a period of years.

Successful and profitable control of plant pests depends upon the application of one or more of the four basic principles of control: exclusion, eradication, protection, and resistance. Chemical pesticides are one of the methods of applying the principles of control. They are available as fungicides, nematicides, insecticides, and herbicides and formulated as solutions, emulsifiable concentrates, wettable powders, granules, pellets, or fumigants. Each formulation is toxic and requires care in handling. Procedures vary with the formulations and specific products.

Integrated pest management incorporates biological control methods and cultural manipulation to supplement pesticides in a multifaceted approach to pest control.

Achievement Review

A. SHORT ANSWER

Answer each of the following questions as briefly as possible.

1. Assume that you are a grower of nursery and greenhouse stock, and your suppliers have the following problems with their propagative stock. Which ones could become a problem for you if you allowed their plants into your operation?
   a. fungal leafspot on roses
   b. weeds under the greenhouse bench
   c. whiteflies on mum cuttings
   d. galls on woody material

e. soil-persistent fungi
f. sun scald on tender cuttings
g. rabbit damage
h. nitrogen deficiency
i. blight of flower buds
j. spray injury to crop

2. Match the branches of science with their definitions.

a. the study of plant diseases
b. the study of viruses
c. the study of fungi
d. the study of insects
e. the study of nematodes
f. the study of bacteria

1. mycology
2. nematology
3. entomology
4. plant pathology
5. bacteriology
6. virology

3. Describe the insects using the following headings.
a. phyllum classification
b. class name
c. type of appendages
d. body symmetry
e. type and location of skeleton

f. three body divisions
g. number of legs
h. number of pairs of wings
i. number of antennae
j. types of eyes
k. location of the reproductive organs
l. method of breathing

4. List the six most common insect mouthparts and at least one insect having each type.
5. List the four stages of complete metamorphosis in the order of their occurrence.
6. Define the following terms common to incomplete metamorphosis.
a. instar
b. molting
7. In which stage of metamorphosis do insects usually do the greatest amount of economic damage?
8. Compare bacteria, fungi, nematodes, and viruses by placing X's in the following chart to identify their characteristics.

Characteristic	Bacteria	Fungi	Nematodes	Viruses
Members of the plant kingdom				
Members of the animal kingdom				
Only single-celled in size				
Multi-celled in size				
Nonchlorophyllous				
Reproduce only in a living host				
Reproduce by cell division				
Reproduce by sexual and asexual means				

9. Complete the crossword puzzle using words related to the life cycle of fungi.

ACROSS

1. The vegetative body of a fungus.
3. The product of sexual fertilization.
5. The pairing of two compatible gametes that marks the start of the diploid phase.
6. Individual threadlike filaments that make up the mycelium.
7. A multicellular, filamentous thallus.

DOWN

2. The stage achieved by the fusion of two sexual nuclei.
4. The condition of the nucleus in the gametophytic stage.
8. Asexually produced spores.

10. Indicate whether each of the following is a form of inoculum (A), a site of infection (B), or an agent of dissemination (C).
    a. animals
    b. spores
    c. wounds
    d. splashing water
    e. bacterial ooze
    f. wind
    g. hyphal strands
    h. stomates
    i. blossoms
    j. virus particles

11. Name the symptom most likely to result from the following situations.
    a. Bacteria destroy the cells of the fruit

of a flowering tree. Cellular fluids are released.

b. A localized swelling appears in the stem of a plant following invasion and egg laying by an insect.

c. A bench of plants in a greenhouse turn yellow despite the sunny weather.

d. The presence of bacteria near the crown of a tea rose causes the tissue to respond with hypertrophy and hyperplasia in a localized region.

e. The presence of a virus in a plant stimulates hypoplasia and hypotrophy in the tissue.

f. Leaf miners in a birch tree chew their way between the upper and lower epidermal tissue of the leaves.

g. Borers feed in the stem tissue of woody shrubs.

h. Insects feed on leaves in localized areas of leaf tissue.

i. Insects feed on the roots of a plant, creating a symptom like that created when pathogens plug a plant's vascular system.

12. What is usually the final symptom of a symptom complex?

13. Define partial control, absolute control, and profitable control of plant pests.

14. Indicate which principle of pest control is being applied when the following measures are taken.

a. A grower plants only resistant varieties.

b. A sod farmer plants only certified seed.

c. Plants are not watered after 2 P.M. to allow them to dry off before evening.

d. A landscaper hauls away all debris from a planting bed before winter sets in.

e. Quarantines are placed upon all plant materials arriving from a country where infected plant material is common.

f. A diseased street tree is cut down and hauled away before the disease spreads to other trees nearby.

g. Growers fund research at a nearby university to speed development of a new resistant variety.

h. Healthy nursery plants are sprayed in the spring prior to onset of the rainy season.

i. Tulip bulbs, stored in a bag where diseased bulbs were found, are dusted before planting.

j. Hardwood cuttings taken from a tree in the woods are dipped in an antibiotic solution before placement into a pasteurized propagation medium.

15. Match the definitions with the pesticide or formulation they describe.

a. a chemical applied after the pest has arrived at the plant

b. a pesticide in the form of a poisonous gas that is most useful in enclosed areas

c. a pesticide whose active ingredients and carrier are in a homogeneous physical mixture

d. a pesticide applied to the host before the pest arrives

e. a pesticide that requires continuous agitation to assure uniform application

f. a chemical that kills some kinds of green plants and not others

g. a chemical that kills all green plants

h. pesticides formulated as coarse, solid particles

i. a pesticide applied via a water carrier despite its insolubility in water, using emulsifiers

1. fumigant
2. granules
3. solution
4. eradicant
5. protectant
6. selective herbicide
7. nonselective herbicide
8. emulsifiable concentrate
9. wettable powder

16. List in order of increasing toxicity, the three signal words that indicate the toxicity of pesticide products. Place an A by each signal word that necessitates protective rubber boots and gloves. Place a B by each signal word that necessitates, in addition, rubber suit, hat, and mask.

17. Explain integrated pest management as an alternative to traditional chemical pesticide use.

B. MULTIPLE CHOICE

From the choices given, select the answer that best completes each of the following statements.

1. Insect legs are attached to the body part known as the _____.
   a. head
   b. thorax
   c. abdomen
2. Insect wings are attached to the _____.
   a. head
   b. thorax
   c. abdomen
3. The type of legs, mouthpart, antennae, and wing venation pattern are important in _____ of a particular insect.
   a. control
   b. reproduction
   c. dissemination
   d. identification
4. An insect digests food in the _____.
   a. fore-intestine
   b. mid-intestine
   c. hind-intestine
5. An insect's nervous system is made up of groups of cells termed _____.
   a. spiracles
   b. tracheae
   c. ganglia
   d. nerves
6. Female insects possess _____ for laying eggs.
   a. an ovipositor
   b. a clasper
   c. a depositer
   d. a spiracle

C. TRUE/FALSE

1. Indicate if the following statements are true or false.
   a. Weeds are parasites.
   b. Weeds may serve as hosts for insects or pathogens.
   c. Weeds can reduce the carbon dioxide content of greenhouses.
   d. Weeds can reduce the heat efficiency of a greenhouse.
   e. Weeds are visible problems only. They do not persist unseen in the soil.
   f. Weeds do not have botanical names because they are not of economic importance.
   g. All weeds are annuals.
   h. Dormant weed seeds can persist in the soil for several years.
   i. Uncultivated land is likely to have more weed seeds in the soil than cultivated land.
   j. Weed seeds can pass through the digestive tract of animals.

SECTION II

THE CRAFTS OF ORNAMENTAL HORTICULTURE

7
Floral Design

Objectives

Upon completion of this chapter, you will be able to

- list and describe the materials needed by a floral designer.
- care for cut flowers properly to prolong their life.
- categorize arrangement materials into basic forms.
- define the principles of design.
- distinguish between different patterns of arrangements.
- use color to advantage in floral design.
- wire flowers correctly.
- make bows, puffs, corsages, table arrangements, and wreaths.

THE VALUE OF FLOWERS

The use of flowers in gardens and homes is well documented through most of recorded history. As civilizations have evolved, flourished, and floundered, the use of flowers has appeared to parallel people's attitudes toward civility, taste, art, and religion. Ancient Romans rewarded their national heroes with wreaths and garlands of flowers and laurel leaves. The Victorians in England, confident of their leadership in the world and buoyed by a century of British rule, used flowers in complex masses in their gardens' perennial borders and in their homes. Japan, isolated for centuries from the civilizations and religions of the Western world, regarded flowers as materials for spiritual expression. The oriental use of flowers is probably the most personal and intimate that the world has known.

Americans began by copying the uses familiar to their emigrant ancestors. Today, however, a distinctively American style of flower arranging has evolved, which if analyzed would probably represent a mixture of the styles of many nations. Selecting the best from an assortment of influences, American floral design is now the world-wide pacesetter for stylized arrangements. We use flowers to express our joys and our sympathies.

Flowers highlight the key moments in our lives—the birth of a child, the holidays, the first dance, weddings, anniversaries, illnesses, and deaths. Such uses have been traditional for many years. In instances like these, most people turn to a professional florist to arrange the flowers. At other times, we arrange our own flowers. When a selection of blossoms is cut from the garden, brought into the house, and placed into a vase, or a single bloom is selected to wear in the lapel of a suit or in the hair, the use of the flowers becomes much more personal. Flowers today serve as gifts to friends and business associates. They accompany our entertainments and reward our heroes, whether stars of theater, opera house, battlefield, or race track. They permit the most inarticulate among us to convey our feelings when words may be hard to find.

FLORAL DESIGN—ART OR CRAFT?

The terms *art* and *science* are widely misused. Tractor repair is ennobled as an industrial art, and lawn mowing is glorified as turfgrass science. Equally tedious is the insistence by some purists that only they know the distinctions that separate true art and science from all pretenders. So it is with the terms *art* and *craft*. What separates the two? Which is floral design? Floral design is a craft; it can be learned by nearly anyone who wishes to take the time. However, the level of expertise and degree of creativity expressed by flower arrangers increase as their appreciation and knowledge of fine art increase.

If art is the creation of new and harmonious relationships among lines and forms, then floral design may qualify as an art in the hands of someone who can use living materials in the same way that others use paint or stone. Commercial florists seldom acquire the knowledge of fine art necessary to create truly original relationships with their materials, although very often their work is beautiful, imaginative, and skillfully executed. Their work is usually a copy or modification of styles created by others, sometimes centuries earlier. Accordingly, theirs can be regarded as the craft of flower arranging.

Being a craft, floral design can be taught. Yet, as with any skill, some practitioners will develop a greater proficiency than others because of the time they spend practicing and their level of personal interest. To become a superior designer, florists must first learn the mechanics of their trade: how to prepare flowers for use in designs, how to position them in arrangements, and how to package the finished work for delivery. In addition, superior florists will study *design* of all types from classical to contemporary, the *natural world* with its endless combinations of line and form, *literature* and the *social sciences* to better understand the many shades of human behavior and emotions, and the *fine arts*, including painting, sculpture, and music. All too often, florists try to substitute personality and pretentions for a solid foundation of knowledge and training. In floral design as elsewhere, however, there is no substitute for hard work and education.

MATERIALS NEEDED TO ARRANGE FLOWERS

Professional florists select their tools from a wide assortment of products available. Following are the major types of tools needed for professional floristry, Figures 7-1 to 7-4.

Cutting Tools

1. *Knife:* Select a short blade made of quality steel that will hold an edge. The short blade will help prevent cut fingers.
2. *Florist shears:* They have short blades with serrated edges and are used for cutting both herbaceous and woody plant material.
3. *Pruning shears:* The best ones have two cutting blades of quality steel. They are used for cutting woody materials.
4. *Ribbon shears:* These are similar to ordinary scissors in appearance.
5. *Styrofoam cutter:* A widely serrated cutting edge makes it suitable for cutting and shaping blocks of styrofoam.

Wiring Materials

1. *Florist wire:* Wire is used to support weak flower stems and to hold curved lines in arrangements. It varies according to length and thickness (gauge). The heaviest gauge is number 18 and the finest is number 32.

FIGURE 7-1.
Florist cutting tools (left to right): florist knife, florist shears, pruning shears, ribbon shears, styrofoam cutter

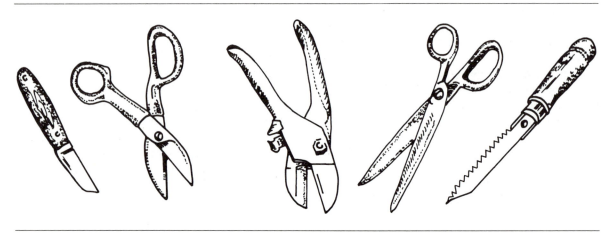

As the weight of flowers or foliage increases, the weight of the wire needed to support them also increases.

2. *Chenille stems:* Chenille stems are like pipe cleaners with extra flocking. They are used both for decorative purposes and to provide water to corsage flowers.

3. *Spool wire:* A heavy gauge wire is needed to wire arrangements such as wreaths and door swags.

4. *Twistems:* These are short pieces of wire enclosed in paper. They fill a number of roles where support or repair is needed.

5. *Wooden picks:* Green wood picks with a length of thin wire attached can be fastened to stems or other materials and inserted into a holding substance. They are widely used with nonliving arrangements.

6. *Metal picks:* These are similar in use to wooden picks and are dispensed from a machine.

Adhesive Materials

1. *Floral tape:* This tape is similar to a waxed crepe paper. It has a stretching quality and is used to wrap flower stems when wire is needed for their support. It comes in assorted colors to blend unseen into the arrangement.

2. *Waterproof tape:* Its most common use is to hold stem support material, such as floral foam, in the container.

3. *Florist clay:* A waterproof material, it is used to hold pinpoint stem holders in place within a container.

4. *Styrofoam glue:* Some glues dissolve styrofoam. This material does not, yet it is a strong adhesive.

5. *Hot glue gun:* The glue is inserted as a solid stick into the gun. It is heated electrically and dispensed in a liquid state that quickly cools and hardens.

Stem Support Materials

1. *Water-holding foams:* Several excellent products are available to commercial florists. The foams are sold shredded and as dry, pressed blocks. When filled with water, they support the flower and foliage stems inserted into them while providing the water needed for freshness.

FIGURE 7-2.
Floral wiring materials: (front left to right) Enameled wire, spool wire, twistems, chenille stems; (rear) wooden picks and a steel pick machine

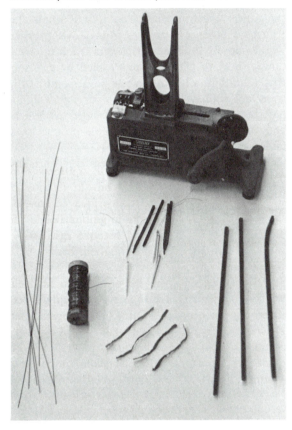

FIGURE 7-3.
Floral adhesives: (front) waterproof tape and florist clay; (rear) floral tape, styrofoam glue, and a hot glue gun

2. *Shredded styrofoam:* Styrofoam does not hold water. It is commonly used to hold the stems of dried silk and plastic flowers and foliage, especially in tall containers. If fresh materials are used, water must be added to fill in between the particles of styrofoam.

3. *Block styrofoam:* An excellent base for dried and permanent arrangements, it cannot be used with fresh flowers.

4. *Chicken wire:* This wire has a 1-inch mesh and can be rolled loosely into a ball and inserted into the container. Stems can then be inserted into the folded mesh.

5. *Pinpoint holders:* These are used more commonly by homeowners than by professional florists because of their expense. They are held in the container with florist clay. Stems are then pressed onto the pins. They are reusable.

CONTAINERS FOR FLORAL ARRANGEMENTS

Among the changes that have elevated public appreciation of floral arrangements has been the imaginative use of containers. From antique vases to beer mugs, anything that can hold stem support material is now fair game for the creative florist. The container may set or at least contribute to the theme of the arrangement. If it can hold water, it can be used for fresh flowers. If it cannot hold water, it

FIGURE 7-4.
Assorted stem support materials: (left) water-holding foams; (center) shredded styrofoam and pinpoint holders; (right) chicken wire and block styrofoam

may still hold water-filled foam wrapped in waterproof foil and fixed in place with wire, tape, or clay. Containers that do not hold water may also be used successfully with permanent materials.

An important quality of all containers is that they not detract from the overall arrangement and the beauty of its plant materials. Such is often the case when a designer selects a novelty container for the sheer cleverness of doing so. A flashy container can never improve a mediocre design, but it can definitely hurt a good one.

Containers should be carefully selected as one element of the total arrangement. They must first serve the functional needs of the arrangement—be appropriate for the shape of arrangement desired and be able to hold the proper stem support material. Designers can then give their imagination full reign in selecting the appropriate container. You should know the basic types first and how to arrange in them; then try your hand with more inventive containers, Figure 7-5.

CARE OF CUT FLOWERS

Cut flowers come to the florist shops of America from all over the nation and the world. Although some retail florists may order flowers direct from the grower, most purchase from a local wholesaler who buys in large quantities from flower growers worldwide. By the time cut flowers reach the local flower shop, they may have traveled by airplane, truck, and bus, all within a few days. Some of the transport will have been refrigerated; some will not have been. The flowers must be attended to immediately upon arrival or the perishable product may be lost, and with it the florist's investment.

Flowers arriving from a wholesaler are usually bound in bunches or clusters. The blossoms may be wrapped in waxed paper sleeves to prevent damage, and the clusters may be shipped in special boxes that reduce crushing and permit some air circulation. If you recognize the logic behind them, the steps to take when flowers arrive at the flower shop are easy to remember. The cut flowers must receive:

1. *nutrition* for continued good health
2. *water* to prevent wilting
3. *cool temperatures* to slow their metabolic activity and prolong their life

Thus, cut flowers are first unpacked carefully. Their bases are then recut on an angle to expose fresh vascular tissue for maximum water uptake. (*NOTE:* In addition, gladiolus and chrysanthemum stem ends may be crushed with a hammer to facilitate water absorption.) The freshly cut stems are immediately placed into disinfected containers filled approximately one-third full with fresh water containing a flower preservative. The preservative contains nutritional sugar plus an antibiotic to prevent bacterial plugging of the vascular tissue. Flower containers should not be overfilled with water or the stems may become waterlogged or even start to rot. Warm water is preferable to cold because the flowers can absorb it more quickly. After several hours in water at room temperature the flowers should be turgid and ready to be cooled. They are then placed into the cooler, usually kept between 38° and 40°F.

FIGURE 7-5.

Common container types for floral arrangements

TYPE	DESCRIPTION	TYPICAL SHAPE
VASES	THE HEIGHT OF THE CONTAINER IS GREATER THAN ITS WIDTH. IT IS NOT PEDESTALED BUT IS USED FOR VERTICAL DESIGNS.	
BOWLS	THE WIDTH OF THE CONTAINER IS GREATER THAN THE HEIGHT. IT IS OFTEN USED FOR TABLE ARRANGEMENTS.	
PEDESTALS	THE CONTAINER IS ELEVATED ON A BASE THAT MAY BE SHORT- OR LONG-STEMMED. IT IS USEFUL WHEN A TALL ARRANGEMENT IS NEEDED YET THE MASS OF A VASE CONTAINER IS NOT DESIRED.	
BASKETS	THE STYLES AND MATERIALS VARY, AND THEY ARE USED FOR BOTH VERTICAL AND HORIZONTAL DESIGNS. THE DESIGNS ARE USUALLY INFORMAL IN STYLE.	
NOVELTY	THESE ARE LIMITLESS IN POSSIBILITIES. CARE SHOULD BE TAKEN TO ASSURE THAT THEY DO NOT OVERPOWER THE DESIGN.	

FORMS OF ARRANGEMENT MATERIALS

Most of the materials that are visible in an arrangement are flowers and foliage. As dried arrangements and novelty designs increase in popularity, additional material may be considered for use. Neverthe-less, all of these materials can be categorized into four basic types, Figure 7-6.

Some flowers and nonflower materials may play more than one role. For example, gladiolus blossoms are often removed from their long, vertical stem and clustered at the center of an arrangement, changing them from line forms to mass forms. Roses used while tight buds on long stems will be strong

FIGURE 7-6.

Types of arrangement materials

TYPE	TYPICAL SHAPE	FLOWER EXAMPLES	NON-FLOWER EXAMPLES
LINE • THIN • VERTICAL • TAPERED • USED TO CREATE THE BASIC SHAPE OF THE ARRANGEMENT • MOST EFFECTIVE AT OUTER EDGES OF THE ARRANGEMENT		GLADIOLUS, SNAPDRAGONS, STOCK, DELPHINIUM, LONG-STEMMED FLOWERS, SUCH AS IRIS AND ROSE, BRANCHES OF FLOWERING SHRUBS, SUCH AS FORSYTHIA, SPIREA, AND PUSSY WILLOW	CATTAIL, SCOTCH BROOM, SNAKE PLANT, TWIGS, EAR OF CORN, TAPERED CANDLE, THISTLES ON LONG STEMS, EUCALYPTUS
MASS • ROUNDED • USED AT THE CENTER OF THE ARRANGEMENT	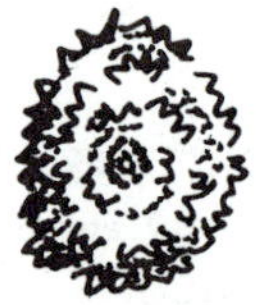	OPEN TULIP AND ROSE, ZINNIA, GLOXINIA, MARIGOLD, CARNATION	CHRISTMAS ORNAMENTS (MILLIMETER BALLS), PINE CONES, SEED PODS, FRUIT
FORM • UNCOMMON SHAPES • UNUSUAL SILHOUETTES • SELDOM MIXED WITH OTHER FORM MATERIALS		ORCHID, LILY, IRIS, BIRD-OF-PARADISE, ANTHERIUM	FIGURINES, CANDLES, DRIFT-WOOD, CERTAIN FRUITS
FILLER • USED TO FILL IN BETWEEN LINE AND MASS MATERIALS • OFTEN HAS MANY BLOSSOMS OR SMALL LEAVES ON A SINGLE STEM		BABIES' BREATH, SPRAY MUMS, HEATHER, STATICE	ASSORTED FOLIAGE, SUCH AS ASPARAGUS FERN, HUCKLE-BERRY, IVY, DRIED LEAVES

line forms. When used as fully opened flowers on short stems, they function as mass forms. Spray mums are regarded as filler flowers in a large arrangement; yet in a small container they can serve as mass forms.

PRINCIPLES OF FLORAL DESIGN

Because floral design, like all design, is personal, it is difficult to evaluate. If a furniture designer creates a chair with one leg shorter than the other three, most people will agree that the design is unsatisfactory because it overlooks one of the most obvious functions of the chair. With a floral design, such clear-cut cases of right and wrong are usually lacking. The florist who designs an arrangement can be expected to like it. Another florist or a customer may or may not agree. They will all be confident that they are right, but their judgements are usually more subjective than objective. Nowhere do opinions come into conflict more frequently than in a classroom. If teachers and learners are to discuss design based upon logic rather than personal opinion, they need a common ground where both master and apprentice can tread with confidence and mutual respect. That common ground is created by the *principles of design* that guide all creative endeavors. Whether sculpted in marble, woven into tapestries, or arranged in a vase, a design can be judged on the basis of how closely it adheres to these principles.

Simplicity

Even the most imaginative, stylish designs are based upon uncomplicated themes and draw upon few elements for their composition. Designs should be limited to one distinctive line, readily apparent to the viewer's eye. Colors should be limited to a few that contrast or blend harmoniously. The container should contribute to the total arrangement and not attract too much attention to itself. The kinds of flowers and foliage used should also be limited; too many shapes and textures can overcomplicate the design. Finally, the setting and background for the arrangement need to be considered. A heavily patterned wallpaper or dress can add complexity to the floral design.

Focalization of Interest

Each arrangement benefits by having only one center of attention for the viewer's eye. Such a focal point becomes the visual center of the design. It is usually the point where the major lines of the design converge—at the center of the container and just above its edge. There are several ways to focus a design. One way is to concentrate the mass form

flowers at the center of the design. Another way is to use larger flowers at the center. Still another way is to use more eye-attracting colors at the focal point. The placement of a figurine, bow, candle, or other nonflower element at the design's visual center can also emphasize the focal point.

Scale and Proportion

There are three size relationships to be considered in a floral arrangement.

1. The *relationship among the flowers and other materials* used as components of the arrangement. To illustrate, the arrangements shown throughout this chapter have the smallest flowers at the top and edge of the design; intermediate sizes follow; and the largest flowers are at the center or focal point. Such a use of flower sizes follows the sequence by which blossoms unfold in nature and is termed *transition*. Regardless of whether the flowers and foliage are of the same species throughout the arrangement, the transitional sequencing of flower and leaf sizes is an important concept if scale and proportion are to be maintained.
2. The *relationship between the flowers and their container*. Neither should overpower or be dwarfed by the other. There is one widely accepted measure of correct scale, Figure 7-7. If the container is taller than it is wide, the height of the arrangement should be at least 1 1/2 but no more than 2 1/2 times the container's height. If the container is wider than it is tall, the arrangement height should be at least 1 1/2 but no more than 2 1/2 times its width.
3. The *relationship between the finished arrangement and the situation in which it will be used*. A dinner table arrangement should not encroach on the guests' plates with its width, nor be so tall that guests cannot see over it. A corsage should not be so large that it pulls the fabric of a gown with its weight. An arrangement of violets can appear to advantage on a bookshelf

FIGURE 7-7.

How to determine the correct height of an arrangement

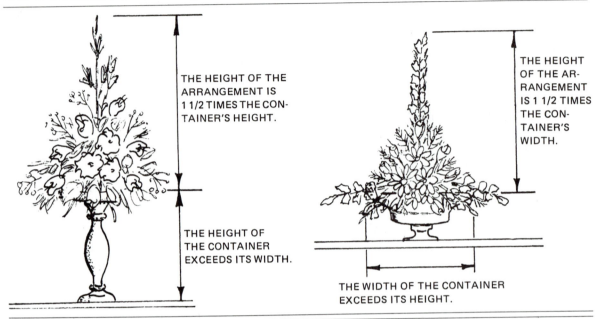

THE HEIGHT OF THE ARRANGEMENT IS 1 1/2 TIMES THE CONTAINER'S HEIGHT.

THE HEIGHT OF THE CONTAINER EXCEEDS ITS WIDTH.

THE HEIGHT OF THE ARRANGEMENT IS 1 1/2 TIMES THE CONTAINER'S WIDTH.

THE WIDTH OF THE CONTAINER EXCEEDS ITS HEIGHT.

in a small room but go unnoticed in a large entrance hall. Conversely, an arrangement designed to fill a window seat in a Victorian home would nearly fill a small dining room in an apartment.

Balance

In floral design, balance is more visual than actual. The viewer needs to sense the same amount of importance on each side of center. Center may or may not be the midpoint of the container; it should be regarded as the point of convergence of the design's horizontal and vertical lines. When correctly balanced, a design will not appear to lean forward or backward or to the left or right.

Two types of balance are common to floral design: symmetrical and asymmetrical. A *symmetrical* arrangement, if bisected by a line (axis) running from the vertical tip through the base of the container, has flowers and foliage in almost exactly the same places on opposite sides of the line, Figure

7-8. An *asymmetrical* arrangement has an axis also, but it may or may not bisect the container equally. It also extends from the vertical tip of the design through the point where the arrangement's lines converge. The collective visual weight of the flowers, foliage, and container on one side of the axis should equal that on the opposite side; yet one side is not a mirror image of the other, Figure 7-9. The majority of modern designs are asymmetrical because they permit greater creativity by the designer. Still, symmetrically balanced arrangements are not uncommon, and they are a good place for new designers to begin.

Rhythm and Line

A design should impart a sense of frozen motion. It should also lead the viewer's eye in a deliberate direction. The viewer's eye should begin at the edge of the design, move to the point of greatest interest, and either remain there or pass on through the design, exiting at the opposite edge. The placement

FIGURE 7-8.
A symmetrically balanced arrangement is identical on opposite sides of a line running from the vertical tip through the point of stem convergence.

FIGURE 7-9.
An asymmetrically balanced arrangement has differing sizes, numbers, or placement of flowers on opposite sides of a line running from the vertical tip through the point of stem convergence.

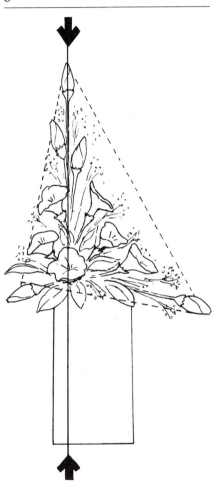

and repetition of selected elements, such as seed pods in a dried autumn arrangement, can create a sense of rhythm. The use of transition described under scale and proportion is also important to the development of rhythm and line. The simple, free-flowing lines of nature as observed in a meandering stream, a gently curled leaf, and a drift of snow or sand are good lines to duplicate within a floral arrangement. Equally interesting can be lines that converge to a single point, like highways radiating to and from a city. Gradations of size, color, and texture will create these lines and promote the essential sense of frozen motion.

PATTERNS OF ARRANGEMENTS

Three standard geometric shapes lend themselves naturally to flower arrangements: the circle, the triangle, and the rectangle. All lend themselves to the creation of symmetrical, formal arrangements. All are easily modified into informal arrangements as ovals and asymmetrical triangles, Figure 7-10.

FIGURE 7-10.

Standard patterns of floral arrangements

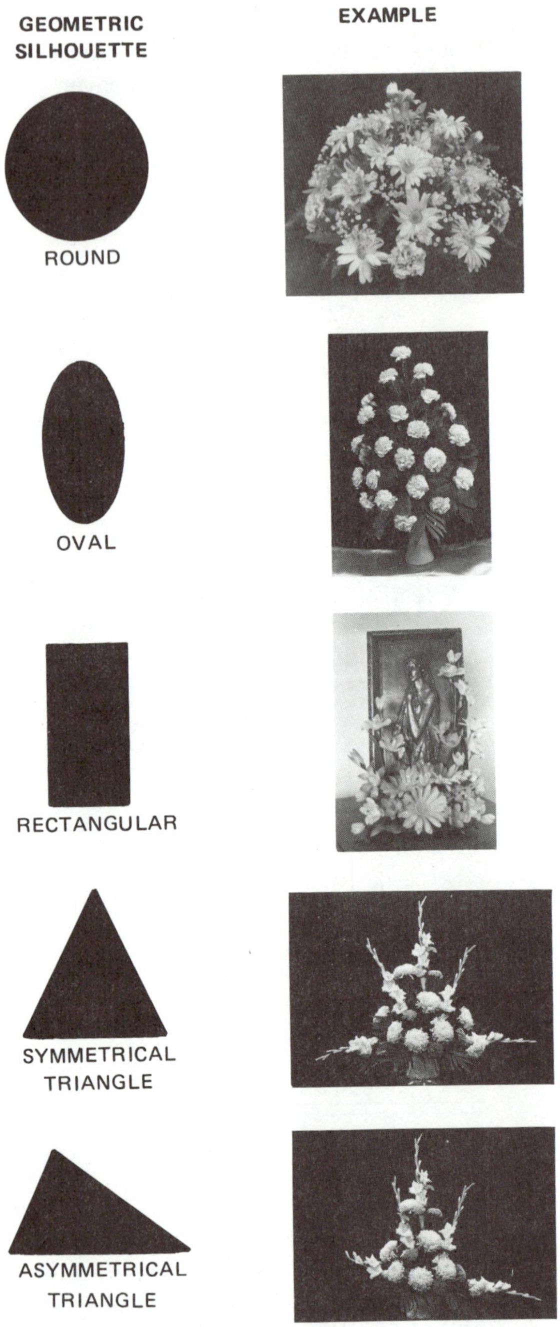

GEOMETRIC SILHOUETTE	EXAMPLE

ROUND

OVAL

RECTANGULAR

SYMMETRICAL TRIANGLE

ASYMMETRICAL TRIANGLE

These *five standard patterns* form the foundation of American floral design. The refinement of these geometric forms has been the major contribution America has made to worldwide floristry. The patterns are made up of *six basic lines,* which are universal in their origins and applications. While limitless variations of these lines exist, the aspiring floral designer needs to begin with a knowledge of the basics, Figure 7-11.

Once the basic patterns and lines are understood, it is easy to analyze the composition of most floral arrangements. For example, the lines of a symmetrical triangle arrangement are seen to be that of a balanced, inverted T. An asymmetrical triangle has an L line. Rectangular arrangements employ either the vertical or horizontal lines. Circular or oval patterns are built around the crescent or S-curve lines.

You should now apply this information by analyzing the arrangements pictured in Figures 7-10 and 7-11. Determine which basic lines are shaping the patterns. Note the forms of the flowers used, their sizes and numbers. Note also their placement within each arrangement. Even though the photos are black and white, the principles of design should be apparent. Find the focal point of each design; assess its type of balance. How have simplicity, scale and proportion, and rhythm and line been achieved?

USING COLOR TO ADVANTAGE

The topic of color, its many contributions, and how best to take advantage of its values is integral to the study of floral design. It is also a topic that can be overintellectualized, thereby destroying the fun of studying it. Although the subject does lend itself to study and research by scientists and artists alike, you may take comfort in knowing that nearly everyone possesses a natural ability to use color correctly. Therefore, a study of color can be regarded mostly as a deciphering of *why* certain colors go together pleasantly or unpleasantly.

Given a selection of painted blocks or swatches of cloth, most people could select and arrange colors in combinations that they and most viewers would find attractive. The precise nature of the color combinations might vary depending upon the setting,

FIGURE 7-11.
Basic lines of floral arrangements

LINE

EXAMPLE

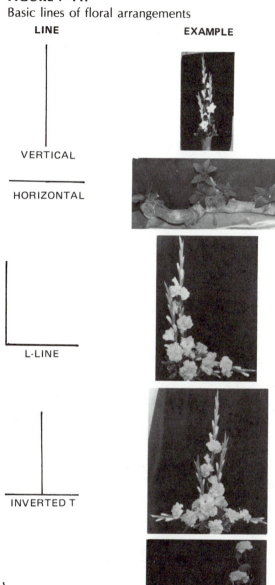

VERTICAL

HORIZONTAL

L-LINE

INVERTED T

CRESCENT

S-CURVE

mood, time of day, season, and assorted other influences. In floral design, the colors chosen for an arrangement may be based upon:

- the occasion that the flowers acknowledge
- the room where they will be placed
- the background against which they will be placed
- the light under which they will be viewed
- the season of the year
- the preferences of the person for whom the flowers are intended
- the preferences of the purchaser
- the preferences of the floral arranger

Color Terminology

To understand color, you should understand the terminology used to describe its qualities and its variations.

- *Color families* are the six major groupings of colors visible when white light is passed through a prism or when a rainbow is seen after a shower. The six color families are red, orange, yellow, green, blue, and violet.
- *Hue* is the quality from which the family name is derived. It is the color in its most brilliant and unaltered state.
- *Tint* is the hue lightened by the addition of white. In flowers, a tint appears as a pastel color.
- *Shade* is the hue darkened by the addition of black. Dark reds or bronzes are examples of shades in flower colors.
- *Tone* is the hue grayed by the addition of both white and black. Certain flowers are distinguished by their smokey tones in contrast to their more brilliant counterparts that remain closer to the original hue.
- *Intensity* is the quality of visual strength or weakness that characterizes a color. Both tints and shades may appear weak if the white or black dilutants dominate the original hue. Where the hue predominates, the color appears stronger.

- *Weight* is also a visual quality based upon the amount of white, black, or gray in the color. White gives a lightness to colors, whereas black creates a sense of heaviness.
- *Luminosity* is the quality of certain colors that allows them to be seen under dim light. Tints, because of the white in them, have a higher luminosity than tones or shades.
- *Warmth* and *coolness.* We frequently associate colors with temperatures. Reds and oranges are considered warm colors because we associate them with fire. Blue and green are considered cool colors because of their association with water and shade.
- *Movement* is the ability of colors to appear closer or farther away. In general, intense colors appear to advance and less intense colors appear to recede.

FIGURE 7-12.
The color wheel

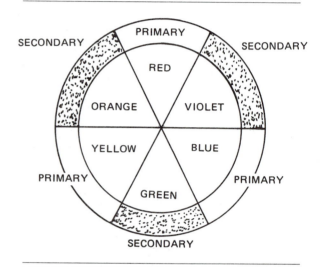

Color Schemes

Certain groupings of colors (color schemes) have gained sufficient understanding and acceptance to justify describing them. All are based upon the color wheel, Figure 7-12. A color scheme may be related or contrasting depending upon the location of the colors on the wheel. Related color schemes may be monochromatic or adjacent. Contrasting color schemes may be complementary, triadic, or polychromatic.

Related Color Schemes

- *Monochromatic* color schemes utilize one color in its many related values. The one color may incorporate assorted tints, shades, and tones in the flowers, foliage, and container.
- *Adjacent* color schemes use one of the primary colors (red, yellow, or blue) with other colors derived from that primary. For example, browns, oranges, and golds have in common the primary color of yellow and work well together in an autumn arrangement. Adjacent colors are so named because of their close proximity on the color wheel.

Contrasting Color Schemes

- *Complementary* color schemes use colors that are opposites or near opposites on the color wheel. The colors do not share a common primary color relationship. Examples include red and green, blue and orange, and yellow and violet.
- *Triadic* color schemes use three colors that are spaced equidistantly on the color wheel. It may be necessary to use some variations of tints, tones, or shades to find the right color combination.
- *Polychromatic* color schemes use all of the hues together. The brilliance can often be too intense, so the floral designer may prefer a modified polychromatic scheme that tones it down.

These are not the only possible or desirable groupings of colors. As stated at the outset, most people have an intuitive ability to combine colors attractively. The selection of colors for a floral arrangement should be based upon logic but also upon that natural intuition.

In the use of color, professional florists are often restricted to the colors of the flowers currently in the cooler. They are also restricted by an inability to dictate the setting or background that will display the arrangement to best advantage. A polychromatic arrangement for a church may be set on a white linen communion table or against a stained glass window. One setting could enhance the flowers and the other absorb them completely. A white corsage worn on a white dress will not have the visual impact that it will have if worn against a darker color or pattern.

Within any given arrangement, darker shades and tones are best used in lesser quantities than lighter tones and tints. The dark shades and tones are most effectively used at the focal center of the arrangement, with lighter colors used for line and filler flowers.

WIRING FLOWERS FOR DESIGN

It is common practice among florists to wire certain flowers prior to their use in designs. Fresh, dried, and permanent (artificial) flowers and foliage may all be wired. The wire must not show after the flowers are arranged, so it is concealed with foliage and floral tape.

Flowers are wired for several reasons:

- to support stems weakened by age or poor production
- to straighten crooked stems
- to hold stems in an intentional curve for a design
- to extend stem length for large arrangements
- to replace a bulky stem, permitting use of the flower in a corsage
- to combine single blossoms for a mass effect

The technique of wiring depends upon the flower being wired. More precisely, it depends upon the type of stem and flower head involved, Figure 7-13.

After wiring, the wires may require taping to cover them or disguise them to resemble stems. To apply the tape, hold the wired flower and stem in one hand while the other hand applies the floral tape from the base of the calyx downward, Figure 7-14. The tape is stretched tight as the flower is twirled.

MAKING BOWS AND PUFFS

Two of the mainstays of floral designs are bows and puffs. *Puffs* are small clusters of netted fabric (sometimes called tulle) that are used as background and lightweight filler for corsages and bouquets. The netting is available in assorted colors and patterns. *Bows* are usually constructed from fabric ribbon, although plastic ribbons are sometimes used for outdoor pieces. Small bows are used for corsages and other arrangements that are to be worn. Larger bows are used in table arrangements and on potted plants. Still larger bows are used on wreaths, door swags, and on wedding, funeral, and grave pieces. A florist must be able to make a respectable bow and puff quickly and repeatedly. Many florists use slack periods to make them up in quantity. *NOTE:* It is the handcrafted nature of floral materials such as bows, puffs, and wiring that distinguishes floral products and justifies their cost to the consumer.

The Floral Bow

Ribbon used by florists varies in both material and width. Every few years certain materials capture the fancy of florists and consumers. Burlap ribbons are a recent example of the faddish materials that enjoy popularity for awhile at least. Velvet-like flocked ribbons remain popular for Christmas arrangements and weddings, as do satin ribbons. Even metallic ribbons have been tried. Regardless of the material, the sizes manufactured are standardized. The most commonly used are:

- 4 1/2 inches wide (No. 120): used for funeral work, arm bouquets, wreaths, and door swags
- 2 7/8 inches wide (No. 40): used for funeral sprays, wedding work, wreaths, and swags
- 5/8 inch wide (No. 3): used for corsages
- 1/4 inch wide (No. 1): used for corsages, as streamers for bridal bouquets, and for head pieces

After practicing the steps illustrated, you will be able to produce a simple and attractive bow, Figures 7-15A through 7-15L.

FIGURE 7-13.
Techniques for wiring flowers

Wiring Technique	Examples of Flowers
Wire inside a hollow stem 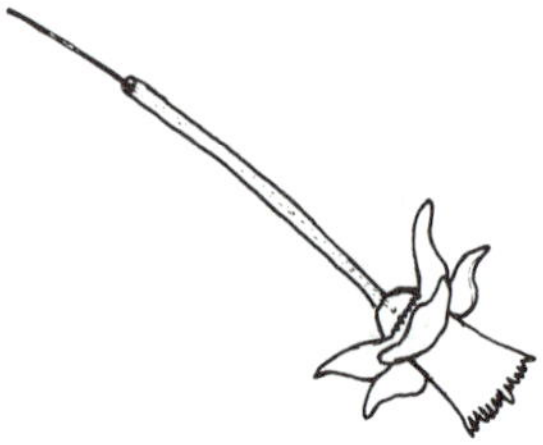	Bulb flowers such as tulips, daffodils, hyacinths, and gladioli
Wire in the calyx and around the stem 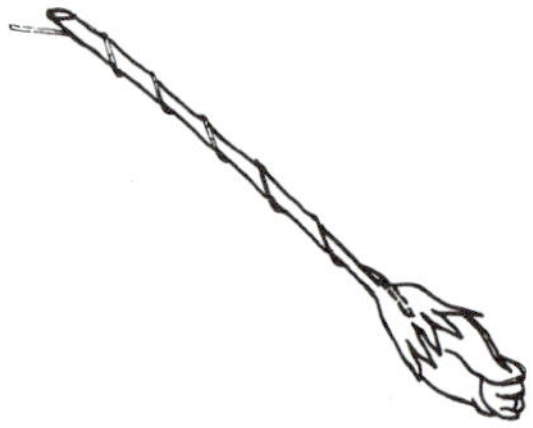	Roses, carnations, large mums
Wire looped through the neck of a blossom and around a wad of moistened cotton 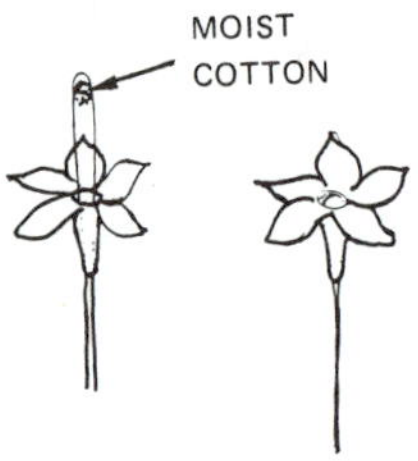	Tubular, fragile flowers such as stephanotis
Wire through the calyx and the ends bent to form a replacement stem 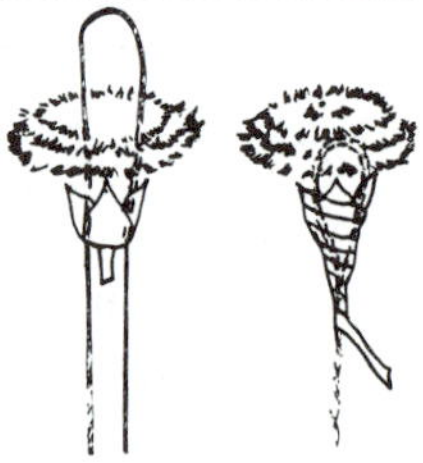	Flowers with a thick, strong calyx such as carnations and roses
Wire through the calyx, formed into a hook, and pulled back to center 	Flowers with weak stems and flattened heads such as daisies, pom pom mums, black-eyed susans

FIGURE 7-14.
Floral tape is applied after the flower has been wired. It is stretched tightly to assure a secure fit.

FIGURE 7-15A.
Steps in the construction of a florist bow

FIGURE 7-15B.

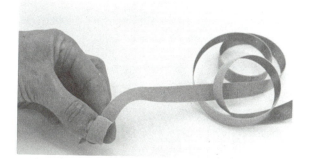

Step 1: Hold the ribbon in your nondominant hand with the finished side toward you, Figure 7-15A.

Step 2: Make a loop over the thumb, and twist the ribbon beneath the thumb so that the finished side stays upward, Figure 7-15B.

Step 3: Make the first loop (smaller than later loops will be) and gather the ribbon beneath the thumb, Figure 7-15C.

Step 4: Make a second loop of equal size to complete the first pair of loops, Figure 7-15D. Twist the ribbon beneath the thumb to keep the finished side upward.

Step 5: Add a second pair of loops using the same techniques. These should be the largest loops of the bow, Figure 7-15E.

Step 6: Add a third and fourth pair of loops. They should be of equal size and intermediate between the size of pairs one and two. Remember to keep twisting the ribbon as it passes beneath the thumb, Figure 7-15F.

FIGURE 7-15C.

Step 7: Cut the ribbon free from the spool, leaving a 3- to 4-inch streamer, Figure 7-15G.

Step 8: Loop the streamer under the bow, keeping the finished side outward, Figure 7-15H.

Step 9: Insert a thin wire through the center loop and beneath your thumb to secure the ribbon, Figure 7-15I.

Step 10: Grasp the bow firmly and pull the wire tightly. Twist it to secure all loops in their proper place, Figure 7-15 J.

Step 11: Cut the last large loops, on an angle, to create streamers for the bow, Figure 7-15K.

Step 12: The finished bow, Figure 7-15L.

FIGURE 7-15D.

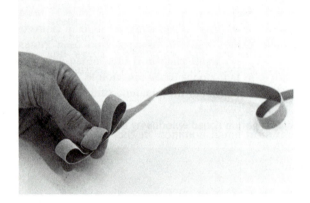

FIGURE 7-15E.

FIGURE 7-15F.

FIGURE 7-15G.

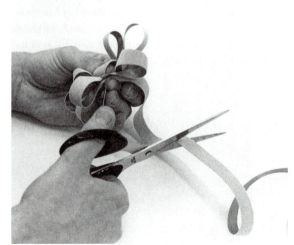

The Corsage or Bouquet Puff

Puffs fill out a corsage or bouquet without adding weight or much additional expense. The netting material comes on rolls like a wide ribbon and ranges from 2 7/8 to 8 inches in width. The netting is cut into squares and gathered with a wire to create

FIGURE 7-15H.

FIGURE 7-15J.

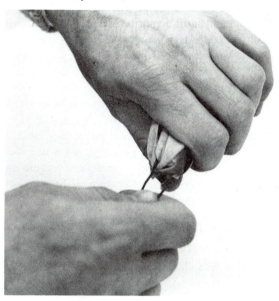

FIGURE 7-15I.

FIGURE 7-15K.

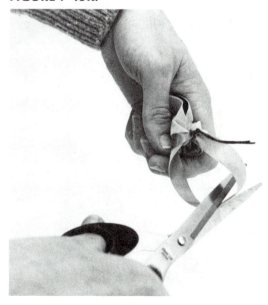

the puff. If gathered in the center, a butterfly-shaped puff results. If gathered at the base, a fan-shaped puff is formed. Once completed, the wires are covered with floral tape, Figure 7-16.

ASSEMBLING THE ARRANGEMENTS

The technique of arranging floral materials rapidly and skillfully into a corsage, table arrangement, wreath, or larger piece requires time to develop. Just as knowing how to read a recipe does not guarantee a successful result, neither will the first floral designs be flawless or easily accomplished. However,

FIGURE 7-15L.

FIGURE 7-16.
Two styles of puffs: fan (left) and butterfly (right)

by following the assembly instructions and studying the illustrations, you can begin to develop proficiency. From this beginning point, further practice, added study, and experience will refine your construction skills and your eye for style.

Corsages

Corsages are most commonly pinned to clothing at the shoulder or banded to the wrist. On occasion they may be pinned in the hair or carried as a nosegay. Teenage girls are the most frequent recipients of wrist corsages and nosegays. Adult women generally prefer to use corsages as accents to their personality, clothing, or a special occasion. Whenever possible, the florist needs some questions answered before the corsage can be designed correctly.

- *What color is the garment to be worn?* The flowers should complement the outfit, not compete with it. It is usually impossible to match the color of the flowers to the color of the garment, so contrasting colors are best used.
- *What is the neckline of the garment?* A sharply vertical neckline such as a V-neck is best accented by a corsage that is strongly vertical also. A scooped neckline looks best with a corsage whose line is gently curved.
- *What is the occasion?* Since corsages can be either formal or informal in style, it is helpful to know.
- *What is the size of the wearer?* Small women look best wearing a small corsage. Larger women need a larger corsage to retain the proper proportionate relationship between flowers and wearer.
- *What is the skin tone of the wearer?* Dark skinned women look better with pastel toned corsages. Fair skinned women often need more vivid colors to brighten their appearance.
- *Does the wearer have a preference for or an allergy to certain flowers?*

Obviously, the professional florist may not have all the information needed to create a totally customized corsage. However, too many florists ask only, "What color?" and overlook other equally vital information.

To prepare flowers for assembly into a corsage, all or most of the stems are removed and replaced with wired and taped stems. A moistened wad of cotton or chenille stem may be used to prolong the life of blossoms that wilt easily. The reason for removing the stems is to lighten the weight of the corsage and decrease the mass. Modern corsages must be lightweight so as not to pull and tear synthetic fabrics but solidly constructed to stand up to today's active life-style.

Although corsages can be made from one or two flowers, corsages made with an odd number of blossoms are most common and most interesting. Accordingly, that is the type to be described here, Figure 7-17.

To construct a corsage of three, five, or seven blossoms:

Step 1: Select flowers that have small blossom heads and are in varying stages of maturity. Wire and tape flowers.

Step 2: In one hand, place the smallest, most tightly closed blossom. This will be the uppermost flower of the corsage and determine the direction of wearing, since it always points upward.

FIGURE 7-17.

Construction of a three-blossom corsage

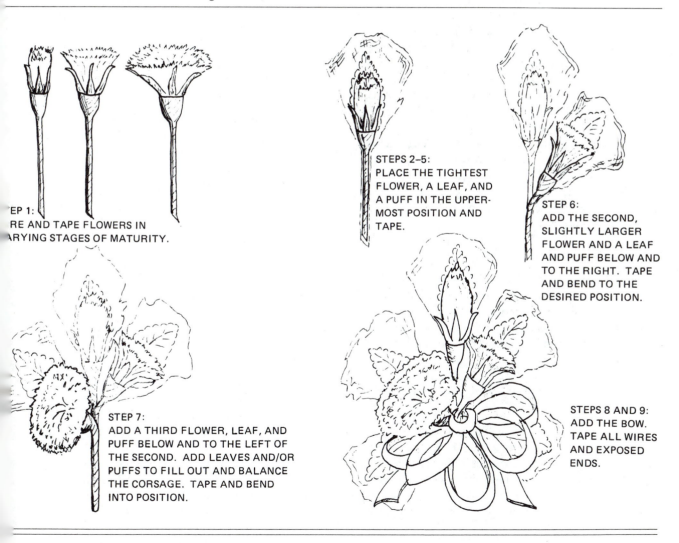

STEP 1:
WIRE AND TAPE FLOWERS IN VARYING STAGES OF MATURITY.

STEPS 2–5:
PLACE THE TIGHTEST FLOWER, A LEAF, AND A PUFF IN THE UPPERMOST POSITION AND TAPE.

STEP 6:
ADD THE SECOND, SLIGHTLY LARGER FLOWER AND A LEAF AND PUFF BELOW AND TO THE RIGHT. TAPE AND BEND TO THE DESIRED POSITION.

STEP 7:
ADD A THIRD FLOWER, LEAF, AND PUFF BELOW AND TO THE LEFT OF THE SECOND. ADD LEAVES AND/OR PUFFS TO FILL OUT AND BALANCE THE CORSAGE. TAPE AND BEND INTO POSITION.

STEPS 8 AND 9:
ADD THE BOW. TAPE ALL WIRES AND EXPOSED ENDS.

Step 3: Behind the blossom, place an artificial or fresh leaf, allowing it to extend about an inch above the first flower.

Step 4: If net puffs are being used, one may be added at this point, behind the flower and leaf.

Step 5: Tape the stems together tightly.

Step 6: Add a second and slightly larger flower below the first and to the right. Positioning is accomplished by bending the wired stem as each flower, leaf, or puff is added.

Step 7: Add a third flower slightly lower and to the left of the second. Continue in this manner until the desired number of flowers have been added. (Reserve the most widely opened flowers for the base of the corsage.) Leaves and puffs should also be added where needed to provide shape and background. Do not twist the wired stems around each other. Instead bind them with tape after every third stem is added to the corsage. Each flower, leaf, or puff counts as a stem.

Step 8: Add the bow to complete the corsage. Depending upon the style of the corsage, the bow may be at the base, mixed with the base flowers, or at the side of the arrangement.

Step 9: Tape all wires and exposed ends. Add a corsage pin and place in a corsage bag or box. Store in a cooler.

Table Arrangements

Table arrangements can range from a simple bud vase on a bedroom nightstand to large buffet designs. As discussed here, though, they will be regarded as the moderately sized and moderately priced arrangements commonly used on dinner tables, coffee tables, or end tables.

As with corsages, the florist needs to get answers to some questions before designing a table arrangement.

- *What is the purpose of the arrangement?* A dinner table centerpiece must be kept low to permit guests to see over it. Other functions may permit a stronger vertical line to the design.
- *Will it be visible from all sides?* One-sided arrangements require fewer flowers and are suitable if the arrangement is to be placed against a wall. However, dinner and coffee table arrangements must be designed for viewing from all sides.
- *What colors predominate on the table or in the room?* The colors of the arrangement should complement the decor, not compete or clash with it.
- *Is there a theme to the occasion?* This is especially important if the arrangement is for a party or holiday event.
- *How long is the arrangement to last?* While reputable florists always use fresh flowers, certain species have a longer life-span than others. Weekend house guests need to be celebrated with a more durable arrangement than would a five-hour dinner party.
- *Does the client have a preference for colors, flowers, or styles?*

Before beginning the assembly of table arrangements, you should refresh your memory about the container types, forms of arrangement materials, patterns, and basic lines of floral arrangements. Then assembly can begin.

Symmetrical Triangle Arrangement, One-Sided

- *Pattern:* triangular
- *Line:* inverted T
- *Container:* pedestal
- *Stem support:* water-holding foam block (for fresh flowers) or block styrofoam (for permanent flowers)
- *Flower shapes:* all one type of flower with some having long stems *or* a combination of line, mass, and filler flowers

Step 1: Begin by securing the stem support material in the container, Figure 7-18. The water-holding foam or styrofoam should be cut to fill the container completely and extend above the rim. If water-holding foam is used, it should be presoaked to assure that it contains adequate water. The block of material is secured with waterproof tape applied as two cross-strips from side to side of the container and over the block of foam.

FIGURE 7-18.

Construction of a symmetrical triangular arrangement

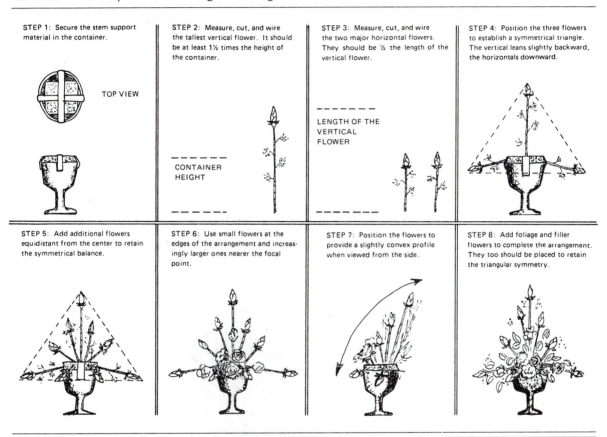

STEP 1: Secure the stem support material in the container.

TOP VIEW

STEP 2: Measure, cut, and wire the tallest vertical flower. It should be at least 1½ times the height of the container.

CONTAINER HEIGHT

STEP 3: Measure, cut, and wire the two major horizontal flowers. They should be ½ the length of the vertical flower.

LENGTH OF THE VERTICAL FLOWER

STEP 4: Position the three flowers to establish a symmetrical triangle. The vertical leans slightly backward, the horizontals downward.

STEP 5: Add additional flowers equidistant from the center to retain the symmetrical balance.

STEP 6: Use small flowers at the edges of the arrangement and increasingly larger ones nearer the focal point.

STEP 7: Position the flowers to provide a slightly convex profile when viewed from the side.

STEP 8: Add foliage and filler flowers to complete the arrangement. They too should be placed to retain the triangular symmetry.

Step 2: Measure, cut, and wire the tallest vertical flower. Its length should be at least 1 1/2 times the height of the container.

Step 3: Measure, cut, and wire the two major horizontal flowers. They should be 1/2 the length of the vertical flower.

Step 4: Place the three flowers into the container, pressing them far enough into the foam to secure them. The vertical line flower should be in the center of the container and angled back just slightly. The two horizontal line flowers should be in the center of the container and angled slightly downward. These three flowers establish the size and shape of the design, an equilateral triangle.

Step 5: Add additional flowers, staying within the triangular line. Each pair of flowers should be positioned at equal points right and left of center to retain the symmetrical balance of the design.

Step 6: Flowers should be small at the edges of the arrangement and increase in size and mass as they approach the focal point.

Step 7: Place the flowers so that the arrangement has a slightly convex appearance when viewed from the side.

Step 8: Foliage and filler flowers can be added to complete the arrangement. They too should respect the symmetry and shape of the design.

This arrangement can be varied by keeping the triangle symmetrical but not equilateral. A candle or figurine can be used as the focal point instead of massing flowers. Also, the container can be changed.

Symmetrical Triangle Arrangement, Two-Sided

- *Pattern:* triangular
- *Line:* inverted T
- *Container:* a low, wide bowl
- *Stem support:* water-holding foam block or block styrofoam.
- *Flower shapes:* all one type of flower or a combination of line, mass, and filler flowers.

Step 1: Proceed as for the one-sided symmetrical triangle arrangement, but place the tallest vertical flower directly in the center of the container and perfectly upright, not leaning, Figure 7-19.

Step 2: Unlike the one-sided triangle, the two-sided arrangement must have depth as well as width. Measure, cut, and wire two flowers of the same length but approximately 1/3 shorter than the other two horizontal flowers. Place them into the foam at the base and at 90° angles to the two horizontals that established the width of the arrangement. The four base flowers will form a diamond when viewed from above. The vertical flower will extend upward from the diamond's center.

Step 3: As flowers and foliage are added to complete the arrangement, keep turning the container to assure that it is developing equally well on all sides but that no flowers extend outside the trian-

FIGURE 7-19.
For a one-sided symmetrical triangle arrangement (left), the three major flowers are angled slightly backward. For two-sided viewing, the major flowers are positioned without angle (right).

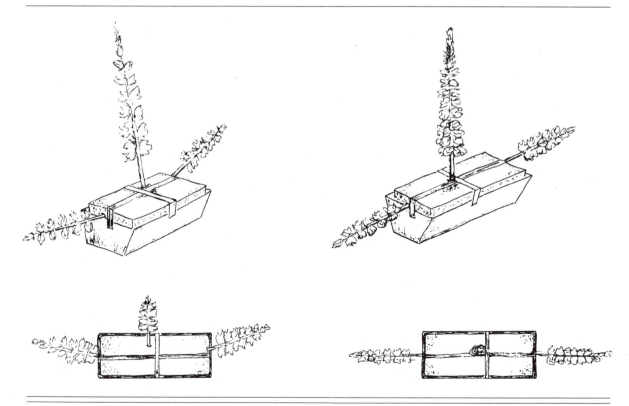

gular lines. The focal point will be replaced by a *focal area* of larger and/or more brilliant blossoms.

Asymmetrical Triangle Arrangement, One-Sided

- *Pattern:* Triangular.
- *Line:* L-line.
- *Container:* A low, wide bowl.
- *Stem support:* Water-holding foam block or block styrofoam.
- *Flower shapes:* All one type or a combination of line, mass, and filler.

This arrangement can be varied considerably and still fit the triangular pattern. If constructed as a right triangle, it follows the L-line. If constructed as neither a right triangle nor an equilateral triangle, the line is best described as a lopsided inverted T. In either case, it possesses the three sides of a triangle but is not symmetrically balanced. The assembly directions that follow are for an asymmetrical triangular arrangement having a strong L-line.

Step 1: Tape the foam into the container to the left of center, Figure 7-20. This helps to balance the design visually in relation to the container.

FIGURE 7-20.

Construction of an asymmetrical triangular arrangement

STEP 1: Secure the stem support material in the container. If the container is larger than the support medium, place the medium off-center.

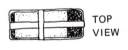

TOP VIEW

STEP 2: Measure, cut, and wire the tallest vertical flower. Its length can be as much as twice the container's length.

CONTAINER LENGTH

STEP 3: Measure, cut, and wire the two major horizontal flowers. One should be approximately ¾ the length of the tallest vertical. The other should be approximately 1/3 the length of the tallest vertical.

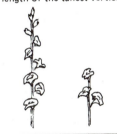

STEP 4: Position the three flowers to establish the asymmetrical triangle. The vertical leans slightly backward. The right horizontal is at 90°. The left horizontal is slightly acute.

STEP 5: Add additional flowers that stay within the triangular shape. No two should be directly opposite. The size should enlarge as the focal point is approached.

STEP 6: Add foliage and filler flowers to complete the arrangement while retaining the asymmetrical triangular shape.

Step 2: Measure, cut, and wire the tallest vertical flower. Its length can be twice the container's width without being too long. Place it into the foam to the rear and left of center. Allow it to lean slightly backward.

Steps 3 and 4: Measure, cut, and wire the two major horizontal flowers that will form the remaining points of the triangle. One should be approximately 3/4 the length of the vertical. It should be placed at the base (right) and at a 90° angle to the vertical. The other should be 1/3 the length of the vertical and placed opposite its longer horizontal counterpart, angled slightly upward.

NOTE: The vertical and longest horizontal flowers should have smaller, more tightly closed blossoms to signify their distance from the design's focal center. The smaller horizontal flower can be larger or more open since it is close to the design's focal center (the point where the three stems appear to converge).

Steps 5 and 6: Subsequent flowers and foliage added should stay within the triangular shape. Flowers used at the focal point should be larger and more eye attracting. As each flower is added, lengths should be adjusted so that no two blossoms align exactly, lest the asymmetry be lost.

The arrangement can be varied by reversing the L-line and moving the foam and focal point to the right of center. Also, foliage can be given a greater role in creating the triangle's major lines, with flowers used principally at the focal center. This is helpful when flowers are limited or when no good line flowers are available.

An S-Curve Arrangement, One-Sided

- *Pattern:* oval
- *Line:* S-line, sometimes called the Hogarth Curve after the English artist, William Hogarth
- *Container:* a tall pedestal container or tall vase
- *Stem support:* water-holding foam block or block styrofoam
- *Flower shapes:* round mass forms for the focal point; pliable line forms (flowers or foliage) for the S-shape

This is one of the loveliest formal design forms and one of the most difficult to create. The two half ovals can be of the same size, creating a perfect S-shape; or the top oval can be larger than the bottom oval for a variation of the style. Flowers and foliage used to create the S-shape should be chosen for their natural curvature when possible to reduce the amount of wiring needed.

Step 1: Tape the water-holding foam or styrofoam into or on top of the container, Figure 7-21.

Step 2: Measure two line flowers (or foliage) to be at least 1 1/2 times as long as the container's longest dimension. Wire and tape the stems with wire of a gauge sufficient to hold the desired curvature.

Step 3: Insert the top flower or foliage form into the left side of the foam, angled upward and slightly backward. The tip of the curved stem should reach toward the vertical center of the arrangement but not quite touch it.

Step 4: Insert the lower stem into the right side of the foam, angled downward and slightly forward. The reverse curvature of the lower stem should equal that of the upper stem.

Step 5: Add flowers and foliage, increasing the size and coarseness of the texture as the focal point is approached.

NOTE: As the arrangements are completed, the back sides should be finished as well, with foliage used to fill in blank areas and conceal the foam base. All fresh stems must be firmly implanted in the moist foam to prevent premature wilting of the arrangement.

Wreaths

For centuries, the Christmas holiday season has been celebrated with festive wreaths. In recent years, wreaths have been used to commemorate many additional occasions and seasons. Although the evergreen Christmas wreath is still most common, wreaths can be made of cones, straw, grapevines, fruit and nuts, and assorted permanent materials.

Lightweight permanent materials often have a circle of styrofoam as their base. The material is attached to wooden or metal picks and stuck into position. Heavier fresh materials may require a frame that can be filled with spagnum moss or other

FIGURE 7-21.
Construction of a one-sided, S-curve arrangement

STEP 1: Secure the stem support material in the container. If the opening in the container is too small, attach the material to the top of the container. 	STEP 2: Measure, cut, and wire the two line flowers or foliage to be at least 1½ times as long as the container's greatest dimension. 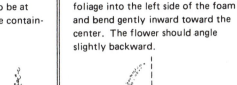 	STEP 3: Insert the top flower or foliage into the left side of the foam and bend gently inward toward the center. The flower should angle slightly backward. 
STEP 4: Insert the lower flower or foliage into the right side of the foam and bend gently to equal the curvature of the top flower. The flower should angle forward slightly. 	As viewed from the side, the two flowers should appear as a continuous line. 	STEP 5: Add additional flowers and foliage while retaining the S-curvature. Increase both size and coarseness as the focal point is approached. 

water-retaining material. Christmas wreaths are usually supported on a heavy metal ring, the thickness of which is determined by the size of the wreath. It is essential that the support ring be strong lest the wreath sag into an oval shape when hung.

Wreaths can be made entirely by hand or with the aid of a wreath machine. The directions that follow are for a traditional evergreen Christmas wreath assembled by hand.

Step 1: Using hand pruners, cut several large evergreen boughs into a pile of smaller pieces, 4 to 6 inches in length and as fully needled as possible. Use of fresh boughs will help assure that the wreath remains green through the season.

Step 2: With the wreath ring resting on the table, grasp a cluster of three or four evergreen branches and wire them tightly to the ring. A spool of medium-gauge wire works well for this purpose.

Step 3: Wire subsequent clusters of greenery into place, overlapping the previous clusters to conceal their wiring. All greenery should be oriented in the same direction around the wreath, Figure 7-22.

FIGURE 7-22.
Hand wiring a wreath

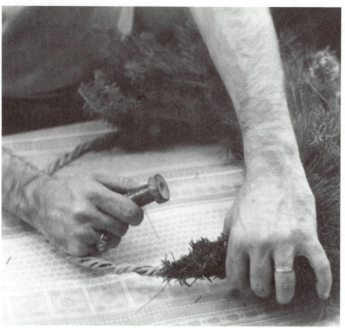

Step 4: The wreath may be developed with one side or two sides. The former is more common than the latter.

Step 5: Wire a loop for hanging onto the back side of the wreath.

Step 6: Once the foliage has been completed, a bow can be added to finish it, Figure 7-23. Other additions can include decorative balls (called millimeter balls), pine cones, birds, or other ornaments.

Retail florists often purchase undecorated wreaths in quantity, already assembled. The time pressures of a busy holiday season may prevent each wreath from being made to order. The preassembled wreaths are made by hand but using a wreath machine, Figure 7-24. The wreaths are usually made well in advance of Christmas and so dry out and shed sooner than those made with fresher materials.

SUMMARY

Our use of flowers is rooted deep in history, and much of what is being done with flowers today is merely a copying of what has been done for centuries. Still, America has made significant contributions to modern flower arranging and is now setting the standards for professional flower arranging worldwide.

Floral design is a craft that requires its practitioners to be knowledgeable not only about the construction and care of floral products but also about design of all types, the natural world, literature, the social sciences, and the fine arts.

The materials of a professional florist include assorted cutting tools, wiring materials, adhesives, and stem support materials. Becoming familiar with their uses and differences is an important first step

FIGURE 7-23.
A completed wreath of evergreen materials, finished with a bow

FIGURE 7-24.
A wreath machine frees both hands to collect and hold the evergreen boughs in place while the foot powers the machine that wires the boughs and secures them.

in a florist's training. Equally important is knowing how to care for fresh cut flowers in the shop. They must receive nutrition for continued good health, water to prevent wilting, and cool temperatures to slow their metabolic activity and prolong their life.

The materials used in floral arrangements, whether fresh, dried, or permanent, can be categorized into basic form types as: line, mass, form, or filler elements. Depending upon how they are used and the size of the arrangement, flowers and nonflowers may fill different form roles.

Liking or disliking a design is a subjective judgement. However, a floral design can be evaluated more objectively if the principles of design are applied to the judgement. The principles are: simplicity, focalization of interest, scale and proportion, balance, and rhythm and line.

Using the five principles of design, florists often arrange the flower and foliage form types into five standard patterns: round, oval, rectangular, symmetrical triangle, and asymmetrical triangle. These five standard patterns form the foundation of American floral design and have been the major contribution of American florists to the worldwide profession.

Six basic lines make up the five standard patterns. The lines are the vertical, horizontal, L-line, inverted T, crescent, and S-curve. Once the basic patterns and lines are understood, it becomes easy to analyze the composition of most floral arrangements.

Color is one of the features of flowers and floral arrangements that makes them so attractive. A florist needs a good color sense and an understanding of color terminology. Careful study of a color wheel is a good place to begin.

More mechanical than a good color sense, but just as necessary, is knowledge of how and when to wire the flowers and foliage of an arrangement. Wire may need to be added if the stems are weak or crooked, if a certain curvature is desired, if the stem length needs to be extended, or if the stem is too bulky and a new one needs to be created. The exact technique of wiring varies with the type of stem and flower head.

This chapter also covered the construction of bows and puffs, a corsage, table arrangements (one-sided and two-sided symmetrical triangles, a one-sided asymmetrical triangle, and an S-curve), and finally a wreath.

Achievement Review

A. TRUE/FALSE

1. Indicate if the following statements are true or false.
   a. The use of flowers is a recent event in recorded history.
   b. Flowers are often an expression of our sentiment or emotions.
   c. American use of flowers began as a copying of styles developed in Europe and elsewhere in the world.
   d. At present, the American style of floral design is a standard for the rest of the world.
   e. The European use of flowers is probably the most personal and intimate that the world has known.
   f. Floral design is more often a craft than an art.
   g. The natural world serves as an important teacher for the floral designer.
   h. Formal education can aid the development of a floral designer.

B. SHORT ANSWER

Answer each of the following questions as briefly as possible.

1. Assume that you are assembling the tools and materials you need to begin in the floral design profession. List as many examples as you can recall from each of the following categories.
   a. cutting tools
   b. wiring materials
   c. adhesive materials
   d. stem support materials

2. Label each of these container types.

3. Following are the steps taken by a florist when an order of fresh flowers arrives. For each step that contains an error, write the correct procedure.
   a. The flowers are carefully unpacked and unwrapped if necessary.
   b. Stem bases are cut squarely across to improve water uptake.
   c. The stems are placed into unwashed containers filled 1/3 full with fresh, cold water.
   d. A floral preservative is added.
   e. The flowers are placed immediately into the cooler at 38° to 40° F.

4. Indicate whether the following are most typical of line flowers, mass flowers, form flowers, or filler flowers.
   a. uncommon shapes and unusual silhouettes
   b. thin and vertical in shape
   c. rounded and often bulky
   d. form the basic shape of the arrangement
   e. used in the open areas between line and mass materials
   f. exemplified by zinnias, marigolds, large mums, and carnations

g. exemplified by orchids, iris, and lilies
h. exemplified by babies' breath and spray mums
i. exemplified by gladiolus and snapdragon

5. Match the definition or example with the correct principle of design.

   a. proper size relationship between flowers and container
   b. main height components equal to 1 1/2 times the container width
   c. the use of one distinctive design line and a few harmonious colors
   d. frozen motion
   e. equal importance on either side of the point of line convergence
   f. selection of a container that does not overpower the flowers
   g. use of mass form flowers at the point of line convergence

   1. simplicity
   2. focalization of interest
   3. scale and proportion
   4. balance
   5. rhythm and line

6. List the five major patterns of floral arrangements.
7. List the six basic lines of floral arrangements.
8. List the six major reasons that flowers are wired.
9. List the word that completes each of the following definitions.
   a. Color families are the _____ major groupings of colors visible when white light is passed through a prism.
   b. The brilliant and unaltered state of color from which a family name is derived is termed its _____.
   c. Tint is the hue lightened by the addition of _____.
   d. A hue darkened by the addition of black is termed _____.
   e. A hue that has been grayed by addition of both black and white is termed _____.
   f. Colors that are more visible under dim light than other colors have a high _____.
   g. Reds and oranges are _____ colors.
   h. Blue and green are _____ colors.
   i. Groupings of colors are termed _____.

C. MULTIPLE CHOICE

From the choices given, select the answer that best completes each of the following statements.

1. A color scheme that uses colors that are opposite or nearly so on the color wheel is _____.
   a. monochromatic
   b. complementary
   c. adjacent
   d. triadic
2. A color scheme that uses one primary color and other colors derived from that same primary is _____.
   a. monochromatic
   b. complementary
   c. adjacent
   d. triadic
3. A color scheme of yellow, green, and orange is _____.
   a. monochromatic
   b. complementary
   c. adjacent
   d. triadic
4. A color scheme that uses many variations of a single color is _____.
   a. monochromatic
   b. adjacent
   c. triadic
   d. polychromatic
5. A color scheme whose three colors are equidistant on the color wheel is _____.
   a. monochromatic
   b. adjacent
   c. triadic
   d. polychromatic
6. A polychromatic color scheme contains _____ hues.
   a. three
   b. four
   c. five
   d. six

D. DEMONSTRATION

To demonstrate your competence, prepare:
   a. five butterfly-shaped puffs and five fan-shaped puffs
   b. a corsage bow
   c. a corsage of five flowers
   d. a table arrangement, the style to be determined either by you or your instructor
   e. a one-sided door wreath, the materials to be decided by your instructor depending upon the season and availability

8
The Interior Uses
of Plants

Objectives

Upon completion of this chapter, you will be able to

- discuss the current status of the interior foliage plant industry.
- list problems unique to the interior use of plants.
- describe the role of light duration, quality, and intensity.
- list the characteristics of a good growing medium.
- describe steps in installation, watering, and drainage.
- describe the working relationship between architects, landscape architects, and maintenance professionals.

BACKGROUND AND STATUS UPDATE

The use of containerized plants is not new. Evidence of the use of potted plants is found in ancient Chinese artifacts, on the tomb walls of the Egyptian pharaohs, and in the ruins of Pompeii. Containerized ornamentals adorned the palaces of European nobility for centuries and filled the parlors of middle-class Victorians and Americans more recently. A pot of ivy on the windowsill and a fern in the entry hall have been standards in U.S. homes for many years. The rubber plant enjoyed several years as the reigning houseplant until other species caught the consumers' favor.

An amateur psychologist can wade deep in conjecture as to the meaning of all this. Historically, such use may represent an attempt to display dominion over nature. In recent years, the use of foliage plants may represent the desire of urbanites to bring back into their lives the greenery lost as our cities and highways have sprawled their paved surfaces across the landscape.

Potted foliage plants (*houseplants*) have had their place in homes for many years. Larger specimens are now appearing in shopping centers, office buildings, hospitals, schools, churches, airports, restaurants, and cocktail lounges. In response to positive public reaction, architects and interior decorators are using plants as architectural elements in buildings. These plants are as important to the decor as the wall coverings, furniture, and carpeting.

So explosive has been the demand for interior plantings in public buildings that the technology necessary to assure the plants' survival has not always been able to keep pace. Some of the early efforts, heralded for their aesthetics, became industry horror stories a few months later when lush greenery turned to chlorotic and necrotic stalks. An unsatisfactory outcome is predictable when architects and interior decorators try to work with plant materials whose qualities and maintenance requirements are unknown to them.

The demand for maintaining plants indoors has created a new profession within the larger field of ornamental horticulture. The term coined to describe it is *interior plantscaping,* a variation on the term *landscaping.* Since *plantscaping* has a distinctly contrived sound to it, the term may not survive. *Interior landscaping* may suffice to distinguish it from exterior landscaping. Semantics aside, the new profession is fresh and exciting and not overcrowded. For now, the term *interior plantscaping* will be used to describe the work.

Tropical foliage plants have proven to be the most successful indoor plants because they do not require the period of cool temperature dormancy that often makes temperate zone plants unsatisfactory. We recognize the desirability of such plants in interior design, and there is a sufficient body of knowledge to permit their widespread production in the nurseries and greenhouses of Florida, California, Texas, Ohio, Pennsylvania, and Latin America. There is far less knowledge of how to move these plants from their natural habitat or production facility to an office lobby and assure their successful transplanting and maintenance. It is not an exaggeration to state that the profession of interior plantscaping is still in its infancy. As a profession it deals with the *design, installation, and maintenance of plants in interior locations.* In scope, it includes but reaches far beyond conventional plant production skills.

UNIQUENESS OF INTERIOR PLANTSCAPES

Imagine how difficult it would be for humans, after countless centuries of living in sheltered dwellings, suddenly to be required to live permanently outdoors. Unlikely as the example may seem, it illustrates the difficulty of relocating plants from their accustomed and optimal growing sites to indoor locations.

When plants are used indoors, they must adjust to numerous changes. Among them:

- a drastic reduction in the quality and intensity of the light
- reduction and constriction of the plant's root system
- the replacement of natural rainfall by dependency upon humans for correct watering
- a reduction in nutrient requirements and a potential for buildup of soluble salts (fertilizers)
- a lack of air movement and rainfall, allowing dust to accumulate on the leaves, often plugging stomata and reducing photosynthesis
- potential damage by air conditioners, central heating systems, cleaning chemicals, water additives, and other irritants

Under these conditions, plants sustain themselves but seldom grow.

LIGHT AND INTERIOR PLANTINGS

Common sense tells us that plants need light to survive and grow. Architects who have envisioned living greenery throughout their building interiors have been disappointed when all the leaves dropped several days after installation, returning slowly or not at all. These are the questions that need answers.

- How much light do interior plantings need?
- What kind of light do they need?
- What sources of light exist indoors?

One final question must be answered if everyone connected with the plantscape is to have realistic expectations: How long will plants live indoors? The answer depends on the plant, of course, but the general answer is that in interior design, as elsewhere, nothing lasts forever. Carpets wear thin; furniture must be replaced; walls require fresh paint. Plants must be regarded as perishable furnishings as well. If installed correctly into a properly designed setting and maintained properly, they will serve satisfactorily for a time period that will unquestionably justify their cost. Then they will require replacement.

Light Intensity

Human activities do not require as much light as plants require for growth. No matter how many windows and skylights are designed into a home or office building, the light intensity inside will never equal that outside. Even unshaded greenhouses filter the sun's light and reduce it by at least 15 percent.

To understand light intensity requires knowing how light is measured. Light intensity is expressed in the units of *lux* or *footcandle*. A *lux* is the illumination received on a surface that is 1 meter from a standard light source known as unity. A lux is an international measurement comparable in use to the metric system. In the United States, the footcandle unit is more commonly used and understood. One *footcandle* (fc) is equal to the amount of light produced by a standard candle at a distance of 1 foot. Direct-reading meters are manufactured that measure light intensity in footcandles up to 10,000 (the illumination on a typical clear, sunny, summer day). A light meter is the only way to measure light intensity accurately and should be the first piece of equipment purchased by a beginning plantscape designer.

The challenge of bringing plants accustomed to outdoor light intensities into a home or shopping mall is best appreciated through several examples. The average residential living room has a light intensity of 10 to 1,000 f.c. by day and as few as 5 f.c. by night. A good reading light provides 20 to 30 f.c. A typist may have 40 to 50 f.c. of illumination on the typewriter surface. The average shopping mall provides 20 to 30 f.c. of light in pedestrian circulation areas and up to 100 f.c. in sales areas. How is a plant grown in 10,000 f.c. of light to survive?

The keys to a plant's survival are acclimatization and maintenance of the minimum light intensity required for its survival in an attractive and healthy condition. Note that *survival* and *attractive appearance* are the maintenance objectives, not growth.

Acclimatization is the adjustment of an outdoor plant to interior conditions. It involves both morphological and physiological changes in the plant and takes time to occur. The *minimum light intensity* is the level of illumination necessary to allow the acclimatized plant to produce new leaves at a rate equal to or slightly greater than the rate at which old leaves *senesce* (age) and *abscise* (fall off). Both the time required for acclimatization and the minimum light intensity will vary depending on the plant species.

There is a great lack of carefully researched and comprehensive data on which an interior plantscaper can rely. To the credit of forward-looking industry professionals, the Associated Landscape Contractors of America (ALCA) and the Florida Foliage Association (FFA) continue to promote research and publication of research data needed by members of the interior plantscaping industry. Although much is still not known about the needs of plants indoors, more is being learned every day.

The procedures necessary for successfully transplanting a tropical foliage species to an interior locale must begin during production of the plant. The growers can start reducing the light of selected plants once contracts for their purchase are signed. Note that acclimatization also involves the plant's adjustment to reduced water in the soil and in the surrounding air. Based upon current knowledge and technology, plants are usually acclimatized in the following way.

Light-intensity Acclimatization Light intensity is reduced gradually over a period of several weeks or months. Each change reduces the light by 50 percent until the desired intensity (usually 100 to 200 f.c.) is reached. The acclimatization process cannot be rushed without a severe reaction by the plant; that is, defoliation or death. Once indoors, the reduced light provided to the plants must be of sufficient duration to permit the plants' slowed photosynthetic processes to manufacture adequate food. Most plants require at least twelve hours of continuous light every day, including weekends. A reliable timed-lighting system is essential.

Nutrient Acclimatization The high fertility level of soil necessary for maximum plant growth during production is unnecessary and even life-threatening to the indoor plant, whose use of soil nutrients is greatly diminished. Thorough soil leaching at the beginning of the acclimatization period and occasionally afterwards will prevent a buildup of soluble salts.

Moisture Acclimatization The frequency of watering is reduced during acclimatization to prepare plants for their more stressful interior locale. The high humidity levels of the production area are also gradually reduced to ready plants for the drier air of home and building interiors.

Temperature Acclimatization Production area temperatures are usually higher than human comfort levels in order to promote more rapid plant growth. During acclimatization, temperatures are gradually reduced to the range common to most interior areas, 65° to 75° F.

Light Quality

Once acclimatized to the reduced light intensity of the interior, the plantscape may still prove unsat-

isfactory if the light quality is incorrect. *Light quality* is the color of light emitted by a particular source. The sun emits all colors of light, some of which the human eye can perceive and others that are imperceptible to humans but beneficial to plants. The green-yellow light most comfortable for humans is of little use in photosynthesis by plants. They depend on light from the blue and red bands of the visible light spectrum. Visible light is only a narrow region of the radiant light spectrum, Figure 8-1. The unit of measurement for light wavelengths is the *nanometer*.

As long as both humans and plants can derive their light energy from the sun, the needs of each are satisfied. Indoors, however, where light energy is usually created by artificial means, the quality of the light can vary considerably. Light preferred by an interior plant specialist concerned with the health of plantings may cause the skin tones of human beings to appear ashen and deathly. In similar fashion, an interior decorator may specify a quality of lighting that gives the human complexion a healthy glow while making nearby plantings appear brown and dead. Clearly, someone seeking a career in interior plantscaping needs to know the types

FIGURE 8-1.

Electromagnetic spectrum and spectral distribution of visible light (From J. Boodley, *The Commercial Greenhouse,* © 1981 by Delmar Publishers Inc.)

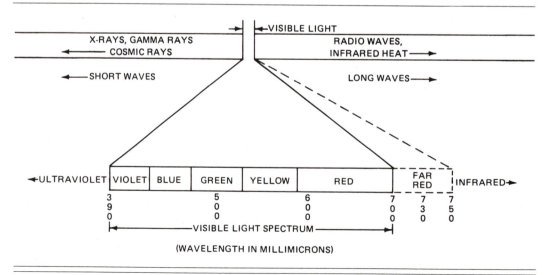

of lamps currently available and the quality of light that they provide. Categories and examples of lamps that have some use in interior plant illumination are shown in Table 8-1. Table 8-2 compares the lamps in all areas important to interior plant survival. You should study these tables thoroughly before proceeding further.

Selection of the proper lamp for the illumination of an interior planting will depend upon the answers to several questions:

- How extensive is the planting?
- Are the plants to be encouraged to grow or merely to be maintained at their current size?
- Will the plants receive any sunlight? If so, how much and for how long?
- How far will the artificial light source be from the plants?
- What types of lamps are being used for general lighting of the area and what is the intensity of surface illumination provided?

For example, consider the plantscape of a typical office. The plants may be permanently located or

TABLE 8-1.
Lamps for Interior Plant Illumination

1. **Tungsten filament incandescent lamps**
   - Standard (the familiar household lightbulb)
   - Reflector (spot or flood lights)
   - Parabolic aluminized reflector (a weather-resistant type of floodlight with a more precise beam)
   - Incandescent plant lamps (not proven to be any better than the standard incandescent)

2. **Fluorescent lamps**
   - Cool white
   - Warm white
   - Plant lamps
   - Wide spectrum plant lamps

3. **High-intensity discharge lamps**
   - Mercury
   - Metal halide
   - High-pressure sodium

in movable planters. Their functions may be to serve as room dividers, establish a mood, or relieve a cluttered desk top. The ceiling may be 8 or 10 feet high. Side windows or a skylight may admit some natural light. In such a setting, cool white fluorescent lighting would be ideal for both general lighting and the growth of the plants. People, plants, and furnishings look natural beneath cool white light due to its excellent color rendition, and the plants receive the right quality of light for photosynthesis. If additional task lighting is needed, small desk lights should be used. Special effects such as shadows or textural highlights can be created with incandescent lights installed beneath the plants and directed upward. (These are called *uplights.*) Some benefit will accrue to the plants from the addition of lighting at the base. However, if supplemental lighting is needed for photosynthesis, it is most efficient when applied from overhead because chloroplasts are concentrated in the upper leaf surface.

A shopping mall presents different problems. The corridors may have ceilings too high to permit the use of fluorescent lamps, which are not good for illumination when the ceiling is much beyond 10 feet. There may be skylights as well as decorative architectural lighting. Overall illumination by mercury or metal halide lamps would be best. As in an office, supplemental or decorative lighting might also be desirable for special effects or the health of the plants. When uplights are used, they should be installed directly into the planters and waterproofed. When supplemental lights are added for overhead illumination, they should be positioned to light the plants fully without shining in the eyes of viewers.

Natural Light

Most important of all light sources for interior plantscapes is natural sunlight when it can be planned for and depended upon. Each footcandle of illumination that nature provides is one less that has to be provided and paid for with artificial lighting. In our energy-conscious society, such savings are worth planning for. However, knowledge of how to maximize the benefits of natural light is vital;

TABLE 8-2.

A Comparison of Artificial Lighting Sources for Interior Plantscapes

Lamp Type	How Light Is Produced	Quality of Light Produced	Percent of Visible Light Radiation	Color Rendition	Initial Cost	Operating Cost
Incandescent (all types)	Current flows through a tungsten filament heating it and making it glow.	High in red light; low in blue light	7–11	Good	Low	High
Cool White Fluorescent	Phosphor coating inside the glass tube is acted upon by radiation from a mercury arc.	High in blue and yellow-green light; low in red light	22	Good (blends with natural daylight	Moderate	Moderate
Warm White Fluorescent	Phosphor coating inside the glass tube is acted upon by radiation from a mercury arc.	Low in blue and green light; more yellow and red light	22	Poor (blends with incandescent light)	Moderate	Moderate
Fluorescent Plant Growth Lamps	Same as other fluorescents. Special phosphors transmit most light energy in blue and red light regions of the spectrum.	High in red and blue light; low in yellow-green light	22	Average (enhances red and blue colors; darkens green colors)	Moderate	Moderate
Wide Spectrum Plant Growth	Same as other fluorescents. Special phosphors transmit most light energy in blue and red light regions of the spectrum.	Less blue and red than standard plant growth lamps; more far-red and yellow-green light	22	Average (favors red and blue colors; darkens green colors)	Moderate	Moderate

Life of the Lamp	Placement Height Above Plants	Plant Responses	Major Advantages	Major Disadvantages
750 to 2000 hours	At least 3 feet to avoid foliage burn	Plants become long and spindly with pale foliage. Flowering is promoted and senescence is accelerated.	• Good for special lighting effects • Compact source of light • Simple installation	• Energy inefficient; too much lost as heat • Light does not distribute evenly over a surface. • Glass blackens with time and light output is reduced. • Frequent replacement is needed.
Up to 20,000 hours	10 feet or less	Plants stay short and compact. Side shoots develop. Flowering extends over a longer period.	• Energy efficient • Heat is radiated over the length of the lamp, allowing closer proximity to plant foliage. • Light distributed more evenly over a flat surface	• Light does not focus well. • They are difficult to start when line voltage drops or humidity is high. • Installation is expensive. • Special fixtures are needed.
Up to 20,000 hours	10 feet or less	Same as CW fluorescent	• Same as CW fluorescent	Same as CW fluorescent
Up to 20,000 hours	10 feet or less	Rich green foliage color. Large leaf size. Side shoots develop. Plants stay short. Flowering is delayed.	• Same as CW fluorescent • Light emission is from the region of the spectrum most important to photosynthesis.	• Same as CW fluorescent • Greater expense with little increase in benefit to the plants
Up to 20,000 hours	10 feet or less	Stems elongate. Side shoots are suppressed. Flowering is promoted. Plants age rapidly.	• Same as CW fluorescent	• Same as CW fluorescent • Growth may not be desired. • Poor color rendition on nonplant materials

TABLE 8-2.

A Comparison of Artificial Lighting Sources for Interior Plantscapes (continued)

Lamp Type	How Light Is Produced	Quality of Light Produced	Percent of Visible Light Radiation	Color Rendition	Initial Cost	Operating Cost
Mercury (Deluxe white model, for interior plants)	An electric arc is passed through mercury vapor.	High in yellow-green light; less red and blue light, but still usable for plant growth	13	Poor (favors blue and green colors)	High	Moderate
Metal Halide	Similar to mercury lamps but with metal and gas additives to produce a different spectrum	High in yellow-green light; less red and blue light, but usable for growth	20–23	Good (similar to CW fluorescent)	High	Low
High Pressure Sodium	Sodium is vaporized into an arc.	High in yellow-orange-red light	25–27	Poor (similar to WW fluorescent)	High	Low

otherwise more heat energy is lost through inefficient windows than is gained in light energy.

Natural light is most helpful when it offers high levels of illumination throughout most of the year. Traditionally sunny areas like the Southwest can make better use of natural light than areas like the Northeast, where clouds block the sun and snow blankets the skylights through much of the winter season.

Sunlight entering from overhead is of greater use in the illumination of interior plantings than light entering from the side, although both are helpful. In neither situation will the natural light be as in-tense as outside light. It will be significantly reduced by the glass glazing through which it passes and the distance it travels between the point of entry and the leaf surface. Little usable light passes more than 15 feet beyond glass, so skylights in high lobby areas or shopping malls are of no benefit to plantings beneath them. Nevertheless, they can be of great benefit in single-story buildings with lower ceilings. In a similar fashion, a large interior plantscape may derive little benefit from side lighting, since usable light enters at a 45° angle and plants must be placed within that narrow beam if they are to benefit, Figure 8-2. They cannot be too close to the glass or

Life of the Lamp	Placement Height Above Plants	Plant Responses	Major Advantages	Major Disadvantages
Up to 24,000 hours	10–15 feet or more	Plants respond in a manner similar to CW fluorescent.	• Long life; useful for inaccessible fixtures • Medium energy efficiency	• Not interchangeable with other lamps • Warm-up time required
Up to 20,000 hours	10–15 feet or more	Plants respond in a manner similar to CW fluorescent.	• High energy efficiency, surpassing the mercury lamp • Good for both plant and general lighting	• Warm-up time required • Color and light quality change with operating hours.
Up to 24,000 hours	10–15 feet or more	Typical red-light plant responses; similar to fluorescent plant growth lamps when compared on equal energy	• High energy efficiency. When combined with blue light sources (such as metal halide), they provide good lighting for plants and people. • Long life	• Yellow color makes them unsatisfactory for general indoor lighting by themselves.

the foliage may burn, however. In a smaller room, as in a residence, natural side light can be of great value.

When skylights are used, they must be designed to permit the most light to enter while insulating against as much winter heat loss as possible. Within the limits allowed by heating and structural engineering, the ceiling well through which the light passes should be as wide and shallow as possible, with the sides painted a reflective white and beveled outward at 45° angles, Figure 8-3. As skylights become more narrow and the ceiling well deeper, there is less area through which the sunlight can enter and a narrower focus of illumination on the surfaces below. The difference between wide, shallow skylights and narrow, deep ones is similar to the difference in illumination between a floodlight and a spotlight. With natural light and interior plantings, the wide floodlight effect is most desirable.

Selecting the Correct Lighting

In summary, no single recipe for correct lighting can be given. There will be varied settings, needs, and objectives to accommodate. Plants will seldom be the only consideration in the selection of lamps and the quality of illumination. When both plants and people are to be considered, a lamp should be selected that provides the yellow-green visible light needed to render human complexion, clothing, and furnishings attractive, while still providing sufficient blue and red light to allow photosynthesis to exceed respiration in the plants. The cool white fluorescent

FIGURE 8-2.
Interior plantings receive the best quality of natural light within a 45° arc of the side windows.

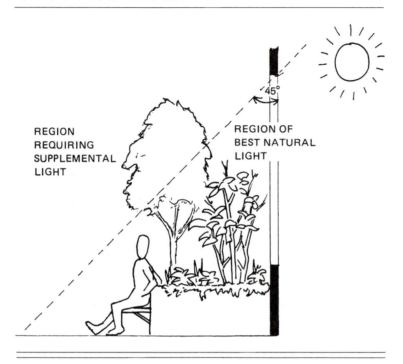

REGION REQUIRING SUPPLEMENTAL LIGHT

REGION OF BEST NATURAL LIGHT

45°

lamp is ideal for such a situation provided that the ceiling is not too high and that growth of the plants is not an objective. If growth is desired, additional incandescent lighting can be focused directly on the plants. The use of more expensive growth lamps is unnecessary since they have not been proven superior to the ordinary cool white fluorescent in maintaining plant health, and they do not render color well. Any natural light that can be used advantageously will reduce the cost of lighting the interior planting.

As for light intensity, not all plants available from growers have been carefully studied to determine the minimum at which they will survive attractively. The ALCA, the FFA, and the colleges of agriculture in the major foliage production states should be queried by anyone responsible for creating an interior plantscape. The many books that purport to give lighting specifications and that stock the shelves

of libraries nationwide are satisfactory for homeowners but not for professional use. Their information is often dated, usually based upon greenhouse lighting, and may be questionable in its accuracy. Recommendations should be sought that give lighting minimums in footcandles, not in general terms such as high, medium, and low. When plants are installed without documented knowledge of their lighting requirements, the situation is risky at best, and the plants must be watched carefully to determine if additional lighting is needed.

To assure that lighting is of the right intensity and duration, a simple timing device may be necessary. The lights must shine on the leaves for enough time each day (twelve hours minimum) to allow adequate photosynthesis to occur. Should the hours of lighting have to be reduced for some reason, the intensity of the lighting must be increased to compensate, Figure 8-4. It does not seem to matter

FIGURE 8-3.
Cross-sectional views of skylights, and the influence of their design on the amount of usable light available to plants

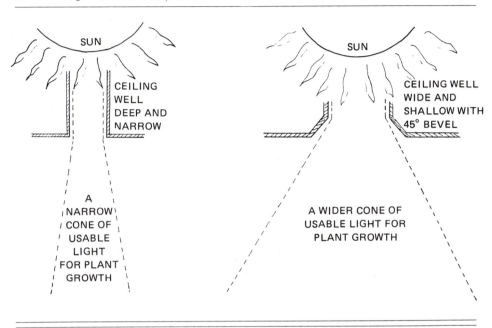

whether foliage plants receive their needed lighting over a short or long period as long as the cumulative photosynthetic activity balances and slightly exceeds respiration. With flowering plants, day length often plays a critical role in determining if and when

FIGURE 8-4.
A timing system can be used to provide supplemental light for an interior plantscape.

the blossoms will appear. Since interior plantings are currently valued most for their foliage, day length is of limited importance.

THE GROWING MEDIUM

No less important than light to the successful acclimatization of interior plants is the provision of a proper growing medium. The roots must be placed in an environment that provides structural support, allows the roots to absorb water, and provides essential minerals. Further, the growing medium must allow rapid drainage of water past the root zone and provide the correct pH for growth.

The medium that serves when plants are growing in a nursery field or production container is likely to be inappropriate for an interior installation. Natural field soil may be:

- too heavy to permit rapid drainage
- too heavy for the floor to support if the container is large

• inconsistent in composition, making standardized maintenance of separate planters difficult
• infested with insects, pathogens, or weeds

Although pasteurized natural soil may be a component of the growing medium for interior plantings, additives will probably be needed. It is even possible that the growing medium selected for the interior plantscape will have no natural soil in it for reasons of improved drainage, hygiene, pH balance, or nutritional consistency. The use of synthetic soils, whose composition is as controlled as a cake recipe, is becoming the rule rather than the exception.

Cornell University and the University of California have been leaders in the formulation of synthetic soils for interior plant production. The Cornell mixes are made from vermiculite or perlite and sphagnum moss. The University of California (U.C.) mixes are made from fine dune sand and sphagnum moss. The ratio of components varies with the species of plants and the maintenance program to be followed.

In addition to the U.C. and Cornell mixes (termed the peat-lite mixes) there are bark mixes, composed of pine bark, sand, and sphagnum moss. These are especially acidic growing media and may require buffering (with dolomitic and hydrated lime) to sustain healthy plants.

All of the synthetic soils are mixed and sold commercially. In large interior installations, it may be more economical to mix the medium at the site rather than purchase the premixed commercial products. All decisions about the growing medium should be made before the planters are filled and the plants installed. Once planted, errors in the medium's composition are difficult to correct without removing the plant.

Because the artificial media have low nutritional value for the plant, complete fertilizers are required as well as periodic application of the minor elements. Regular soil testing is necessary to assure that the fertilization program is correct.

Installing the Plants

The acclimatization of a plant's root system involves establishment of the correct relationship between roots and foliage. Outdoors, a sizeable root system is needed to supply adequate amounts of water and minerals to the leaves for near-maximal photosynthesis. Indoors, photosynthesis is reduced to survival-maintenance levels, so the root system need not be as large. Thus, one of the first steps at the time of installation is to remove the production medium from around the roots and prune away excess roots, Figure 8-5.

Several methods can be used for setting plants into an interior plantscape, depending upon whether the plants are to be placed in ground beds or raised planters. The anticipated frequency of plant replacement will partially determine the method of installation. For example, large trees are often permanently planted in the floor of an enclosed shopping mall because they are too large and expensive to replace, Figure 8-6. The growing medium around them must permit good drainage and retain nutrients, while drainage tiles beneath the plant carry away excess water. Plants in such a setting must not be subjected

FIGURE 8-5.
Excess roots are pruned off the foliage plant before transplanting to establish a more balanced relationship between the foliage and the root mass.

FIGURE 8-6.
The grate around this permanently installed fig tree allows watering and fertilizing while protecting the root system from being trampled or compacted.

to detergents and waxes used for floor maintenance, so special design provisions, such as raised edging, may be necessary.

Plants that are in raised planters or that require frequent replacement are usually not installed permanently. Instead, they are often planted in a growing medium within a nursery container having good drainage, Figure 8-7. The containerized plant is then placed into a support medium of peat, perlite, sand, or similar well-drained material. The support medium serves to retain the plants as well as permitting excess water to drain away from the root zone and insulating the root system against abrupt temperature or moisture changes. A separating sheet (usually of fiber glass or a rot-resistant fine mesh material) is placed between the support medium and the coarse gravel lining the bottom of the planter. The gravel is needed to facilitate drainage, and without the separator the growing medium and support medium would gradually wash into the gravel and

plug it. A plastic tube inserted into the planter permits dip-stick testing to determine if the planter is being overwatered. It also provides a means of pumping out excess water if the planting is endangered by overwatering. Mulch on the surface of the planter serves to discourage moisture loss, provides a decorative appearance, and conceals the rims of the plant containers.

The same techniques used for the planting of containerized plants can be applied to in-ground plantings. If the planter drains directly into the building's drainage system, then the vertical plastic tube is unnecessary. All else remains the same except that there is no outer container.

WATERING AND DRAINAGE

As noted above, the ability of the growing medium to drain off excess water is critical. More interior plant deaths result from overwatering than for any other reason. All planters must permit the removal of standing water. The layer of coarse gravel already referred to is one method. Other methods include setting containerized plants on top of inverted pots within the larger planter and incorporating drains and spigots into planter bottoms. Where drainage from the planter base is planned, additional planning must assure that carpets, tiles, and other floor surfacings are not damaged by the runoff water. Siting the plants on gravel beds into which the water can drain and then evaporate is one method.

Hand-in-hand with planning for drainage goes provision for proper watering. Some interior plantings require continuous moisture; others do better if permitted to dry out between regular waterings. Obviously the two types of plants would not coexist compatibly in the same planter.

An interior plantscape must be watered according to a schedule, and not according to the judgement of a custodian or other unqualified individual. The right watering frequency is most easily determined in controlled environments such as enclosed shopping centers. It is most difficult in locations where environmental variables are not stable, such as by open windows or in drafts near doors.

FIGURE 8-7.

Installation of a containerized plant within an indoor planter

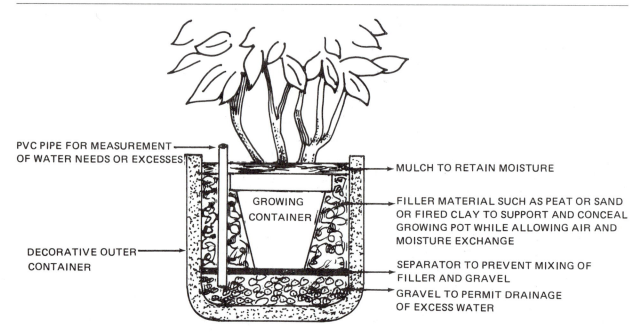

PVC PIPE FOR MEASUREMENT OF WATER NEEDS OR EXCESSES

GROWING CONTAINER

DECORATIVE OUTER CONTAINER

MULCH TO RETAIN MOISTURE

FILLER MATERIAL SUCH AS PEAT OR SAND OR FIRED CLAY TO SUPPORT AND CONCEAL GROWING POT WHILE ALLOWING AIR AND MOISTURE EXCHANGE

SEPARATOR TO PREVENT MIXING OF FILLER AND GRAVEL

GRAVEL TO PERMIT DRAINAGE OF EXCESS WATER

The need to water a planting can be determined by feeling the soil and observing its color. A gray surface color and failure of soil particles to adhere to the fingers indicate dryness. Moisture meters are also available for a more carefully controlled reading of the growing medium's water content.

Although some automated watering systems exist, there is a surprisingly limited use of them in large installations. Nationwide, there are interior plantscapes utilizing hundreds of plants that are all watered by hand. Most definitely, the technology of watering is still in the developmental stages.

When water is applied, it must be in a quantity adequate to wet the soil deeply, not shallowly. Shallow watering encourages shallow rooting and increases the vulnerability of the plants to damage from drying. Deep watering promotes deep and healthy rooting while providing the soil leaching necessary to prevent soluble salt buildup.

The quality of water used on the interior planting may vary with the location. The most likely source

will be municipal water lines. Most public drinking water contains chlorine and often fluoride as germicides and tooth-decay deterrents respectively. Neither additive will harm plants under normal conditions. Although chlorine is potentially harmful, the amounts used in drinking water are dissipated by aeration as the water bubbles from the faucet or hose nozzle. More heavily chlorinated swimming pool or fountain water can damage plants and should never be used as a watering source. *NOTE:* Plants grown around enclosed pools require good air exchange in the room or the chlorine gas from the pool may damage them.

Water is a source of soluble salts. Water that has been softened by means of cation-exchange softeners may be dangerously high in sodium, which can be toxic to plants. In buildings with such water softeners, alternate sources of water should be sought for the plantings. In regions of the country where the need for water conservation causes recycled water to be used on interior plantings, a chemical analysis of the water should be made to determine if any toxic chemicals are present that could damage the planting.

OTHER CONCERNS

As with any planting, certain routine procedures are needed to keep the plants healthy and attractive. The problems that are common to plants grown outdoors are similar to those that trouble plants grown indoors. In addition, indoor plantings often encounter stresses not common in outdoor landscapes.

Fertilization

Fertilization is needed to provide the mineral elements required for photosynthesis. Interior plantings need a complete fertilizer, but not as often as outdoor plantings, because the rate of plant growth is greatly reduced indoors. The ratio of the complete fertilizer should be fairly uniform; for example, 1-1-1 or 2-1-2. Too much nitrogen may lead to spindly, succulent, and unattractive vegetative growth. Trace elements will also need to be applied, especially if the growing medium contains no real soil. Frequent soil testing is necessary, regardless of whether the growing medium is true soil or synthetic. Excessive fertilizing follows only slightly behind overwatering as a major reason why interior plants fail to survive satisfactorily.

Either organic or inorganic fertilizers will work well, but the organics are generally slower in releasing their minerals for use by the plants. This means that soluble salt buildup is less frequently a problem. However, the odor of decomposing organic fertilizers may not be welcomed in such settings as shopping malls or library lobbies. Therefore, the use of organic fertilizers may not be practical in all situations.

Inorganic fertilizers are applied to large interior plantscapes in liquid form. This is faster and easier than applying dry granulars to each pot. When there are hundreds of pots and planters to maintain, speed is vital.

Humidity

Humidity is seldom a problem for plants growing outdoors, but it can cause problems for interior plantscapes. Because of the drying effects of central heating and air conditioning, interior plantings must adapt to an air environment that may contain half or less of the relative humidity outside. Interior humidities of 40 percent or less are common.

Preparation for the dry air must begin during the acclimatization process. Gradual drying of the plant's atmosphere will usually allow the plant to survive after transplanting indoors. In some cases, attempts to increase the humidity around plantings by misting the foliage during the day have proven to be of little or no value. Misting may also cause damage to carpeting or furnishings. Proper acclimatization is the best solution at present.

Air Pollution

Air pollution cannot be escaped by bringing the landscape indoors. Pollutants in the exhaust of cars and trucks, the smoke from cigarettes, the chlorine gas escaping from swimming pools, and the chemical soup which passes for air in our major metropolitan areas are all harmful to plants when sufficiently concentrated. Some plant species are more susceptible than others.

Good ventilation is an important element in the health of interior plantings. Proper ventilation will carry away chlorine vapors, fumes from smokers and chemical cleaning agents, or the ethylene which may be present if a building is heated by some form of hydrocarbon combustion (fossil fuel).

Ethylene can also damage plants as they are shipped to the site. This toxin is present in vehicle exhaust and if permitted to seep into the cabin where plants are stored may result in injury. If packed tightly or in restrictive packaging, the plants also may injure themselves, since their own tissue produces ethylene. Because of the potential harm from ethylene, plants should be unpacked and ventilated immediately upon arrival. At the time the plants are purchased, the grower or shipper should guarantee that the transport vehicles will be ventilated and sealed against exhaust fumes.

Dust

Dust is an air pollutant different from the others in that it is a particulate, not a vapor. When the leaves of a plant are coated with dust, they are not only unattractive, but gas exchange may be reduced due to plugged stomata. Air filtration reduces the amount of dust. Regular cleaning of the plants can also prevent dust buildup. Most plants in a residential interior can be rinsed off under the shower or set outside during a rainfall to wash away the dust. Commercial plantings can be kept clean with regular feather dusting and periodic washing. Cleaning should be a regular task within a total maintenance program for the interior plantscape.

Pruning

Pruning will not be extensive in an interior planting because of the plants' reduced rate of growth. Most pruning will be done to keep the plants shaped for an attractive appearance. Broken or damaged branches will also require removal. If the plants are not intended to grow, in order to avoid crowding, the roots as well as the foliage must be pruned back. Excessive root growth in containers can result in strangulation of the root system.

For pruning the plantings, hand pruners will be suitable for most of the herbaceous material and much of the woody material. Lopping shears will be helpful with larger materials, and pruning saws may be needed for indoor trees. Methods of pruning are described later in the text.

Interior plants should be pruned to a shape that suggests the appearance of a full canopy. Due to the restricted lighting in an interior planting, a full canopy seldom forms, however, and the plant's full branching structure may be visible. The pruning must enhance and take advantage of this sparse foliage covering.

Repotting

Repotting of plants is necessary in plantings where growth is allowed. Containers need to be removed from the planter and from the plant, the next larger size selected, and the excess space filled with growing medium. If roots are matted, they should be loosened before repotting. If they have started to grow around themselves, the large and excess roots should be pruned away. The pot should not be filled to the rim with the growing medium. An inch of unfilled space below the pot rim will give water room to flow into the container.

Insects and Diseases

Insects and diseases are not as common to interior plantings as to exterior ones, but they do occur. Insect problems are more common than diseases.

The initial pest presence may be introduced by the plants themselves as they arrive from the grower. Insects or pathogenic inoculum may be present in the foliage, roots, soil, or containers. All should be checked carefully upon arrival, at an area away from the installation site. The same careful check should be made each time replacement plants arrive. Obviously a reputable grower is a first defense against pests.

The most common pest problems of interior foliage plants are:

- aphids
- mealybugs
- spider mites
- white flies
- scale
- thrips
- nematodes
- root mealybugs
- root rots
- leaf spots
- anthracnose
- mildews
- blights

These pests and others are not uncontrollable when modern pesticides and methods of application can be used against them. However, the interior location and the presence of people make the use of sprays, dusts, and fumigants difficult. Whenever there is the possibility of people making contact with the pesticide, it is dangerous to use it. Even furnishings and carpeting can be damaged by many of the corrosive or oil-based chemicals. Where practical, plants can be wrapped in loose plastic bags and sprayed within the bag. This helps reduce the drift and the danger to people and furnishings.

Control measures are restricted to:

- chemical pesticides approved for application indoors
- removal of infected or infested plant parts
- washing away of insects and inoculum from plant foliage
- replacement of plants with healthy new ones

Vandalism and Abuse

Vandalism and abuse to interior plantings may necessitate replacement earlier than expected. Certain locations will bring certain predictable types of abuse. Plantings in cocktail lounges or nightclubs may have alcoholic drinks poured into their soil. Those in college snack bars and dorms may have cigarette holes burned through their leaves. Plantings in shopping malls may be mutilated by home propagators who take cuttings faster than the plants can replace them. Sometimes entire plants are stolen from the planters. The planters themselves are used as litter bins for paper cups, cigarettes, chewing gum, and assorted other debris.

The only real defense against such damage and abuse is public education and cooperation. By keeping the plantings attractive and well maintained, they are less likely to be deliberately damaged. Replacing or repairing damaged plants as soon as they are noticed displays the concern of the owners for an attractive planting that all can enjoy.

Grouping Compatible Species

Grouping compatible species simplifies the maintenance of the interior plantscape. Within any one planter the species selected should have the same requirements for light, moisture, fertilization, and

soil mix. While a large plantscape may effectively combine tropical species, desert species, and sometimes even temperate species, they should be in separate planters. When plant replacement is necessary, the new plants should be compatible with the existing ones.

THE INTERDISCIPLINARY TEAM

The popularity of interior plantings has grown faster than the ability of any one profession to stay abreast of it all. A successful interior planting in a commercial building requires the expertise of interior plantscapers, plant growers, interior decorators, landscape architects, architects, maintenance professionals, and building management. Each professional brings his or her point of view to the project. The architect sees the plants as architectural features, not as living organisms. Building management personnel want the plants to attract customers and please employees but they too do not have a horticultural sense of plant needs. Maintenance professionals regard the plants as something to be dusted, watered, and fertilized. They are concerned about the proximity of water, the difficulty of changing lamps, and the ease of reaching the plants. Each has concerns and contributions to the planning process that the others need to know.

The success of the plantscape is measured by its appearance and health. The interior plantscaper should be consulted while the building is still in the planning stage, as errors in lighting quality and intensity can be difficult to correct later. Drainage of planters directly into the building's drainage pipes requires that plantings be sited permanently and near the pipes. Watering of the plantings necessitates a nearby supply of water to which a hose can be attached. A shopping mall or office lobby should have appropriate water outlets every fifty feet along the wall and preferably within each planter. Otherwise hoses will be stretched across walkways, endangering pedestrians and creating puddles. Maintenance of the plantings should be the responsibility of a contracted professional plant maintenance firm, which should work closely with the architect to assure sufficient water outlets, storage space, accessibility for equipment, and so on. Failure to involve the maintenance firm in the early stages of planning can result in an unattractive plantscape shortly after installation. The managers of the building must be made to understand why specific lamps are needed for plant survival even if less expensive ones are available.

In short, fewer problems will develop for the interior planting if all professionals whose work impacts upon it work together from the beginning. At present, such interdisciplinary cooperation is the exception rather than the rule. The failure to use the team approach is often a case of architects not realizing their own limitations with horticultural materials.

THE FUTURE

No career field in ornamental horticulture holds greater promise than interior plantscaping. The current technology and knowledge of growing plants indoors is comparable to that when Henry Ford and the Wright Brothers began working in their professions. There is a need for fresh approaches to move us beyond *Ficus* trees and hanging baskets. More research is needed to introduce new species and varieties suitable for interior use. Also needed is better data on how to acclimatize plants with less foliage drop and transplant shock. Further study of soil mixes, fertilizer needs, and lighting requirements is needed to replace the guesswork of today. Finally, more professional plant maintenance firms are needed. They need dependable studies of such matters as the time required to dust and clean various leaf sizes and textures. Buildings need to be designed as carefully for their planted occupants as for their human ones. Automated watering and fertilization systems are needed to ease the current labor-intensive methods. There is ample work for young people who wish to train themselves for it.

SUMMARY

The use of plants indoors is not a new idea. Nevertheless, so explosive has been the recent demand for interior plantings in public buildings that the technology necessary to assure their survival has not kept pace.

In general, tropical foliage plants have proven better suited than temperate zone plants for indoor use because they do not require a period of cool temperature dormancy. Nevertheless, when plants are used indoors, a number of problems can occur. The procedures necessary for the successful transplant of tropical foliage plants to an interior locale include acclimatization to reduced light intensity, reduced nutrients, greater moisture stress, and lower temperatures.

Although light acclimatization allows the plant to survive at a reduced light intensity, it is still vital that the proper quality of light be provided. Thus the designer of an interior plantscape must understand the differences in light quality and the lamps that provide them.

The provision of a proper growing medium is equally important to the successful acclimatization of interior plants. The medium must provide structural support, water absorption, essential nutrients, proper pH, and good drainage. It is most often a mixture of natural soil and additives.

Plants may be set into an interior plantscape in ground beds or raised planters. The method of installation will be determined partly by the anticipated frequency of plant replacement. Whether planting in ground or in planters, allowance must be made for the removal of standing water from the planters. More interior plantings die from overwatering than for any other reason.

Watering the plantings, like other procedures, needs to be done as part of a carefully planned program of maintenance. Scheduled maintenance, rather than impromptu attention, is necessary to assure optimum satisfaction with the plantscape.

The most successful interior plantings require the expertise and cooperation of interior plantscapers, plant growers, interior decorators, landscape architects, architects, maintenance professionals, and building management. Each brings a distinct point of view to the project from which all others, and the plantscape, can benefit.

Achievement Review

A. TRUE/FALSE

Indicate if the following statements are true or false.

1. Interior use of plants is a new concept.
2. Architects and interior decorators are trained to write correct specifications for interior plantscapes.
3. Tropical plants have proven to be better suited for interior use than temperate plants.
4. Interior plantings require less fertilizer than outdoor plantings.
5. Soluble salts originate from fertilizers and water.
6. Clear glass transmits 100 percent of the sunlight that shines upon it.
7. Light intensity is measured in nanometers.
8. Light wavelength is measured in footcandles.
9. Most plants will survive after being acclimatized to a light intensity of 100 to 200 f.c.

B. MULTIPLE CHOICE

From the choices given, select the answer that best completes the following statements.

1. The most efficient natural light for interior plantscapes enters from _____.
   a. the sides of buildings
   b. skylights 20 feet overhead
   c. skylights 15 feet or less overhead
   d. incandescent bulbs
2. The most efficient skylight has _____.
   a. a shallow, wide well with vertical sides
   b. a shallow, wide well with 45° beveled sides
   c. a deep, narrow well with vertical sides
   d. a deep, narrow well with 45° beveled sides

3. In most single-story settings, the best lamp for lighting an indoor planting is _____.
   a. an incandescent
   b. a cool white fluorescent
   c. a warm white fluorescent
   d. a fluorescent plant growth lamp
4. Interior plantings illuminated with 100 to 200 f.c. of light require _____ hours of lighting each day.
   a. eight
   b. ten
   c. twelve
   d. fourteen
5. The most important reason for using a synthetic soil in an interior planting rather than natural soil is _____.
   a. drainage
   b. nutrient content
   c. structural support
   d. cost
6. An interior planting that reflects seasonal changes through use of such plants as daffodils, mums, and poinsettias, would be best if designed _____.
   a. as an in-ground planting
   b. as a containerized planting
7. Deep watering of interior plantings will _____.
   a. encourage deep rooting
   b. leach away soluble salts
   c. do both of these
   d. do none of these
8. Watering of interior plantscapes is best done _____.
   a. irregularly
   b. weekly
   c. when convenient
   d. according to an established schedule

9. The two major reasons why interior plants die are overwatering and _____.
   a. too much light
   b. too much fertilizer
   c. too little fertilizer
   d. incorrect pH
10. Chlorine gas, dust, and ethylene are all possible _____ affecting interior plantings.
   a. pesticides
   b. air pollutants
   c. pathogens
   d. toxins

C. SHORT ANSWER

Answer each of the following questions as briefly as possible.

1. Place X's where appropriate to compare the different lamp types.

Characteristic	Incandescent	Cool White Fluorescent	Fluorescent Plant Lamps	Mercury	Metal Halite	High Pressure Sodium
High in red light						
Low in red light						
High in blue light						
Low in blue light						
Good color rendition						
High initial cost						
Low initial cost						
Moderate initial cost						
Low operating cost						
Moderate operating cost						

2. Label the parts of this containerized plant.

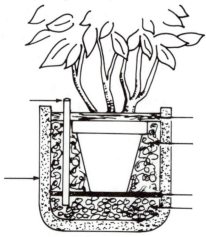

3. Indicate whether the following are characteristic of interior or exterior plantings.
   a. Rapid growth is encouraged.
   b. Pruning is minimal.
   c. Dusting of foliage is needed.
   d. High-analysis fertilizers are used regularly.
   e. Ventilation of the planting site is needed.
   f. Plants must be acclimatized to low humidity.
   g. Plants are most affected by insects and diseases.
   h. Plants suffer most from vandalism and abuse.

9
Landscape Design

Objectives

Upon completion of this chapter, you will be able to

- list three influences on contemporary landscape design.
- define the objectives of contemporary landscaping.
- describe the outdoor room concept.
- select plants to fill a role in the landscape.
- select enrichment items and construction materials for the landscape.
- recognize graphic tools and techniques used to illustrate a landscape plan.
- create a residential landscape plan that incorporates the principles of design.

THE NEED FOR GARDENS

Gardens are the pleasure grounds of civilization. They date back to the ancient Egyptians, Greeks, and Romans and perhaps earlier. Historically, gardens have been as much an expression of their owners' social status and political power as they have been places for recreation. Nobility throughout Europe, Asia, and Africa built gardens on a grand scale, often measured in square miles, as a way of exhibiting their domination over a hostile natural world. In other times and places, gardens were developed as tributes to religious figures: Monastic gardens honored Christian martyrs throughout Europe, while Far Eastern cultures built gardens to their deities as well.

Through much of history, gardens have belonged to the privileged, not the common person. Today, however, gardens grow in the middle-class suburbs, city centers, and on farms. They fill public plazas, shopping malls, and parks. And the American approach to landscaping is giving new direction to an old profession.

Three major influences have shaped American attitudes toward landscaping:

1. the formal tradition of sixteenth and seventeenth century Europe
2. the naturalism of eighteenth century England
3. the nature symbolism of the Orient

The sixteenth century Italian renaissance gardens and the seventeenth century French baroque gardens exemplify the formal tradition in landscape design, Figure 9-1. They are characterized by sym-

FIGURE 9-1.
A garden in the Italian Renaissance style

metrical layout patterns, severe shearing of plants (topiary pruning, described in Chapter 12), the use of ornate fountains and sculpture, and an overall architectural quality. The gardens left no doubt about the importance of the owners and their ability to dominate the natural world.

England, in the eighteenth century, found its royal court life less formal than that of continental Europe. Also the general attitude of the people toward the natural world was less fearful and more conciliatory. As a result, the formal garden fell from fashion and a more natural appearance became popular, Figure 9-2. *Natural* was defined by the landscape designers, however, and the style called naturalism required planning as meticulous as that of the most formal gardens. The designers of these English Naturalism gardens tore out the formal, symmetrical gardens with zeal and replaced them with lakes and islands, rolling hills, pastoral lawns, grazing animals, and added such picturesque touches as manufactured "ruins," dead trees, hermits' cottages, and vistas of distant temples. Grass was brought right up to the walls of the stately homes, a departure from the embroidery-like flower beds that surrounded them in formal gardens. Straight lines were practically nonexistent, in the belief that nature's lines are always curvilinear.

Naturalism not only reshaped the landscape of England but reached across the Atlantic to influence the public park movement in the United States. When Frederick Law Olmstead, the dean of American landscape architects, joined Calvert Vaux in

FIGURE 9-2.
Romanticized landscapes typified eighteenth century English gardens. Carefully composed natural scenes were intended to evoke pleasurable emotions.

the planning of New York City's Central Park, the design concepts of naturalism found new form. Central Park was the first important urban garden created for the pleasure of ordinary people. It was and it remains the flagship of the American landscape movement. It is a movement that now sets the pace for much that is done worldwide.

The other major influence on America's landscape style came from the Orient, principally from China and Japan. For centuries, Eastern and Western cultures were isolated from each other and developed distinctly different garden styles. Even today, full appreciation of a genuine oriental garden is beyond the comprehension of most Europeans and Americans, who are unfamiliar with the religious and spiritual beliefs of the orient. The style of these gardens is based upon miniaturization of the natural world. Within a small garden area, carefully screened from the world outside, mountains, rivers, islands, clouds, animals, and forests may all be found. Components of the natural world are symbolized with materials including sand and stones, special pruning techniques, and clever land forms, Figure 9-3. Animal forms may be shaped with stones; rivers and pools may be represented by raked sand; ripples may be fixed in time as concentric

FIGURE 9-3.
There is a strong Japanese influence in the use of stone and water for this garden.

Americans have been instrumental in restructuring the landscape profession in a fashion that sets trends worldwide. Our democratic system recognizes no privileged class of people, so gardens are available for everyone, in addition to the financial resources to support them. More inherent to America is the single-family dwelling with its front, back, and side yards that provide the space for garden development. The opportunity for ordinary citizens to own land and to develop it as they choose is unparalleled in most other nations outside of North America. It has moved landscape design from a qualitative level to a quantitative level. Yet, since no two properties have the same combination of physical characteristics and user needs, the mass reproduction of look-alike landscapes is not as common as might be feared.

By definition, modern *landscape design* is the arrangement of outdoor space in a way that serves the needs and desires of people without damage to natural ecological relationships. Dissecting the definition a bit shows that the profession is more service- than product-oriented. The landscape design must first satisfy the people who will use the land. Therefore, the designer must first carefully ascertain the needs of the user population. Yet in the larger sense, ownership of land is temporary, and our stewardship of the land mandates that we not make changes that will damage it irreparably and restrict its future use. A concern for the environmental impact of a proposed landscape change is a recent and important development in the evolution of the landscape design profession. It has added an ethical obligation to a profession whose past concerns have been largely aesthetic and technical.

rings in a pool of sand; mountains may be rolling mounds of soil; and clouds may be tufted shrubs.

Unlike Western gardens, usually planned for recreation and entertainment, Eastern gardens are planned for contemplation and meditation. Their influence on American gardens can be seen in the use of space and form, in pruned yet naturally shaped trees and shrubs, in stone as a sculptural form, and in certain furnishings.

LANDSCAPING TODAY

According to countless landscape designers and theorists, there is nothing truly new in landscape design. Our most modern landscapes are eclectic conglomerations of ideas from earlier times and distant places.

THE OUTDOOR ROOM

To serve the needs and desires of clients, the landscape designer must take an inventory of their characteristics and their attitudes toward the landscape site. Necessary information for a residential design might include the clients':

- composition of the family, that is, number, gender, and ages

- hobbies and special interests
- frequency and style of outdoor entertainment
- attitudes toward outdoor living
- attitudes toward landscape maintenance
- attitudes toward privacy
- attitudes toward their neighbors
- pets
- outdoor service needs (such as garbage cans, dog yard, clothes lines, storage for garden tools)
- special preferences for plant species and colors
- use of swimming pool, barbecue area, music system, night lighting
- special needs because of age (young children, elderly residents) or physical disabilities
- budget (for annual and total expenditures)

The site of the proposed landscape may or may not support all of the uses desired by the clients. It may also offer possibilities not recognized by the clients. Therefore, its initial assessment by the designer should be independent of client input. The *site analysis* may include:

- dimensions of the lot
- topography (elevation variations)
- geology (rock formations, including the bedrock base and rock outcroppings)
- hydrography (surface and underground water and drainage patterns)
- existing vegetation
- existing buildings
- location of utility lines and pipes above and below ground
- soil analysis (structure, type, fertility, and pH)
- historic importance of the site
- directional orientation
- established circulation patterns
- proximity to roads, public transportation, or waterways
- views from the site (pleasant and unpleasant)
- views toward the site
- problem areas (such as wet spots, low spots, wells)
- prevailing wind direction and velocity

Most of this information can be gathered by the designer, although some may require the assistance of other agencies and professionals. For example, a complicated terrain may require that an engineering firm be hired to produce a detailed survey. Many designers use a checklist to summarize the characteristics of each site, Figure 9-4.

Once the needs of the client and the capabilities of the site are determined, the two must be coordinated by the designer. In doing so, it is helpful to apply knowledge and experience gained from interior design. Known as the *outdoor room concept*, the approach is based upon two notions:

1. People live differently in different rooms of their home and can be expected to do the same in different spaces of the landscape.
2. The components of an indoor room (walls, ceiling, and floor) are also present in the outdoors, thereby permitting outdoor spaces to be perceived as rooms.

Use Areas

The average home is divided into four areas, each having a different function.

1. *The public area* is where guests are received. It includes the front door and entrance hall. In larger homes, it may also include a reception room.
2. *The general living area* is where most of the daily family interaction occurs and where friends are entertained. It includes the living room, dining room, and family or game room.
3. *The service area* is where the utilitarian functions of the family are carried out. It includes the laundry room, sewing room, workshop, and kitchen.
4. *The private living area* is used for personal aspects of family life. It includes the bathrooms, bedrooms, and dressing rooms, Figure 9-5.

Similarly, the average American residential landscape has three or four different areas to accommodate different uses.

FIGURE 9-4.
Site analysis checklist (From J. Ingels, *Landscaping: Principles and Practices,* 2nd edition, © 1983 by Delmar Publishers Inc.)

Site Analysis for the Property of

Client's Name *Mr. & Mrs. John Doe*
Client's Address *1234 Main Street*
Tucson, Arizona
Taken By *JCd* Date Taken *April 12, 1982*

Site Characteristics	Physical Importance	Visual Importance	Pos. +	Neg. −	Neutral ?
NATURAL FACTORS					
Existing vegetation	2 SHADE TREES IN RA. 1 FL. TREE OFF S.W REAR CORNER	ALL IN GOOD HEALTH AND ATTRACTIVE	✓		
Stones, boulders, rock outcroppings	NONE	NONE			
Wind, breezes	WESTERLY / GUSTY			✓	
Surface water features	NONE	NONE			
Groundwater	TOO DEEP TO BE USABLE				✓
Soil conditions	CALICHE LAYER			✓	
Birds and small game	✓	✓	✓		
Large game	NONE				
Existing shade	FAIRLY GOOD		✓		
Turf plantings	NONE OF QUALITY				✓
Terrain features	LEVEL				✓
Direct sunlight	LOTS OF IT / ALL DAY				✓
Off-site views		MOUNTAINS IN DISTANCE	✓		
Others					
Hardiness zone 9					
Soil pH 8.4					
Soil texture SANDY DOWN TO THE CALICHE					
MAN-MADE FACTORS					
Architectural style of building(s)		SPANISH			✓
Presence of outbuildings	NONE				✓
Existing patios	10x15 CONCRETE	NOTHING SPECIAL			✓
Existing walks, paths, steps, ramps	DRIVEWAY IN /CONCRETE WALKS IN		✓		
Swimming pool	YES	OVAL/ 30' LONG	✓		
Fountains, reflecting pools	NONE	NONE			
Statuary	NONE	NONE			
Fences, walls	NONE	NONE			
Existing lighting	NONE	NONE			
Off-site features		HOMES ON ALL SIDES			✓
Others FIRE HYDRANT	FRONT CORNER OF LOT			✓	
CULTURAL FACTORS					
Power lines (aboveground)	YES	UNATTRACTIVE		✓	
Power lines (belowground)	NO	NO			
Telephone lines	YES			✓	
Water lines	INSTALLED				✓
Historical features	NONE	NONE			
Archaeological features		ATTRACTIVE SPANISH STYLING	✓		
Nearby roadways	IN FRONT OF HOUSE	ELDERLY ON N SIDE			✓
Neighbors	GOOD RELATIONSHIP	FAMILIES + KIDS ON E+W	✓		
Off-site benefits		GOOD DISTANT VIEWS	✓		
Off-site nuisances	WIND IS TOO STRONG			✓	
Zoning regulations	YES/ PROTECTIVE		✓		
Nearby public transportation	NO				✓

FIGURE 9-5.

Examples of use area categories within the home (From J. Ingels, *Landscaping: Principles and Practices,* 2nd edition, © 1983 by Delmar Publishers Inc.)

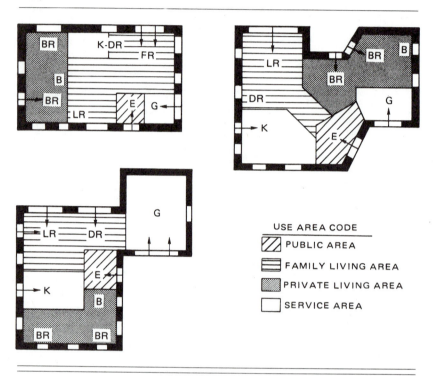

1. *The public area* is between the house and the street. It is the area everyone crosses to approach the house. The public area serves two major functions: to direct traffic to the entrance of the home and to place the house in an attractive setting when viewed from the street.
2. *The general living area* is usually at the rear of the house and often to the sides as well. It is the area where games are played, where the patio and pool may be found, where friends are entertained and barbecues held, and where the family members relax.
3. *The service area* contains the trash and garbage cans, utility sheds for storage of gar-den tools, compost piles, vegetable gardens, clothes lines, and similar items. It is always screened from view.
4. *The private living area* is not found in all landscapes. Where it does occur, it is the site for activities such as quiet conversation, sun bathing, and hot tubs. It is usually limited to the use of family members.

For the outdoor use areas to function most effectively, they must be properly oriented to the sun and must adjoin their interior counterparts directly. The public area is at the front of the property and should be just large enough to serve its purpose, Figure 9-6. The ideal general living area is at the rear of the house, has a southern exposure (to maxi-

FIGURE 9-6.
Assigning landscape use areas is easier on some lots than on others (From
J. Ingels, *Landscaping: Principles and Practices,* 2nd edition, © 1983
by Delmar Publishers Inc.)

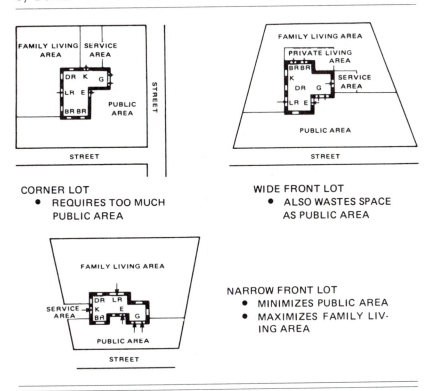

CORNER LOT
- REQUIRES TOO MUCH PUBLIC AREA

WIDE FRONT LOT
- ALSO WASTES SPACE AS PUBLIC AREA

NARROW FRONT LOT
- MINIMIZES PUBLIC AREA
- MAXIMIZES FAMILY LIVING AREA

mize its use), and adjoins the living room, family
room, and/or dining room of the house. The service
area should be near the kitchen or a utility room
so that trash, garbage, and laundry can be moved
conveniently. The area should also be close to the
point of municipal trash pickup so that heavy containers do not have to be carried far. The private
living area is usually a morning use area. It should
be oriented to the east or south to catch the morning
sunshine. Also, it must have a direct linkage with
the bedrooms of the house. Its purpose as a private
area will be lost if users have to pass through other
areas of the house and landscape to reach it.

The Walls, Ceilings, and Floors of a Landscape

Interior rooms are defined by their walls, ceilings,
and floors. In like manner, outdoor spaces can be
interpreted as having walls, ceilings and floors. The
functions and materials of these components of the
outdoor room are shown in Table 9-1.

Materials should be selected for the outdoor
room with an idea of the function they will serve
and the way they will fit in with the others being
used. Improper design decisions can often be
avoided by application of the outdoor room concept.

TABLE 9-1.

Components of an Outdoor Room

Outdoor Room Component	Function	Materials Commonly Used
Walls	• Define the limits and shape of the out-door room • Direct traffic through the landscape • Provide full or partial privacy • Provide security	• Shrubs • Low-branching trees • Fences • Stone or brick walls • Groundcover plantings • Flower plantings • Bodies of water • Exterior walls of buildings
Floors	• Define the base plane of the room • Absorb the impact of traffic	• Turf • Hard pavings • Soft pavings • Groundcovers • Flowers • Water
Ceilings	• Define the upper limits of the room • Provide full or partial shade • Provide privacy from overhead viewers	• Trees • Canopies and awnings • Vines on an overhead trellis • Building overhangs

For example, shrubs will not be planted in the middle of a lawn if the action is compared to placing a piece of wall in the center of the room instead of at the edge, where walls are meant to be.

The suitability of certain materials will become clearer as their functions are more precisely defined. For example:

- Is the wall material to block a view, frame a view, divert the wind, dilute the wind, separate areas but not block the view, or offer security?
- Is the ceiling to offer full or partial shade, block a view from above, create an intimate or a lofty effect, offer protection from rainfall, or merely to suggest the upper limits of the room?
- Is the floor to support vehicle traffic, pedestrians, pets, or no traffic? Will it be next to an exit or in areas of less traffic? Will it be subject to staining, heat, or ice and snow?

Other questions will need to be answered as the design progresses:

- What colors or textures are important to match or complement as the outdoor room materials are selected?
- How available are the desired materials and at what price?
- How immediate is the effect to be?
- How much maintenance is required and how much is the client prepared to do?

Begin to study outdoor rooms to determine the wall, ceiling, and floor components, Figures 9-7 to 9-9.

THE PRINCIPLES OF DESIGN

The creation of usable, imaginative landscapes depends upon how skillfully the designer applies the five basic principles that guide all creative planning.

FIGURE 9-7.
This public area blends patterned paving, flowers, and a containerized tree, all elements of an outdoor room.

FIGURE 9-8.
The recessed court at this shopping mall has all of the elements of an outdoor room. The planter walls form the enclosure; the surfacing is durable to withstand use; and the building overhang defines the ceiling. (Courtesy United States Department of Agriculture)

FIGURE 9-9.
Wall, ceiling, and floor elements are easily distinguished in this patio development. Note also the comfortable furnishings and provision for grilling.

Simplicity

Simplicity in the landscape can be either visual or physical and is usually both. Simplicity usually dictates that separate elements be given less emphasis than massed elements, fussy bedlines be replaced with gentle curves or straight lines, and maintenance-free materials be used. By massing plant species into groups, attention is not diverted to each individual plant, Figure 9-10. In a similar fashion, the massing of colors is preferable to a salt-and-pepper mixture. Simplicity results when pools, patios, walks, and other constructions avoid complex shapes or forms that conflict with the overall shape of the outdoor room.

FIGURE 9-10.
Simplicity results from massing a limited selection of species in gently curvilinear beds.

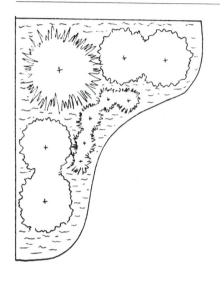
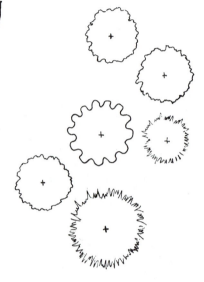

- SIMPLE, FLOWING BEDLINE
- MASSED PLANTS
- LIMITED NUMBER OF SPECIES

- FUSSY BEDLINE
- SEPARATED PLANTS
- TOO MANY SPECIES

- NO UNIFYING BEDLINE
- SEPARATED PLANTS
- TOO MANY SPECIES

Rhythm and Line

Rhythm and line bring a feeling of continuity to the landscape. As bedlines flow between different use areas, the viewer's eye should follow them comfortably, without meeting jarring changes. By repeating the basic layout lines, colors, and materials throughout the landscape, the principle of rhythm and line is applied.

Balance

Balance is another visual quality of the landscape's design. In theory, it puts the landscape on a seesaw and requires that each side have the same visual weight. There are three types of balance, Figure 9-11:

1. *symmetrical,* in which one side of the landscape is an exact duplicate (mirror image) of the other

2. *asymmetrical,* in which one side of the landscape has the same visual mass as the other but does not duplicate it
3. *proximal/distal,* in which the on-site landscape is developed to counterbalance the off-site landscape

Balance cannot be developed from only one or two points within the landscape. It must exist from every important vantage point off the site and within it. For example, when viewed from across the street, the public area of a residence should frame the building, with the house as the center of attention. As the viewer gets closer, the entry door must be balanced with foundation plantings that warrant equal attention on both sides. Then as the viewer stands at the door and looks back toward the street, the off-site view must balance against the on-site landscape.

FIGURE 9-11.
Three types of balance

SYMMETRICAL
BALANCE

ASYMMETRICAL
BALANCE

PROXIMAL/DISTAL
BALANCE

Focalization of Interest

In the application of this principle, a landscape designer either creates focal points for the landscape or enhances them where they already exist. Each major use area of a landscape should have one focal point. Only where the use areas are further subdivided (for example, a pool area within the larger family living area) should additional focal points be considered.

Some focal points are obvious. For example, in the public area the house entrance is the point to which the viewer's eye should be directed, Figure 9-12. Examples of focal points commonly used in other areas are specimen plants, statuary, ponds or pools, birdbaths, and sundials.

Proportion

The principle of proportion should guide the size relationships between elements of the landscape, Figures 9-13 and 9-14. The designer must envision a mature landscape with comfortable, non-threatening size relationships between:

- plants and buildings
- plants and other plants
- plants and people

FIGURE 9-12.
Focalization: Plants are arranged in an asymmetrical manner to move the viewer's eye toward the entrance, the focal point. (From J. Ingels, *Landscaping: Principles and Practices,* 2nd edition, © 1983 by Delmar Publishers Inc.)

FIGURE 9-13.

Proportion: Each element of the landscape must be in the proper size relationship with all other elements. (From J. Ingels, *Landscaping: Principles and Practices,* 2nd edition, © 1983 by Delmar Publishers Inc.)

Occasionally a landscape is larger than human scale and we are comfortable in the setting, as when we picnic in a large park beneath a canopy of towering forest trees. But, in general, the landscape should be designed to be in proportion with the people who will use it and the dwellings they build within it.

SELECTING PLANTS FOR LANDSCAPES

Many influences guide the choice of plants for use in landscapes. Too often, personal preference, sentimentality, or an unfamiliarity with plants prevail over more logical reasons for selection. The best reasons for selecting one species of plant over another are that:

- the plant fills the role assigned it in the design
- it will survive the growing conditions of the site
- it is affordable by the client

If several species meet these specifications, then personal preference and sentimentality can be applied unabashedly.

Plants are selected for certain roles based on architectural, engineering, and aesthetic considerations. For example, when plants are used to shape the outdoor room, frame a view, shade a patio, or soften a brick wall their function is *architectural.* When they solve a problem , such as directing traffic, reducing wind velocity, or absorbing dust and noise, they are *engineering* elements. If valued essentially for their appeal to the senses, through fragrance, sound, and color, or other visual attributes, they are *aesthetic* contributors.

When comparing plant species, landscape designers look at characteristics that include the following:

- hardiness (ability to survive the winter)
- mature size (height and width)
- flowering qualities (blossom color and fragrance)
- fruiting qualities (color, edibility or toxicity)
- foliage and bark color
- foliage and bark texture
- rooting system (effective or ineffective in erosion control, moisture tolerance, drought resistance)
- foliage silhouette, Figures 9-15 and 9-16
- deciduous or evergreen foliage
- presence or absence of thorns
- brittleness of the wood
- rate of growth and length of life
- ability to attract wildlife
- soil preferences (nutrients, composition, and pH)

FIGURE 9-14.
The height of the outdoor ceiling can create an intimate feeling for a patio or a cathedral effect in a park or forest.

- susceptibility to pests
- frequency of pruning required
- availability
- cost

The relationship between rate of plant growth and brittleness of the wood is worth noting. In general, rapidly growing trees have more brittle wood. That is why silver maples, cottonwoods, and willows drop branches all over the yard after a windstorm. They grow quickly and break up easily. It also explains the greater strength of slower growing trees like the oaks.

Different plants possess different combinations of the selection factors and plants can only be judged superior or inferior within the role requirements established by the landscape designer. An immediate need for shade might make fast-growing species preferable to slower-growing ones. Both might be planted, one for immediate effect and the other for long-term benefit. Location may determine the quality of a species: Certain weed trees of the Northeastern and Central United States are the premier lawn trees of the Southwest, where they surpass less adaptive native Eastern trees. The plant lists given in Tables 9-2 to 9-6 represent many such lists given in countless texts and journals worldwide. In the lists, plants are described using some of the selection factors mentioned earlier. Other specialized lists, (for example, plants for seashores, plants for containers, plants for street trees) are often available from the local Cooperative Extension Service.

FIGURE 9-15.

Typical tree silhouettes, characteristics, and landscaping uses. (From J. Ingels, *Landscaping: Principles and Practices*, 2nd edition, © 1983 by Delmar Publishers Inc.)

Silhouette and Examples	Characteristics	Possible Landscape Uses
wide-oval Flowering crabapple Silk tree Cockspur hawthorn Flowering dogwood	• spreads to be much wider than it is tall • often a small tree • horizontal branching pattern • branches low to the ground	• focal point plant • works well to frame and screen • can be grouped with spreading shrubs beneath
vase-shaped American elm	• high, wide-spreading branches • majestic appearance • usually gives excellent shade • an uncommon tree shape	• excellent street trees • allows human activities underneath • frames structures • used above large shrubs or small trees • *note:* the American elm is easily killed by Dutch elm disease; this limits its use
pyramidal Pines Fir Spruce Hemlock Filbert Sweetgum Pin oak Sprenger magnolia	• pyramidal evergreen trees are geometric in early years • pyramidal deciduous trees are less geometric • pyramidal shape is less noticeable as the trees mature	• accent plant • large, high-branching trees allow human activity beneath • save older trees for their irregular shapes • *note:* avoid planting large trees near small buildings

Silhouette and Examples	Characteristics	Possible Landscape Uses
round Shinyleaf magnolia Cornelian cherry dogwood American yellow wood Norway maple	• width and height are nearly equal at maturity • usually dense foliage • if the tree is large, a heavy shade is cast	• lawn trees • mass well to create grove effect • larger growing species may be used for street plantings • smaller growing species can be pruned and used as patio trees
 columnar  Columnar Norway maple Columnar Chinese juniper Fastigiata European birch	• somewhat rigid in appearance • much taller than wide • branching strongly vertical	• useful in formal settings • accent plant • group with less formal shrubs to soften its appearance • frames views and structures
 weeping 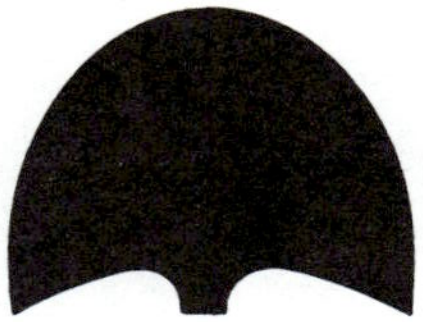 Weeping willow Weeping hemlock Weeping cherry Weeping beech	• very graceful appearance • branching to the ground • easily attracts the eye • grass or other plants cannot be grown beneath them	• focal point plant • screens • attractive lawn trees • *note:* avoid grouping with other plants

FIGURE 9-16.
Shrub silhouettes (From J. Ingels, *Landscaping: Principles and Practices*, 2nd edition, © 1983 by Delmar Publishers Inc.)

Shrub Silhouette and Examples	Characteristics	Recommended Landscape Uses	Shrub Silhouette and Examples	Characteristics	Recommended Landscape Uses
globular  Brown's yew Globe arborvitae Burford holly Globosa red cedar	• as wide as it is tall • geometric shape • attracts attention • does not mass very well	• accent plant • use several with a single pyramidal shrub for strong eye attraction • avoid overuse	**pyramidal** Upright yew Pyramidal junipers False cypress Arborvitae	• taller than it is wide • rigid and stiff • attracts attention • geometric shape • usually evergreen	• accent plant • focal point • use to mark entries and at incurves • group with less formal spreading shrubs
low and creeping  Andorra juniper Bar-Harbor juniper Cranberry cotoneaster Prostrate holly	• low growing • much wider than it is tall • masses well • irregular shape • loose, informal shape	• use to edge walks • cascades over walls • controls erosion on banks • grown in front of taller shrubs	**upright and loose**  Smoke bush Lilac Rose of Sharon Rhododendron	• taller than it is wide • loose, informal shape • usually requires pruning to prevent leggy growth	• closely spaced for privacy • use to soften building corners and lines • useful for screening and framing views
spreading Hetz junipers Pfitzer junipers Spreading yew Mugo pine	• wider than it is tall • medium to large shrub • masses well • usually dense foliage	• use at outcurve • place at corners of buildings • useful for screening, privacy, and traffic control	**columnar**  Hicks yew Italian cypress Arizona cypress	• width is about half the height • geometric, flat topped, and dense	• accent plant • foundation plantings • closely spaced for hedges • mass closely when a solid wall is desired
arching Forsythia Beautybush Vanhoutte spirea Large cotoneaster	• wider than it is tall • prevents the growth of other plants beneath itself • graceful silhouette • usually requires yearly thinning	• provides screening and dense enclosure • softens building corners and lines • background for flowers, statuary, fountains			

TABLE 9-2.

A Guide to Landscape Trees

Tree	Evergreen	Deciduous	Maximum Height 10'-25'	Maximum Height 25'-60'	Maximum Height 60' and up	Time of Flowering Early Spring	Time of Flowering Late Spring	Time of Flowering Early Fall	Fruiting Time Late Summer	Fruiting Time Early Fall	Fruiting Time Late Fall	Good Fall Color	Hardiness Zone Rating	Comment
Almond		X	X							X			8	edible fruit
Amur corktree		X	X							X			4	does not fruit in warmer zones
Apples														
Summer		X		X		X			X				4	
Fall		X		X		X				X			4	
Winter		X		X		X					X		4	
Apricot		X		X		X			X				5	edible fruit
Arborvitae	X			X									3	prunes well
Ash														heavy seed formation
Arizona		X		X								X	5	
Green		X		X			X			X			2	
White		X			X		X			X			3	
Flowering		X		X			X			X			5	
Beech														
American		X			X							X	3	low branching
European		X			X							X	4	does not do well in city air
Birch														
Canoe		X			X	X						X	2	attractive but often short lived because of certain insect damage
European		X		X		X						X	2	
Sweet		X			X	X						X	3	
Cherry		X		X		X			X				4	edible fruit attracts birds
Chinese chestnut		X		X			X			X			4	disease resistant edible fruit
Crabapple														
Flowering		X	X				X					X	3	edible fruit
Fruiting		X	X				X			X		X	4	attractive flowers
Crape Myrtle		X	X					X					7	difficult to transplant
Cypress														
Italian	X				X								7	pyramidal growth habit
Monterey	X				X								7	
Sawara false	X				X								3	
Dogwood														
Flowering		X	X			X						X	6	good patio tree
Douglas fir	X				X								4	dense foliage pyramidal shape
Elm		X			X								?	susceptible to several diseases
Fig	X			X		X			X				6	edible fruit
Fir														
Balsam	X				X								4	
White	X				X								4	
Fringe tree		X	X				X			X			4	may also be grown as a shrub.
Ginkgo		X			X							X	4	plant only male trees good for city conditions

TABLE 9-2.
A Guide to Landscape Trees (continued)

Tree	Evergreen	Deciduous	Max Height 10'-25'	Max Height 25'-60'	Max Height 60' and up	Flowering Early Spring	Flowering Late Spring	Flowering Early Fall	Fruiting Late Summer	Fruiting Early Fall	Fruiting Late Fall	Good Fall Color	Hardiness Zone Rating	Comment
Goldenchain		X	X										5	
Goldenrain tree		X	X				X				X		5	coarse textured
Hawthorn														avoid using near children's area; plants are thorny
Cockspur		X		X			X			X		X	4	
English		X	X				X			X			4	
Washington		X		X			X			X		X	4	
Hemlock	X				X								3	grows best in full sunlight
Holly														male and female plants are needed for fruit set
American	X			X									5	
English	X				X						X		6	pyramidal silhouette
Honeylocust		X			X					X			4	good city tree
Hornbeam														
American		X		X					X			X	2	
European		X		X					X			X	5	
Larch		X			X							X		a deciduous, needled conifer
Linden														good street trees
American		X			X								2	
Little leaf		X			X								3	
Silver		X			X								4	
Magnolia														large, showy flowers
Saucer		X	X			X				X			5	
Southern	X				X		X			X			7	
Star		X	X			X				X		X	5	
Maple														has spectacular autumn foliage color where autumn days are cool and crisp
Amur		X	X			X			X			X	2	
Hedge		X	X			X			X				5	
Japanese		X	X			X			X			X	5	
Norway		X			X	X			X				3	
Red		X			X	X						X	3	
Sugar		X			X	X			X			X	3	
Mountain ash (European)		X		X			X		X			X	3	susceptible to borer insects
Oak														strong trees used widely as lawn and shade trees good for attracting squirrels
Live	X			X									7	
Pin		X		X								X	4	
Red		X			X							X	3	
Scarlet		X			X							X	4	
Peach		X		X		X			X				5	edible fruit, but may not bear if winters are too warm
Pear		X		X		X			X				4	edible fruit
Pecan		X			X		X			X			5	may not bear mature fruit outside of southeastern states.

185

TABLE 9-2.
A Guide to Landscape Trees (continued)

Tree	Evergreen	Deciduous	Maximum Height 10'-25'	25'-60'	60' and up	Time of Flowering Early Spring	Late Spring	Early Fall	Fruiting Time Late Summer	Early Fall	Late Fall	Good Fall Color	Hardiness Zone Rating	Comment
Pine														
Austrian	X				X								4	good for use as windbreaks
Red	X				X								2	most effective when used
Scotch	X				X								2	in groups of three or more
White	X				X								3	
Plum														
Fruiting		X	X			X			X				4	
Purple		X	X			X			X				4	attractive, delicate flowers
Redbud		X		X		X							4	
Russian olive		X	X				X			X			2	silver foliage color
Sapodilla	X			X		X				X			10	
Silk tree		X		X				X					7	wide spreading delicate foliage
Spruce														
Black Hills	X				X								3	rigid, pyramidal trees
Blue	X				X								2	
Colorado	X				X								2	
Norway	X				X								2	
White	X				X								2	
Sweet gum		X			X					X		X	5	spectacular fall color mixed shades
Sycamore		X			X		X				X		5	white peeling bark
Tulip tree		X			X		X		X				4	needs room to grow and spread
Walnut														
Black		X			X					X			4	edible fruit roots of black walnut
English		X			X					X			6	kill most plants growing beneath the tree
Willow														
Babylon weeping		X		X		X						X	6	grows quickly
Pussy		X	X			X							4	weak wood
Thurlow weeping		X		X		X						X	4	
Zelkova (Japanese)		X			X							X	5	good substitute for American elm

TABLE 9-3.

A Guide to Landscape Shrubs

Shrub	Evergreen*	Deciduous	Mature Height 3'-5'	Mature Height 5'-8'	Mature Height 8' and up	Season of Bloom** Early Spring	Season of Bloom** Late Spring	Season of Bloom** Early Fall	Light Tolerance Sun	Light Tolerance Semi-shade	Light Tolerance Heavy Shade	Good Fall Color	Zone of Hardiness	Comment
Almond, Flowering		X	X			X			X				4	very showy blooms
Azaleas														requires an acidic soil condition and often iron chelate fertilizers
Gable		X	X				X		X	X			6	
Hiryu		X					X			X			7	
Indica	X			X		X				X			8	
Kurume	X		X				X			X			7	
Mollis		X	X				X		X	X			6	
Torch		X	X				X			X		X	6	
Barberry														good plants for traffic control; thorny
Japanese		X		X			X		X	X		X	4	
Mentor	semi			X			X		X	X			5	
Red leaved		X		X			X		X	X			4	
Bayberry	semi				X				X	X			2	fragrant leaves and fruit good for seashore areas
Boxwood														prunes well; good for formal hedges
Common	X				X				X	X			5	
Little leaf	X		X						X	X			5	
Camellia	X				X	X				X	X		7	blooms from late October to April
Coralberry		X	X					X		X		X	2	good on banks for erosion control
Cotoneaster														fall color comes from bright red fruit
Cranberry		X	X				X		X	X		X	4	
Rockspray	semi		X				X		X	X		X	4	
Spreading		X		X			X		X	X		X	5	
Deutzia, slender		X	X				X		X				4	white flowers
Dogwood														red twig is very good for erosion control. All dogwoods have good fall color
Cornelian cherry		X			X	X			X			X	4	
Grey		X			X	X			X			X	4	
Red twig		X			X	X			X			X	2	
Firethorn														fall color comes from brightly colored fruit
Scarlet	semi			X			X		X			X	6	
Formosa	X				X		X		X			X	8	
Forsythia														bright yellow flowers cascading branching patterns
Early		X		X		X			X				4	
Lynwood		X			X	X			X				5	
Showy border		X			X	X			X				5	

*Semi-evergreen indicates that the plants retain their leaves all year in warmer climates, but drop them during the winter in colder areas.

**Where no rating is given, flowers are either not produced or are not of importance.

TABLE 9-3.
A Guide to Landscape Shrubs (continued)

Shrub	Evergreen*	Deciduous	Mature Height 3'-5'	Mature Height 5'-8'	Mature Height 8' and up	Season of Bloom** Early Spring	Season of Bloom** Late Spring	Season of Bloom** Early Fall	Light Tolerance Sun	Light Tolerance Semi-shade	Light Tolerance Heavy Shade	Good Fall Color	Zone of Hardiness	Comment
Gardenia	X		X				X	X		X			8	very fragrant flowers
Hibiscus														
Chinese	X								X				9	
Shrub althea		X					X	X	X				5	
Holly														
Chinese	X				X				X				7	fruit color is most attractive in the fall
Inkberry	X				X				X				3	
Japanese	X				X				X				6	
Honeysuckle														
Blue leaf		X		X			X		X	X			5	
Morrow		X			X		X		X	X			4	
Tatarian		X					X		X	X			3	
Hydrangea														
Hills of Snow		X	X				X		X				4	coarse-textured shrubs
Oak leaf		X		X				X	X			X	5	
Pee gee		X			X			X	X			X	4	
Jasmine														
Common white	semi				X		X		X				7	
Italian	X				X		X		X				8	
Primrose	semi				X		X		X				8	
Juniper														
Andorra	X		X						X				2	grows well in hot, dry soil
Hetz	X		X	X					X				4	
Japanese garden	X			X					X				5	
Savin	X								X				4	
Pfitzer	X				X				X				4	
Lilac		X			X		X		X				3	large, fragrant flowers
Mahonia														
Leatherleaf	X				X	X				X			6	holly-like foliage
Oregon grape	X		X			X				X			5	bluish, grapelike fruit
Mock orange		X			X	X	X		X				4	creamy white fragrant flower
Nandina	X			X				X	X	X		X	7	very attractive in flower and fruit stage
Ninebark		X			X		X		X			X	2	
Pieris (Andromeda)														
Japanese	X					X			X	X			5	
Mountain	X			X		X			X	X			4	
Pine, Mugo	X			X					X	X			2	
Poinsettia	X				X			late fall	X				9	long-lasting blooms
Pomegranate		X			X		X		X	X		X	8	colorful in both spring and fall
Potentilla (Cinquefoil)		X	X				X	X	X				2	produces yellow flower all summer

*Semi-evergreen indicates that the plants retain their leaves all year in warmer climates, but drop them during the winter in colder areas.
**Where no rating is given, flowers are either not produced or are not of importance.

TABLE 9-3.
A Guide to Landscape Shrubs (continued)

Shrub	Evergreen*	Deciduous	Mature Height 3'-5'	Mature Height 5'-8'	Mature Height 8' and up	Season of Bloom** Early Spring	Season of Bloom** Late Spring	Season of Bloom** Early Fall	Light Tolerance Sun	Light Tolerance Semi-shade	Light Tolerance Heavy Shade	Good Fall Color	Zone of Hardiness	Comment
Privet														
Amur		X			X		X		X	X			3	prunes well
California	semi				X		X		X	X			5	popular hedge plants
Glossy	X				X			X	X				7	
Regal		X		X			X		X	X			3	
Quince, Flowering														
Common		X	X			X			X	X			4	densely branched, thorned plants
Japanese		X		X		X			X	X			4	good for traffic control
Rhododendron														
Carolina	X			X			X						5	
Catawba	X			X			X						4	
Rosebay	X				X		X						3	
Rose, Hybrid tea		X	X				X	X	X				varies	very diversified group of plants special culture required
Spirea														
Anthony Waterer		X	X				X		X	X			5	very attractive when
Bridal wreath		X		X			X		X	X		X	4	blooming
Billiard		X			X		X		X	X			4	most are resistant to
Frobel		X	X				X		X	X			5	disease and insect pests
Thunberg		X	X			X			X	X		X	4	
Vanhoutte		X		X			X		X	X		X	4	
Viburnum														
Arrowwood		X			X		X		X	X		X	2	attractive spring flowers
Black haw		X			X		X		X	X		X	3	good fall foliage color
Cranberrybush		X			X		X		X	X		X	2	many are good as wild-life food
Doublefile		X			X		X		X	X		X	4	
Fragrant		X			X		X		X	X		X	5	
Japanese snowball		X			X		X		X	X			5	
Leatherleaf	X				X		X		X	X			5	
Sandankwa	X			X			X		X	X			9	
Weigela		X			X		X		X				5	blooms late
Winged Euonymus		X		X					X			X	3	spectacular crimson fall color
Wintercreeper	X		X							X		X	5	excellent for foundation plantings and hedges
Yew														
Spreading Anglo-Japanese	X				X				X	X			4	
Upright Anglo-Japanese	X				X				X	X			4	prunes well
Spreading Japanese	X				X				X	X			4	long lived
Upright Japanese	X				X				X	X			6	will not tolerate poorly drained soil
English	X				X				X	X			6	
Canada	X		X						X	X			2	

*Semi-evergreen indicates that the plants retain their leaves all year in warmer climates, but drop them during the winter in colder areas.

**Where no rating is given, flowers are either not produced or are not of importance.

TABLE 9-4.

A Guide to Groundcovers

Groundcover	Evergreen	Deciduous	Height	Optimum Spacing	No. Needed to Plant 100 sq. ft.	Light Tolerance	Hardiness Zone Rating	Flower or Fruit Color and Time of Effectiveness
Ajuga or bugle		X	5"	6 inches	400	sun or shade	4	blue or white flowers in summer
Cotoneaster, creeping		X	12"	4 feet	10	sun	4	pink flowers, red fruit in summer and fall
Cotoneaster, rockspray	semi		18" plus	4 feet	10	sun	4	pink flowers, red fruit in summer and fall
Euonymus, big leaf wintercreeper	X		18" plus	3 feet	14	sun or shade	5	orange fruit in fall
Euonymus, purple leaf wintercreeper	X		18"	3 feet	14	sun or shade	5	not of significance
Honeysuckle, creeping		X	12"	3 feet	14	sun	5	pale yellow flowers in spring; red fruit in fall
Ivy, Baltic English	X		8"	18 inches	44	shade	4	none
Mondo	X		12"	10 inches	144	partial shade	8	white or pink flowers in spring
Myrtle or Periwinkle	X		8"	12 inches	92	shade	4	blue flowers in spring
Oyster Plant		X	12"	12 inches	92	sun or shade	9	not of significance
Pachysandra	X		12"	12 inches	92	shade	4	white flowers in spring
Sarcococca	X		tall – requires shearing	3 feet	14	sun or shade	7	white flowers and scarlet berries in fall
Wandering Jew	X		6"	12 inches	92	shade	9	red-purple flowers in spring and summer
Weeping lantana	X		18" plus	24 inches	25	sun	9	lavender flowers all year
Yellowroot		X	18" plus	18 inches	44	sun	5	brown-purple flowers in spring

190

TABLE 9-5.
A Guide to Vines

Vine	Broad-leaved Evergreen	Deciduous	Height	Clinging	Twining or Tendrils	Light Tolerance	Hardiness Zone Rating	Flower or Fruit Color and Time of Effectiveness
Actinidia, bower		X	30'		X	full sun or semishade	4	white flowers in spring
Actinidia, Chinese		X	30'		X	full sun or semishade	7	insignificant
Akebia, fiveleaf	semi		35'		X	full sun or semishade	4	purple flowers in spring
Ampelopsis, porcelain		X	20'		X	semishade	4	multicolored fruit in fall
Bignonia (or crossvine)	X		60'		X	full sun or semishade	6	orange-red flowers in spring
Bittersweet, American		X	20'		X	sun or semishade	2	yellow and red fruit in fall and winter
Boston ivy		X	60'	X		sun or shade	4	insignificant
Bougainvillea	X		20'	X		full sun	7	multicolored in summer
Clematis		X	3' to 25'*		X	full sun or semishade	4 to 7*	many colors of flowers in late spring
Euonymus, evergreen bittersweet	X		25'	X		sun or shade	5	yellow and red fruit in fall and winter
Fig, creeping	X		40'	X		sun or shade	9	insignificant
Honeysuckle, trumpet		X	50'		X	full sun or semishade	3	orange flowers in summer; red fruit in fall
Hydrangea, climbing		X	75'	X		full sun or semishade	4	white flowers in summer
Ivy, English	X		70'	X		semishade	5	insignificant
Kudzu vine		X	60'		X	sun or shade	6	insignificant
Monks hood vine		X	20'		X	semishade	4	yellow-orange fruit in fall
Rambling roses		X	10' to 20'		support needed	sun	5	flowers of many colors in spring and summer
Trumpet vine		X	30'	X		sun	4	orange flowers in summer
Virginia creeper		X	50'	X		sun or shade	3	insignificant
Woodbine, Chinese		X	50'		X	shade	5	yellow flowers in summer; red fruit in fall

*Dependent upon the actual species selected

TABLE 9-6.
A Guide to Selected Southwestern Plants

Plant	Growth Habit	Mature Height							Season of Bloom					Special Use in the Landscape
		1' or less	2'-5'	6'-9'	10'-15'	15'-30'	30'-50'	Over 50'	Early Spring	Late Spring	Summer	Fall	Winter	
Ash														
Arizona	T						X		NS					shade tree
Modesto	T						X		NS					shade tree
Citrus trees	T				X				varies with the variety					excellent for containers
Coral tree	T					X			X					brilliant flowers
Crabapple, flowering	T					X				X				specimen plant
Cypress, Arizona	T						X		NS					screens and windbreaks
Elderberry, desert	T					X			NS					screens and windbreaks
Elephant tree	T					X			NS					
Elm														
Chinese	T						X		NS					shade tree
Siberian	T						X		NS					windbreak
Eucalyptus	T						X		varies with the variety					many species prized for flower and/or foliage
Hackberry, netleaf	T						X		NS					shade tree
Honeylocust														
Shademaster	T					X			NS					good in dry, desert conditions
Sunburst	T					X			NS					
Thornless	T						X		NS					
Ironwood, desert	T			X						X				specimen tree
Jujube, Chinese	T					X				X				very salt tolerant
Locust														
Black	T							X		X				frequent pruning makes these attractive flowering trees
Idaho	T						X			X				
Pink flowering	T						X			X				
Magnolia, southern	T							X			X			lawn tree
Mesquite														
Honey	T					X			NS					shade trees and windbreaks
Screwbean	T					X			NS					
Mulberry, white	T							X	NS					shade tree

192

TABLE 9-6.
A Guide to Selected Southwestern Plants (continued)

Plant	Growth Habit	Mature Height							Season of Bloom					Special Use in the Landscape
		1' or less	2'–5'	6'–9'	10'–15'	15'–30'	30'–50'	Over 50'	Early Spring	Late Spring	Summer	Fall	Winter	
Myrtle	S		X								X			prunes and shapes well
Ocotillo	S				X					X				specimen plant
Oleander	S				X						X	X		does well in heat and poor soil
Photina	S					X			X					screens
Privet														all species can be pruned to lower heights
California	S				X				X					
Glossy	S					X			X					
Japanese	S				X					X				
Texas	S			X						X				
Rose, floribunda	S		X							X				massing effects
Silverberry	S				X				NS					good for containers
Sugar bush	S				X				X					
Bougainvillea	V			X							X			very colorful
Ivy														
Algerian	G	X							NS					
Boston	V					X			NS					
Jasmine, star	V					X					X			very fragrant
Lavender cotton	G	X									X			effective as edging
Periwinkle	G	X								X				
Trumpet creeper	V					X						X		
Virginia creeper	V					X			NS					
Wisteria	V				X					X				may be trained as shrubs and weeping trees

T Trees
S Shrubs
V Vines
G Ground covers
NS Flowers are not showy.

193

TABLE 9-6.
A Guide to Selected Southwestern Plants (continued)

Plant	Growth Habit	Mature Height							Season of Bloom					Special Use in the Landscape
		1' or less	2'-5'	6'-9'	10'-15'	15'-30'	30'-50'	Over 50'	Early Spring	Late Spring	Summer	Fall	Winter	
Olive, European	T					X			NS					good multi-stemmed tree
Pagoda tree, Japanese	T					X					X			lawn tree
Paloverde														specimen trees
Blue	T					X				X				
Little leaf	T					X				X				
Mexican	T					X				X				
Pine														
Aleppo	T						X		NS					grows well in poor soil
Digger	T						X		NS					specimen plant
Italian stone	T							X	NS					good in desert conditions
Japanese black	T					X								good in planters; prune well
Pinyon	T				X									multi-stemmed effects
Pistache, Chinese	T							X	NS					good patio tree / good fall color
Poplar														
Balm-of-Gilead	T						X		NS					narrow columnar form
Bolleana	T						X		NS					windbreaks
Cottonwood	T							X	NS					
Lombardy	T							X	NS					
White	T						X		NS					
Silk tree	T						X				X			showy shade tree
Smoke tree	T				X					X				
Sycamore														excellent street trees
American	T							X	NS					
Arizona	T							X	NS					
California	T							X	NS					
Tamarisk														
Athel tree	T						X				X			wind, drought, and salt resistant
Salt cedar	T					X					X			
Umbrella tree, Texas	T						X			X				shade tree
Willow														
Babylon	T						X		NS					
Globe Navajo	T							X	NS					
Wisconsin	T						X		NS					

194

TABLE 9-6.
A Guide to Selected Southwestern Plants (continued)

Plant	Growth Habit	Mature Height							Season of Bloom					Special Use in the Landscape
		1' or less	2'-5'	6'-9'	10'-15'	15'-30'	30'-50'	Over 50'	Early Spring	Late Spring	Summer	Fall	Winter	
Zelkova, sawleaf	T						X		NS					windbreak
Abelia, glossy	S			X							X			
Apache plume	S		X						X					
Arborvitae, Oriental	S				X				NS					
Barberry														barrier plantings
Darwin	S			X						X				
Japanese	S		X						NS					
Bird of paradise	S			X							X			
Brittlebush	S		X							X				
Butterfly bush	S			X							X			vigorous growth
Cherry laurel, Carolina	S					X			X					screens and hedges
Cotoneaster, silverleaf	S			X						X				wind screen
Crape myrtle	S					X					X			very colorful
Creosote bush	S			X							X			screens and hedges
Firethorn, Laland	S			X					X					espaliers well
Hibiscus														
Perennial	S			X							X			
Rose of Sharon	S				X						X			
Holly														Wilson and Yaupon clip and shade well
Burford	S			X					NS					
Wilson	S			X					NS					
Yaupon	S					X			NS					
Hopbush	S				X				NS					screens
Jojoba	S		X						NS					hedges
Juniper														
Armstrong	S		X						NS					
Hollywood	S				X				NS					
Pfitzer	S			X					NS					
Lysiloma	S				X					X				good for transition between garden and natural landscape

DESIGNING PLANTINGS FOR LANDSCAPES

While no two properties are exactly alike, some methods of grouping plants are common to most residential designs.

Corner Plantings

Defining the corners of the outdoor room are the *corner plantings.* Depending upon how much privacy is desired, the corner plantings may or may not be connected to the line plantings that make up the walls of the outdoor room.

A corner planting has two parts, the *incurve* and the *outcurves,* Figure 9-17. The incurve is the most desirable location for an attractive specimen plant because it is a natural focal point. The plants in the outcurves should be selected and placed to direct attention even more strongly to the incurve, Figure 9-18.

The incurve plant is usually the tallest plant in the bed. If the corner planting is not the major focal point of the outdoor room, an accent rather than a specimen plant can be selected for the incurve. An accent plant will attract the eye more than the outcurve plants, but not as much as the focal point.

Many variations are possible with a corner planting, Figure 9-19. Shorter plants can be placed in

FIGURE 9-17.

Parts of the corner planting bed: the incurve and the outcurves (From J. Ingels, *Landscaping: Principles and Practices,* 2nd edition, © 1983 by Delmar Publishers Inc.)

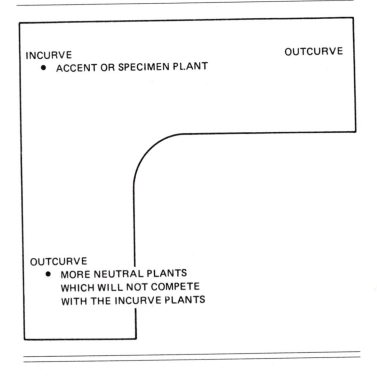

INCURVE
• ACCENT OR SPECIMEN PLANT

OUTCURVE

OUTCURVE
• MORE NEUTRAL PLANTS WHICH WILL NOT COMPETE WITH THE INCURVE PLANTS

FIGURE 9-18.
In a corner planting, attention is drawn from the outcurves to the incurve by stair-stepping plants. (From J. Ingels, *Landscaping: Principles and Practices,* 2nd edition, © 1983 by Delmar Publishers Inc.)

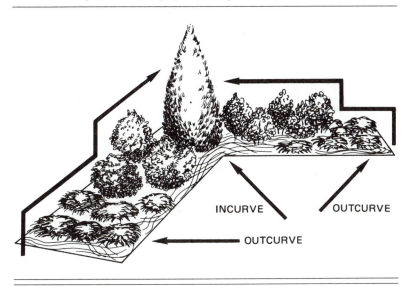

INCURVE   OUTCURVE

OUTCURVE

front of taller plants, and a statue or bench can be used instead of a plant at the incurve. Whatever is done, it is important to keep the design simple. Only three or four species should be used unless the bed is exceptionally large.

Line Plantings

Line planting is the basic method of forming outdoor walls with plants. Depending upon the species and their arrangement, a line planting can accomplish many purposes. It can provide full, partial, or no privacy. It can also modify the climate, block or frame a view, and serve as a backdrop for flowers, Figure 9-20.

Like a corner planting, a line planting should be in a cultivated bed to separate it from the lawn and provide for mulching and easier maintenance. The height and thickness of the planting will depend on the size and number of plants used. As plants are chosen, the designer must know their mature size and space them far enough apart so that they

can grow to maturity, otherwise they will have to be pruned frequently to prevent crowding.

Skillful design of a line planting requires practice and experience. If too few species are used, the planting is monotonous; if too many species are used, chaos results, Figure 9-21. Interest can be created even with a limited number of species by grouping them into masses and staggering their placement to enhance depth perception by the viewer, Figure 9-22. For example, placing lower shrubs in front of taller ones creates a stepped effect, Figure 9-23.

Foundation Plantings

Houses built during the last century usually had foundations of stone or concrete blocks. The foundation planting was developed in an attempt to hide this unsightly base. Modern houses frequently do not have exposed foundations. Nevertheless, the belief persists among many homeowners that a foundation planting in its 1940s form is still the measure

FIGURE 9-19.

Three variations of the corner planting. Note that in each example the eye is drawn from the ends of the planting to the center and from the front to the rear. (From J. Ingels, *Landscaping: Principles and Practices,* 2nd edition, © 1983 by Delmar Publishers Inc.)

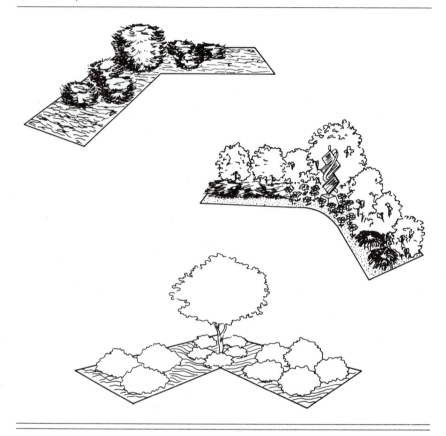

of a properly landscaped home, Figure 9-24. Characteristic of that style were rigid upright forms accenting the corners and entrance, with spreading forms in between. In contrast, the modern foundation planting, Figures 9-25 and 9-26:

- uses softer, less geometric plant forms
- combines deciduous and evergreen species for greater seasonal change and interest
- reaches outward and forward to tie the house more intimately with the rest of the garden

- accents the house rather than presenting a solid mass of plant materials
- focuses attention on the entrance by the use of flowers and other eye-attracting features

ENRICHMENT ITEMS

When the indoor room has been constructed, the final step is to add those things that make it usable and personal. Furniture, pictures, a stereo, and lamps are just a few of the things needed to complete

FIGURE 9-20.
Some functions of line plantings (From J. Ingels, *Landscaping: Principles and Practices,* 2nd edition, © 1983 by Delmar Publishers Inc.)

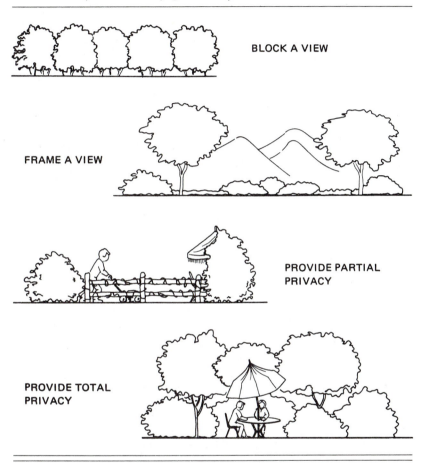

BLOCK A VIEW

FRAME A VIEW

PROVIDE PARTIAL PRIVACY

PROVIDE TOTAL PRIVACY

an indoor room. Additions are also needed for an outdoor room. They are termed *enrichment items.* These are elements of the landscape that do not function primarily as wall, ceiling, or floor, Figures 9-27 through 9-30.

Outdoor enrichment items may be classified as natural or man-made in origin and as tangible or intangible in character, Table 9-7.

There are two pitfalls in the use of landscape enrichment items. One is that the quality of the items selected may not match the quality of the overall landscape. The other is that the items may be used to excess. To avoid the overuse of enrichment items, a good test is to remove an item, then stand back and decide if it is missed. If the design does not suffer by its absence, it probably was not needed.

Determining the quality or worth of an enrichment item is more difficult because it is so subjective. In general, the types of outdoor furniture usually sold at local discount centers, such as aluminum frame lawn chairs with plastic webbing, are too

FIGURE 9-21.

Common mistakes in landscape design (From J. Ingels, *Landscaping: Principles and Practices,* 2nd edition, © 1983 by Delmar Publishers Inc.)

A monotonous view results when there is not enough variation in plant height or texture.

The simplicity of a line planting is destroyed by too many plant species and too much variation in height.

FIGURE 9-22.

An effective line planting consists of a few species, in massed groupings, with staggered placement. (From J. Ingels, *Landscaping: Principles and Practices,* 2nd edition, © 1983 by Delmar Publishers Inc.)

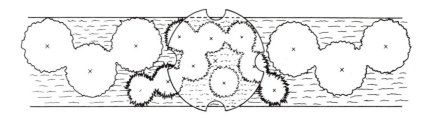

gaudy to fit tastefully into a well-designed outdoor room. As for gazing globes, pink flamingos, plaster ducks, and impish gnomes, they are better suited to carnivals than landscapes. Much of what is available in garden statuary is mass production of sculpture created for sixteenth and seventeenth century gardens. As such, it is anachronistic in most modern gardens. Whenever possible, invest a bit more money for enrichment items that are more natural or original. Even without much money, a good designer can incorporate many tangible and intangible natural enrichment items into the landscape. Thoughtful selection of plant species can attract interesting wildlife to the garden. Pleasant fragrances are also merely a matter of selecting the proper plants.

FIGURE 9-23.
The placement of low shrubs in front of taller ones adds a stepped effect to the line planting. (From J. Ingels, *Landscaping: Principles and Practices,* 2nd edition, © 1983 by Delmar Publishers Inc.)

FIGURE 9-24.
A common, unimaginative foundation planting. Upright shrubs at the corners of the house and spreaders beneath the windows create a rigid appearance. (From J. Ingels, *Landscaping: Principles and Practices,* 2nd edition, © 1983 by Delmar Publishers Inc.)

SELECTING CONSTRUCTION MATERIALS

Comparable in importance to the selection of plants is the choice of nonplant construction materials for the outdoor room. Essential for architectural and engineering functions, the construction materials available are nearly as diverse as the plant materials that a designer must choose from. A good designer must know the characteristics, potential uses, and availability of various woods and pavings, of materi-als that may be precast, quarried, or prefabricated. Landscape designers can also benefit from the personal experience of working with construction materials in order to understand the difficulties encountered when they are specified for particular uses. Several excellent construction specification manuals are available for reference. These are updated regularly to keep professional designers abreast of current materials, their dimensions, strength, potentials, and costs. The following comparison of some common construction materials will serve as a simplified introduction to a complex subject.

FIGURE 9-25.
The foundation planting reaches forward from the house, does not hide the costly stonework, and does not block the view from the bay window.

FIGURE 9-26.
The foundation plants reach outward from the corner of the house, forward from the entry, and stay below the window levels.

FIGURE 9-27.
Water, in natural or directed forms, ranks as the favorite enrichment item in any landscape. Its movement, sound, and sparkle are without equal.

FIGURE 9-28.
Lamps, fountains, and planters enrich this urban plaza. (Courtesy United States Department of Agriculture)

Enclosure Materials

Enclosure materials form walls for the outdoor room. As such, they function in many ways like the walls of indoor rooms. In addition, they often do things that indoor walls do not do. The particular enclosure material selected will depend upon whether the purpose is to offer:

- *strength*, as with a security fence or a retaining wall

- *privacy*, requiring a height of at least 6 feet and the properties of solidity and nontransparency

- *climate control*, which may necessitate the strength to withstand and divert strong winds or high snowdrifts

- *space articulation*, which may require height and screening or nothing more than a suggestion of the room's limits, permitting attractive off-site features to be seen

FIGURE 9-29.
Classic sculpture such as this requires a setting receptive to its formality.

FIGURE 9-30.
This patio is enriched by an umbrella table with matching chairs. Also note the telephone for additional convenience.

Much of the enclosure material available today is prefabricated. It may also be constructed on-site, using materials such as wood or stone, poured or precast concrete, brick or railroad ties. Figure 9-31 compares commonly used landscape enclosure materials and several representative styles.

Surfacing Materials

Surfacing materials form floors for the outdoor room. While top-quality turfgrass would be the first choice of most people as the most attractive and comfortable surfacing, it will not withstand concentrated foot traffic and tolerates very little vehicular traffic. Groundcover and flowers are also attractive surfacings, but their use is aesthetic; they will tolerate no traffic of any type. Thus the need for constructed surfacings is firmly established in the landscape industry.

Depending upon their degree of durability and their physical structure, constructed surfacings are categorized as either hard pavings or soft pavings. The terms have little to do with the comfort of walking on them.

TABLE 9-7.
Landscape Enrichment Items

Natural	
Tangible	**Intangible**

• Rocks, boulders, and natural outcroppings
• Specimen plants
• Waterfalls, streams, or natural pools
• Animals
	• Plant fragrance
	• Sounds:
	  • Wind in the pines
	  • Birds
	  • Water
	  • Crickets and other musical insects

Man-Made	
Tangible	**Intangible**

• Outdoor furniture
• Sculpture, murals, and other art
• Lighting systems
• Pools and fountains
• Music systems
	• Night lighting
	• Sounds:
	  • Wind-chimes
	  • Music

FIGURE 9-31.
A comparison of enclosure styles and materials

Style	Material	Security at 6-Foot Height	Privacy at 6-Foot Height	Noise Reduction at 6-Foot Height	Wind Deflection at 6-Foot Height	Grading Structure	Useful for Raised Beds	Comments
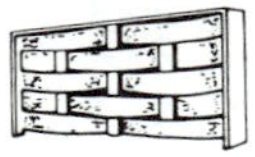 BASKETWEAVE FENCE	Wood	Yes	Yes	Moderate	Moderate	No	No	Available in prefabricated sections, attractive on both sides
 BRICK WALL	Brick	Yes	Yes	Good	Good	Yes	Yes	An ideal material for free-standing retaining or seat walls; width will vary with height and function
 CHAINLINK FENCE	Steel	Yes	No	None	None	No	No	Good for use around pet areas or to safeguard children's play areas
 CONCRETE BLOCK WALL	Concrete	Yes	Yes	Good	Good	Yes	Yes	Less expensive than poured concrete, stone, or brick
 GRAPE STAKE FENCING	Wood	Yes	Yes	Moderate	Moderate	No	No	A rustic style that weathers to an attractive gray; also comfortable in urban settings.
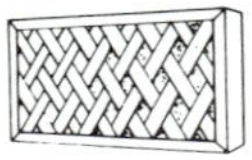 LATTICE FENCE	Wood	No	Variable	No	Limited	No	No	Effectiveness as a screen depends upon how closely the lattice is spaced.
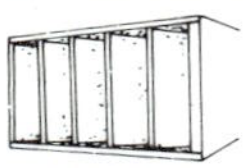 LOUVERED FENCE	Wood	Yes	Yes	Moderate	Moderate	No	No	Louvers may be vertical or horizontal and are angled to provide privacy.
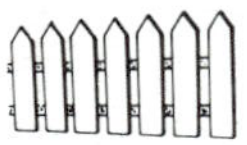 PICKET FENCE	Wood or Iron	Depends on the height	No	No	No	No	No	High maintenance costs because of frequent painting needed
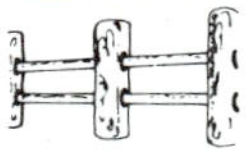 POST-AND-RAIL FENCE	Wood	No	No	No	No	No	No	Degree of formality varies with finish of lumber; a style valued for aesthetics more than for security or privacy
 POURED CONCRETE WALL	Concrete	Yes	Yes	Good	Good	Yes	Yes	Can be smooth or textured, colored or inset with materials to add interest to the surface; requires reinforcing if it is to provide strength.
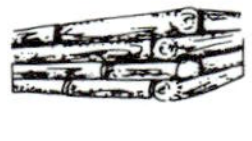 RAILROAD TIES	Wood	No	No	No	No	Yes	Yes	Ideal for rustic, natural enclosures; widely used for soil retention; care should be taken to assure that the ties have not been preserved with a phytotoxic.
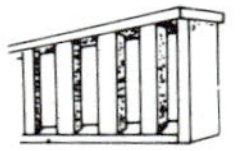 SLAT FENCE	Wood	Yes	Variable	Moderate	Moderate	No	No	Effectiveness as screen or wind deflector depends upon how closely slats are spaced.
 SOLID BOARD FENCE	Wood	Yes	Yes	Moderate	Good	No	No	Expensive but the best for security and privacy combined; maintenance easier than with other styles
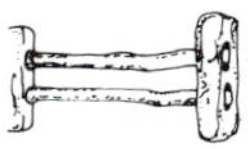 SPLIT RAIL FENCE	Wood	No	No	No	No	No	No	A rustic style best used in rural settings; lumber is rough and unfinished
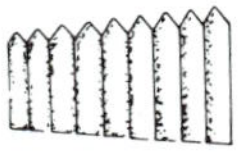 STOCKADE FENCE	Wood	Yes	Yes	Moderate	Good	No	No	A variation of solid board fencing
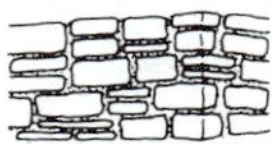 STONE WALL, ASHLAR	Stone	Yes	Yes	Good	Good	Yes	Yes	The stone is cut, usually at the quarry. Stones vary in their smoothness and finish. They are laid in a horizontal and continuous course with even joints.
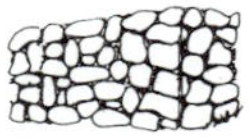 STONE WALL, RUBBLE	Stone	Yes	Yes	Good	Good	Yes	Yes	The stone is not cut. No course is maintained. Small stones are avoided. Larger stones are used at the base of the wall.
 WOOD RETAINING WALL	Wood	No	No	No	No	Yes	Yes	Wood must be preserved to avoid rapid decay; the preservative should not be phytotoxic; reinforcement necessary to assure strength
 WROUGHT IRON GRILLS	Iron	Yes	No	No	No	No	No	Expensive and used mostly for aesthetics; grills may be continuous or used as baffles separately.

Hard pavings are either poured or set into place as modular units and become solid when installation is completed. Examples include decking, poured concrete, precast concrete slabs, brick, cobblestone, flagstone, and an ever-expanding selection of precast interlocking units that permit the creation of highly decorative patterned pavings.

Soft pavings are loose aggregate materials, sometimes finely particulate like sand, sometimes coarsely particulate like crushed stone or brick chips. They may be hard and virtually indestructible or may shatter, pulverize, and even decay. Examples include sand, crushed stone, brick chips, marble chips, stone dust, wood chips, tan bark, sawdust, and asphalt. All are temporary, either because of deterioration or because they are easily carried away on shoes and tires. They may be dusty when the weather is dry, sticky in the case of asphalt, and tracked into the house if used near doorways. Soft pavings are seldom satisfactory as surfacings for areas that receive intensive use, such as patios or entrance walks. However, they are well suited to areas where grass is impractical yet the expense of hard surfacing is unwarranted, such as secondary walks, dog yards, service areas, and children's play areas.

Table 9-8 compares commonly used landscape surfacing materials.

MATERIALS AND METHODS

A landscape design begins as an idea in the mind of a landscape architect or designer. As the needs of the client are matched with the capabilities and limitations of the site, new needs may be generated as the client more fully understands the property's potential. Eventually the abstract ideas in the designer's mind must be given a form that will permit subsequent ideas to be related to the earlier ones. Still later, the more developed proposals of the designer must be presented in a form that the client can understand. Finally, assuming acceptance by the client, the designer's plan must be read by the landscape contractor who will implement the design.

The landscape designer communicates in the language of *graphics*. The graphic art techniques used by designers are directed toward helping themselves, their clients, and their landscape contractors understand how the design will appear after it is installed and mature. The landscape designer uses many of the same tools used by a drafter, as illustrated in Figure 9-32.

While these tools are used by all designers at all levels, many other tools are used in accordance with the preferences of the designer, the scope of the project, the purpose of the drawings, and the money the designer chooses to invest in graphic gadgetry. Various templates and lettering guides are available. Copy machines reproduce positive images of drawings in seconds. Color can be applied in numerous ways and special effects created by use of different papers.

Drawing Surfaces

A variety of drawing surfaces are used by landscape design professionals. Some have a short life. Others must be stronger because the drawings are meant to be permanent or must endure repeated handling and folding. Some of the surfaces are for original drawings, while others are for copies. Some are best suited to pencil; others are more suitable for ink, felt pen, colored pencils, or other media. A brief summary of the drawing surfaces and their uses follows.

Drawing Paper   A heavy, opaque surface, it is used for sketching and for one-of-a-kind drawings. It takes pen and pencil equally well. Markers are likely to bleed and run on it. The smoother finishes are less absorbent.

Vellum   A translucent paper surface, vellum is manufactured in several weights and qualities. Vellum is a surface on which drawings are originated or traced. The most lightweight vellum has a limited life. It is used for first-draft drawings and sketches. Heavier grades are used to trace designs that are to be presented to the client or contractors. Being translucent, vellum can be copied in a diazo duplicator.

TABLE 9-8.

A Comparison of Surfacing Materials

Material Description	Hard Paving	Soft Paving	Modular	Continuous and Solid
Asphalt: A petroleum product with adhesive and water-repellant qualities. It is applied in either heated or cold states and poured or spread into place.	Semihard; allows weeds to germinate and grow through it			X
Asphalt Pavers: Asphalt combined with loose aggregate and molded into square, rectangular, or hexagonal shapes. They are applied over a base of poured concrete, crushed stone, or a binder.	Semihard if not applied over concrete		X	
Brick: A material manufactured of either hard baked clay, cement, or adobe. While assorted sizes are made, the standard size of common brick is 2 1/4 x 3 3/4 x 8 inches.	X		X	
Brick Chips: A byproduct of brick manufacturing. The chips are graded and sold in standardized sizes as aggregate material.		X	X	
Carpeting, Indoor/Outdoor: Waterproof, synthetic fabrics applied over a concrete base. They are declining in popularity. Their major contribution is to provide visual unity between indoor and outdoor living rooms.				X
Clay Tile Pavers: Similar to clay brick in composition, but thinner and of varying dimensions (most commonly 3 x 3-inch, or 6 x 6-inch squares). They are installed over a poured concrete base and mortared into place.	X		X	

Slippery When Wet	Permeable to Water	Suitable for Vehicles	Suitable for Walks	Suitable for Patios
		X	X	Certain formulations are suitable. Others may become sticky in hot weather. *Note:* The application of a soil sterilant before applying the asphalt can eliminate the weed problem in walks, drives, and patios.
	If installed over crushed stone		X	X
	If installed in sand	X	X	X
	X		Edging needed to hold them in place	
	Provision must be made for surface water drainage or the carpeting becomes soggy.			X
X			X	X

TABLE 9-8.

A Comparison of Surfacing Materials (continued)

Material Description	Hard Paving	Soft Paving	Modular	Continuous and Solid
Concrete: A versatile surfacing that can be made glassy smooth or rough. It can also be patterned by insetting bricks, wood strips, or loose aggregates into it. Concrete is a mixture of sand or gravel, cement, and water. It pours into place, is held there by wood or steel forms, then hardens.	X			X
Crushed Stone: Various types of stone are included in this umbrella term: limestone, sandstone, granite, and marble. Crushed stone is an aggregate material of assorted sizes, shapes, and durability.		X	X	
Flagstone: An expensive form of stone rather than a kind of stone. Flagstone can be any stone with horizontal layering that permits it to be split into flat slabs. It may be used as irregular shapes or cut into rectangular shapes for a more formal look. It is usually set into sand or mortared into place over a concrete slab.	X		X	
Granite Pavers: Granite is one of the most durable stones available to the landscaper. The pavers are quarried cubes of stone, 3 1/2 to 4 1/2 inches square, that are mortared into place. Various colors are available.	X		X	
Limestone: A quarried stone of gray coloration. Limestone can be cut to any size. It adapts to formal settings.	X		X	

Slippery When Wet	Permeable to Water	Suitable for Vehicles	Suitable for Walks	Suitable for Patios
Only when smoothly fin- ished		X	X	X
	X	X	Edging needed to hold the mate- rial in place	Limited use except beneath picnic tables where stains might spoil hard paving
Depends upon the rock used and how smooth the surface is			X	X
		X	X	Too rough
		X	X	X

TABLE 9-8.

A Comparison of Surfacing Materials (continued)

Material Description	Hard Paving	Soft Paving	Modular	Continuous and Solid
Marble: An expensive quarried stone of varied and attractive colorations. It has a fine texture and a smooth surface that becomes slippery. Its use as surfacing is limited. It can be inset into more serviceable surfaces such as poured concrete.	X		X	
Marble Chips: A form of crushed stone, marble chips are more commonly used as a mulch than a surfacing. They are expensive compared to other loose aggregates; still they enjoy some use as pavings for secondary walks and areas that are seen more than walked upon.		X	X	
Patio Blocks: Precast concrete materials available in rectangular shapes of varied dimensions and colors. Limitless patterns can be created by combination of the sizes and colors. The blocks are set into sand or mortared over concrete.	X		X	
Sandstone: A quarried stone composed of compacted sand and a natural cement such as silica, iron oxide, or calcium. Colors vary from reddish brown to gray and buff white. The stone may be irregular or cut to rectangular forms.	X		X	
Slate: A finely textured stone having horizontal layering that makes it a popular choice for flagstones. Black is the most common color, but others are available.	X		X	

Slippery When Wet	Permeable to Water	Suitable for Vehicles	Suitable for Walks	Suitable for Patios
X				Best used in dry climate where slipperiness will not be a frequent concern
	X		X	
			X	X
		X	X	X
X			X	X

TABLE 9-8.
A Comparison of Surfacing Materials (continued)

Material Description	Hard Paving	Soft Paving	Modular	Continuous and Solid
Stone Dust: A by-product of stone quarrying. Stone dust is finely granulated stone, intermediate in size between coarse sand and pea gravel. It is spread, then packed down with a roller. The color is gray.		X		X
Tanbark: A by-product of leather tanning. The material is processed oak bark. It has a dark brown color and a spongy soft consistency. It is ideal for children's play areas.		X	X	
Wood Chips: A by-product of saw mills, wood chips are available from both softwoods and hardwoods. The latter decompose more slowly than the former. Wood chips have a spongy soft consistency. They are often used as mulches.		X	X	
Wood Decking: Usually cut from softwoods, the surfacing can be constructed at ground level or elevated. The deck is valuable as a means of creating level outdoor living spaces on uneven terrain. Space should be left between the boards to allow water to pass through and the wood to dry quickly. Use of a wood preservative will slow decay.	X		X	
Wood Rounds: Cross-sections of wood cut from the trunks of trees resistant to decay, such as redwood, cypress, and cedar. The rounds are installed in sand. Individual rounds are replaced as they decay.	X		X	

Slippery When Wet	Permeable to Water	Suitable for Vehicles	Suitable for Walks	Suitable for Patios
	X		X	
	X		Edging needed to hold the material in place	
	X		X	
	X		X	X
	X		X	X

FIGURE 9-32.
A guide to landscape design tools

INSTRUMENT		DESCRIPTION AND USES
DRAFTING BRUSH		A small brush with fine bristles to remove drafting powder and bits of eraser from the tracing surface.
ERASER		Several types of erasers are used in the profession. • Plastic: to remove heavy pencil lines • Kneaded: to clean away smudges • Ink: to remove ink from vellum or film surfaces
ERASURE SHIELD		A small piece of sheet metal with various shapes punched out. When an error is made, the shield is placed over the lines to be removed before erasing. Smudging is minimized.
DRAWING BOARD	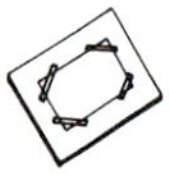	A working surface with an ungrained top (usually wood) and parallel sides. A good drawing table with room for frequently used tools is vital to good quality graphic work.
T-SQUARE		A two-piece instrument that derives its name from its shape. It is a straight edge for supporting the pencil or pen tip. When held tightly against the drawing board, it permits parallel lines to be drawn either horizontally or vertically.
TRIANGLES		Three-sided flat instruments used in several ways. Constructed with angle combinations of $45°-45°-90°$ and $30°-60°-90°$, they can be used to duplicate those angles. Also, they are used as straight edges and with the T-square to create parallel perpendicular lines.
PENS, PENCILS, AND LEAD HOLDERS		Drawing instruments used according to the designer's preference and the importance of the work. Pencils and lead holders use graphite leads. Technical pens use ink. Pencils and lead holders are less expensive than pens, but graphite can smudge. Pens are better for work that requires greater permanence.
DRAFTING TAPE		A paper tape that resembles masking tape but has a lower adhesive quality. It is used to secure paper, vellum, film, and other tracing surfaces to the drafting board. It releases easily and without damage to the drawing's surface.
DRAFTING POWDER		A soft, finely granulated material that is sprayed or dusted over the surface of a drawing to keep the T-square, triangles, and designer's hands and sleeves from smearing the graphite. It can also be used to remove a graphite-gray cast from a drawing.
SCALE		A tool for measuring and dimensioning. It has three sides with six edges. Two types exist: an *engineer's scale* and an *architect's scale*. They permit proportional reduction of actual dimensions to drawing size. This allows the designer to create plans indoors that can be installed outdoors with precision.
PROTRACTOR	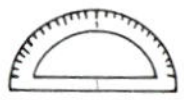	A device for measuring angles. The base of the protractor is aligned with the bottom of the angle. Where the angle's other line intersects the protractor's gauge, the angle is read in degrees. A protractor may also be used to construct an angle.
COMPASS	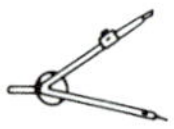	A tool for creating circles. One of the two legs has a metal point that locates the center of the circle. The other leg has a graphite lead or ink tip that draws the circle to the desired radius.
FRENCH CURVES		Pencil or pen supports for the creation of curvilinear lines. They may be used to generate new shapes and lines as well as to trace and define lines from a rough sketch.

Mylar Film   A translucent plastic surface, it is used like heavyweight vellum, to make final tracings for presentation and duplication. Mylar film is more expensive than vellum. It is frosted on one or both sides, which is necessary for ink or graphite to adhere to the surface.

Drafting Cloth   A translucent surface of plastic with a thin layer of linen bonded to it. The cloth side is the one drawn on. Cloth is the most expensive drafting surface. It is used where strength is important.

Reproductive Paper, Film, and Cloth   Sensitized materials that permit positive duplication of drawings done in pencil or ink on vellum, film, or cloth. Use of the paper requires a copying machine like the diazo duplicator. Special-effect papers are available as well.

Types of Landscape Drawings

While there are many types of mechanical drawings and ways to illustrate objects, the landscape designer most frequently utilizes the *plan view*. The plan view assumes a vantage point directly above the proposed landscape and looking down perpendicular to the ground, Figure 9-33. The plan view is a collection of *symbols* that represent the plants and construction materials to be used in the landscape, Figure 9-34. Symbols are drawn to a scale that permits the property to fit onto the drawing surface and labeled adequately to permit the client to visualize the project when installed. Plants are scaled to their mature size to avoid overcrowding.

The advantages of plan views are that they:

- are easily drawn to scale
- efficiently combine the concepts of the design with necessary mechanics, such as plant spacing, mulch depth, and concrete thickness
- can be easily adapted from presentation tracings into working construction plans

The major disadvantage of plan views is that some clients find it difficult to visualize the symbolic, two-dimensional drawing in three dimensions.

Two other types of illustration, the elevation and the perspective, are used as sales tools to help clients visualize the completed landscape. *Elevations* are two-dimensional views of the front, rear, or side of a landscape as seen from ground level, Figure 9-35. *Perspectives* simulate three-dimensional views, with the vantage point at ground level or slightly above, Figure 9-36.

CREATING A RESIDENTIAL PLAN

The development of a residential landscape plan merges the arrangement of plants and other materials with the graphic skills necessary to symbolize them. It requires that the principles of design be applied to an entire property, not just sections, and that the final design possess a unity which integrates all use areas into a total plan, Figure 9-37.

The development is best approached logically and in a sequential, orderly manner, Figure 9-38.

Step 1: Analyze the characteristics of the site.

Step 2: Determine the needs of the clients.

Step 3: Match client needs and site capabilities as closely as possible, allowing new needs to be suggested by the site's potential for development.

Step 4: Select a scale that allows the site to be reproduced on paper, including lot, buildings, and existing features.

Step 5: Assign use areas to appropriate regions of the property. Lay out the areas with wide angles to avoid the creation of narrow, tight, and impractically shaped spaces.

Step 6: Select focal points and locate them within the use areas.

Step 7: Shape each use area in a way that directs attention to the focal point and relates it to adjoining use areas. You may have to try numerous possibilities before you are satisfied. Lightweight vellum can be taped over the drawing to eliminate the need to erase as ideas are tried and rejected.

Step 8: Retaining the shapes from the step above, convert the lines to planting beds and other outdoor wall elements. It is important to determine the location of the walls before considering specific plants or other materials.

FIGURE 9-33.

A plan view (From J. Ingels, *Landscaping: Principles and Practices,* 2nd edition, © 1983 by Delmar Publishers Inc.)

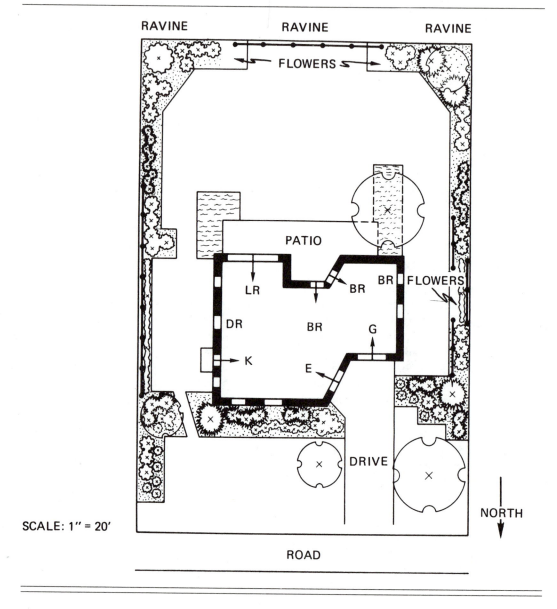

Step 9: Define the function to be served by each plant in the design. The function directs and restricts the choice of species.

Step 10: Select plant species that will fill the roles defined for them and symbolize them with their mature size and plant type.

FIGURE 9-34.
Typical plan-view symbols used by landscape designers

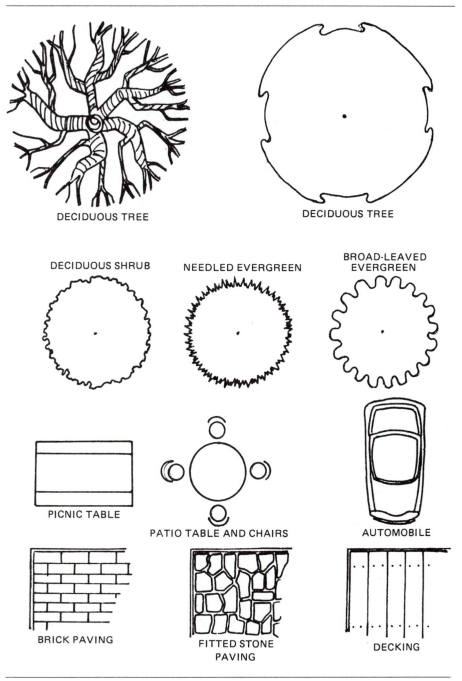

DECIDUOUS TREE

DECIDUOUS TREE

DECIDUOUS SHRUB

NEEDLED EVERGREEN

BROAD-LEAVED EVERGREEN

PICNIC TABLE

PATIO TABLE AND CHAIRS

AUTOMOBILE

BRICK PAVING

FITTED STONE PAVING

DECKING

FIGURE 9-35.
An elevation view permits the height and width to be drawn to scale.

FIGURE 9-36.
A perspective view permits easier visualization of how the proposed design will appear.

PERSPECTIVE DRAWING by MARTIN BOZAK 1977

Step 11: Select enclosure, surfacing, and enrichment items that complement the total design.

Step 12: Label all elements of the plan. Keep the lettering on or near the symbols.

Step 13: Compile a list of all plant species used and the total number needed of each.

Step 14: Trace the design, eliminating all unnecessary guidelines, on heavy vellum, film, or cloth, using a hard lead pencil or technical pen.

SUMMARY

The making of landscape gardens dates back to the ancient Egyptians and perhaps before. These early gardens were as much an expression of status, power, and dominance as they were places for recreation and personal enrichment. Modern landscape designers attempt to arrange outdoor space in a way that serves the needs and desires of the people who use it without damage to natural ecological relationships.

In developing a landscape, a designer must first inventory the characteristics of both the clients and the site. Client interviews and site analysis, with possible subcontracting of services from other professionals, are means of acquiring the necessary data.

FIGURE 9-37.
A professional landscape plan prepared as a plan view

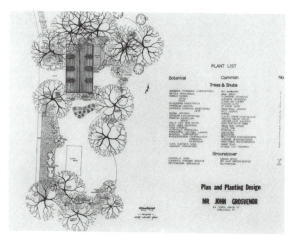

Recognizing that people use various areas of the landscape for different purposes, designers plan distinct regions (use areas) into each site. As many as four different use areas may be found in a residential landscape: public area, private living area, general living area, and service area. Each of these areas is then developed as an outdoor room, with wall, ceiling, and floor elements selected from the rich

FIGURE 9-38.
Steps in the layout of a residential landscape plan

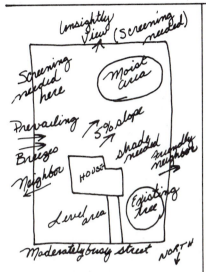

STEP 1. ANALYZE THE SITE'S CHARACTERISTICS.

STEPS 2 AND 3. DETERMINE THE CLIENT'S NEEDS AND MATCH THEM WITH THE SITE'S CAPABILITIES.

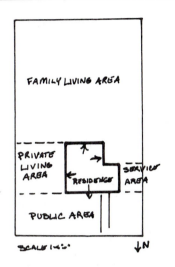

STEPS 4 AND 5. CHOOSE A SCALE THAT ALLOWS THE SITE TO FIT ONTO PAPER. ASSIGN APPROPRIATE USE AREAS.

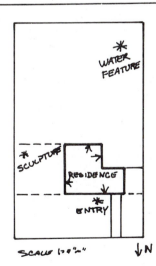

STEP 6. SELECT FOCAL POINTS AND THEIR LOCATIONS WITHIN THE USE AREAS.

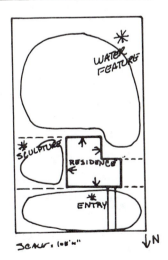

STEP 7. ROUGHLY SHAPE EACH USE AREA TO DIRECT ATTENTION TO THE FOCAL POINTS AND DEVELOP UNITY BETWEEN AREAS.

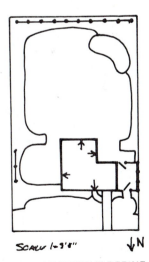

STEP 8. ACCURATELY DEFINE THE SHAPES AND DIMENSIONS OF THE OUTDOOR WALLS.

FIGURE 9-38.
Continued.

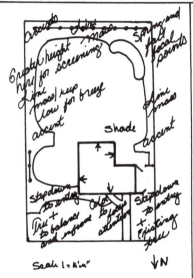

STEP 9. DEFINE THE FUNCTIONS TO BE SERVED BY THE PLANTS.

STEP 10. SELECT PLANTS THAT CAN FILL THE ROLES PRESCRIBED FOR THEM.

STEP 11. SELECT NONPLANT ITEMS THAT COMPLEMENT THE TOTAL DESIGN.

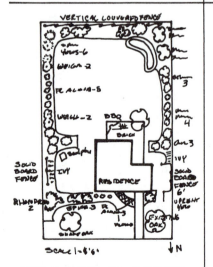

STEP 12. LABEL ALL ELEMENTS OF THE PLAN.

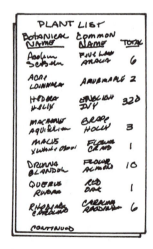

STEP 13. COMPILE A LIST OF ALL PLANT SPECIES USED. ALPHABETIZE BY GENERIC NAME. TOTAL THE NUMBER USED.

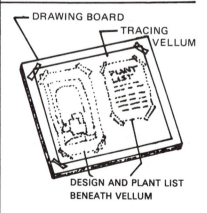

STEP 14. OVERLAY THE DESIGN AND PLANT LIST WITH VELLUM. TRACE ALL SYMBOLS AND LETTERING. DO NOT TRACE GUIDE LINES. LABEL WITH CLIENT'S NAME AND ADDRESS.

and varied array of natural and man-made materials available.

In creating usable and attractive outdoor living spaces, landscape designers apply the five basic principles of design that guide all creative effort: simplicity, balance, focalization of interest, proportion, and rhythm and line. Judging how effectively these principles are applied aids the objective evaluation of a landscape.

It is equally important that the selection of plants for the landscape be objective. Personal preference must be put aside in favor of choices that contribute needed architectural, engineering, or aesthetic qualities. The qualities of plants can only be judged superior or inferior within the role requirements established for a property by the landscape designer.

Enrichment items may also be selected to fill a need. Their contributions supplement the wall, ceiling, and floor elements in the outdoor room and make the landscape more usable and personal. The landscape's enclosure and surfacing materials are also selected by matching their characteristics with the role they must fill in the landscape design.

The landscape designer communicates through the language of graphics. The graphic art techniques used by designers are directed toward helping the designers, their clients, and their contractors understand how the design will appear when it is installed and mature. The tools of the landscape designer are similar to those used by a draftsman, although the techniques are often quite different. Illustration forms include plan views, elevations, and perspectives.

This chapter also included instructions for the creation of corner plantings, line plantings, and a total residential design.

Achievement Review

A. SHORT ANSWER

Answer each of the following questions as briefly as possible.

1. Based on the characteristics of gardens, in the left column, indicate their style, from the right column.

   a. designed from a romanticized concept of how nature should appear
   b. represents the natural world in miniature
   c. symmetrical layouts
   d. designed to display dominance over nature
   e. influenced the American park movement

   1. nature symbolism of the Orient
   2. formal tradition of sixteenth and seventeenth century Europe
   3. eighteenth century English naturalism

   f. spiritual appreciation of the garden may be lacking in American and European observers

2. Define contemporary landscape design.
3. Indicate if the following information, needed to develop a landscape, would be obtained from the client (A), the site analysis (B), or both (C).

   a. off-site views
   b. attitudes toward outdoor living
   c. preferences for certain plants
   d. problem areas such as wet spots
   e. direction of prevailing winds
   f. needs of the handicapped
   g. existing vegetation
   h. topography
   i. budget
   j. hydrography

4. List and describe the four major use areas in a residential landscape.

5. Indicate if the functions listed below are performed by outdoor walls (A), outdoor floors (B), or outdoor ceilings (C).
   a. define the base plane of the outdoor room
   b. define the shape of the outdoor room
   c. provide privacy from overhead viewers
   d. absorb the impact of user traffic
   e. direct traffic through the landscape

6. Which principle of design is being applied in each of the following situations?
   a. One side of the landscape has the same visual weight as the side opposite.
   b. Attention is directed to the house entrance.
   c. Trees are selected that will not dwarf the nearby house.
   d. Bedlines are designed to be gently curving, not fussy and intricate.
   e. Rectangular lines in the house and in the public area are repeated in the general living area.

7. Indicate if the following functions of plants are architectural (A), engineering (B), or aesthetic (C).
   a. contribute fragrance to a private living area
   b. cast needed shade across a concrete patio
   c. block pedestrian traffic
   d. absorb nearby highway noise
   e. attract attention to the incurve with their colorful flowers
   f. trace a pattern of green against a long, unbroken concrete wall
   g. reduce the velocity of wind blowing across a patio
   h. establish a windbreak for a farm in Kansas

8. Classify the following examples of enrichment items as natural (N) or man-made (M) and tangible (T) or intangible (I).
   a. outdoor furniture
   b. birds
   c. the sound of birds
   d. the fragrance of flowers
   e. large boulders
   f. sculpture
   g. reflecting pool
   h. the sound of a waterfall

B. MULTIPLE CHOICE

From the choices given, select the answer that best answers each of the following questions.

1. Which fencing style would *not* provide privacy for a patio area?
   a. basketweave
   b. grapestake
   c. horizontally louvered
   d. chainlink

2. Which enclosure would *not* deflect the wind?
   a. 6-foot brick wall
   b. 6-foot concrete wall
   c. picket fence
   d. 6-foot solid board fence

3. Which enclosure material would be most suitable for a children's play area where security and visibility into the area were desired?
   a. chainlink fencing
   b. brick wall
   c. stockade fencing
   d. railroad ties

4. Which fencing style would require the greatest maintenance?
   a. chainlink
   b. picket
   c. solid board
   d. split rail

5. Which surfacing material would *not* be suitable for parking cars?
   a. asphalt
   b. brick
   c. slate
   d. concrete

6. Which surfacing would not be suitable for paving the patio of a senior citizens' center?
   a. brick
   b. concrete
   c. indoor-outdoor carpeting
   d. marble
7. Which soft paving is most likely to deteriorate due to microbial activity?
   a. brick chips
   b. wood chips
   c. crushed stone
   d. stone dust

C. DEMONSTRATION

Select a property whose residents are agreeable to serving as clients, and follow the steps outlined in this chapter to create a total landscape design. For this initial effort, the following suggestions are offered:

1. Select a site whose dimensions and house location are either easily measured or can be taken from an existing property map.
2. Consult with the clients several times during the development of the design to assure that their wants and needs are being satisfied.
3. Work out the first draft on drawing paper and trace the final form onto heavy vellum using your best drafting skills.
4. Have a professional landscape designer evaluate your work and offer a critique.

10
Installing
Landscape Plants

Objectives

- identify the tools used in the installation of landscape plants.
- condition soil used in the installation of landscape plants.
- describe the advantages and disadvantages of bare-rooted, balled-and-burlapped, and containerized plant material.
- select the best season for transplanting.
- outline procedures for the installation of trees, shrubs, groundcovers, bedding plants, and bulbs.
- describe the advantages and disadvantages of organic and inorganic mulches.
- explain the benefits of antitranspirants.
- describe installation problems unique to the American Southwest.

THE IMPORTANCE OF PROPER INSTALLATION

High-quality landscapes begin with top-quality plant materials. Both depend upon careful installation techniques to assure the survival and growth of the transplanted stock. Landscape contractors joke about $25 dollar plants set into $75 dollar holes. In fact, a great deal of labor and materials are often needed to prepare a hostile planting site for a new plant. Few sites offer a perfect combination of proper soil texture, fertility, and pH with correct drainage and optimum water and humidity throughout the post-transplant period. All are necessary for successful transplanting.

THE NECESSARY TOOLS

Since the plant material to be installed includes seeds, bulbs, bedding plants, groundcovers, and trees and shrubs of all sizes, a wide range of tools must be available to accomplish the installation. The hand tools most commonly used are shown and described in Figure 10-1.

As plant materials increase in size, hand tools must be supplemented with power tools. The tree spade makes possible the successful transplant of large trees and shrubs, Figure 10-2. It operates on a hydraulic system, with each movement controlled from a set of levers that permits the machine to be operated by anyone, regardless of physical

FIGURE 10-1.
Tools for landscape installation

Tool	Function
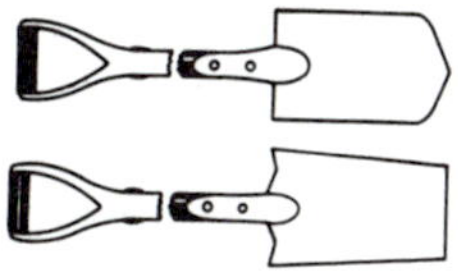 **Spades**	The spade performs obvious service in digging. It has a flat back and flat or pointed end. It has no sides and is not effective as a scoop. Spades are useful in the installation of all types of landscape plants.
 **Spading fork**	Used for turning over the soil when it is not too hard or compacted, the spading fork is useful in preparing planting beds that need not be too deep, as for flowers, bulbs, and groundcovers.
Rounded Shovel Square	Used for cleaning loose soil from planting holes and for other scooping uses, a shovel has sides where a spade does not. The blade may be rounded or square. The handle may be long or short.
 **Spading shovel**	A combination tool similar to both the spade and the shovel, it can be used for digging as well as scooping. It is useful in the installation of all types of landscape plants.
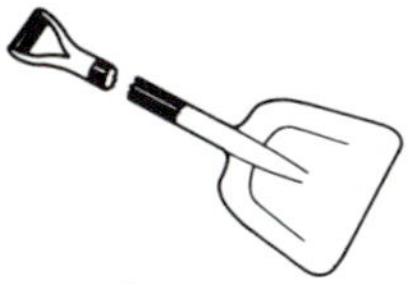 **Scoop**	Good for moving loose materials such as crushed stone, peat moss, and soil, a scoop has high sides. It is not used for digging. Aluminum scoops are available for lightweight materials. Steel scoops are used for heavy materials.
 **Manure fork**	The best tool for moving coarse, lightweight materials, such as straw and wood chips.
 **Bulb planter**	Used to install flowering bulbs, it is pressed into the soil, removing a plug of soil that is replaced after the bulb is inserted.

FIGURE 10-1.
Continued.

Tool	Function
 **Hand trowel**	Used to install bedding plants, groundcovers, and bulbs
 **Transplanting hoe**	Its uses are similar to those of a hand trowel. It is less adaptable to other types of digging.
 **Garden hoe**	Widely used for breaking up the soil prior to planting, it is helpful in the installation of all types of landscape plants.
 **Grading hoe**	Used to loosen hard or compacted soil during preparation for planting, it has a sharpened flat end.
 **Cutter mattock**	Stronger than a grading hoe, its uses are similar. It has two flat ends.
 **Pick**	Used for breaking up hard rocky soil, it has two pointed ends for gouging into the soil. A variation is the pick mattock, which has one pointed end and a flat mattock on the other end.
 **Wheelbarrow**	The wheelbarrow has a multitude of functions. It can haul materials to and from the planting site. It can also be used at the site to hold soil conditioners for mixing.
 **Ball cart**	Used to move balled-and-burlapped plants to the planting site.

FIGURE 10-2.
Large trees can be moved successfully with powered transplanters. There is limited disturbance to the root system. (Courtesy Big John Tree Transplanter Mfg.)

strength. Other power tools helpful in the installation of landscape plants include the power auger, power tiller, tractor, and front-end loader.

THE SOIL FOR INSTALLATION

The soil removed from a planting hole is seldom satisfactory in its unaltered state for use as replacement (or backfill) soil after the new plant is set. It may be too heavy with clay and need the addition of sand and peat moss to provide better aeration and flocculation (aggregation of soil particles). There may be too much sand, requiring additions of humus to retain moisture around the new plant's roots. Landscapers may have to deal with the hard caliche layer if in the Southwest, salt-saturated soil

next to roadways and walks, construction debris buried by builders, and the natural stoniness of rocky regions. Rich loam with ideal pH and good drainage is not common on most sites, where the landscaper is usually one of the last developers to be called in.

The soil to be filled into the planting hole must provide a medium in which the root system of the new plant can resume growth and develop fibrous root hairs to absorb water and nutrients for the new plant. If the soil does not drain adequately, the new plant may die from a lack of oxygen. If the backfill is too sandy, the new roots of the plant will stay within its soil ball and not grow out into the new soil.

Correct soil for plant installation has these qualities:

- loamy texture (near equal mix of sand, peat, and soil)
- good drainage
- suitable pH
- balanced nutrients

These qualities can only be attained by carefully blending conditioners such as sand, peat moss, compost, leaf mold, and manure with the soil before backfilling. In some situations, the original soil may be so unsatisfactory that it must be completely replaced, although recent research suggests that this is seldom necessary. In large landscape installations, the conditioning needs should be determined in advance so that necessary quantities of the additives can be ordered and available at the planting site. All members of the planting crew should be instructed in how to prepare the backfill mix. Periodic checks by the crew supervisor will assure that the new soil is mixed correctly and uniformly.

ROOT FORMS OF LANDSCAPE PLANTS

Landscape plants are available in a variety of root forms. Bedding plants and groundcovers are usually grown in pressed peat pots or plastic packets that permit the root system to be transplanted intact. Trees and shrubs can be purchased as bare-rooted, balled-and-burlapped (b & b), or containerized

FIGURE 10-3.
Root forms of landscape plants

Root Form	Advantages	Disadvantages
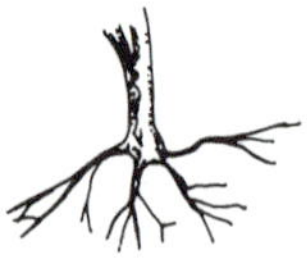 Bare-rooted	• Comparatively inexpensive • Lightweight and easy to transport • Dormant at the time of planting	• Severely reduced root system • Transplant season limited to early spring • Usually small, requiring time to mature
 Balled-and-burlapped	• Larger material can be transplanted • Less damage to the root system • Can be transplanted throughout spring and fall	• Usually the most expensive • Soil ball adds weight and bulk • For large plants, costly installation equipment is required
 Containerized	• Less expensive than b & b material • Root system intact • Can be transplanted throughout spring, summer, and fall	• Seldom available in large sizes • Can become root-bound if kept in containers too long

plants. The advantages and disadvantages of each are compared in Figure 10-3.

Which root form is best to use depends upon the season of the year, the availability of stock, the size of the plants at the time of installation, and the budget of the project. Bare root is a common root form for deciduous shrubs and a few trees that develop new roots quickly after transplanting. Ever-greens are most often balled and burlapped or containerized. Deciduous trees and shrubs may also be obtained in b & b or containerized forms. Vines are usually containerized.

THE TIME TO TRANSPLANT

The best season for transplanting depends upon the type of material being planted. Usually, the prime objective is to transplant at a time that will permit good root growth before shoots and leaves develop. For most plants in most parts of the country, that time is early autumn. Then the roots can grow as long as the soil remains unfrozen, while the cool air temperatures encourage the above-ground parts to go dormant. Early spring, when root growth exceeds shoot growth, is the second best season. Summer is not a good season unless containerized material is used, with its intact root system. Winter is not a good season for transplanting in northern regions because the roots cannot grow. In regions where the winter temperatures are milder and the ground doesn't freeze, winter can be a satisfactory alternate season for transplanting.

Flowering bulbs have definite transplant seasons. Hardy bulbs, which bloom in the spring, must be planted in the fall. Tender bulbs, which flower in the summer and will not survive the winter, are planted in the spring, dug up in the fall, and stored indoors over winter.

Annual flowers, purchased as bedding plants, are transplanted in the spring after all danger of frost is past.

METHODS OF INSTALLATION

The root form of the transplant will determine how the plant is installed. In certain cases, the size of the root mass may necessitate specialized equipment, but the principles behind the installation will remain the same.

Balled-and-Burlapped Plants

Plants with roots in a rounded soil ball wrapped in burlap are installed in a flat-bottomed, straight-sided hole that is as much as 50 percent wider and deeper than the ball. The extra volume of the hole leaves room for conditioned backfill soil to be added around the soil ball, Figure 10-4. Large plants need a smaller hole to help support the plant and prevent tipping.

Deciduous trees and shrubs may benefit from several handfuls of a slow-release, complete fertilizer mixed into the soil. This type of fertilizer provides nitrogen, phosphorus, and potassium gradually throughout the first year.

Application of fertilizer at the time of transplanting is seldom recommended for evergreens. Their roots will be severely harmed by the chemicals unless the plant has had at least a year to become established. If the plant appears to need fertilization, a slow-release fertilizer may be used; but it should be applied on the surface and beneath the mulch, not around the roots. Excessive amounts of any fertilizer applied to new plantings can cause serious damage to their root systems.

The conditioned soil should fill the hole sufficiently to raise the plant to the level at which it grew in the nursery. That level is marked by the top of the soil ball. Before the backfill soil is placed around the ball, the burlap should be loosened but not removed; then the soil can be added and tamped down with the feet to drive out the air pockets. *CAUTION:* If burlap has been replaced by plastic mesh or a wire basket, they should be removed before the backfill is added as they can stunt the development of the root system. Also, burlap that is left on should be thoroughly buried to hasten its decay and prevent it serving as a wick for moisture loss.

After the soil has been added, a mounded ring of compacted soil should be formed around the edge of the hole to catch and hold water. In large landscapes where frequent watering may not be possible (such as golf courses), the presence of the soil ring may determine whether the transplant survives during the first year, when moisture is especially critical.

Staking may be necessary to hold the plant straight until the roots become established in the new location. A year or longer may be required. Depending upon the size of the plant, one of two techniques of staking may be used. Smaller trees may be supported with stakes driven parallel to the

FIGURE 10-4.
The finished planting (From J. Ingels, *Landscaping: Principles and Practices,* 2nd edition, © 1983 by Delmar Publishers Inc.)

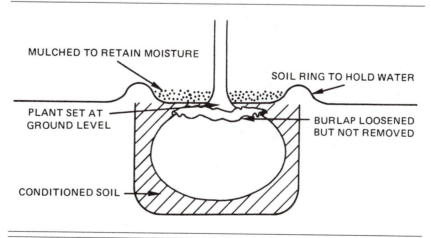

trunk, Figure 10-5. Short wires wrapped in sections of hose brace the tree at or near the first set of branches.

Larger trees, having more sizeable foliage canopies, need stronger bracing. The technique utilizes three stakes driven into the ground beyond the soil ball. Wires extend upward to brace the tree. The wires should be wrapped in protective hose sections to prevent damage to the tree. Twisting the wires to keep them taut will assure that the tree stays straight and secure. To prevent someone from tripping over them, the wires should be flagged with reflective tape or cloth, Figure 10-6.

Following the transplanting, the trunks of deciduous trees can be wrapped with paper or cloth tree wrap to reduce desiccation from wind or direct

sunlight, Figure 10-7. The tree wrap should be applied from the base of the trunk to the top of the trunk where the branching begins. The wrap may remain on the tree for a year or until it decomposes, whichever comes first. Plants may also benefit from the application of an antitranspirant before and after transplanting. Antitranspirants are discussed later in this chapter.

Containerized Plants

The same type of large, flat-bottomed planting hole used for balled-and-burlapped plants is needed for containerized plants. The depth of planting, backfill with conditioned soil, tamping and watering to drive out air pockets, and formation of a ring of soil are the same for containerized plants as for b and b stock. Two additional steps are necessary in the installation of containerized plants. First is the removal of the container, whether it be a clay

FIGURE 10-5.
This two-stake technique is good for small trees. It is not a trip-hazard, but it is not as sturdy as the three-stake method.

FIGURE 10-6.
This three-stake method of securing a new transplant is strong and necessary for large trees. However, the wires are a potential trip-hazard in populated areas.

FLAGGING STRIPS
ADDED FOR SAFETY

FIGURE 10-7.
Tree wrap is applied from the base of the tree to the first set of branches. Each coil should slightly overlap the preceding one to seal out precipitation and insects.

or plastic pot, wooden basket, or other form of production holder. Second is preparation of the roots for normal orientation. Too much time in the container may have caused the roots to encircle themselves in a condition often termed *pot-bound.* If the roots cannot be easily unwound, it is best to make vertical cuts through them at 2-inch intervals around the root mass, Figure 10-8. From these cuts, new roots will develop, growing naturally downward. Not making the cuts will allow the roots to continue their encircling growth even after the plant is removed from the container, eventually killing the plant.

Bare-Rooted Plants

If bare-rooted plants were placed in a flat-bottomed hole, the roots would be wadded and dis-

oriented and a stunted plant could result. The hole must have a mounded bottom to permit the plant's roots to spread over it and to encourage their natural orientation downward, Figure 10-9. Broken and damaged roots should be pruned away before planting.

As with other trees and shrubs, the planting hole for bare-rooted material should be up to 50 percent wider and deeper than the root system it is to receive. As the backfill soil is added, there is a greater possibility of air pockets forming with bare-rooted material than with other root forms. Special care must be taken to prevent their formation.

Groundcovers and Bedding Plants

Both groundcovers and bedding plants are commonly sold in strips of plastic or pressed peat moss pots or in flats. Certain plants such as geraniums or flowering perennials may also be marketed in small clay or plastic pots. Peat pots need not be removed, but other containers must be. Remember that the rim of the peat pot must be buried in the soil to prevent the wick-drying effect.

To install these plants, the entire bed is prepared rather than individual holes. This allows the groundcovers or flowers to be planted with a hand trowel or hoe rather than a spade, Figure 10-10. The soil must provide good drainage and nutrients. It should also be as weed-free as possible at the time of planting. For large areas, a garden tiller may be used to loosen the soil and incorporate the necessary conditioners. The use of a preemergence herbicide prior to planting can reduce some of the maintenance requirements of the planting.

The spacing of groundcover plants depends upon the species and the speed of coverage desired. Naturally, closer spacing will result in more rapid coverage, but it will also increase the installation cost. Table 9-4 in Chapter 9 lists some of the most common species of groundcovers, their optimum spacing, and selected other characteristics. To assure even and maximum coverage, groundcovers should be installed in a staggered planting pattern, Figure 10-11.

FIGURE 10-8.
Containerized plants can become pot-bound. Before transplanting, they should be removed from the container and the root mass cut vertically at 2-inch intervals to promote new root growth that is oriented properly.

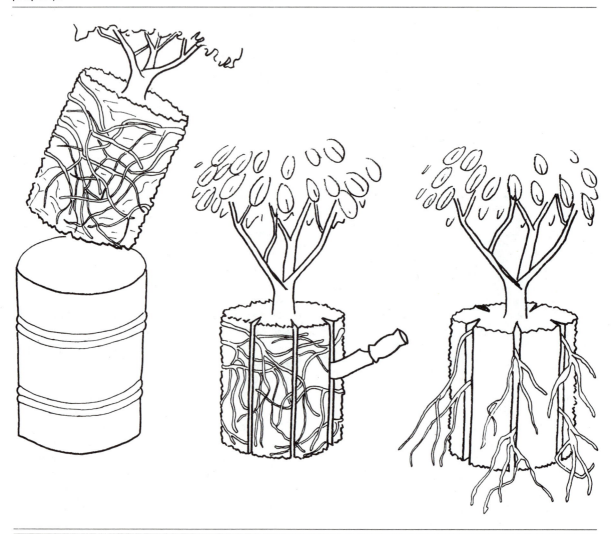

Due to the shallow root system of groundcovers and bedding plants, the plants can dry out easily. Prior to transplanting, each plant must be watered thoroughly. As each is set into the soil, it should be watered again. Thereafter, the new planting must be watered frequently and deeply to establish the plants successfully.

Groundcovers will benefit from mulching to reduce weeds and, most importantly, aid in preventing alternate freezing and thawing of the ground. Such ground activity can result in *heaving* the groundcovers to the surface where their roots are exposed to the cold and drying air. To be successful, mulch should be applied to a new groundcover planting

FIGURE 10-9.
Different forms of nursery stock require different shapes of planting holes. (From J. Ingels, *Landscaping: Principles and Practices,* 2nd edition, © 1983 by Delmar Publishers Inc.)

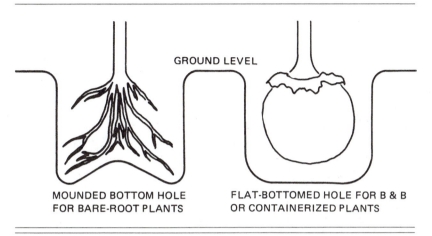

MOUNDED BOTTOM HOLE
FOR BARE-ROOT PLANTS

FLAT-BOTTOMED HOLE FOR B & B
OR CONTAINERIZED PLANTS

FIGURE 10-10.
With its plastic pot removed and the bed properly conditioned, this geranium is easily installed using a hand trowel.

after the ground has frozen to prevent premature thawing.

Bulbs

Flowering bulbs require a rich, well-drained soil. They are planted either in flower beds and borders or as masses in the lawn. They may be gently tossed by the handful into open, turfed lawn areas to be planted wherever they land, in an irregular spaced pattern. Bulbs are planted at differing depths and spacings depending upon their species. Table 10-1 lists the depths and spacings of some common bulbs.

Bulbs are always set into the ground with the base oriented downward and the shoot pointed upward, Figure 10-12. Many can be installed with a bulb planter.

Other bulb-like structures, called tubers and rhizomes, are installed in mounded holes that permit the structure to be oriented horizontally and the roots directed downward, Figure 10-13. As always, the backfilling step should be done carefully to assure that no air pockets form. Water collecting around the bulbs in air pockets can promote rotting.

MULCHING

All plants benefit from mulching after installation. Mulching refers to the application of loose aggregate materials to the surface of a planting bed. The materials may be organic or inorganic. Examples of both types and their advantages and disadvantages are listed in Table 10-2.

FIGURE 10-11.
Alternating the placement of groundcovers fills space most efficiently. (From J. Ingels, *Landscaping: Principles and Practices,* 2nd edition, © 1983 by Delmar Publishers Inc.)

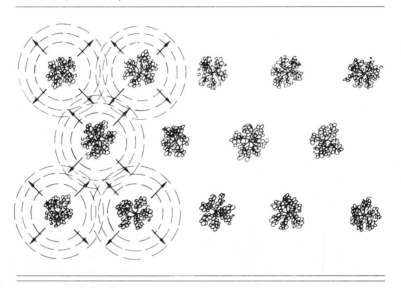

The benefits of mulching a newly installed plant are that:

1. water is retained in the soil around the root system and wilting is avoided
2. weed growth is discouraged
3. the aesthetic appearance of the planting is enhanced
4. soil temperature fluctuation is minimized, preventing winter heaving of bulbs and groundcovers. Repeated freezing and thawing of the soil around a plant's base can also damage the bark and permit the entry of pathogens or insects.

For mulches to be effective, they must be applied 3 to 4 inches deep. A shallow layer of mulch does not reduce sunlight enough to discourage weed seed germination, retain moisture, or prevent changes in the surface temperature of the soil. If a more shallow layer of mulch is desired, weeds can still be controlled and water retained by spreading black plastic around the plant base and adding one or two inches of mulch to weight down the plastic. The plastic must be black to prevent sunlight from penetrating and promoting weed growth. This mulching technique works well on flat land but is not advised for use on slopes, because rainwater tends to wash the mulch off the slick plastic. It must also be used cautiously on heavy clay soils because the plastic prevents evaporation and may cause the soil to hold too much water, drowning the new transplant.

USING ANTITRANSPIRANTS

Antitranspirants, also called *antidessicants,* are chemicals that reduce the amount of water plants lose through transpiration. Antitranspirants are useful because excessive water loss can result in transplant shock. They normally act either to induce closing of the stomata or to cover the stomata with a water-impermeable coating. Several popular brand names are available. The antitranspirants are sprayed onto the plant before and again after trans-

TABLE 10-1.

A Guide to Bulb Installation

Plant	Depth to Top of Bulb	Spacing
Amaryllis	Leave upper 1/3 of bulb exposed	One bulb per pot
Anemones	2 inches	12 inches
Bulbous iris	2 inches	12 inches
Caladium	2–3 inches in North/1 inch in South	As desired for effect
Calla lily	Leave upper 1/3 of bulb exposed	One bulb per pot
Cannas	3–4 inches	1 1/2–2 feet
Crocus	3 inches	2–4 inches
Daffodil	4–5 inches	6–8 inches
Dahlia	5–6 inches	2–3 feet
Elephant ears	Just below soil surface	As desired for effect
Gladiolus	3–4 inches	6–8 inches
Grape hyacinths	3 inches	2–4 inches
Hyacinths	4–6 inches depending on size	6–8 inches
Lilies	Two to three times the thickness of bulb	1 foot
Paperwhite narcissus	Just below soil surface	One or two bulbs per pot
Ranunculus	2 inches	12 inches
Snowdrops	3 inches	2–4 inches
Summer hyacinths	4 inches	6–8 inches
Tuberous begonia	Just below soil surface	6–8 inches
Tulips	4–5 inches	6–8 inches

planting. Since the majority of plant stomata are present in the greatest numbers on the lower surface of leaves, the underside of the canopy should receive the greatest coverage.

Antitranspirants are of greatest benefit in the transplanting of deciduous trees and shrubs that are in leaf. They are also of benefit to evergreens, especially broad-leaved forms. Any evergreen will benefit from antitranspirants if it is transplanted in the fall, right before the dry winter period.

PROBLEMS OF ARID REGIONS

The landscaper installing plants in the American Southwest encounters four distinct problems:

1. The soil quality is generally poor.
2. Irrigation is necessary throughout the year.
3. Higher altitudes can produce extremely hot daytime temperatures and very cool nights.
4. High winds dry out plants quickly and often damage them physically.

Arid soils generally fall into three categories: pure sand or gypsum, adobe, and caliche. Sand lacks both nutrient content and humus. *Adobe* is a heavy, clay-like soil that holds moisture better than sand but needs humus to lighten it and improve its aeration. *Caliche* soils are highly alkaline due to excessive lime content. They have a calcareous hardpan deposit near the surface that blocks drainage, making plant growth impossible. The hardpan layer may lie right at the surface or from several inches to several feet below ground level. The deposits may occur as a granular accumulation or as an impermeable, concrete-like layer.

FIGURE 10-12.
Bulbs are installed either by hand (as shown) or with a bulb planter. They are set into the ground "noses up" and are covered with soil pressed firmly to prevent air pockets. (Courtesy United States Department of Agriculture)

FIGURE 10-13.
This iris rhizome is installed by spreading its roots evenly over a mound of soil, then backfilling to cover the roots while allowing the leaves and top of the rhizome to remain exposed.

Generally, these are the characteristics of arid soils:

- lack humus
- require frequent irrigation
- are nutritionally poor; nutrients are continually leached out by the irrigation water
- are highly alkaline (pHs of 7.5 to 8.5 and higher)

TABLE 10-2.
Characteristics of Common Mulches

Organic Mulches (peat moss, wood chips, shredded bark, chipped corncobs, pine needles)	Inorganic Mulches (marble chips, crushed stone, brick chips, shredded tires)
• reduce soil moisture loss • often contribute slightly to soil nutrition • may alter soil pH • are not a mowing hazard if kicked into the lawn • may be flammable when too dry • may temporarily reduce nitrogen content of soil • require replacement due to biodegradation • may support weed growth as they decompose	• reduce soil moisture loss • do not improve soil nutrition • seldom alter soil pH • are a hazard if thrown by a mower blade • are nonflammable or fire-resistant • have no effect upon nitrogen content of soil • do not biodegrade

- are low in phosphate; phosphate may be rendered unavailable by the high pH
- lack iron or contain it in a form unavailable to plants
- have a high soluble salt content resulting from alkaline irrigation waters and from manures and fertilizers that do not leach thoroughly

When installing plants in the Southwest, the landscaper must add organic matter to the soils to improve their structure. Organic matter improves the water retention capability of light sandy soils and breaks up heavy adobe soils. The only way to improve the drainage of caliche soil is to break through it and remove the impermeable layer. The excavated soil can be replaced with a conditioned mix that will support healthy plant growth.

To catch and retain the water so vital yet so limited in arid regions, the planting beds should be recessed several inches below ground level to create a catch basin, Figure 10-14. This method traps and holds applied water, preventing loss through runoff. In addition, organic mulches should be applied to a depth of 4 inches or more to slow moisture loss and create a cooler growth environment for the roots. Trunk wraps and whitewash paint are also applied to the trunks of trees to prevent water loss through their thin bark due to sun scald.

Cactus plants are sufficiently different from other plants to warrant special mention. They can be transplanted successfully by following these steps.

1. Before transplanting, mark the north side of the cactus. Orient this side of the plant to the north in its new location. The plant will have developed a thicker layer of protective tissue on its south side to withstand the more intense sunlight.

FIGURE 10-14.
A recessed planting bed creates a catch basin for moisture. (From J. Ingels, *Landscaping: Principles and Practices,* 2nd edition, © 1983 by Delmar Publishers Inc.)

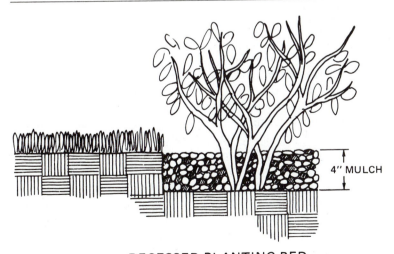

RECESSED PLANTING BED

2. By trenching around the cactus, lift as much as possible of the root system.

3. Brush soil from the roots and dust them with powdered sulfur.

4. Place the cactus in a shaded area where air circulates freely and allow the roots to heal for a week before replanting.

5. Plant the cactus in dry, well-drained soil. Stake the plant if necessary.

6. Water the plant in three or four weeks, *after* new growth starts. Thereafter, apply water at monthly intervals.

Whenever possible, native or naturalized plants should be selected for Southwestern landscapes. They have a better chance of surviving the transplant and they keep maintenance costs down. In situations where the soil is especially unsuitable for planting, there may be little choice but to install the plants above ground in planters.

SUMMARY

High-quality landscapes require high-quality plant materials, properly installed. Many different tools are available for use in the installation of landscape plants. Some have multiple uses; others are more specialized. Knowledge of these tools and their functions is important to the landscaper who seeks to use time and work crews most efficiently.

Properly conditioned soil is also important if new plants are to have the best opportunity for successful growth. Conditioners such as sand, peat moss, compost, leaf mold, and manure can give the planting soil loamy texture, good drainage, a suitable pH, and balanced nutrient levels. In large landscape installations, the conditioning needs should be determined in advance so that necessary quantities of additives can be ordered and available at the planting site.

Landscape plants are commonly available in bare-root, balled-and-burlapped, or containerized forms. Each root form has its advantages and disadvantages. The form selected usually depends upon the type or size of plant, the budget for the project, and the season of the year.

The best time to transplant is when good root growth can occur prior to the development of shoots and leaves. For most plants in most parts of the country, early autumn is the best season to transplant. However, there are good alternative planting seasons throughout the nation depending upon the climate.

How the plant is installed is determined by the root form. Balled-and-burlapped or containerized plants require a flat-bottomed, straight-sided hole. Bare-root material requires a mounded bottom in the hole to permit proper orientation of the roots before backfilling. The size of the hole can be as much as 50 percent wider and deeper than the root mass it is to receive. Larger soil balls need a smaller planting hole for better plant support. After backfilling with conditioned soil, the soil is tamped down, water is applied, and a mounded ring of soil is formed to catch and hold water around the plant base. Staking, mulching, the wrapping of deciduous tree trunks, and perhaps use of an antitranspirant complete the transplant operation. Groundcovers, bedding plants, and bulbs require equally conscientious care in their installation; however, the entire planting area is usually conditioned rather than individual planting holes, and the methods of installation are frequently more labor intensive.

All plants benefit from mulching after installation. Mulches aid in moisture retention, reduce soil temperature fluctuation, and improve weed control around new or established plantings. Mulches may be either organic or inorganic; both have advantages and disadvantages. The choice of one over the others may be based merely on its appearance.

Because the arid landscape of the Southwest is so different from other parts of the country, this chapter also included a brief discussion of its unique problems.

Achievement Review

A. SHORT ANSWER

Answer each of the following questions as briefly as possible.

1. Identify the following tools used in the installation of landscape plants.

a.

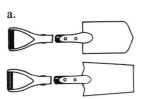

b.

c.

d.

e.

f.

g.

h.

i.

n.

j.

o.

k.

l.

2. List the four qualities of good backfill soil.
3. Indicate whether the following are characteristic of bare-rooted (A), balled-and-burlapped (B), or containerized (C) plants.
   a. lightweight and easily transported
   b. retain the entire root system
   c. severely reduced root system
   d. usually the most expensive
   e. plants may become pot-bound
   f. permits large plants to be transplanted
   g. transplant season limited to early spring
   h. allows transplanting in any season
4. What is the primary objective when deciding the timing of a plant transplant?
5. Why is early autumn the most desirable transplanting season for most plants?

m.

6. Indicate whether the following are characteristic of inorganic (A) or organic (B) mulches or both (C).
    a. reduce water loss from the soil
    b. may alter soil pH
    c. may temporarily reduce nitrogen content of soil
    d. hazardous if thrown by a mower blade
    e. do not biodegrade
    f. sometimes are slightly nutritional
7. What benefit is gained by applying antitranspirants to new transplants?

B. TRUE/FALSE

Indicate if the following statements are true or false.

1. The root form of the transplant determines how the plant is installed.
2. Balled-and-burlapped plants are installed in a hole with a mounded bottom.
3. Burlap can be left around the ball of earth, but plastic mesh or wire baskets should be removed before planting.
4. Earthen rings to retain water and staking if necessary to assure alignment are used regardless of the root form.
5. Containerized plants are installed in a hole with a mounded bottom.
6. Plants are installed at the depth at which they were growing before relocation.
7. Bare-rooted plants are installed in a hole with a mounded bottom.
8. The planting hole for tree and shrub transplants may be as much as 50 percent wider and deeper than the root system it is to receive.
9. For rapid and maximum coverage, groundcovers should be installed in staggered, alternating rows.

C. ESSAY

Installation of plants in the American Southwest is complicated by poor soil and the lack of adequate natural water supplies. What must a landscaper do to counter these threats to plant survival?

11
Maintaining Landscape Plants

Objectives

Upon completion of this chapter, you will be able to

- water, fertilize, edge, and mulch tree and shrub plantings.
- prune trees and shrubs correctly.
- maintain annual and perennial flower plantings.
- winterize landscape plants.

SUSTAINED CARE OF PLANTINGS

As important as the design and proper installation of landscape plantings is their ongoing maintenance. Many gardens do not attain the appearance envisioned by the landscape designer until the plants have time to mature. Aiding those plants in their healthy maturation requires attentive and knowledgeable maintenance. Also, some designs require specific plant effects such as clipped formal hedges or espalier training to create the garden as envisioned. In both instances, the value of skillful landscape maintenance is apparent. A maintenance program generally will include the following tasks: watering, fertilizing, mulching, edging, pruning, pest control, and winterization.

Watering

Depending upon the region of the country, supplemental watering may be an infrequent task or so regular that it requires an automatic irrigation system. The need for a good water supply to the plantings has been discussed in earlier chapters. Here the subject is the objectives of proper watering, its frequency, and its quantity.

Initially, watering must promote deep root development by the plant to establish it securely in its location. Later, watering must keep the plant healthy and growing actively even during dry summer weather. Much winter damage to evergreens can be avoided if the plants are kept well-watered throughout the summer and autumn.

Not all plants require the same amounts of water. Neither will all plant root systems grow to the same depth in the soil. While nearly all landscape trees and shrubs will die if kept in either arid or water-logged soil for long, certain species are especially sensitive to sites that are too dry or too wet.

Infrequent and deep watering is preferable to frequent, shallow watering. Enough water should be applied to wet the soil to a depth of 12 to 16 inches. In the Southeast and Southwest, supplemental water may be required nearly every day. In other regions, supplemental watering may be weekly or even less frequent.

Fertilization

Trees will grow without fertilization in most soils once they are established. However, they will grow with greater health and vigor if they are fertilized annually. Shrubs will respond to proper fertilization with lush growth, greater resistance to pests, and less winter damage.

For shrubs growing in cultivated beds, fertilizer may be applied in late March or early April. Depending upon the plants involved, each 100 square feet of bed area should receive between 1 and 3 pounds of a low-analysis, complete fertilizer. The fertilizer should be distributed uniformly over the soil beneath the shrubs, with most of the fertilizer under the outer edge of the shrub where the fibrous roots that absorb the nutrients are located. The fertilizer should not be allowed to touch the foliage, or *foliar burn* (a reaction to the chemicals) may result. If the soil is dry, the fertilizer should be worked into the soil with a hoe. If the weather has not been abnormally dry, the fertilizer can be left untilled and the next rainfall or irrigation will wash it into the soil.

Traditionally, trees have been fertilized in a somewhat similar fashion by drilling holes (2 inches in diameter and 12 inches deep) at intervals of 18 to 24 inches around the tree beneath the drip line of the canopy. Into these holes is poured a dry mixture of 50 percent high-analysis, complete fertilizer and 50 percent sand as a carrier. The holes can

be drilled by hand or with a powered auger. In recent years, tree spikes have been promoted, especially to homeowners. They are complete fertilizers manufactured in torpedo shapes and sold at a price that is prohibitive if used in large quantities. They are pounded into the ground, eliminating the need to drill holes. For use with a few small trees, they are convenient.

Landscape maintenance specialists most commonly fertilize trees using an automated process that drills the holes and injects the fertilizer quickly and neatly. Recently, a method for injecting fertilizer (and pesticides) directly into trees' vascular systems has been introduced to the industry. It is rapidly gaining favor as more people become familiar with the technique and skilled in its use, Figure 11-1.

The objectives of fertilization change as plants, especially trees, mature. Initially, plants are fertilized to promote their health and growth. Later, as the plants approach or attain maturity, the purpose of fertilization is not to enlarge them but to keep them healthy, attractive, and strong.

FIGURE 11-1.
The patented Mauget system releases liquid pesticides and/or fertilizers into trees through tubes injected into the vascular system. (Courtesy J. J. Mauget Company)

The one most common mistake in the fertilization of landscape plants is the application of fertilizer too late in the growing season. The result is often a flush of vegetative growth in response to the nitrogen that leaves the plant ill-prepared for winter. Great damage can result to plants from well-meaning but ill-timed fertilization.

Mulching

The objectives, advantages, and disadvantages of mulching, and examples of commonly available products were outlined in the preceding chapter. For extended maintenance of landscape plantings, mulches require replacement. Organic mulches decompose, forming humus *and* an ideal medium for the germination of newly deposited weed seeds unless mulch is replaced annually. Inorganic mulches do not decompose but decline in appearance if not freshened periodically. *Caution:* Old mulch should be removed before new is added. Otherwise, the soil level over the roots is gradually deepened and the plants may die.

One important use of mulch is as a protective divider to prevent lawnmower damage to the base of plants, Figure 11-2. A gouge from a mower creates a site for pathogen or insect invasion of the plant. If hit repeatedly, a tree may become partially or completely girdled. In large landscapes with numerous trees and an understaffed maintenance crew who rely on riding mowers for grass cutting, frequent injury to trees results from attempts to mow as close as possible, thereby eliminating hand trimming. A square of mulch around the base of each tree can protect the tree, speed the overall mowing time, and create a neat appearance, Figure 11-3.

Edging

The term *edging* has a double use. As a verb, *edging* refers to cutting a sharp line of separation, usually between a planting and the adjacent lawn. Reference may be made to edging a bed or edging the lawn along a walk. The term can also be used to describe a product, usually a steel or plastic strip that can be installed as a physical separator between

FIGURE 11-2.
Typical damage resulting from a lawn mower when grass is allowed to grow next to a tree.

FIGURE 11-3.
The mulched area around the tree protects the plant from mower injury. The square shape is preferable to a round shape because it makes mowing easier.

planting beds and lawns or between lawns and paved areas.

To edge a bed requires an edging tool or a flat-back spade. The edger or spade is dug into the ground to a depth of 6 to 8 inches and a wedge of sod removed, Figure 11-4. The process continues along the edge of the bed or the paved area. Cutting through the sod sharply and vertically discourages the roots of the turf from growing into the bed. If the landscape is large, edging can be accelerated by using a power edger.

The installation of edging material against the cut edge further discourages the horizontal spread of grass roots into the bed. It also retains a sharp turf line along walks and drives. The best edging materials are firm, not easily bent and crushed. The corrugated foil edging promoted to the home garden market is not satisfactory for professional use and should be avoided. Heavy-gauge steel and polyvinyl-chloride edging is available from numerous manufacturers. Satisfactory edges can also be created with wood, bricks, and other modular manufactured materials.

In regions where winter heaving of the soil is common, it is best to select an edging material that can be anchored, Figure 11-5. Most anchoring tech-

FIGURE 11-4.
A sharply beveled edge around a planting bed will slow rhizome grasses and stop bunch grasses.

FIGURE 11-5.
Plastic edging should be anchored at the time of installation to prevent winter heaving. Here a metal rod is driven through the edging into the soil and clipped over the lip of the edging.

niques are only moderately effective, however, and resetting heaved edging is a common spring activity in northern landscapes. It is worth the effort, considering the advantages that the material offers in retention of the bed line and separation of the mulch from the lawn mower.

PEST CONTROL

Chapter 6 dealt with plant pests and the methods used to control them. Here the subject is how to choose a chemical product, and in what formulation, what strength, and when to apply it. Remember, chemical control of plant pests is not the only or even the best approach to control; nevertheless, it is one that may be determined to be appropriate.

Regardless of the state of the union in which the landscaper practices, certification is required of all who would purchase or supervise the application of any chemical registered with the United States Environmental Protection Agency (EPA). A few pesticides in dilute formulations are available for general purchase, but most of those needed by professional landscapers are restricted in use.

Every state has a college of agriculture that is responsible for approving the use of every pesticide within the state. Scientists at the colleges of agriculture are also responsible for determining the appropriate product and formulation for use against a particular disease, weed, or insect on a particular host at a particular time of year or stage of development. Their recommendations are updated and published each year for use by the states' professionals. Copies of the recommendations are available by direct mail from the university and often from the local Cooperative Extension office. Landscapers should realize that although pest problems may be much the same in different states and in different regions of a single state, the environmental conditions can vary in subtle ways not always apparent. Therefore, if a pesticide is not approved for use within a state or region, it is dangerous, unethical, and illegal to purchase the product elsewhere and use it in the restricted zone. It is equally dangerous and illegal to increase the dosage or frequency of application beyond those recommended. It is these abuses of pesticide application that the certification program of the EPA seeks to eliminate.

PRUNING TREES AND SHRUBS

Pruning is the removal of a portion of a plant to improve its appearance and health and to control its growth and shape. It is easily done, but not so easily done correctly. Each time a bud or branch is removed from a plant, it has both a short-term and long-term effect. The short-term effect is the way the plant looks immediately after pruning, and perhaps through the remainder of the growing season. The long-term effect is the way the plant appears after several seasons of growth without the part that has been pruned.

Parts of a Tree

The following parts of a tree are important to an understanding of proper tree pruning, Figure 11-6. The *lead branch* of a tree is its most important branch. It is dominant over the other branches, called the *scaffold branches*. The lead branch usually cannot be removed without destroying the distinctive shape of the tree. This is especially true in young trees.

The scaffold branches create the *canopy*, or foliage, of the tree. The amount of shade cast by the canopy is directly related to the number of scaffold branches and the size of the leaves. When it becomes necessary to remove a branch from a tree, removal usually occurs at a *crotch*, the point at which a branch meets the trunk of the tree or another, larger branch.

It is always desirable to leave the strongest branches and remove the weakest. Where the crotch union is wide (approaching a right angle), the branch is strong because there has been no crowding and pinching of the new wood produced each year by the cambiums of the trunk and branches. Where the crotch union is narrow, the branch is weak due to a pinch point forming where the expanding trunk meets the expanding branch, Figure 11-7. Growth in that area becomes compressed and dwarfed and the branch may snap off at that point during a heavy

FIGURE 11-6.
The parts of a tree (From J. Ingels, *Landscaping: Principles and Practices,* 2nd edition, © 1983 by Delmar Publishers Inc.)

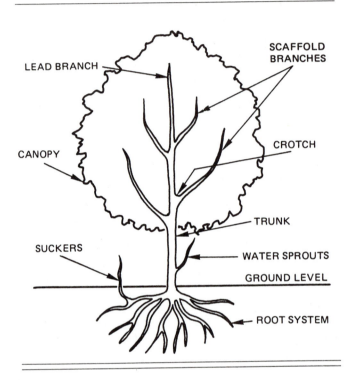

wind or in response to the weight of a person climbing on it.

Two other types of branches often found on trees are suckers and water sprouts. *Suckers* originate from the underground root system. *Water sprouts* develop along the trunk of a tree. Neither is desirable for an attractive tree and both should be removed.

Parts of a Shrub

A shrub is a multistemmed plant, Figure 11-8. Its branches and twigs differ in age, with the best flower and fruit production usually on the younger branches. The younger branches are usually distinguished by a lighter color, less bark, and smaller diameter. The point at which the branches and the root system of a shrub meet is the *crown.* New branches originate at the crown, causing the shrub to grow wider. New shoots, called *stolons,* may spread underground from existing roots to create new shrubs from the parent plant. In some grafted plants, the *graft union* may be seen at or near the crown. (Grafting is described in Chapter 14.) Shoots originating from the *stock* (or root portion) of a grafted plant are cut away since the quality of their flowers, fruit, and foliage is inferior. Only shoots originating from the *scion* (or shoot portion) are allowed to develop.

The Proper Time to Prune

Landscapers who design and install as well as maintain landscapes usually prefer to prune when

FIGURE 11-7.
Tree crotch structure (From J. Ingels, *Landscaping: Principles and Practices,*
2nd edition, © 1983 by Delmar Publishers Inc.)

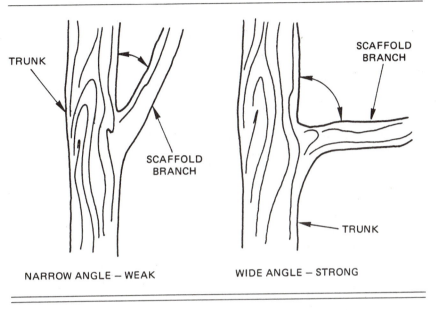

FIGURE 11-8.
The parts of a shrub (From J. Ingels, *Landscaping: Principles and Practices,*
2nd edition, © 1983 by Delmar Publishers Inc.)

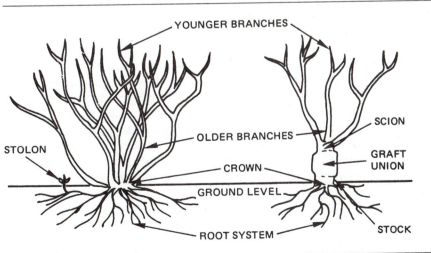

they have little other work. This distributes their work and income more evenly throughout the year. Some plants can accept this off-season attention and remain unaffected by it. Other species accept pruning only during certain periods of the year.

There are advantages and disadvantages to pruning in every season. Since seasons vary greatly from region to region, the following can be used only as a general guide to the timing of pruning.

Winter Pruning   Winter pruning gives the landscaper off-season work. It also allows a view of the plant unblocked by foliage. Broken branches are easily seen, as are older and crossed branches. The major disadvantage of winter pruning is that without foliage it is difficult to detect dead branches. Because of this, plants can become seriously misshapen if the wrong branches are removed. An additional disadvantage is the damage that can be done through cracking frozen plant parts.

Summer Pruning   Summer pruning also provides work during a slower season for the landscaper. An advantage of summer pruning is that it allows time for all but very large wounds to heal before the arrival of winter. The major limitation of summer pruning is that problems of plants may be concealed by their full foliage. Branches that should be removed are often difficult to see. Especially with trees, it is difficult to shape the branching pattern unless all the limbs are visible.

Autumn Pruning   Pruning during the autumn season may conflict with more profitable tasks for the landscaper. In terms of the health of the plant, autumn pruning is acceptable as long as it is done early enough to allow cuts to heal before winter. Autumn pruning should not be attempted on plants that bloom very early in the spring, however. These early bloomers produce their flower buds the preceding fall. Thus, fall pruning cuts away the flower buds and destroys the spring show. Autumn pruning should be reserved for plants that bloom in late spring or summer, producing their buds in the spring of the year.

Spring Pruning   Since spring is the major planting season, most landscapers do not welcome pruning requests unless maintenance is their principal business. However, most plants are most successfully pruned during the spring. As buds begin to swell, giving evidence of life, it is clear which are the live and dead branches. Furthermore, there is little foliage to block the view of the complete plant. Spring pruning provides the plant with maximum time for wounds to heal. In addition, the unfolding leaves conceal the fresh cuts from the viewer's eye.

If the plant is an early spring bloomer, it is best to prune it immediately after flowering. Plants that have a high sap pressure in the early spring, such as maples, birches, walnuts, and poinsettias, should not be pruned until summer or fall, when the sap pressure is lower. Otherwise, the excessive exudation becomes unsightly.

Parts of the Plant to Prune

The reason for pruning will determine the limbs and branches to be removed from a tree or shrub. If the objective is to remove diseased portions, the cut should be made through healthy wood between the trunk or crown and the infected part, Figure 11-9. The cut should never be made through the diseased wood or very close to it. This contaminates the pruning tool, which may transmit the pathogen to healthy parts pruned later.

If the objective is to improve the overall health and appearance of the plant, branches growing into the center of the plant should be removed. Limbs and twigs that grow across other branches can crowd the plant and cause sites of infection to form by rubbing abrasions through the bark. If more than one limb originates at a tree crotch, the strongest should be left and the others removed, Figure 11-10. Major structural limbs and twigs must be left so that no holes appear in the plant. Often overlooked is the fact that many secondary branches can stem from one older branch. The removal of one branch from a young tree can result in an older tree missing an entire side.

If, the purpose of the pruning is to create denser foliage, as with evergreens, the center shoot is shortened or removed. This encourages the lateral buds to grow and create two shoots, where there had been only one, Figure 11-11.

FIGURE 11-9.

The correct way to remove diseased limbs or twigs (From J. Ingels, *Landscaping: Principles and Practices*, 2nd edition, © 1983 by Delmar Publishers Inc.)

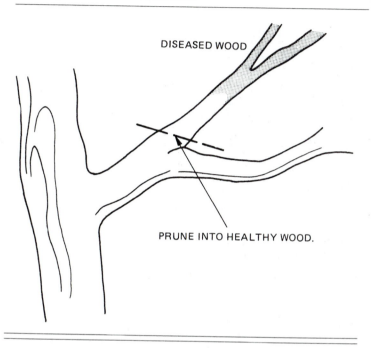

DISEASED WOOD

PRUNE INTO HEALTHY WOOD.

Pruning Methods

The method of pruning a tree or shrub depends upon the size and number of branches to be removed. Limbs are pruned from trees with a technique called *jump-cutting*. This method allows a scaffold limb to be removed without taking a long slice of bark with it when it falls. A jump-cut requires three cuts for the safe removal of a limb, Figure 11-12. The final cut should remove the stub of the limb as close to the trunk as possible. The wound may then be covered with a wound paint to make it less conspicuous until the plant has time to heal. Wound paints are available in aerosol cans for small-scale use and in quart and gallon cans for professional use. There is little evidence to support the claims of manufacturers that some formulations offer greater protection than others. Some recent research has suggested that wound paints may

actually delay healing of plant tissue. Where this is suspected, the landscaper may choose not to use it.

When shrubs are pruned, one of two techniques is used. *Thinning out* is the removal of a shrub branch at or near the crown. It is the major means of removing old wood from a shrub while retaining the desired shape and size. *Heading back* is the shortening, rather than total removal, of a twig. It is a means of reducing the size of a shrub. In cases where shrubs have become tall and sparse, a combination of thinning out and heading back can rejuvenate an old planting, Figures 11-13 and 11-14.

In heading back, the location of the cut is important, Figure 11-15. If too much wood is left above the bud, the twig will die from the point of the cut back to the bud, but the cut may not heal quickly enough to prevent insect and pathogen entry. Also,

FIGURE 11-10.
Selecting branches to be pruned (From J. Ingels, *Landscaping: Principles and Practices,* 2nd edition, © 1983 by Delmar Publishers Inc.)

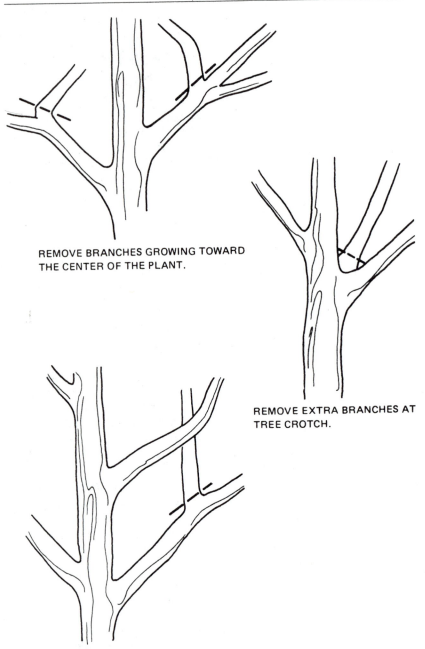

REMOVE BRANCHES GROWING TOWARD
THE CENTER OF THE PLANT.

REMOVE EXTRA BRANCHES AT
TREE CROTCH.

REMOVE BRANCHES GROWING ACROSS
OTHER BRANCHES.

FIGURE 11-11.
Evergreens are pruned in the spring if denser foliage is desired. (From J. Ingels, *Landscaping: Principles and Practices,* 2nd edition, © 1983 by Delmar Publishers Inc.)

DORMANT WINTER BUDS

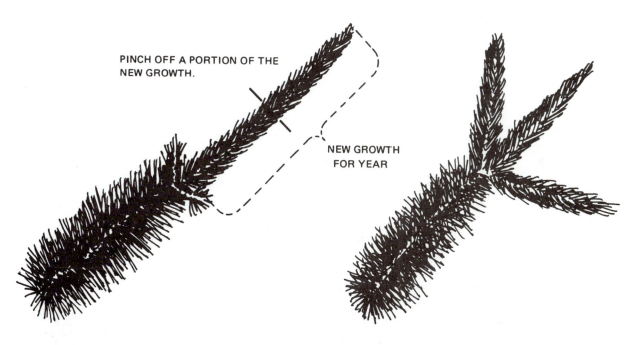

PINCH OFF A PORTION OF THE NEW GROWTH.

NEW GROWTH FOR YEAR

IN THE SPRING, THE DORMANT CENTRAL BUD HAS THE GREATEST GROWTH UNLESS PINCHED BACK.

AFTER PINCHING, ALL BUDS ARE ABLE TO GROW. THE RESULT IS A FULLER PLANT.

FIGURE 11-12.
The removal of large limbs using the technique of jump-cutting. The cut at A allows the limb to snap off after a cut at B without stripping bark from the trunk as it falls. The final cut at C removes the stub. (From J. Ingels, *Landscaping: Principles and Practices,* 2nd edition, © 1983 by Delmar Publishers Inc.)

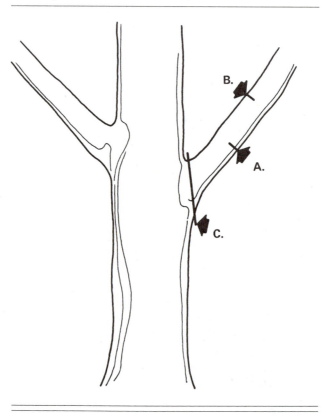

the woody stub itself may decay later. A cut below the bud will cause the bud to dry out and possibly die. The cut should be made just above the bud and parallel to the direction in which the bud is pointing. The cut should be close enough to the living tissue to heal over quickly but not so close to the bud that it promotes drying. The direction in which the branch of a plant grows can be guided by good pruning techniques. Branches growing into the plant can be discouraged by the selection of an outward-pointing bud when heading back, Fig-

ure 11-16. If the twig has an opposite bud arrangement, the unnecessary one is removed.

How to Prune Hedges

The creation of a hedge requires close spacing of the shrubs at the time of planting and a special type of pruning. The landscaper must shear the plant so that it becomes as dense as possible. This is usually done with hedge shears. The hedge shears easily cut through the soft new growth of spring,

FIGURE 11-13.

The techniques of thinning out and heading back (From J. Ingels, *Landscaping: Principles and Practices,* 2nd edition, © 1983 by Delmar Publishers Inc.)

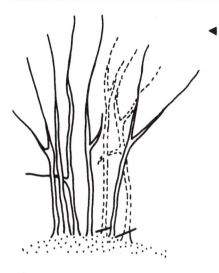

THINNING OUT. As its name implies, this method involves selection of an appropriate number of strong, well-located stems and removal at the ground level of all others. This is the preferred method for keeping shrubs open and in their desired shrub size and form. With most shrubs, it is an annual task; with others, it is required twice a year.

HEADING BACK. This method involves trimming back terminal growth to maintain desired shrub size and form. It encourages more compact foliage development by allowing development of lateral growth. This is the preferred method for controlling the size and shape of shrubs and for maintaining hedges.

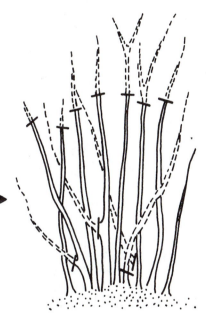

the season when most hedges are pruned. For especially large hedges, electric or gasoline powered shears are available. However, practice and skill are required for the satisfactory use of power shears. Damage can occur quickly if the landscaper does not keep the shears under control.

A properly pruned hedge is level on top and tapered on the sides, Figure 11-17. It is important that sunlight be able to reach the lower portion of the hedge if it is to stay full. Otherwise it becomes leggy and top-heavy in appearance.

FLOWER PLANTINGS

No components of the landscape require more maintenance time than flower plantings. This is why

FIGURE 11-14.
Two techniques used to rejuvenate old shrubs (From J. Ingels, *Landscaping: Principles and Practices,* 2nd edition, © 1983 by Delmar Publishers Inc.)

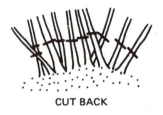

CUT BACK

SELECT SIX OR MORE WELL-PLACED VIGOROUS SHOOTS

HEAD BACK

GRADUAL RENEWAL. This pruning method involves removal of all mature wood over a 3-to 5-year period. Approximately one-third of the mature wood is removed each season. This is the preferred method for shrubs that have not been recently pruned and are somewhat overgrown.

COMPLETE RENEWAL. This method involves complete removal of all stems at the crown or ground level. Two to three months later the suckers or new growth that emerges is thinned to the desired number of stems. These, in turn, are headed back to encourage lateral branching. Unpruned, seriously overgrown, or severely damaged shrubs are prime prospects for this treatment.

landscapes with a low budget for maintenance must minimize the use of flowers. Reasons for the high maintenance costs are that:

- Weeds must be pulled by hand or controlled with costly selective herbicides.
- Flowers are more susceptible to insects and diseases that mar their aesthetic appearance, therefore requiring additional expenditure for pesticides.
- Flowers tend to go to seed or get leggy with age if not pinched back frequently during the growing season. (This subject is discussed later in the chapter.)
- Many perennials bloom only once a year but must be cared for throughout the growing sea-

FIGURE 11-15.

Where to prune the twig (From J. Ingels, *Landscaping: Principles and Practices,* 2nd edition, © 1983 by Delmar Publishers Inc.)

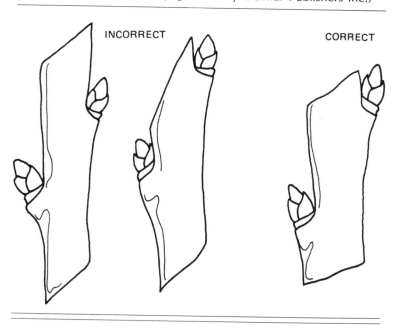

INCORRECT   CORRECT

son to assure a good flower display the following year.

- Some perennials, notably the tender bulbs, must be dug up, put into storage for the winter, and set out again each spring.

Weed Control

Mulches help control weeds in flower plantings and retain surface moisture for the planting as well. Due to the short height of some flowers, the thickness of the mulch may be reduced to 3 inches rather than 4. Black plastic beneath the mulch can be beneficial in annual beds. The plants can be set into the soil through holes in the plastic and mulch put on top of the plastic. In perennial beds, where the flowers go dormant each fall and revive in the spring, the plastic could do more harm than good by trapping new flower growth beneath it.

Several good herbicides are approved for use in flower plantings. If applied for pre-emergence *and* post-emergence control, a weed-free flower planting can be attained. One danger of herbicides in flower beds is that the chemicals are always selective; that is, they kill grasses or broadleaved weeds, but seldom both. Since flower plantings can have characteristics of both, damage can be done to some flowers even when others are uninjured.

Most herbicides used in flower plantings lose their effectiveness if the soil is disturbed after application. In such cases, landscapers must avoid cultivating the soil surface if they wish the full benefit of the herbicide.

Watering

All flowers should be watered frequently and deeply during dry periods. Their shallow roots quickly react to drought conditions and they reach a critical wilting point much sooner than the woody plants of the landscape. Flowers planted beneath trees and shrubs must compete with the woody

FIGURE 11-16.
Twigs should be pruned to leave an outward-pointing bud. (From J. Ingels, *Landscaping: Principles and Practices,* 2nd edition, © 1983 by Delmar Publishers Inc.)

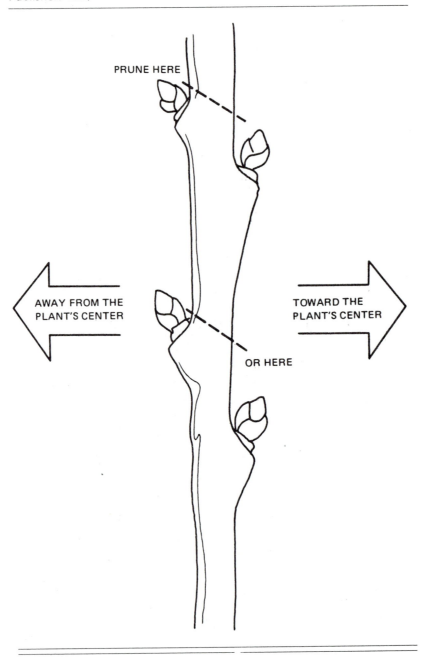

FIGURE 11-17.
Good and poor hedge forms as seen from the side (From J. Ingels, *Landscaping: Principles and Practices,* 2nd edition, © 1983 by Delmar Publishers Inc.)

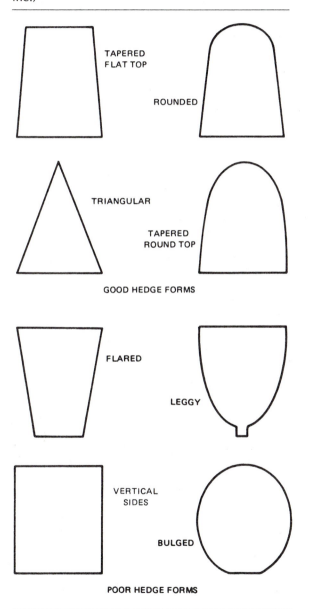

TAPERED
FLAT TOP

ROUNDED

TRIANGULAR

TAPERED
ROUND TOP

GOOD HEDGE FORMS

FLARED

LEGGY

VERTICAL
SIDES

BULGED

POOR HEDGE FORMS

plants for surface water, and they will dry out faster than flowers not in such competitive locations.

Fertilization

Annuals can be fertilized in midsummer with a low-analysis fertilizer to keep them lush and healthy. Bulbs should be fertilized immediately after flowering with a high-phosphorus fertilizer such as bone meal. Nonbulbous perennials grow best if fertilized in the early spring. Summer or fall fertilization of perennials can harm the plants by keeping them too succulent as winter approaches.

Pinching

Annuals are most likely to benefit from pinching back, but certain perennials such as hardy mums will also do better if pinched. *Pinching* removes the terminal shoot on each branch of the flower and allows the lateral shoots to develop, thereby creating a fuller plant. *Soft pinching* is done with the thumb and forefinger and removes the terminal bud or, at the most, the terminal bud and first set of laterals. *Hard pinching,* may shorten each stem by one-third or more. Most flowers benefit by a hard pinch soon after being set out, followed by one or two soft pinches during the summer. With perennials such as mums whose flower bud initiation is tied to a photoperiod response, the last pinch should not be after mid-July if a good flower display is to be seen in the autumn.

Flowers such as petunias or rose moss that are valued for their profuse blossoms can be kept from getting leggy and going to seed by severely cutting them back about midsummer. A tool as indelicate as a pair of grass clippers can be used to provide a hard pinch if the planting is extensive. A period of several weeks with few flowers will follow until new reproductive growth begins. However, the fresh look and new flowers that result will carry the annuals right into the fall season.

WINTERIZING THE LANDSCAPE

Winter injury is any damage done to the landscape during the cold season of the year. The injury may be due to natural causes or to human error. It may be predictable or totally unexpected. At times, winter injury can be avoided, while at other times it can only be accepted and dealt with. It affects most elements of the outdoor room: plants, paving, steps, furnishings, and plumbing are all susceptible to damage. This discussion will center around the types of winter injury common to plants. However, it is important for the landscaper to guard against winter injury to all elements of the garden.

There are two general categories of winter injury: natural and human-induced. The examples that follow are common to most landscapes where the winter is severe.

Natural Injuries

Windburn Windburn can result when evergreens are exposed to strong prevailing winds throughout the winter months. The wind dries out the leaf tissue, and the dehydrated tissue dies. Windburn causes a brown to black discoloration of the leaves on the windward side of the plant. Very often, leaves further into the plant or on the side opposite the wind show no damage. Broadleaved evergreens are highly susceptible to windburn because they have a great amount of leaf surface. Responding to dropping temperatures, many broadleaved evergreens roll their leaves in the winter to reduce the amount of exposed surface area, Figure 11-18.

Damage from Temperature Extremes Temperature extremes are injurious to plants that are at the limit of their hardiness (termed *marginally hardy*). Damage may range from stunting (when all of the previous season's young growth freezes) to death of the plant, if the severely cold weather is prolonged.

After an especially cold winter, certain plants may show no sign of injury except that their spring flower display is absent. This happens if the plant's flower buds have frozen while the leaf buds have not. The injury is common to forsythia and certain spireas in the northern states.

Unusually warm weather during late winter can also cause plant damage. Fruit trees may be encouraged to bloom prematurely, only to have their flowers killed by a late frost. As a result, the fruit harvest can be greatly reduced or even eliminated. Spring-flowering bulbs can also be disfigured if forced into bloom by warm weather followed by freezing winds and snow.

Sun Scald Sun scald is a special type of temperature-related injury. It occurs when extended periods of warm winter sunshine thaw the above-ground portions of a plant, but the period of warmth is too brief to thaw the root system. Above ground, the thawed plant parts require water, which the frozen roots are unable to provide. Consequently, the tissue dries out and a scald results.

Sun scald is especially troublesome on evergreens planted on the south side of a building. It also occurs on newly transplanted young trees in a similar location. The thin young bark scalds easily, and the moisture content of the tissue is already low because of the reduced root system.

FIGURE 11-18.
This rhododendron has begun to roll its leaves as a means of reducing the exposed surface area during the winter.

Heaving When the ground freezes and thaws repeatedly because of winter temperature fluctuations, turfgrass, bulbs, and other perennials can be forced up from the soil. When their roots are exposed to the drying winter wind, they are killed.

Damage from Ice and Snow The sheer weight of snow and ice on plant limbs and twigs can cause breakage and result in permanent destruction of the plant's natural shape. Evergreens are most easily damaged because they hold heavy snow more readily than leafless deciduous plants, Figure 11-19. Snow or ice falling off a pitched roof can split foundation plants in seconds, Figure 11-20. Plants that freeze before snow settles on them are even more likely to be injured, because freezing reduces plant flexibility, causing twigs to snap rather than bend under added weight. Unfortunately, the older and larger a plant becomes, the greater the damage resulting from heavy snowfalls and ice storms. There have been accounts of severe winter storms destroying the entire street trees of cities.

Damage from Animals Lacking other food in winter, small animals often feed on the tender twigs and bark of plants, especially shrubs. Bulbs are also susceptible. Entire floral displays can be destroyed by the winter feeding of small rodents. Shrubs can

FIGURE 11-19.
Heavy snow can break the limbs of evergreens. Snowplow damage can also result if trees are planted too close to the road edge.

be distorted and stunted by the removal of all young growth. In places where the plant becomes girdled, it is unable to transport nutrients and will eventually die.

Human-Induced Injuries

Certain types of injury are inflicted by human beings during wintertime landscape maintenance. Some types of injury result from carelessness on the part of groundskeepers. Other types are the predictable result of poor landscape design. A large number occur because the landscape elements are hidden beneath piles of snow.

Damage from Salt Salt is toxic to nearly all plant life. When salt is applied to melt winter ice;

FIGURE 11-20.
Falling ice can split and break shrubs planted beneath the overhang of a roof. (Courtesy Rodney Jackson, Photographer)

it can harm trees, shrubs, bulbs, and lawns as well as paving. Often the damage does not appear until long after the winter season has passed. Thus the cause of the injury may go undiagnosed. The injury may appear as strips of sterile, barren soil paralleling walks where salt was applied. Burn injury can also be seen on the lower branches of evergreens. Some sensitive plants weaken and die after several winters of excessive salt runoff.

Damage from Snowplows Snowplows can damage plant and construction materials in several ways. A plow operator may be careless or unfamiliar with the landscape and push snow onto a planting or a bench. On the other hand, poor design may have placed plants and other vulnerable elements too close to walks, parking areas, or streets, where they interfere with winter snow removal. Damage to lawns can result when the plow misses the walk and actually plows the grass. The turf may not survive if this occurs repeatedly.

Rutting of Lawns When the surface layer of the soil thaws but the subsoil remains frozen, surface water is unable to soak in. Users of the landscape accustomed to finding the ground firm may be unaware of the damage they cause by parking their vehicles on temporarily soft lawns. The soil can become badly compacted, resulting in unsightly ruts.

Reducing Winter Injury

Some types of winter damage can be eliminated by winterizing the landscape during the preceding autumn. Other types can be reduced by better design of the grounds. Still other winter injuries can only be minimized, never totally eliminated.

Windburn Windburn can be eliminated in the design stage of landscaping by selecting deciduous species, rather than evergreens, for windy corners and other exposed locations. If evergreens are important to the design or already exist in the garden, windburn can be reduced by:

- wrapping the plants in burlap during the winter
- applying antitranspirants in the autumn and again in late winter to reduce water loss from the tissue

The use of antitranspirants is more practical for large evergreens than wrapping. In very cold regions, both techniques may be applied to certain plants.

Damage from Temperature Extremes Temperature extremes can only be partially guarded against. Where wind chill is a factor, plants should be located in a protected area. Burlap wrapping also helps, especially with hardy species that have tender flower buds.

Certain plants, such as roses, can be cut back in the autumn and their crowns mulched heavily to insulate them against the effects of winter. Similarly, any plant that can be damaged by freezing and thawing of the soil should be heavily mulched after the ground has frozen to insulate against premature thawing.

To prolong the life of annual flowers in the autumn, when a frost is forecast, the foliage can be sprinkled with water prior to nightfall. The water will give off heat as it changes from liquid to ice and can keep the plant tissue from freezing if the temperature does not go too low for too long.

Sun Scald Sun scald of transplants lessens as plants grow older and form thicker bark. Wrapping the trunks of young trees with paper or burlap stripping can provide an effective protection during the first year or two after transplanting. For other types of sun scald, such as that which affects broadleaved evergreens, the same remedies practiced for windburn and temperature extremes are effective.

Some sun scald can be avoided by the landscape designer. Vulnerable species of plants should not be placed on the south side of a building, nor should they be placed against a white wall that will magnify the sun's effect on the above-ground plant parts.

Heaving Heaving of the turf is impossible to prevent completely. The best defense against it is to encourage deep rooting through deep watering, properly timed fertilization, and correct cutting. Bulbs, groundcovers, and other perennials can be protected from heaving by application of a mulch after the ground has frozen. The mulch acts to insulate the soil against surface thawing.

Damage from Ice and Snow Ice and snow damage to foundation plants can be avoided if the designer is careful not to specify plants beneath the overhanging roof line of buildings. If the groundskeeper must deal with plants already existing in an overhang area, the use of hinged wooden A-frames over the plants can help to protect them, Figure 11-21. As large pieces of frozen snow and ice tumble off the roof, the frame breaks them apart before they can damage the plants.

To aid plants that have been split or bent by heavy snow accumulation, the groundskeeper must work quickly and cautiously. A broom can be used to shake snow off weighted branches, but this must be done gently and immediately after the snow stops. If the branches are frozen or the snow has become hard and icy, removal efforts may do more harm than good.

If plant breakage occurs during the winter, the groundskeeper should prune the damaged parts as soon as possible. This reduces the possibility of further damage to the plants during the remainder of the winter.

Large and valuable plants in the landscape can be winterized in the autumn by tying them loosely with strips of burlap or twine. (Do not use wire.) When prepared in this manner, the branches cannot be forced apart by heavy snow, and splitting is avoided.

Damage from Animals   Animal damage can be prevented either by eliminating the animals or by protecting the plants from their feeding. While rats, mice, moles, and voles are generally regarded as offensive, plantings are damaged as much or more by deer, rabbits, chipmunks, and other more attractive kinds of animals. Certain rodenticides may be employed against some of the undesirable animals that threaten the landscape. In situations where the animals are welcome but their winter feeding dam-

FIGURE 11-21.
Hinged wooden A-frames protect foundation plantings from damage caused by sliding ice and snow

FIGURE 11-22.
A plastic coil around the trunk of a young tree can protect it against girdling by rabbits and rodents.

age is not, a protective enclosure of fine mesh wire fencing or a hard plastic coil around the plants helps to discourage animals from feeding there, Figure 11-22. Nontoxic liquid repellants can also be applied to the base of trees and shrubs in the fall to discourage animal feeding.

Damage from Salt Salt injury to turf adjacent to walks can be minimized if caution is exercised by the groundskeeper. Salt mixed with coarse sand does a better job than either material used separately and reduces the total quantity of salt applied. The sand provides traction on icy walks, and a small amount of salt can melt a large amount of ice.

Damage from Snowplows Snowplow damage is to be expected if plants are too close to walks and roadways. Therefore, when planning landscapes for snowy climates, the designer should avoid placing shrubs near intersections or other places where snow is likely to be pushed. Another type of damage results from the plow driver's inability to see objects beneath the snow. If possible, all plants and other objects that cannot be removed for winter should be marked with tall, colored poles that can be seen above the snow. Whenever possible, snow blowers should be used instead of plows. These machines are much less likely to cause damage.

Rutting of Lawns Rutting of lawns usually results from permitting cars to be parked there. The best solution to the problem is to avoid the practice. Otherwise, sawhorses or other barriers offer a temporary solution.

SUMMARY

As important as the design and proper installation of landscape plantings is the ongoing maintenance of the plantings.

Watering must promote deep root development to establish new transplants successfully. Later, watering must be sufficient to keep the plant healthy throughout the growing season. Not all plants require the same amount of water. Infrequent, deep watering is preferable to frequent, shallow watering.

Fertilization of trees and shrubs promotes healthier and more vigorous growth. Generally, they are best fertilized in the spring when they can take full advantage of the nutrients and still harden off before winter arrives.

Mulching, edging, pruning, and pest control are also important parts of a professional maintenance program. Pruning is the removal of a portion of a plant for better appearance, improved health, controlled growth, or attainment of a desired shape. To prune properly, landscapers must understand the structure of trees and shrubs, when to prune to accomplish the desired objective, what and where to cut, and the appropriate tools to use.

Flower plantings are labor intensive, requiring much hand work to keep them attractive. Herbicides, black plastic, and mulch can help control weeds. Frequent watering, midsummer fertilization, and pinching are also necessary.

Winter injury can be due to either natural causes (windburn, temperature extremes, sun scald, heav-ing, ice and snow, or animals) or human error (salt-ing, snowplow damage, or rutted lawns). Certain types of injury can be reduced or eliminated by properly winterizing the landscape during the pre-ceding autumn.

Achievement Review

A. MULTIPLE CHOICE

From the choices given, select the answer that best completes each of the following statements.

1. The main objective when watering trees and shrubs is to _____.
   a. keep the humidity high
   b. promote deep root development
   c. keep the foliage clean and moist
   d. prevent wilting
2. The proper time to fertilize shrubs is in _____.
   a. early spring
   b. late summer
   c. early fall
   d. winter
3. Trees should be fertilized _____.
   a. near their anchor roots
   b. at the nursery before digging
   c. near their base
   d. beneath the drip line of the canopy
4. The purpose of edging a bed is to _____.
   a. divert surface water
   b. discourage turf roots from entering the bed
   c. hold the mulch within the bed
   d. improve drainage

B. SHORT ANSWER

Answer each of the following questions as briefly as possible.

1. You discover a leaf spot disease on the English ivy plantings that are the major groundcover of a large landscape. How can you determine what control to use, what products to select, and when and how much to apply?
2. Complete the following statements.
   a. Removing a portion of a plant for better appearance, improved health, controlled growth, or attainment of a desired shape is termed _____.
   b. The most important branch on a tree is the _____.
   c. Scaffold branches create the _____ of the tree.
   d. The point at which a branch meets the trunk or a larger branch is termed the _____.
   e. Undesirable branches originating from the root system of a tree are termed _____.
   f. Undesirable branches originating from the trunk of a tree are termed _____.
   g. The point of a shrub at which the branches and root system meet is termed the _____.
   h. Generally, the younger branches in a shrub will have a _____ color.
   i. Shoots originating from the stock of a grafted plant are _____ compared to those originating from the scion.
   j. Flower and fruit production occur most profusely on _____ branches.

3. Place X's in the following chart to compare the advantages and disadvantages of pruning during different seasons.

Characteristic	Spring	Summer	Autumn	Winter
A full view of the plant's branching structure is possible.				
View of the plant's branching structure is blocked.				
Dead branches cannot be detected.				
The timing may conflict with more profitable activities.				
The most time is allowed for cuts to heal.				
This is the best season to prune for the good of most plants.				
This is the best season to prune shrubs that flower in early spring.				
Plants may be damaged by unintentional breakage.				

4. Which three branches shown in this drawing should be removed, and why?

5. What is the difference between thinning out and heading back in the pruning of shrubs?

6. Indicate a maintenance technique that will reduce or eliminate the following types of winter injury.
   a. windburn on coniferous evergreen trees
   b. a flowering shrub that fails to bloom in the spring
   c. roses that exhibit long dead canes every spring
   d. young trees that show sun scald in the spring
   e. bulbs and groundcovers that heave to the surface and dry out during the winter
   f. tall shrubs that are split by falling ice
   g. young trees girdled by rabbits
   h. small shrubs broken by a snowplow because the driver was unable to see them

C. ESSAY

Write a short essay on the maintenance of flower plantings. Outline the care needed by both annuals and perennials to keep them healthy and attractive throughout the growing season. Assume the environmental conditions to be those in your own area.

12

Special Training Techniques for Plants

Objectives

Upon completion of this chapter, you will be able to

- define vines, espalier pruning, topiary pruning, and bonsai.
- explain how each specialized plant form is used.
- train plants for use in these specialized forms.

SELECTING AND USING VINES IN THE LANDSCAPE

Vines are plants with a vigorous central lead shoot and a long, linear growth habit. They do not grow naturally into clumped, shrub-like forms and are seldom able to grow upright without support. They may be either woody or herbaceous. Vines are among the most versatile plants that can be selected for use in the landscape. Given the opportunity to climb, they form interesting walls for the outdoor room. Allowed to trail, they carpet the outdoors with rich and varied textures. If trained to grow on a trellis or along a fence, they can be controlled to create various effects envisioned by a designer. Left to grow naturally, they lend charm and softening qualities to a landscape that might otherwise appear too contrived. Vines are especially useful

where height is needed from a plant, but a narrow planting space makes trees or shrubs impossible. Table 9-5 in Chapter 9 lists some of the more common vines available for use in landscapes.

Vines climb in one of three ways depending upon the species, Figure 12-1.

1. Some climb by *twining* themselves around a trellis, fence, or another plant.
2. Others produce fine *tendrils* that wrap around the supporting structure and allow the vine to climb.
3. Still other vines produce *holdfasts* that permit the plant almost to glue itself to the support.

A landscaper should be aware of the climbing methods of each species being considered for use in the landscape before making a final selection.

FIGURE 12-1.
How vines climb (From J. Ingels, *Landscaping: Principles and Practices,* 2nd edition, © 1983 by Delmar Publishers Inc.)

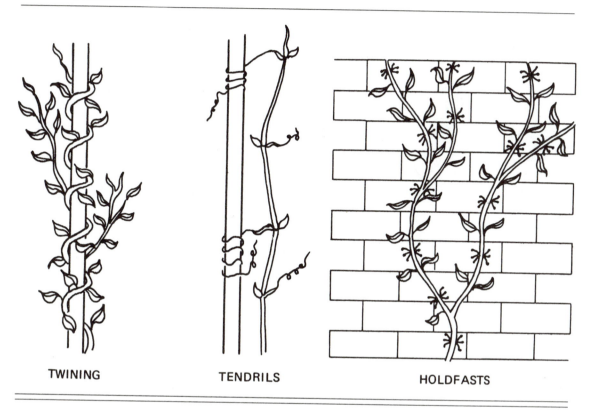

TWINING   TENDRILS   HOLDFASTS

Damage to buildings can result from excessive moisture retention caused by vines growing against an unsuitable surface or from pitting of the surface by the vine's holdfasts.

As with trees and shrubs, vines can be deciduous or evergreen, flowering or nonflowering, dense or sparse. Many people avoid using vines because of their maintenance requirements. They do require some special attention if maximum flowering is to be obtained, or if vigorous growth is to be kept under control. However, to avoid their use is to deprive the garden of one of its most contributive elements, Figure 12-2. Factors to consider when choosing vines include:

- *Hardiness:* Does the vine need a sheltered location or can it survive an open location?

- *Sun orientation:* Although many flowering vines will survive on a north-facing wall, they do not flower satisfactorily and their vegetation may be sparse.
- *Amount of coverage desired:* It is unwise to select a vigorous species where only a tracework of vine is desired.
- *Amount and type of support required:* Will a trellis be required or does the species climb unaided?

INSTALLING AND TRAINING VINES

The installation and training of vines will be discussed in three stages: planting, initial training, and developmental pruning.

FIGURE 12-2.
Vines contribute an architectural quality to this building. Regular maintenance is required to keep the windows and shutters from being overgrown.

Planting

Vines are usually sold as containerized plants. The method of installation described for containerized trees and shrubs is also applicable to the installation of vines.

Points specific to vine planting have to do with the vine's need for support. Even if the vine is to grow vertically, it should not be placed directly against the support. The soil next to a building wall or even around a fence post is often drier than the surrounding soil. Therefore, a foot of space should be left between the support structure and the new vine.

If the vine is a species that climbs by holdfasts, it will need no additional support. However, if the vine climbs by twining or tendrils, it will need support in the form of a trellis or wire fencing. To climb an open style of fencing, support assistance is seldom needed after the initial training.

Where a trellis or support fence is required, it must be in place before the vine is planted. The support must be strong enough to hold the mature vine, with its fruit if necessary, or when it is covered with heavy snow. The support must also be securely anchored so that it does not become a sail when the wind blows.

Initial Training

The initial training of a vine is intended to direct and accelerate its coverage. This may involve tying the shoots of twining and tendril species to the support; holdfast-producing vines will not need to be tied. Shoots that grow outward from the wall or support should be pruned off. Only lateral shoots growing parallel to the support should be permitted to develop.

If the young vine does not have several vigorous shoots at its base at the time of planting, it should be pruned back to the lowest pair of healthy buds. From these buds several vigorous shoots will develop, which can then be tied or otherwise directed.

Application of a complete, low-analysis fertilizer to the vine during its first spring of new growth, and deep watering to promote deep rooting, will assure a successful beginning for the vine.

Developmental Pruning

Developmental pruning is usually necessary if the vine's full potential is to be realized. The amount of work required depends upon the species. If the vine climbs by holdfasts, and total wall coverage is the only objective, little attention is needed except the removal of dead or damaged shoots and those extending too far outward from the wall.

If the pattern of growth is to be more restricted and defined, the vine is likely to require regular pruning once or twice annually. If profuse flowering is desired, young wood must be encouraged and unruly, older, nonflowering shoots must be pruned away. Severe pruning can cut away key tendrils or entwined branches that support the vine, however. Even vines that climb easily may need to be tied to the support in strategic places to guard against falling after a heavy pruning.

Since vines, like trees and shrubs, flower at different times of the year, they cannot all be pruned at the same time if the flower display is to be encouraged. Table 12-1 provides a general timetable for pruning.

TABLE 12-1
A Guide to Pruning Flowering Vines

Time of Flowering	Season Flower Buds Produced	When to Prune
Spring (April to June)	Preceding summer	Cut the flowering shoots back to within several inches of the main framework branches *immediately after flowering.*
Early summer (May to July) and late summer (August to September)	First flowers are from old wood of the previous year. Second flowers are on young shoots of the current season.	Cut back each time after flowering by removing one-third of the shoot. Fertilize after the first summer pruning.
Summer and autumn (June to September)	Current season on new growth	Prune back all the growth from the previous year *during the winter* and mulch the crown, or prune in the *early spring* before shoot growth begins.

ESPALIERS

Among the horticultural contrivances of gardeners over their centuries of cultivating ornamental plants are espaliers. *Espaliers* are trees and shrubs that are allowed to develop only two dimensionally: they have height and width, but hardly any depth, Figure 12-3. In appearance, they may suggest a vine to the viewer because of their flatness. Whereas vines are a natural growth form of plants, however, espaliers are entirely created by gardeners and must rely totally upon continuing human attention to exist in their unusual form.

Espaliers were originally developed as a way to incorporate into gardens desirable fruiting species, such as apples, pears, and peaches, without giving them a lot of space. The trees were trained against courtyard walls and along fences. They were even developed inside early orangeries (see Chapter 20) and in greenhouses. Later, other species were tried and found usable solely for their decorative effect.

Espaliers are most commonly grown against walls or fences for two reasons.

1. The support is convenient and necessary for initial training of the young plant.
2. The wall provides a good background for viewing the plant and its branching pattern. Although a decorative branching pattern is not necessary, it is traditional, and over the years some attractive and clever styles have been developed. It is possible for certain tree espaliers to be free-standing however, Figure 12-4.

As elements of the landscape, espaliers serve as novelties, since they are not commonplace plant

FIGURE 12-3.
A row of espaliered trees covers this brick wall.

forms. If used correctly, they can function as focal points or as accent features against a garden wall. Their uses are similar to those of decorative vines, yet they have a distinctively different appearance.

INSTALLING AND TRAINING ESPALIERS

The espalier begins its training as a one-year-old tree or shrub with no developed lateral shoots at its base. It may be planted against a wall initially and a support frame set behind it, or it may be grown in a container for several years with a training frame inserted into the container.

The plant is installed like any other tree or shrub, set about 12 inches from the wall, fence, or lattice that is to support it. At the time of installation, the plant should be dormant, Figure 12-5.

Step 1. Select an unbranched, one-year-old plant and install while it is dormant. Prune back the single stem to a height of 15 to 18 inches. Allow three healthy buds to remain near the top of the cut stem. The lower two must point in opposite directions.

Step 2. After the shoots break dormancy, train the top shoot to grow vertically and the lower two shoots to grow horizontally. The shoots may be

FIGURE 12-4.
This espaliered fruit tree is being trained inside a greenhouse, but it is self-supporting. The wires direct and hold the young branches until they mature.

supported with bamboo stakes tied to the support frame. In the first growing season for each set of horizontal branches, train and tie the branches at 45° angles rather than full 90° angles. Remove any lateral shoots growing outward from the wall and cut back any vertical shoots from the horizontal branches to allow only two or three nodes on each vertical spur.

Step 3. At the end of the growing season, the two horizontal shoots (now branches) should be removed from the bamboo stakes and tied to the frame at 90° angles.

Step 4. After the plant has gone dormant again, prune the central branch back to a point 15 to 18 inches above the horizontal branches. Allow three more healthy buds to remain near the top

FIGURE 12-5.

Steps involved in training an espalier

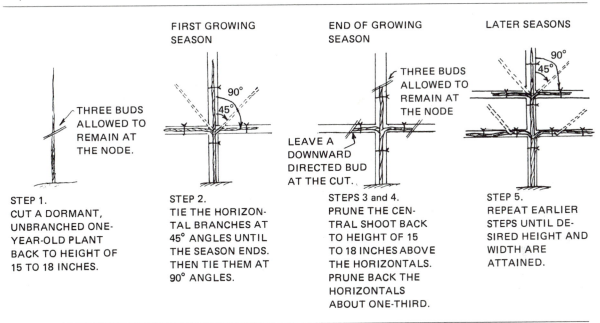

FIRST GROWING SEASON

END OF GROWING SEASON

LATER SEASONS

THREE BUDS ALLOWED TO REMAIN AT THE NODE.

90°

45°

THREE BUDS ALLOWED TO REMAIN AT THE NODE

LEAVE A DOWNWARD DIRECTED BUD AT THE CUT.

90°

45°

STEP 1.
CUT A DORMANT, UNBRANCHED ONE-YEAR-OLD PLANT BACK TO HEIGHT OF 15 TO 18 INCHES.

STEP 2.
TIE THE HORIZON-TAL BRANCHES AT 45° ANGLES UNTIL THE SEASON ENDS. THEN TIE THEM AT 90° ANGLES.

STEPS 3 and 4.
PRUNE THE CEN-TRAL SHOOT BACK TO HEIGHT OF 15 TO 18 INCHES ABOVE THE HORIZONTALS. PRUNE BACK THE HORIZONTALS ABOUT ONE-THIRD.

STEP 5.
REPEAT EARLIER STEPS UNTIL DE-SIRED HEIGHT AND WIDTH ARE ATTAINED.

of the cut stem. The lower two must point in opposite directions. Prune back both of the horizontal branches about one-third. A downward-oriented bud should be left at each cut. Lateral buds along the main trunk should be reduced to three or four, not counting the three at the top, which will form the new branches during the next growing season.

Step 5. Repeat these steps until the plant attains the height and width desired. After that, little or no additional extension of the framework branches should be permitted. They can be kept pruned back to the previous year's terminal points.

Should the support frame or trellis weaken over time, the espalier can be attached to the wall with special nails, part steel and part pliable lead. The steel end is driven into the wall and the pliable portion is hooked around the branch of the plant.

Horticulturists with enthusiasm and opportunity for espalier training have developed numerous complicated style variations, Figure 12-6.

- single, double, and triple U-shapes
- horizontal
- palmette verrier
- palmette oblique
- fan shape
- Belgian fence
- losange
- free-standing spiral
- free-standing pyramidal

Truly espalier training is one of the highest manifestations of the horticulturist's skills.

TOPIARY PRUNING

Equally clever, difficult, and historic is the ancient craft of topiary pruning, Figure 12-7. *Topiary* is a technique of shearing plants into nontypical shapes, usually sculptural in form. Begun centuries ago, when nature had been tamed far less than today, topiary was an attempt to illustrate human dominance over Nature. Once, entire gardens were created where no plants were permitted to develop

FIGURE 12-6.
Eleven different styles of espalier pruning

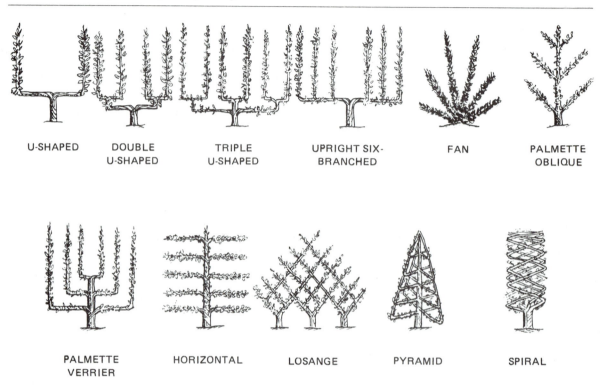

| U-SHAPED | DOUBLE U-SHAPED | TRIPLE U-SHAPED | UPRIGHT SIX-BRANCHED | FAN | PALMETTE OBLIQUE |

| PALMETTE VERRIER | HORIZONTAL | LOSANGE | PYRAMID | SPIRAL |

natural shapes. The English yew was the plant most often selected for topiary shaping and there are historic gardens throughout Britain and Europe whose topiary yews date back hundreds of years. Other species of plants have been tried with varying degrees of satisfaction. Whatever the species used, it must be one that accepts repeated pruning with little or no growth permitted after the desired shape is attained. Species that do not produce new growth on old wood are unsuitable as topiaries.

Topiaries are garden novelties, such as espaliers are. They have been shaped into forms as fantastic as the imaginations of the horticulturists. Corkscrews, chess pieces, chairs, peacocks, tables, and countless animals are but a few of the subjects chosen, Figure 12–8. One famous amusement park has created dozens of modern topiaries in the shapes of characters from animated films.

Historic techniques of topiary pruning required time, patience, and complicated training techniques. If the winter season was harsh, a section of plant could be killed, leaving a peacock without a tail or some similar tragedy. In such cases then as now, there was no alternative to rebuilding the plant from new shoots or by grafting.

With predictable impatience, modern gardeners sometimes wish to create the effects of topiary without the lengthy wait. They use more rapidly growing species such as the privets, which accept severe pruning and recover quickly. They plant the rapidly growing species inside wired forms that are commercially available. As the plant fills the form, the gar-

FIGURE 12-7.
Topiary pruned arborvitaes accent the entry of this home.

FIGURE 12-8.
Specialty nurseries can provide espaliered and topiary pruned plants to be grown further on the landscape site. The espaliers shown here are firethorns, and the topiaries are juniper. (Courtesy Monrovia Nursery, Azusa, CA)

dener shears the plant back to the wire form, creating a fuller plant and a satisfactory topiary in a relatively short period of time with relatively little skill.

Another modern variation on the topiary of old uses a vine such as English ivy and a mold of sphagnum moss. The ivy is encouraged to grow over the form, which rapidly disappears from sight, leaving some curiosity such as an ivy-covered French poodle in its place, Figure 12-9.

Topiary pruning has existed for a long time. It comes in and out of fashion depending upon current attitudes toward formal garden design. It has ties to our modern clipped hedges and to the Japanese craft of cloud pruning (suggesting tufts of clouds in a garden by the manner of pruning certain shrubs), yet it remains a unique and distinctive use of plants.

BONSAI DEVELOPMENT

Bonsai is the ancient Japanese craft of dwarfing trees. Forest giants, such as pines, maples, beeches, and spruces have been successfully maintained in shallow containers for a hundred years or more while restricted to a height of mere inches, Figures 12-10 and 12-11.

In the hands of a skilled bonsaist, the containerized tree acquires a mature appearance but in miniature. The same shaping, molding, and damage by

FIGURE 12-9.
An ivy-covered topiary form

FIGURE 12-10.
This bonsaied black pine is 100 years old. (Courtesy Brooklyn Botanic Garden)

wind and storms that determine the noble, often craggy character of a tree in nature are interpreted in the bonsai. So much of the Japanese concept of landscape gardening is related to the symbolism of nature in miniature (see Chapter 9) that their fondness for bonsai is understandable.

Even more than espalier or topiary, bonsai requires time and patience, Figure 12-12. All bonsais require several years even to begin developing a shape suggestive of a full-sized tree. Many successful bonsais outlive the gardeners who work with them. Thus, the ancient bonsai trees have been developed and cared for by a series of horticulturists, each one appreciative and respectful of the work done on the plant in the past.

Obviously, bonsai development is not easy. Although it has great popular appeal, it has proven too time-consuming for most hobbyists to pursue. Also, many people, believing the bonsai to be a houseplant, have been disappointed when it re-

sponded to its place of honor on the coffee table by dying. Most modern homes are too warm, too dry, and too dimly lit for the traditional woody species of the temperate zone to survive as bonsais. More recently, species from the subtropics and tropics have been tested as bonsai plants and many have proven highly satisfactory, at least for the Western market. Bonsai purists would probably not approve.

Use of traditional, temperate-zone, woody species for bonsais necessitates that they be treated in some ways like other woody plants. They must be kept outdoors most of the time, perhaps on the patio or porch, and only brought in occasionally for indoor display. They may need a period of dormancy during the winter but will not tolerate severely cold weather for extended periods. During their indoor visits, bonsais need plenty of light, cool daytime temperatures (about 68° F) and even cooler temperatures at night (from 62° to 65°F).

Whether temperate-zone trees or houseplants are selected, the bonsai usually begins its development

FIGURE 12-11.
This white pine is estimated to be over 500 years old. (Courtesy Brooklyn Botanic Garden)

FIGURE 12-12.
A bonsaied camellia started in 1964. The extensive wiring indicates that it is still in its training stage.

as a seedling or sapling, although it is possible to use a commercially produced containerized nursery shrub or to collect from the wild. A few nurseries specialize in the production of young plants suitable for bonsai development.

Tools, Containers, and Soil Mix

Special bonsai tools are available, but a beginner can easily assemble the following assortment.

- scissors
- pruning shears
- copper wire
- a wire cutter
- a strong thin stick such as a knitting needle or chopstick

Bonsais can be started in ordinary clay pots or even greenhouse flats. Repotting is one of the requi-

sites of successful bonsai development, so a more stylish container can be selected later. Ultimately, the container will play an important role in complimenting the style of the bonsai and determining its size. Cascading forms will require a taller container to permit the descending branching habit. Forms that are more suggestive of old, upright trees will require a wide, shallow container. In every case, the bonsai container should not be overly decorative, and what decoration exists should suggest the Orient if it is to be in character with the plant. Also, the containers should have drainage holes in the bottom or be able to be drilled without breaking. Unglazed pottery makes an ideal container if all other characteristics are present. To prevent the loss of soil from the container, the drainage holes should be covered with screen wire, metal mesh, or hardware cloth, which can be taped in place with florists' waterproof adhesive tape.

Soil for the bonsai must drain well. A mixture of equal parts of coarse sand, peat moss, and loam will be satisfactory. Each soil component should be put through a fine-mesh sieve before mixing to remove the smallest particles, which can plug the container's drainage holes. The soil should also be pasteurized before use to kill weed seeds and soil-borne pests. The soil is most easily prepared and used while it is dry. Gravel spread on the bottom of the container before adding the soil will aid drainage.

Styles of Bonsai

Since a bonsai symbolizes a larger tree, the most popular styles can be anticipated by recalling how trees grow naturally. Most grow upright, of course, but under various environmental conditions trees are seen to cascade over embankments (perhaps as the soil erodes away from their roots), lean away from a strong wind, or clutch a mountain side. They are also found as single, perfect specimens or in groups where each individual's appearance is lost in the larger shape of the grove, Figure 12-13.

Beginning and Wiring the Bonsai

The plant should be shaped before it is placed in the container. The shallow root system does not permit pruning of the plant until it has time to establish itself in the container.

The bonsaist first selects the style that the plant is to attain. Branches important to the shape are retained; others are pruned away. If cuts are extensive, a wound paint may be used. If the formal, upright style is selected, the branches will need to

FIGURE 12-13.
Typical bonsai styles

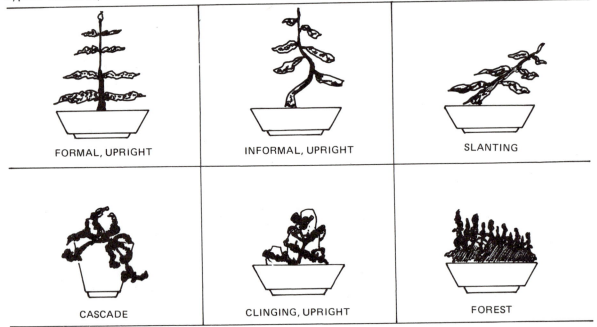

| FORMAL, UPRIGHT | INFORMAL, UPRIGHT | SLANTING |
| CASCADE | CLINGING, UPRIGHT | FOREST |

be evenly distributed around the plant with closely spaced internodes. Branches must be pruned so that they are largest and longest at the base and get shorter and smaller as they approach the top of the plant. The central branch must be straight and strong. In informal uprights, the tapered branch selection is the same, but the central branch can be crooked.

Slanted or cascading styles often have three major branches, representing heaven, earth, and Man, a relationship that is spiritually ingrained in the Oriental tradition. Heaven is always the uppermost branch; earth is the lowest branch; and Man is the middle branch. Other branches may exist on the bonsai, but they are not permitted to compete with the major three. Cascade styles of bonsai are one-sided styles, so the branches should be selected with that in mind. Other styles are not so flat, although many will have a preferred direction or side for viewing, Figure 12-14.

It is unlikely that the plant will have the desired shape, with each branch in its proper place, after the first pruning. Wiring will be needed to direct the plant's growth as the style develops. Copper wire of different gauges is used because copper is pliable and weathers quickly to a neutral, unobjectionable appearance.

Wiring should be started at the bottom of the tree and worked upward, Figures 12-15 and 12-16. Heavy-gauge wire should be used to direct the trunk. The end of the wire should be anchored by insertion several inches into the soil ball. Then evenly spaced and parallel coils of wire should be wrapped around the trunk at approximately one-inch intervals. The trunk wire should stop just short of where the branches meet the trunk to allow the thinner wire used for the branches to be anchored around the trunk. The branch wire should first be anchored to the trunk, then wrapped in coils that match and parallel those on the trunk. The finest gauge wire should be used to direct the thin twigs. It should be anchored to a branch before beginning the coiling. Not all bonsais require the wiring of all three parts: trunk, branches, and twigs.

FIGURE 12-14.
A young bonsai receiving its first pruning

Pruning the Roots and Potting the Bonsai

Before the new plant is placed in its container, and each time it is repotted in the years that follow, the root system must be reduced. Restricting the root system to a size that will sustain the plant without encouraging vigorous growth is critical to the bonsai technique. Such pruning maintains the vital support relationship between the above-ground and below-ground parts of the plant.

Using a long stick, probe the soil ball until the soil has fallen away from the outer roots and they are hanging free. Cut away these roots. Continue until the soil ball and root mass are reduced to a size that will fit into the container chosen. It is easiest to do this when the soil is dry and important

FIGURE 12-15.
The wiring of a new bonsai requires wire of differing gauges. It is wrapped in evenly spaced and parallel coils around the trunk, branches, and twigs.

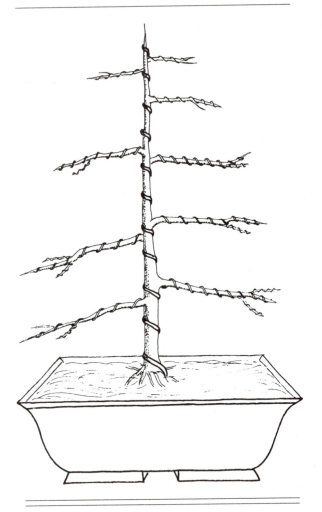

FIGURE 12-16.
Wiring the bonsai requires wire of different sizes

around the roots. Some of the plant's older roots may protrude above the soil. That is acceptable, even desirable, since it adds to the illusion that the tree is old. At least a quarter inch of space should be left below the top of the container for watering.

Follow-Up Care

After planting, cover the surface of the soil with moss or coarse sand to hold it in place and to give the bonsai a finished appearance. Water is applied frequently and from the top. Because the root system is restricted and the soil mass is shallow, the plant dries out quickly. Since it must never be permitted to dry out completely, it may need to be watered as often as once a day, if a houseplant has been used, or several times weekly if a woody plant has been used. If the soil and container have been prepared to drain properly, there is little danger of overwatering, but the plants should not be kept saturated.

not to disturb the soil directly beneath and around the trunk. Young, fibrous roots will develop within the soil ball, Figure 12-17. Each spring, the bonsai can be lifted from its container and a similar treatment given to the root system. This will prevent the plant from becoming potbound and will also permit wires to be removed or added.

The bonsai should be set in its container, off center and to the rear. The dry, pasteurized soil mix can then be added, taking care to avoid air pockets

FIGURE 12-17.
Reducing the root system of a bonsai plant

Fertilization of bonsais is also a frequent necessity due to the rapid leaching of nutrients by the drainage water. Nutrients can be applied several times monthly using a low-analysis fertilizer at half strength. They may also be applied using a pelletized time-release fertilizer that will release small amounts of nutrients over a three- to six-month period. Regardless of the form of fertilizer used, the amount of nitrogen should be limited to avoid the promotion of undesirable growth.

Should the bonsai become top-heavy, it will be necessary to wire it to the container. A heavy-gauge copper wire can be threaded like a shoestring through two drainage holes in the container and bent upward and around or through the soil mass to be tied around the trunk. This wire and the wires used to shape the bonsai should be either removed or loosened and reapplied before they can damage or mark the plant. For branches and twigs, the wire may have served its purpose after a month. Stiffer branches and the trunk may require a year before the shape is permanent.

Pest problems are no different with bonsais than other plants except that woody temperate plants, being displayed both indoors and out, may carry some insects into the home that would otherwise have remained outside. All bonsai plants should be set outside for spraying with pesticides. In the winter, a systemic product can be used. If the species does not accept pesticides without damage, as the Jade plant does not, the foliage can be washed with a dilute soap solution and rinsed off under the shower.

For growth indoors, bonsais need a well-lit location if they are normally outdoor species. Window greenhouses or artificially lit growth racks are ideal. House plant bonsais should receive the same amount of light required by others of their species. Ideally, the temperature will be on the cool side and the humidity will be maintained by frequent misting.

SUMMARY

Vines are plants with a vigorous central lead shoot and a long, linear growth habit. They are versatile landscape plants and generally underused in modern gardens. They are usually sold in containerized form and are planted in a manner similar to other containerized plants. If the vine needs a trellis or other support to aid its climbing, the support should be in place before the vine is planted. Fertilization, deep watering, and developmental pruning will help assure a successful transplant and a satisfactory effect.

Espaliers are trees and shrubs that are allowed to develop only height and width, not depth. In appearance, they suggest a vine. For training purposes and to provide a suitable background, espaliers are usually grown against walls or along fences, although free-standing forms exist. An espalier must be started as a young plant. To develop any of the popular decorative patterns, time, patience, and attention are required.

Topiary is a technique of shearing plants into nontypical shapes, usually sculptural in form. Like espaliers, topiary plants are garden novelties. Also as with espaliers, topiary requires patience and complex training techniques.

Bonsai is the ancient Japanese craft of dwarfing trees. In the hands of a skilled bonsaist, the containerized tree acquires a mature appearance but in miniature. Bonsai development takes the longest time of all the specialized training techniques.

This chapter provided instructions for training plants in these specialized forms.

Achievement Review

A. SHORT ANSWER

Answer each of the following questions as briefly as possible.

1. Define the following terms.
   a. vine
   b. espalier
   c. topiary
   d. bonsai
2. List six common bonsai styles.

B. MULTIPLE CHOICE

From the choices given, select the answer that best completes each of the following statements.

1. The two methods of vine climbing that require supplemental support are _____.
   a. twining and holdfasts
   b. holdfasts and tendrils
   c. tendrils and twining
   d. trellis and lattice
2. The best time to prune vines _____.
   a. is in the spring
   b. is in the early summer
   c. is in the autumn
   d. depends on the species
3. _____ are *not* products primarily of the horticulturist's craft?
   a. vines
   b. espaliers
   c. topiary shrubs
   d. bonsai
4. Vines function in the landscape as _____.
   a. softeners
   b. wall elements
   c. surfacing elements
   d. all of these
5. Espaliers and topiaries function in the landscape as _____.
   a. aesthetic contributions
   b. ceiling elements
   c. support elements
   d. mood evocators
6. How do bonsai plants function?
   a. as hobbies
   b. as patio or houseplants
   c. as expressions of Oriental nature symbolism
   d. all of these
7. In their first season of growth, the horizontal branches of an espalier should be tied _____.
   a. vertically
   b. at a 45° angle
   c. at a 90° angle
   d. loosely
8. The *most* critical relationship in the successful dwarfing of trees is that between _____.
   a. root mass and container size
   b. root mass and top growth
   c. nitrogen content and top growth
   d. wiring and top growth

C. TRUE/FALSE

Indicate if the following statements are true or false.

1. Topiary attempts to suggest a miniaturization of the natural world. It is rooted deep in Japanese garden tradition.

2. Espaliers may be supported or free-standing. If free-standing, they are three-dimensional.

3. Bonsai containers must have drainage holes.

4. Bonsai plants should be shaped and wired before being placed in containers.

5. The copper wire used to shape bonsais should be coiled randomly around each branch and twig.

6. Bonsai plants should be watered on a schedule similar to their outdoor counterparts.

7. Bonsais should be fertilized more frequently than outdoor plants due to nutrient leaching by the drainage water.

8. Wire used to shape bonsai plants can be left on indefinitely.

9. Temperate-zone woody plants are well suited to be indoor bonsais.

10. Bonsai is a technique of landscape pruning.

11. Each year when a developing espalier goes dormant, pruning should cut back the central branch to a point 15 to 18 inches above the horizontal branches and the horizontal branches by one-third.

13

Turf: Selection, Establishment, and Maintenance

Objectives

Upon completion of this chapter, you will be able to

- select appropriate turfgrasses for particular landscape uses.
- explain how grass seed mixtures vary.
- interpret a grass seed analysis label.
- describe common methods of turf installation.
- explain how a spreader is calibrated.
- outline a maintenance program for professional lawn care.

THE MANY USES OF TURFGRASS

Turfgrasses appear at both ends of a landscape time spectrum. They are among the oldest species used by human beings to fashion the outdoors into residential and recreational areas that suit their needs. They are also responsible for one of the newest and most rapidly expanding branches of today's landscape industry. Selected as the most suitable surfacing for historic bowling greens and golf courses, turfgrasses have carpeted the outdoor rooms of cottages and castles for centuries. Today, full-service lawn care corporations are leading profit makers among landscape professionals.

The main reason for the ongoing interest in turfgrass and improved turf care is that nothing has ever surpassed a top-quality lawn as the ideal outdoor surfacing. It is comfortable to walk on, suitable for numerous athletic activities, ideal for picnic blankets and sunning; and it collects and holds dust and dirt from the air, absorbs sound, and produces oxygen.

Following are the major categories of use and a few examples of each.

- *Residential use:* private homes, estates, apartment and condominium complexes
- *Campuses*

- *Commercial use:* corporate landscapes, shopping centers, office buildings
- *Municipal use:* civic buildings, community centers, highway dividers, airports
- *Recreational use:* parks, golf courses, country clubs, stadiums, athletic fields

The diversity of uses is immediately apparent. Turf may function almost solely in an aesthetic capacity, as when it is the setting for a corporate giant. Other uses are utilitarian, as on a football field. No single turfgrass is suitable for all landscapes, geographic regions, soil types, altitudes and environmental conditions. Correct selection of turfgrass for a landscape requires knowledge of all these factors.

COMPARISON OF TURFGRASSES

Turfgrasses are monocotyledonous plants (having only one seed leaf in the embryo) whose growing point is at the crown near the soil. This low and protected location of the growing point permits turfgrasses to be mown and walked upon repeatedly.

Most turfgrasses used in landscapes nationwide are perennial, although there are several annual species of significance. Most species reproduce from seed, although several are propagated vegetatively.

A typical grass plant produces new leaves continuously from its growing point throughout the growing season. It also loses about the same number of older leaves through natural senescence and death. Whether growing naturally or under cultivation, turfgrasses will increase beyond the number of seeds sown. One of the objectives of good turfgrass management is to encourage that growth as quickly and as uniformly as possible.

Growth Habit

Grasses have differing growth habits, resulting from the three different ways that they produce new shoots, Figure 13-1.

1. *Rhizome-producing* (*rhizomatous*): The shoots are produced beneath the soil's surface and send new plants to the surface

FIGURE 13-1.
Growth habits of grasses

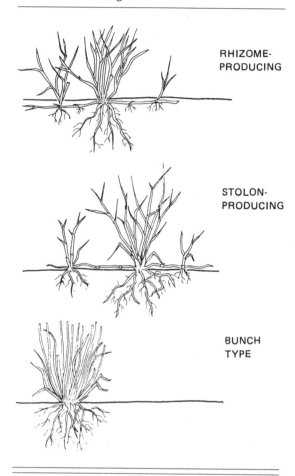

RHIZOME-PRODUCING

STOLON-PRODUCING

BUNCH TYPE

some distance out from the parent plant. The new plant develops its own root system and is independent of the parent plant, although the physical bond through the rhizome may continue.

2. *Stolon-producing* (*stoloniferous*): The shoots are produced and extend outward from the parent plant along the surface of the soil, not beneath it. The new plant develops independently as in rhizome-producing plants.

3. *Bunch-type:* New lateral shoots termed *tillers* are produced from axillary buds within the leaf sheath.

Rhizomatous and stoloniferous grasses tend to reproduce more quickly and evenly than bunch-type grasses. Therefore the bunch-types require more seed and closer spacing in order to cover an area quickly and avoid clumps in the lawn.

Texture, Color, and Density

Leaf *texture* is mostly a measure of the width of the leaf blade: the wider the blade, the coarser the texture. Generally, the fine-textured grasses are regarded as more attractive than the coarse-textured grasses. The color of a grass and its density will also differ among species. Color variance runs the full range of greens from pastel to dark and bluish. *Density* refers to the number of aerial leaf shoots that a single plant or species will produce.

Seed Size

Seed size varies greatly among grass species. Fine-textured grasses have small seeds; coarse-textured grass seeds usually are much larger. Thus a pound of fine-textured grass seed contains considerably more seeds than a pound of coarse-textured grass seed. A pound of fine-textured grass seed will also cover a larger area of land. For example, a pound of fine-textured Kentucky bluegrass contains approximately 2,000,000 seeds. That number of seeds will plant about 500 square feet of lawn. A pound of coarse-textured tall fescue contains about 227,000 seeds. Only 166 square feet of lawn can be planted with this particular seed. To further illustrate the differences in seed sizes, there are as many seeds in 1 pound of bluegrass as there are in 9 pounds of ryegrass; and as many seeds in 1 pound of bentgrass as there are in 30 pounds of ryegrass.

Soil and Climatic Tolerance

Most grasses do best in soil that is moderately fertile, neutral to slightly acidic, well aerated and moist but well drained. Nevertheless, there are sites in every state and county that fall short of those ideal conditions. Some species can adapt to a wide range of soil conditions, while others are very limited in adaptability.

Similarly, some grasses tolerate high humidity and reduced sunlight, others do not. Some prefer droughty climates, most do not. Some thrive in the subtropics and tropics, while others are better suited for temperate and subarctic regions.

Grasses are often grouped into two categories based upon the temperatures at which they grow best:

1. *Cool-season grasses* are favored by daytime temperatures of 60° to 75° F.
2. *Warm-season grasses* are favored by daytime temperatures of 80° to 95° F.

Knowledge of their preferred growing temperatures explains why lawns in northern states are often brown and dormant in midsummer when the days are very warm. In early spring and late fall, when temperatures in the southern states drop, warm-season grasses do not flourish there, Figures 13-2 and 13-3.

Use Tolerance

Under any given conditions of use, some grasses will hold up and others will quickly wear away. Some will accept heavy use and recuperate quickly, while others will rebound much more slowly. Some can accept the compaction of heavy foot traffic and still look good, while others discolor and slow their rate of growth markedly.

Disease and Insect Resistance

As turfgrasses have become more important to people, their growth in large monocultures (as on golf courses or sod farms) has increased the probability of infestation or infection by insects or pathogens. Breeding for resistance is an ongoing effort of horticulture scientists. Certain cultivars enjoy widespread use for a while until their resistance is overcome by a new mutant of an old pest.

In Table 13-1, many of the most commonly used turfgrasses are compared.

FIGURE 13-2.

Relationship of temperature to growth rate for cool-season and warm-season grasses (Courtesy United States Department of Agriculture)

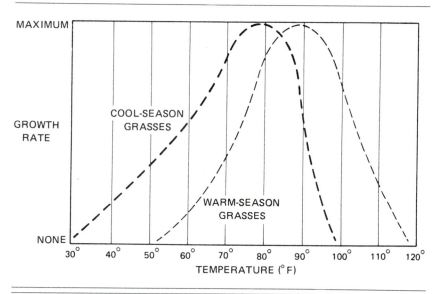

PURCHASING GRASS SEED

Grass seed is sold in small quantities through retail outlets such as garden centers, supermarkets, hardware stores, and department stores. It is also sold in bulk amounts through wholesale suppliers. Professional landscapers, park and recreation specialists, and other industry horticulturists usually purchase seed wholesale. However, because most clients have had the experience of purchasing seed from a retailer, where packaging, product claims, and prices can be very confusing, landscapers need to know how grass seed mixtures differ in order to explain to clients why a seemingly more expensive seed may be needed for the landscape.

The Confusion

Confusion often results from retail packaging with bright colors and names that seem to promise success, such as *Sure-Grow, Easy Lawn, Play Lawn,* and *Shade Mix.* When these are stocked next to

smaller bags of seed that cost twice as much, the untrained consumer predictably will be attracted to the one that seems to offer the most for the least.

The Solution

The key to the mystery is the *seed analysis label.* By law, the seed analysis label must appear on every package of seed sold. If the seed is sold in bulk amounts, the analysis will be printed on a tag tied to the handle of the storage container.

Although legal requirements vary somewhat from state to state, most analysis labels contain the following information:

Purity This is the percentage, by weight, of pure grass seed. The label must show the percentage, by weight, of each species of grass in the mixture. This allows the consumer to compare the permanent with the nonpermanent grasses and the fine-textured with the coarse-textured grasses.

FIGURE 13-3.

Regions of grass adaptations (Courtesy United States Department of Agriculture)

CLIMATIC REGIONS IN WHICH THE
FOLLOWING GRASSES ARE SUITABLE FOR LAWNS:

1. Kentucky bluegrass, red fescue, and colonial bentgrass. Tall fescue, Bermuda, and zoysia grasses in the southern part.
2. Bermuda and zoysia grasses. Centipede, carpet, and St. Augustine grasses in the southern part; tall fescue and Kentucky bluegrass in some northern areas.
3. St. Augustine, Bermuda, zoysia, carpet, and Bahia grasses.
4. Nonirrigated areas: crested wheat, buffalo, and blue grama grasses. Irrigated areas: Kentucky bluegrass and red fescue.
5. Nonirrigated areas: crested wheatgrass. Irrigated areas: Kentucky bluegrass and red fescue.
6. Colonial bentgrass, Kentucky bluegrass, and red fescue.

Percent Germination This is the percentage of the pure seed that was capable of germination on the date tested. The date of testing is important and must be shown. If several months have passed since the germination test, the seed is less likely to germinate at the percentage indicated.

Crop Seed This is the percentage, by weight, of cash crop seeds in the mixture. These are undesirable species for lawns.

Weeds This is the percentage, by weight, of weed seeds in the mixture. A seed qualifies as a weed seed if it has not been counted as a pure seed or crop seed.

Noxious Weeds This is usually the number of noxious weed seeds per pound or ounce of weed seeds. Noxious weeds are those extremely undesirable and difficult to eradicate.

Inert Material This is the percentage, by weight, of material in the package that will not grow. In low-priced seed mixes, it includes materials such as sand, chaff, or ground corncobs. Inert material may be added to make the seed package look bigger, or it may already be present in the seed and not be removed because the cost involved would raise the price of the seed.

Seed Analysis

The percentage of pure live seed (PLS) can be calculated by multiplying the percent purity of desirable species (percent P) and the percent viability (percent V). For example, a seed lot costing $2.50 per pound with 93 percent fine-textured seeds and 85 percent germination (viability) has 79 percent pure live seed (.93 × .85 = .79). If the percent PLS is divided into the cost of 1 pound of the seed and multiplied by 100, the cost per pound of pure live seed is determined. This is a way to compare seeds that look similar in analysis but are dissimilar in price.

$$\frac{\text{cost per pound}}{\text{percent PLS}} \times 100 = \text{cost per pound of PLS}$$

$$\frac{\$2.50}{79} \times 100 = \$3.16/\text{lb PLS}$$

Three sample analyses follow. Study them and determine which mixture would probably cost the most and which the least.

Mixture A

Fine-Textured Grasses

12.76% red fescue	85% germination
6.00% Kentucky bluegrass	80% germination

Coarse-Textured grasses

53.17% annual ryegrass	95% germination
25.62% perennial ryegrass	90% germination

Other Ingredients
  2.06% inert matter
  0.39% weeds—no noxious weeds

Mixture B

Fine-Textured Grasses

38.03% red fescue	80% germination
34.82% Kentucky bluegrass	80% germination

Coarse-Textured Grasses

19.09% annual ryegrass	85% germination

Other Ingredients
  7.72% inert matter
  0.34% weeds—no noxious weeds

Mixture C

Fine-Textured Grasses

44.30% creeping red fescue	85% germination
36.00% Merion bluegrass	80% germination
13.54% Kentucky bluegrass	85% germination

Coarse-Textured Grasses
  None Claimed

Other Ingredients
  5.87% inert matter
  0.29% weeds—no noxious weeds

It is likely that Mixture C would be the most expensive. It contains the highest percentage of fine-textured grasses, no coarse-textured grasses, no annual grasses, and the lowest percentage of weeds. Mixture A would probably cost the least, since it contains a high percentage of coarse grasses and the greatest percentage of weeds. None of the mixtures is very poor in quality, since none contain crop or noxious weed seeds.

MIXTURES VERSUS SINGLE-SPECIES BLENDS

Grass seed is commonly formulated as either a *mixture* or a *blend*. A mixture combines two or more different species of grass. A blend combines two or more cultivars of a single species. Both formulations have their place depending upon the site and circumstances. Mixtures have been most common in temperate zone landscapes; single-species plantings have been more common in subtropical and tropical landscapes.

Mixtures sometimes have the disadvantage of variegated color and texture resulting from the different species. They have the advantages of tolerating mixed environmental conditions and resisting devastation by diseases or insects that might wipe out a single species.

Single-species turf plantings offer a more uniform appearance than mixtures; but they are often victims of their own inability to adjust to varied environmental conditions on a site and to resist diseases and insects.

TABLE 13-1.

A Comparison Chart for Turfgrasses

Grass Species	Cool Season or Warm Season	Growth Habit	Leaf Texture	Mowing Height/Inches	Fertilization Pounds of Nitrogen Per 1000 Square Feet Per Year
Bahiagrass	Warm	Rhizomatous	Coarse	1 1/2 to 2	1 to 4
Bermudagrass	Warm	Stoloniferous and rhizoma-tous	Fine	1 to 2	4 to 9
Bentgrass, Colonial	Cool	Bunch-type (with short stolons and rhizomes)	Fine	1/2 to 1	2 to 4
Bentgrass, Creeping	Cool	Stoloniferous	Fine	1/2 or less	4 to 8
Bentgrass, Redtop (see Redtop) Bentgrass, Velvet	Cool	Stoloniferous	Fine	1/2 or less	2 to 4
Bluegrass, Annual	Cool	Bunch-type or stolonifer-ous	Fine	1	2 to 6
Bluegrass, Canada	Cool	Rhizomatous	Medium	Does not mow well	1 or less
Bluegrass, Kentucky	Cool	Rhizomatous	Fine	1 to 2 1/2	2 to 6
Bluegrass, Rough	Cool	Stoloniferous	Fine	1 or less	2 to 4
Bromegrass, Smooth	Cool	Rhizomatous	Coarse	Does not mow well	1 or less
Buffalograss	Cool	Stoloniferous	Fine	1/2 to 1 1/2	1/2 to 2
Carpetgrass, Common	Warm	Stoloniferous	Coarse	1 to 2	1 to 2
Carpetgrass, Tropical	Warm	Stoloniferous	Coarse	1 to 2	1 to 2

Soil Tolerances	Climate Tolerances	Uses	How Established If Seeded, Pounds Per 1000 Square Feet
Infertile, acidic, and sandy	Subtropical and tropical	Utility turf; good for use along roadways	Seeded at 6 to 8
Does well on a wide range of soils	Warm temperate and subtropical	Sunny lawn areas; good general purpose turf for athletic fields, parks, home lawns	Plugging or seeded at 1 to 1 1/2
Moderately fertile, acidic, and sandy	Temperate and sea-coastal	Areas where intensive cultivation is practical	Seeded at 1/2 to 2
Fertile, acidic, and moist	Subarctic and temperate	Golf greens and other uses where intensive cultivation is practical	Sprigging or seeded at 1/2 to 1 1/2
Moderately fertile, acidic and sandy	Temperate and sea-coastal	Shaded, intensively cultivated areas	Seeded at 1/2 to 1 1/2
Fertile, neutral to slightly acidic	Temperate and cool subtropical	Not planted intentionally; but common in intensively cultivated turfs during spring and fall	Does not apply
Infertile, acidic and droughty	Subarctic and cool temperate	A soil stabilizer	Seeded at 1 to 2
Fertile, neutral to slightly acidic	Subarctic, temperate, and cool subtropical	Sunny lawn areas; good general purpose turf for home lawns, athletic fields, and parks	Seeded at 1 to 2
Fertile and moist	Subarctic and cool, shaded temperate	Some use on shaded, poorly drained sites	Seeded at 1 to 2
Infertile and droughty	Dry and temperate	A soil stabilizer	Seeded at 1 to 2
Does well on a wide range of soils; tolerant of alkaline soils	Dry temperate and subtropical	Useful in semiarid sites as a general purpose lawn grass	Seeded at 3 to 6
Infertile, acidic, and moist	Subtropical and tropical	Utility turf; good for use along roadways and as a soil stabilizer	Seeded at 1 1/2 to 2 1/2
Infertile, acidic, and moist	Humid subtropical and tropical	Utility turf; good for use along roadways and as a soil stabilizer; can be used as a lawn grass in tropics	Seeded at 1 1/2 to 2 1/2

TABLE 13-1.
A Comparison Chart for Turfgrasses (continued)

Grass Species	Cool Season or Warm Season	Growth Habit	Leaf Texture	Mowing Height/Inches	Fertilization Pounds of Nitrogen Per 1000 Square Feet Per Year
Centipedegrass	Warm	Stoloniferous	Medium	1 to 2	1 to 2
Fescue, Chewings	Cool	Bunch-type	Fine	1 1/2 to 2	2
Fescue, Creeping Red	Cool	Rhizomatous	Fine	1 1/2 to 2	2
Fescue, Hard	Cool	Bunch-type	Medium	Does not mow well	1 or less
Fescue, Meadow	Cool	Bunch-type	Coarse	1 1/2 to 3	1 or less
Fescue, Sheep	Cool	Bunch-type	Fine	Does not mow well	1 or less
Fescue, Tall	Cool	Bunch-type	Medium to Coarse	1 1/2 to 3	1 to 3
Gramagrass, Blue	Warm	Rhizomatous	Fine	Does not mow well	1 or less
Redtop (a bentgrass)	Cool	Rhizomatous	Coarse	1 1/2 to 3	1 to 2
Ryegrass, Annual	Cool	Bunch-type	Medium	1 1/2 to 2	2 to 4
Ryegrass, Perennial	Cool	Bunch-type	Fine	1 1/2 to 2	2 to 6
St. Augustinegrass	Warm	Stoloniferous	Coarse	1 to 2 1/2	2 to 6
Timothy, Common	Cool	Bunch-type	Coarse	1 to 2	3 to 6
Wheatgrass, Crested	Cool	Bunch-type	Coarse	1 1/2 to 3	1 to 3

Soil Tolerances	Climate Tolerances	Uses	How Established If Seeded, Pounds Per 1000 Square Feet
Infertile, acidic, and sandy	Subtropical and tropical	Utility turf; also usable as a low-use lawn grass	Seeded at 1/4 to 1/2
Infertile, acidic, and droughty	Subarctic and temperate	Shaded sites with poor soil	Seeded at 4 to 8
Infertile, acidic and droughty	Subarctic and temperate	Shaded sites	Seeded at 3 to 5
Fertile and moist; not tolerant to droughty soil	Moist and temperate	A soil stabilizer	Seeded at 4 to 8
Widely tolerant of all but droughty soils	Moist and temperate	Utility turf; good for use along roadways	Seeded at 4 to 8
Infertile, acidic, well-drained, and droughty	Dry and temperate	A soil stabilizer	Seeded at 3 to 5
Does well on a wide range of soils	Warm temperate and subtropical	Utility turf; good for use along roadways; new cultivars (Brookston, Olympic, and Rebel) good for lawns	Seeded at 4 to 8
Does well on a wide range of soils	Dry and subtropical	Utility turf; good for use along roadways and in arid sites	Seeded at 1 to 2
Does well on a wide range of soils	Subarctic, temperate, and cool subtropical	Utility turf; good for use along roadways and in poorly drained areas	Seeded at 1/2 to 2
Fertile, neutral to slightly acidic and moist	Temperate and subtropical	Useful for quick and temporary lawns in the temperate zone and for winter color in the subtropic zones	Seeded at 4 to 6
Fertile, neutral to slightly acidic and moist	Mild and temperate	Used in mixed species lawns and as an athletic turf	Seeded at 4 to 8
Does well in a wide range of moist soils	Subtropical and tropical seacoastal	A good lawn grass with excellent shade tolerance	Sprigging
Fertile, slightly acidic, and moist	Subarctic and cool temperate	Utility turf; good for athletic fields in cold regions where preferable species won't survive	Seeded at 1 to 2
Does well in a wide range of soils	Subarctic and cool temperate	Useful as a general purpose turf on droughty sites	Seeded at 3 to 5

TABLE 13-1.

A Comparison Chart for Turfgrasses (continued)

Grass Species	Cool Season or Warm Season	Growth Habit	Leaf Texture	Mowing Height/Inches	Fertilization Pounds of Nitrogen Per 1000 Square Feet Per Year
Zoysiagrass (Japanese lawngrass)	Warm	Stoloniferous and rhizomatous	Medium	1/2 to 1	2 to 3
Zoysiagrass (Manilagrass)	Warm	Stoloniferous and rhizomatous	Fine	1	2 to 3
Zoysiagrass (Mascarenegrass)	Warm	Stoloniferous and rhizomatous	Fine	Does not mow well	2 to 3

Blends attempt to retain the advantages of both mixtures and single-species plantings. If the cultivars are carefully selected, the blend will offer uniform appearance plus environmental adaptability, good resistance to wear, compatible maintenance requirements of all component cultivars, and resistance to pests of the area.

METHODS OF INSTALLATION

Depending upon the species of grass, the type of site, and the immediacy of the need for a usable turf, one of four methods is used to install a turfgrass planting:

1. seeding
2. sodding
3. plugging
4. sprigging and stolonizing

Seeding

Seeding is the most common and least expensive method of establishing a lawn. The seed may be applied either by hand or with a spreader on small sites. If the site is extensive, the seed may be applied with a cultipacker seeder (a large seeder pulled by a tractor) or by a hydroseeder (a spraying device that applies seed, water, fertilizer, and mulch simultaneously). The hydroseeder is especially helpful for seeding sloped, uneven areas, Figure 13-4.

Sodding

When a lawn is needed immediately, sodding may be selected as the method of installation. *Sod*

FIGURE 13-4.
Use of the Hydroseeder gives rapid stabilization to this steep embankment. (Courtesy United States Department of Agriculture)

Soil Tolerances	Climate Tolerances	Uses	How Established If Seeded, Pounds Per 1000 Square Feet
Does well in a wide range of soils	Temperate, subtropical, and tropical	Useful as a general purpose turf for home lawns, parks, and golf courses, especially in warmer regions	Plugging
Does well in a wide range of soils	Subtropical and tropical	A good lawn grass	Plugging
Does well in a wide range of soils	Warm subtropical and tropical	A soil stabilizer and groundcover	Plugging

is established turf which is moved from one location to another. A sod cutter is used to cut the sod into strips. These are then lifted, rolled up, and placed onto pallets for transport to the site of the new lawn, Figures 13-5 and 13-6. At the new site, it is unrolled onto the conditioned soil bed. The effect is that of an instant lawn. Sodding is much more costly than seeding, but the immediacy is important to some clients and necessary on certain sites, such as steep slopes and terraces, where seed might wash away.

Plugging

Plugging is a common method of installing lawns in the southern sections of the United States. Certain

FIGURE 13-5.
This sod has been cut and is being folded and stacked for transport to the planting site.

FIGURE 13-6.
After careful preparation of the soil, the sod is unrolled at the site of the new lawn.

grasses, such as St. Augustinegrass and the zoysias, are not usually reproduced from seed. Instead, they are placed into a new or existing lawn as small sections of sod (*plugs*). Because the growing season in the South is longer than elsewhere, the plugs have time to develop into a full lawn. Plugging is a time-consuming means of installing a lawn, and that is its major limitation. However, plugging is necessary for many warm-season grasses that are poor seed producers. On large sites, some mechanization of the process is possible.

Sprigging and Stolonizing

Like plugging, sprigging and stolonizing are vegetative methods of establishing a lawn. Also like plugging, the methods are more common to southern grasses than northern species. A *sprig* is a piece of grass shoot. It may be a piece of stolon or rhizome or even one of the lateral shoots. Sprigs do not have soil attached and so are not like plugs or sod. They are planted at intervals into prepared, conditioned soil. Several bushels of sprigs are required to plant 1,000 square feet. If done by hand, the process is slow and tedious; for large sites, mechanization can lessen the time required.

Stolonizing is a form of sprigging. The sprigs are broadcast over the site, covered lightly with soil, then rolled or disked. Since each sprig is not individually inserted into the soil, this method is faster.

PROPER LAWN CONSTRUCTION

Certain turf installations are unique. A golf green is a highly specialized use of turf that has no counterpart in landscaping. A rocky, steeply sloped highway embankment requires special techniques too. This chapter will only deal with the installation of conventional lawn plantings such as those of the typical home or commercial business site. More specialized references are available for nontypical turfs. Six steps should be followed to assure a beginning lawn every chance for success:

1. Plant at the proper time of year.
2. Provide the proper drainage and grading.
3. Condition the soil properly.
4. Apply fresh, vigorous seed, sod, plugs, or sprigs.
5. Provide adequate moisture to promote rapid establishment of the lawn.
6. Mow the new lawn to its correct height.

Time of Planting

Warm-season grasses grow best when the temperature is from 80° to 95° F. They are best planted in the late spring, just before the summer season. In this way, they have the opportunity to become well established before becoming dormant in the winter.

Cool-season grasses germinate best when temperatures are from 60° to 75° F. The best planting time for them is early fall or very early spring, right before the cool seasons in which they flourish. If cool-season grasses are planted too close to the intensely hot or cold days of summer and winter, they will die or go dormant before becoming well established.

Grading and Draining the New Lawn

Each time the rain falls or an irrigation system is turned on, water moves across the surface of the soil as well as into it. *Grading* (altering level land so that it slopes slightly) can direct the movement of the surface water. Proper *drainage* permits the water to move slowly into the soil to the turf's root system where it can be absorbed, yet pass beyond the root zone before it collects and does harm to the plants.

Even lawns that seem flat must slope enough so that water moves off the surface and away from nearby buildings. If a slight slope does not exist naturally, it may be necessary to construct one. A decline of between 6 inches and 1 foot per 100 feet is required for flat land to drain properly. Failure to grade lawns away from buildings can result in flooded cellars and basements.

Drainage is critical to the survival of the lawn, no matter what the species. Depending upon the soil and site involved, good drainage may require nothing more than mixing organic matter with the existing soil. In cases where the soil is heavy with clay, a system of drainage tile may be necessary.

If drainage tile is needed, it should be installed after the lawn's grade has been established but before the surface soil has been conditioned. Regular 4-inch agricultural tile is normally used, placed 18 to 24 inches beneath the surface. In a typical tile installation, the tile lines are spaced approximately 15 feet apart. Each of the lateral lines runs into a larger main drainage line, usually 6 to 8 inches in diameter. This, in turn, empties into a nearby ditch or storm sewer, Figures 13-7 and 13-8.

Conditioning the Soil

Suitable texture and proper pH are very important in soil that is to become a lawn. Most turfgrasses grow best in a loam or sandy loam soil with a pH that is neutral to slightly acidic (7.0 to 6.5). For a review of soil texture and pH, refer to Chapter 2.

The additives needed to improve the texture, drainage, or pH of the soil can be incorporated best

FIGURE 13-7.
Drainage tile installation (drawing not to scale) (From J. Ingels, *Landscaping: Principles and Practices,* 2nd edition, © 1983 by Delmar Publishers Inc.)

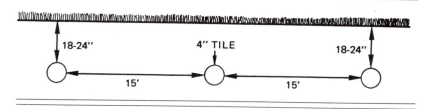

FIGURE 13-8.
The flow of water through tile drainage system (drawing not to scale) (From J. Ingels, *Landscaping: Principles and Practices,* 2nd edition, © 1983 by Delmar Publishers Inc.)

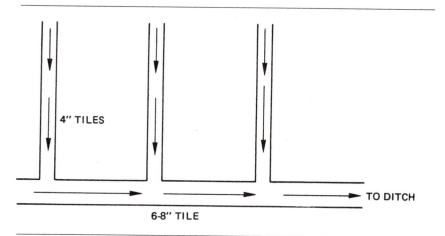

with a rototiller or larger agricultural field equipment, Figure 13-9. The tiller also loosens the soil surface and breaks it into smaller particles.

The tiller may do two additional things: chop up weeds growing on the site and bring rocks to the surface. Both can be bad for the new lawn. The stones mar the even texture of the lawn and, if large enough, can damage mowers and other maintenance equipment. They should be removed with a rake (hand or powered) before seeding. The weeds can cause future problems for the lawn, especially if they are noxious or other perennial types. It is best to eliminate them before installing the grass. A nonselective systemic herbicide or short-term soil sterilant such as methyl bromide can be applied. Both are expensive, widely used, and available under several brand names. Once the soil has been properly conditioned, it is ready to plant.

Planting the Lawn

Seed Seed is applied to the prepared soil in a manner that will distribute the recommended quan-

FIGURE 13-9.
A rototiller is an excellent way to incorporate additives into the soil while loosening it.

tity evenly; otherwise, a patchy lawn develops. When applied with a spreader or cultipacker seeder, the seed is often mixed with a carrier material such as sand or topsoil to assure even spreading. The seed or seed/carrier mix is divided into two equal amounts. One part is sown across the lawn in one direction. The other half is then sown across the lawn in a direction perpendicular to the first, Figure 13-10.

Placing a light mulch of weed-free straw over the seed helps to retain moisture. It also helps to prevent the seed from washing away during watering or rainfall. On a slope that has not been hydroseeded, it is wise to apply erosion netting over the mulched seed to further reduce the danger of the seed washing away, Figure 13-11.

Sod Sod must be installed as soon after it has been cut as possible. Otherwise, the live, respiring grass will damage itself due to the excessive temperatures that build up within the rolled or folded strips of sod. Permitting the sod to dry out while awaiting installation can also damage the grass and result in a weak and unsatisfactory lawn.

The soil should be moist before beginning installation of the sod. The individual strips are then laid into place much as a jigsaw puzzle is assembled. The sod should not be stretched to fit as it will only shrink back later, leaving gaps in the surface. Instead, each strip should be fitted carefully and tightly against the adjacent strips. Using a flat tamper or roller, tamp the sod gently to assure that all parts are touching the soil evenly.

Plugs Plugs are small squares, rectangles, or circles of sod, cut about 2 inches thick. Their installation is similar to that of groundcovers. They are set into the conditioned soil at regular intervals (12 to 18 inches) and in staggered rows to maximize coverage. The top of each plug should be level with the surface of the conditioned soil. The soil should be moist but not wet at the time of installation to prevent some of the plugs from drying out while others are still being installed.

Sprigs Sprigs are planted 2 to 3 inches deep in rows 8 to 12 inches apart. In hand installations, rows are not drawn. Instead the sprigs are distrib-

FIGURE 13-10.
Spreader application: Half the material is applied at a 90° angle to the other half. (From J. Ingels, *Landscaping: Principles and Practices,* 2nd edition, © 1983 by Delmar Publishers Inc.)

FIGURE 13-11.
Erosion netting and mulch permit this newly seeded slope to develop a strong turf planting. Without the mulch and netting, the seed would wash away before becoming established. (Courtesy United States Department of Agriculture)

uted as evenly as possible over the prepared soil surface and pushed down into the soil with a stick. As described earlier, stolonizing uses a top-dressing of soil over the sprigs and eliminates the requirement for individual insertion. The soil should be moist, but not overly wet, when planting begins. If the lawn area is large, planted areas should be mulched and lightly rolled as the installation progresses. To wait until the entire installation has been completed could result in drying out of the sprigs.

Watering

In each of the planting techniques above, mulch is applied immediately after planting to prevent drying of the propagative material.

Once the planting and mulching are completed, water needs to be carefully applied. The soil and developing grasses should not be allowed to dry out until the new grass is about two inches tall.

Nevertheless, the watering cannot be so heavy or continuous that the seedlings or new sprouts drown, the soil erodes, or diseases gain a foothold.

At least a month of watering several times each day is necessary to establish the new lawn properly. Sprinklers are preferable to a garden hose. With a sprinkler, the water can be applied slowly and evenly. If the lawn has been planted for a client and the landscaper or turf specialist cannot be at the site to do the watering, the client must be given explicit instructions about how and when to apply the water. Allowance for naturally occurring rainfall should be included in the written instructions. Client responsibilities, such as watering, should be written into any contract drawn between a professional firm and a consumer.

The First Mowing

The first mowing of a new lawn is an important one. The objective is to encourage horizontal branching of the new grass plants as quickly as possible. This creates a thick (dense) lawn. The first mowing should occur when the new grass has reached a height of 2 1/2 to 3 inches, and the grass should be cut back to a height of 1 1/4 to 1 1/2 inches. Thereafter, different species require differing mowing heights for proper maintenance (Table 13-1). For the first mowing, it is a good practice to collect and remove the grass clippings. After that, clipping removal is usually unnecessary unless the grass has grown so tall between mowings that clumps of grass are visible on the lawn. Note that grasses used for soil stabilization will not require mowing.

CALIBRATING A SPREADER

Two types of spreaders are used to apply seed, fertilizer, or other granular material to lawns. These are the *rotary spreader* and the *drop spreader*. The rotary type dispenses the material from a closed hamper onto a rotating plate that then propels it outward in a semicircular pattern. The drop spreader dispenses the material through holes in the bottom of the hamper as it is pushed across the lawn. In both types, the amount of material applied is controlled by the size of the holes through which the material passes and the speed at which the spreader is pushed. Therefore, use of a spreader requires that it be calibrated to dispense the material at the rate desired. For materials that are applied often and by the same person, the spreader need only be calibrated once and the setting noted on the control for future reference. Different materials usually require different calibrations even when the rate of application is the same.

The object of calibration is to measure the amount of material applied to an area of 100 square feet. A paved area such as a driveway or parking lot is an excellent calibration site. Afterwards, the seed or other material can be swept up easily for future use. Covering the area with plastic is also helpful in collecting the material after calibration. Windy days are not suitable for calibrating.

To calibrate, fill the spreader with exactly 5 pounds of material. Select a spreader setting near the center of the range to begin. Apply the material by walking at a normal pace in a straight line. Shut off the spreader while turning it around. Each strip should slightly overlap the previous one to avoid a streaked appearance. The probability of streaking is greater with a drop spreader than with a rotary spreader. When an area of 100 square feet has been covered once, weigh the material left in the spreader and subtract it from the original 5 pounds. The quantity of material applied per 100 square feet has then been determined. If it is not the distribution desired, the spreader can be adjusted to increase or reduce the rate of application.

MAINTAINING AN ESTABLISHED LAWN

Lawn maintenance requires a great deal of time during the growing season. How the lawn appears is determined by the amount of time and money budgeted by the property owner or manager. Increasing public expectations of turf quality account for the great success of professional lawn maintenance firms nationwide.

Spring Operations

A lawn needs both weekly and seasonal care. In regions of the country where winters are long and severe, spring may find lawns covered with compacted leaves, litter, and partially decomposed organic material termed *thatch*. The receding winter may leave grass damaged by salt, disease, or heaving.

Small areas can be cleaned of debris with a strong-toothed rake. Larger lawns may require the use of power sweepers or thatch removers.

Where heaving has lifted large areas of sod away from direct contact with the soil, a light rolling with a lawn roller is advisable in the spring. The lawn should be rolled first in one direction, then again in a direction perpendicular to the first. Clay soils should not be rolled, since air can easily be driven from such a soil and the surface compacted. No soil should be rolled while wet. The roller can be used properly only after the soil has dried and regained its firmness.

The first cutting of the lawn each spring removes more grass than cuttings that follow later in the summer. The initial cutting to a height of 1 1/4 to 1 1/2 inches has the same objective as the first cutting of a newly planted lawn; that is, to promote horizontal growth of the grass. An additional benefit of the short first cutting is that fertilizer, grass seed, and herbicide can then reach the soil's surface more easily. Cuttings later in the year are usually not as short.

Patching the Lawn

If patches of turf have been killed by diseases, insects, dog urine, or other causes, it may be necessary to reseed them or add new sod, plugs, or sprigs. Widespread thinness of the grass does not indicate a need for patching but a lack of fertilization or improper mowing.

Patching is warranted where bare spots are at least 1 foot in diameter. Seed, sod, plugs, or sprigs should be selected to match the grasses of the established lawn. Plugs can be set directly into the soil using a bulb planter or golf green cup cutter to cut the plug and then to remove the soil where it is to be planted. With seed, sod, and sprigs, it is best to break the lawn surface first with a toothed rake. A reseeding mixture of a pound of seed in a bushel of topsoil is handy for patching where seed is the form of propagation. Mulch and moisture must then be applied as stated earlier.

The timetable for patching is the same as for planting and is related to the type of grass involved.

Aerating the Lawn

Aeration of a lawn is the addition of air to the soil. Air is an important component of the soil and is essential to good plant growth. If the lawn is installed properly, the incorporation of coarse sand and organic matter into the soil promotes proper aeration. However, where traffic is heavy or the clay content is high, the soil may become compacted. The turf specialist can relieve the compaction by use of a powered aerator, Figure 13-12. There are several types of aerators and a range of sizes. All cut into the soil to a depth of about 3 inches and either remove plugs of soil or slice it into thin furrows. A topdressing of organic material is then applied to the lawn and a rotary power mower run over the organic material to blow it into the holes or slits left by the aerator. The plugs of soil left on top of the lawn may be removed by raking or, if not too compacted, broken apart and left as topdressing. Mechanical equipment is made that can aerate and convert the topdressing in a single operation.

Vertical Mowing

Vertical mowing is a technique that can break up the soil plugs left by an aerator or even remove excessive thatch if necessary. It requires a power rake or a mower whose blades strike the turf vertically. It is done when the lawn is growing most rapidly and conditions for continued growth are favorable. For cool-season grasses, late summer or early autumn is the best time; for warm-season grasses, late spring to early summer is best.

FIGURE 13-12.
A small plug aerator, suitable for residential use

The blades of the vertical mower are adjusted to different heights depending upon the objectives of the operation. A shallow setting is used to break up soil plugs; a lower setting gives deeper penetration into the thatch layer, facilitating its removal and relieving compaction of the soil. Deep vertical mowing is only practiced on deep-rooted turfs. Shallow-rooted turfs, often growing mainly in the thatch layer, are harmed more than helped by vertical mowing.

Fertilization

Lawns should be fertilized just before they need the nutrients for their best growth. For example, cool-season grasses derive little benefit from fertilizer applied at the beginning of the hot summer months. Only the weeds benefit from fertilizers that are not timed to coincide with the turf's needs.

Cool-season grasses should be fertilized in the early spring and early fall. This supplies proper nutrition prior to the seasons of most active growth. Late fall fertilization encourages soft, lush growth which is damaged severely during the winter. Warm-season grasses should receive their heaviest fertilization in late spring. Their season of greatest growth is the summer.

A look at Table 13-1 reveals that not all species of grass require the same amounts of fertilizer. As fertilizer costs increase and certain fertilizers become less available, turf managers try to maintain plantings with minimal applications. Increasingly, fertilizers with nitrogen in slowly available forms (such as ureaformaldehyde, sulfur-coated urea, or isobutylidine diurea) are being selected by turf management professionals. Still, the specific amounts needed can vary greatly among species. The most successful mixed species lawns are those that combine species whose fertilization requirements are compatible, such as fescues and bluegrasses.

The amount of fertilizer required is usually stated in pounds of nitrogen per 1,000 square feet per year. The number of pounds of nitrogen in a fertilizer is determined by multiplying the weight of the fertilizer by the percentage of nitrogen it contains. The examples that follow illustrate.

Example 1
Problem: How many pounds of nitrogen are contained in a 100-pound bag of 20-10-15 fertilizer?
Solution: 100 pounds × 20% N = pounds of N
100 × .20 = 20 pounds of N

Example 2
Problem: How many pounds of 20-10-15 fertilizer should be purchased to apply 4 pounds of nitrogen to 1,000 square feet of lawn?
Divide the pounds of nitrogen by the percentage of nitrogen.
Solution: 4 pounds of N desired ÷ 20 percent
= pounds of 20-10-15 fertilizer required
4 ÷ .20 = 20 pounds of 20-10-15 fertilizer required

To obtain the fullest benefit from fertilizer, it should be applied in an amount the turfgrass can use fully before it reaches beyond the root zone. Lawn fertilizer is usually applied in two stages. Half of the recommended poundage for the year can be applied to cool-season grasses in the early spring and the remaining half in the early fall. Warm-season grasses can receive half or more of the recommended fertilizer poundage during the late spring and the remainder in mid-summer. A spreader must be used to assure even distribution of the fertilizer. It is applied in two directions with the rows slightly overlapping.

Watering

Turfgrasses are among the first plants to suffer from lack of water, since they are naturally shallow rooted as compared to trees and shrubs. The groundskeeper should encourage deep root growth by watering so that moisture penetrates the soil to a depth of 8 to 12 inches. Failure to apply enough water will promote shallow root systems that can be easily injured during hot, dry summer weather, Figure 13-13. Infrequent, deep watering is much preferable to daily, shallow watering. The quantity of water applied during an irrigation will depend

FIGURE 13-13.
Deep watering promotes deep, healthy root growth. Shallow watering promotes shallow rooting and leaves the grass susceptible to injury by drought. (From J. Ingels, *Landscaping: Principles and Practices,* 2nd edition, © 1983 by Delmar Publishers Inc.)

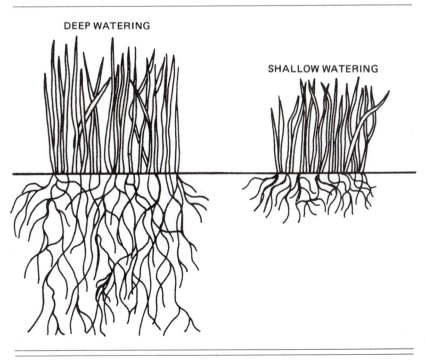

DEEP WATERING

SHALLOW WATERING

upon the time of day and the type of soil. Clay soils allow slower water infiltration than coarser textured sandy soils, but clay soils retain water longer. Therefore less water may need to be applied to clay soils or the rate of application may need to be slower or both. The amount of water given off by a portable sprinkler can be calibrated once and a notation made for future reference. To calibrate a portable sprinkler (Figures 13-14 and 13-15), set several wide-topped, flat-bottomed cans with straight sides (such as coffee cans) in a straight line out from the sprinkler. When most of them contain the amount of water needed to penetrate 8 to 12 inches, shut off the sprinkler and note the time required. Start by applying 1 1/2 to 2 inches of actual water and check its depth of penetration; then adjust accordingly. Permanent and computer-controlled automatic irrigation systems are common in large turf installations. They turn on and off by timers and dispense a predetermined amount of water through valves that pop up for use and disappear

out of sight afterward. Such systems are sophisticated and costly, Figure 13-16.

The best time of day to water lawns is early morning. Watering later in the day is inefficient due to greater evaporation. Watering in the early evening or later should be avoided because of the danger of disease; turf pathogens thrive in lawns that remain wet into the evening. The lawn should have time to dry before the sun sets.

Mowing

Three types of mowers are used to maintain turf plantings: the reel mower, the rotary mower, and the flail mower. The *flail mower* is used for utility and stabilization turfgrasses that are only cut a few times each year. Reel and rotary mowers are used to maintain home, recreational, and commercial lawns. On a *reel mower,* the blades rotate in the same direction as the wheels and cut the grass by pushing it against a nonrotating bedknife at the rear base of the mower, Figures 13-17 and 13-18. The blades of a *rotary mower* move like a ceiling fan,

FIGURE 13-14.
This revolving sprinkler covers a limited area and must be manually moved to each new location. Wind gusts can affect the evenness of the coverage.

FIGURE 13-15.
This oscillating sprinkler casts water in an arch-like pattern. It is suitable for homeowner use and must be moved periodically for full coverage.

FIGURE 13-16.
An automatic lawn irrigation system in operation (Courtesy Weather-Matic Irrigation)

FIGURE 13-18.
Reel mowers provide the best quality cut.

FIGURE 13-17.
A powered reel mower

FIGURE 13-19.
A rotary mower

parallel to the surface of the lawn, cutting the grass off as they revolve, Figure 13-19. Reel mowers are most often used for grasses that do best with a shorter cut, such as bentgrass. Rotary mowers do not cut as evenly or sharply but are satisfactory for lawn grasses that accept a higher cut, such as ryegrass, bluegrass, and fescue. The riding mower, so popular in suburbia, is a rotary mower. Many campuses, parks, and golf course fairways are mown

with a large bank of reel mowers (called a gang mower) pulled behind a tractor that has been fitted with tires that will not rut the lawn.

In every situation, the blades must be sharp to give a satisfactory cut. Dull or chipped mower blades can result in torn, ragged grass blades that will die and give the lawn an unhealthy pallor of gray or brown. The sharp blades of any lawn mower should be respected. When powered, they can cut through nearly any shoe. Workers should never mow unless wearing steel-toed work shoes. Hands should never be brought near the blades while the mower is running. No inconvenience caused by shutting off a mower can equal the instant injury that a power mower can do to a worker's hand or foot.

Table 13-1 illustrates the often wide range of tolerable mowing heights that exists between and within species. Within the range, the mowing height selected often depends upon how much care the lawn can be given and what surface quality is expected. The shorter height will necessitate more frequent mowing, more watering during dry periods, and perhaps greater pest control efforts. In turn, the lawn will respond with greater density and a finer texture due to thinner blades. A taller lawn surface will have a slightly coarser texture (even with grasses classified as fine-leaf forms), take longer to thicken, but need cutting less often. It will also withstand heat and dry periods better, since the extra blade length will cast cooling shadows over the soil's surface. Often, fewer weeds are an added benefit of a taller lawn surface. Since not all species are mowed to the same height, mixed species lawns should be made up of grasses that have similar cutting requirements.

The frequency of mowing is another variable in turfgrass maintenance. Because the rate of growth of a lawn can vary with the temperature and with the moisture provided, the frequency of mowing cannot always be precisely specified. Where an inflexible mowing schedule is followed, the turf planting will be unsatisfactory. Nevertheless, professional firms must submit bids for lawn maintenance and so must have a reasonably accurate idea of how many times during the growing season a lawn will require mowing. Estate gardeners and golf course superintendents have less of a problem. They can mow when the time is right, not when a contract calls for it.

A long-standing and accepted rule of thumb is that mowing should remove about one-third of the length of the grass blade. Thus, if the turf is being kept at 1 1/2 inches, it should be mowed when it reaches a height of 2 1/4 inches. If the grass is permitted to get too long before cutting, the dead clippings can mar the appearance of the lawn. Then the only alternative is to collect the clippings either in a grass-catcher or with a lawn sweeper or rake. If cut properly, the clippings will not be excessive, will decompose rapidly, and will not require collection.

The pattern of mowing should be varied regularly to prevent wheel lines from developing in the lawn and to encourage the horizontal growth of the shoots. The easiest variation is to mow at a 90° angle to the previous mowing. If a lawn is mowed in both directions on the same day with a reel mower, an attractive checkerboard pattern develops. The pattern is not so apparent with a rotary mower, but the practice is just as healthful for the lawn.

Weed Control in Established Lawns

Turfgrass plantings can be marred by the presence of weeds. The weeds may include undesirable grasses and broadleaf plants. They may be annuals, biennials, or perennials. Many are prolific seed producers and infiltrate the lawn each year, often invading from the lawn of a neighbor who is not conscientious about control. Other weeds are indicators that something is wrong with the lawn. For example, certain weeds are predictable where the soil is too compacted for healthy grass growth. Others are found where winter pavement salt has killed off the turf.

For a review of weeds and herbicides, see Chapter 6. Pre-emergence herbicides are used to control many weeds in lawns and must be applied several weeks in advance of the weeds' germination to be effective. The time of application will vary depending upon the region of the country and the weeds involved. Pre-emergence herbicides are the preferred means of controlling annual grass weeds.

Post-emergence herbicides are used against weeds that have already germinated and are visible in the lawn. They are usually applied to the lawn rather than to the soil. They may be formulated as either contact or systemic herbicides, depending upon the product. Post-emergence herbicides are most often used against perennial grasses and broadleaf weeds.

Selectivity is an important characteristic of turf herbicides so that the desirable grasses are not damaged by the control effort. Annual grasses and broadleaf weeds are commonly controlled with selective herbicides. However, perennial grass weeds require the use of a nonselective herbicide (one that kills all vegetation that it touches). Nonselective herbicides must be applied on days without wind and with applicators that control the amount and direction of flow very carefully. After the treated area is free of weeds and time has rendered the herbicide harmless, new patch planting can take place.

For products recommended against specific turf weeds, turf managers can consult their local college of agriculture or Cooperative Extension Service. Because turf weed problems and recommended controls do not vary greatly from state to state, helpful information can also be found in trade journals that serve the turf industry.

Pest Control

Weeds are only one of the pests that affect turf installations. Like all plants, turfgrasses are susceptible to an assortment of pathogen, insect, animal, and human injuries. Table 13-2 is a partial list of the problems that can befall a typical lawn.

With turf, as with any crop, the best defense against most pests is the selection of resistant varieties and the creation of a growth environment that favors the turfgrass more than the pest. For example, watering at night promotes fungal growth of all types and invites disease. Watering should be done earlier in the day to allow the grass to dry before nightfall. Another example: allowing a thick thatch layer to develop provides a good habitat for certain insects. Minimizing the thatch layer reduces

TABLE 13-2.
Common Turf Problems

Turf Insects	Turf Diseases	Other Problems
• Ants	• Anthracnose	• Dogs
• Army worms	• Brown patch	• Gophers
• Bill bugs	• Copper spot	• Ground squirrels
• Chinch bugs	• Dollar spot	• Mice
• Cut worms	• Fairy ring	• Moles
• Grubs	• Fusarium blight	• Human
• Leaf hoppers	• Leaf spots	vandalism
• Mites	• Net blotch	• Vehicles and
• Mole crickets	• Nematodes	equipment
• Periodical cicadas	• Powdery mildew	
• Scale	• Pythium blight	
• Sod webworm	• Red thread	
• Weevils	• Rots	
• Wireworms	• Rusts	
	• Slime molds	
	• Smuts	
	• Snow molds	

these insect populations. When soil insects are reduced or eliminated, the turf is not as attractive to rodents, such as moles, that burrow through the soil to feed on them. Thus, solving one problem may indirectly solve another.

Pest control in turf is not easy or inexpensive, nor can it be approached casually. However, the arsenal of fungicides, insecticides, and rodenticides is adequate to the task of preventing or eradicating most lawn pests. (For a review of the principles of pest control, see Chapter 6.) The local college of agriculture can provide specific recommendations. For the control of people, their vehicles, and their pets, the groundskeeper must rely upon fencing, shrub masses, and walls to direct their movement onto paved areas. Signs requesting cooperation in the preservation of an attractive turf will often help by making people more aware of problems they may cause. After all, not everyone knows the difference between soil aeration and compaction.

Vandalism is impossible to control if the vandals are determined. Turf is not as attractive a target as other components of the landscape, but it gets its share of abuse. Golf courses and parks are most often damaged by vehicles rutting the turf, but the problem is also found on campuses and even in some residential neighborhoods. Locked gates and strategic placement of trees can sometimes help by making vehicular access to the turf difficult for the would-be vandal. Education to increase public awareness of the value of the landscape and the responsibilities of citizens is the only real solution to vandalism. It is also easier to write in a textbook than it is to bring into being.

SUMMARY

Turfgrass is the ideal outdoor surface for residential, academic, commercial, municipal, and recreational landscapes. Since no one grass is suitable for all landscapes, geographic regions, soil types, altitudes, and environmental conditions, the turf professional must have both general and specific knowledge of varieties and their requirements in order to select the appropriate one.

Grasses differ in growth habit, texture, color, density, seed size, soil and climatic tolerance, use tolerance, disease and insect resistance, and maintenance needs.

Grass seed is commonly formulated as either a mixture or a blend. Both formulations have their place depending upon the site and circumstances. Seeding is only one method of installing a lawn, however. Other methods include sodding, plugging, and sprigging or stolonizing.

Warm-season and cool-season grasses require different planting times. Warm-season grasses are best planted just before the summer season; cool-season grasses, in early fall or early spring. For the best lawn possible, the site must first be graded so that surface water will run off. Good drainage must be provided to maximize growth. Suitable conditioning of the soil can improve not only the drainage but texture and pH as well. After planting, mulching and careful watering must keep the new lawn moist until it is well established. The first mowing should be shorter than following ones to encourage horizontal branching and promote greater turf density.

Lawn maintenance requires a great deal of time. Some maintenance, such as mowing, goes on throughout the growing season. In addition, there are distinct spring operations to prepare the turf for the growing season and help it shake off the effects of winter. Other operations occur at various points in the summer and fall depending upon the type of turf and the location.

Turf is mowed with reel, rotary, or flail mowers. Which is used depends upon the type of turf, the size of the installation, and the quality of cut desired. Mowing is potentially hazardous and workers should be especially safety conscious.

Weed control is accomplished with pre-emergence and post-emergence herbicides. Other pests are controlled through the use of resistant varieties, creation of a growth-positive environment, and assorted fungicides, insecticides, nematicides, and rodenticides.

Achievement Review

A. SHORT ANSWER

Answer each of the following questions as briefly as possible.

1. List nine ways in which turfgrasses can vary.
2. Indicate if the following apply to warm-season grasses (W), cool-season grasses (C), or both (B).
   a. favored by temperatures of 80° to 95° F
   b. favored by temperatures of 60° to 75° F
   c. commonly installed by plugging
   d. planted in the late spring for optimum growth
   e. examples are bluegrasses and fescues
   f. planted in the early fall or early spring for optimum growth
   g. require regular watering after planting and for at least a month afterwards
   h. should be fertilized in the late spring
   i. should be fertilized in the early spring and early fall
   j. must be well drained
3. What are the advantages of using a good seed mixture for a home lawn rather than a single species of grass?
4. List four factors that could cause the prices of two 1-pound packages of grass seed to differ.
5. List and define the important terms found on a grass seed analysis label.
6. Match the characteristics on the left with the methods of lawn installation on the right.

   a. a vegetative method that uses pieces of the plant without soil attached
   b. gives the most immediate effect
   c. the most common and least expensive method
   d. a vegetative method that uses small circular or square pieces of sod
   e. a vegetative method that can be accomplished by broadcasting and top-dressing

   1. seeding
   2. sodding
   3. plugging
   4. sprigging/stolonizing

7. Explain how a spreader is calibrated.
8. List the six steps in starting a successful lawn.

B. PROJECTS

1. From Table 13-1, select all the turfgrasses that will survive in your region of the country. Next group them, first on the basis of similarities in use, then on the basis of maintenance requirements. Formulate turfgrass recommendations for a local home lawn, athletic field, and highway soil stabilization project.
2. Select a lawn area with which you are familiar, such as your home lawn or campus or a nearby park. Determine the grasses that make it up. Ask your Cooperative Extension agent for help if you are not certain. Using Table 13-1 and other materials available from your state's college of agriculture, outline an annual maintenance program for the lawn. Include fertilizer needs and time(s) of application, an irrigation schedule, mowing height, spring operations, aeration, type of mower and mowing pattern, and weed and pest control programs. Arrange the maintenance program in chronological order.

14

Techniques of Plant Propagation

Objectives

Upon completion of this chapter, you will be able to

- prepare plant propagation media.
- prepare total environments for plant propagation.
- demonstrate techniques of propagation by the use of cuttings, seeds, plugs, grafting, budding, and layering.
- describe techniques of tissue and organ culture.

SCIENCE AND CRAFT

Among all the crafts of ornamental horticulture, none exemplifies the application of science to the profession better than plant propagation. It is a craft that changes continually in response to new research findings; yet it respects the skills and techniques that only time, practice, and caring can convey.

This is a how-to chapter. It outlines the methods of reproducing plants that are practiced in various ornamental horticulture professions. You should review Chapter 5 as preparation.

Propagation techniques are practiced by all types of ornamentalists. Homeowners are able to establish new plants from seeds and cuttings taken from the plants of friends. Nursery and greenhouse owners can increase their inventories by various propaga-

tion techniques. Some firms become so specialized that they engage solely in plant propagation, providing rooted cuttings, bulbs, or high quality seed to growers nationwide.

THE MEDIA FOR PROPAGATION

To promote the initiation and development of new roots on cuttings, it is necessary to create an environment for the cutting base that will:

- support the cutting
- retain moisture uniformly
- drain away excess water uniformly
- provide adequate aeration
- not support weed seeds and other pests
- pasteurize easily

If seeds are the propagative unit, the medium must also be a source of nutrition for the developing embryo.

The propagating medium will usually contain one or more of the following materials:

- natural soil
- sand
- peat moss
- sphagnum moss
- perlite
- vermiculite
- fired clay

Natural soil is not commonly used for propagation unless mixed with quantities of coarser additives, such as fired clay, sand, or perlite, to improve its drainage and aeration. If pure sand is used for propagation, it must not be too fine or drainage will be poor. The best sand to use is quartz sand, fairly coarse in texture. All the materials used as propagation media should be steam pasteurized before use to eliminate weed seeds, nematodes, fungi, and other soil-borne plant pathogens.

The time required for establishment of new plants will vary from a week to several months. With seed propagation, it will be necessary to transplant the newly rooted plants from the propagating medium to a medium that will provide the nutrients necessary for sustained growth. For nursery crops, the new plants may be set directly into the field. Greenhouse growers may choose to install the plants directly into greenhouse benches or into containers ranging in size from bushel baskets to small clay or peat pots.

Field soils and greenhouse bench soils must be of loamy texture and well drained. Both may require the addition of sand or peat moss for texture, moisture retention, and drainage. One of the soilless mixes may also be used. Tables 14-1 and 14-2 list the components of the (older) University of California and (newer) Cornell mixes that enjoy wide acceptance throughout the ornamental horticulture industry.

THE STRUCTURES FOR PROPAGATION

The structures that serve as production facilities for greenhouse and nursery growers also serve for propagation. Greenhouses, hotbeds, cold frames, and lath houses (see Chapter 19) can all be used directly or in some modified manner to propagate plants.

The propagation structure must possess the following characteristics:

- sufficient light or darkness to permit seed germination and photosynthesis
- high humidity to reduce wilting of the cutting until new roots can form and to promote callus tissue formation in grafts

TABLE 14-1.
One Example of the University of California Soil Mix

Components:	50 percent sand (0.5 to 0.05 mm in diameter)
	50 percent peat moss (moistened before mixing)
MIX WELL WITH:	
Fertilizer	1/4 pound potassium nitrate
Additives (per cubic yard of mix):	1/4 pound potassium sulfate
	2 1/2 pounds single superphosphate
	7 1/2 pounds dolomitic limestone
	2 1/2 pounds calcium carbonate lime
PASTEURIZE WITH STEAM OR CHEMICALS	

NOTE: Additional nitrogen and potassium fertilizers will be needed as the plants grow.

TABLE 14-2.
The Cornell University Peat-lite Mix

Components:	50 percent sphagnum peat moss (moistened before mixing)
	50 percent vermiculite *or* horticultural perlite
MIX WELL WITH:	
Fertilizer	10 pounds ground dolomitic limestone
Additives (per cubic yard of mix):	1 to 2 pounds 20 percent superphosphate
	1 pound potassium nitrate
	5 ounces fritted trace element FTE55
	3 ounces wetting agent (Aquagro® or Surfside®)

NOTE: Additional fertilizer will be needed after two or three weeks of growth.
CAUTION: If vermiculite is used, be careful not to over-apply potassium, since vermiculite
   contains 5 percent available potassium.

- warmth, to accelerate germination or rooting
- ventilation, once roots have been formed, to reduce the risk of disease

Since plant propagation usually requires higher temperature and greater humidity than general crop culture, the grower may designate certain structures for propagation only. Where the quantity of plants propagated does not justify an entire greenhouse or frame, one or two benches may be used as propagation chambers. With a timed mist line to provide the moisture, and a covering of transparent polyethylene film draped on wires over the bench to retain the humidity, an excellent propagation bench can be created, Figure 14-1. The plastic canopy can be folded back at night permitting the cuttings to surface-dry, thus reducing the risk of mildew and other diseases. In such a situation, it is critical that the media used and the benches assigned for propagation be pasteurized and well drained.

PROPAGATION BY SEEDS

Propagation by seeds is practical if the seed is able to produce plants with predictable and desirable features. Seeds are produced and collected by industry specialists who market them through catalogues that illustrate the mature plant to potential buyers. Seed suppliers usually provide the buyer with information about the best time to plant, any pretreatment that may be necessary, and follow-up culture information. A grower who propagates from seeds should select a reputable supplier who will provide seed that is true to species and free of pests, weed seeds, and inert material; it should also have a high germination percentage (as determined by a seed test).

Preconditioning of Seeds

Certain species of plants, particularly those likely to be grown in a nursery, may produce seeds that require preconditioning before germination. As explained in Chapter 5, some seeds possess seed coats that are so hard and impervious to water that they must be altered before water can be absorbed and germination initiated. In their natural habitat, such seeds may require exposure to weathering elements or passage through the digestive tract of animals to break down the seed coat. A commercial grower can overcome such seed coat dormancy by *mechanical scarification* of the seed. Several techniques are commonly used, including:

- abrasion (such as rubbing with sandpaper or tumbling in drums with coarse sand or gravel)
- soaking the seed in hot water and allowing it to cool over a twelve- to twenty-four-hour period
- soaking the seed in concentrated sulfuric acid until the seed coat is paper thin but the embryo not yet affected. This may require from several

FIGURE 14-1.
The mist from the overhead line is retained by the plastic curtain. This is a typical propagation bench. (Courtesy Rodney Jackson, Photographer)

minutes to several hours depending upon the species.

Other species of plants produce seeds that remain dormant until certain physiological changes (termed *after-ripening*) are initiated and completed within the seed. The initiation of the changes is usually temperature-related, beginning when the temperature drops below 50° F for a required period of time. The chilling is termed *stratification,* and it can be duplicated by growers when the needs of a particular species are known. To stratify seeds, they

are first soaked in water for a half day or longer, then mixed with sphagnum moss or another moisture-retaining medium, sealed in polyethylene bags (for oxygen permeability), and stored in a cooler (not a freezer) for the required period of time. If the moisture-retaining medium is not sphagnum moss (which has a natural fungicidal quality), then a seed fungicide should be added for protection.

Some seeds may possess *double dormancy* and require both scarification and stratification before they germinate.

A great amount of research continues to be directed to the use of growth regulators as substitutes for the after-ripening period. Most of the results have yet to prove of commercial value, although gibberellic acid and kinetin are of some use to commercial growers.

Planting the Seeds

The successful propagation of new plants from seeds depends upon careful attention to all of the production factors. Should any one factor be limiting, the survival or quality of the plants can be affected. The factors of good production are:

- good quality seed
- correct propagation medium
- correct planting techniques
- appropriate lighting
- proper watering
- good drainage
- proper temperature
- adequate nutrients

The attributes of high-quality seed and the features of good propagation media have been discussed.

Planting techniques for seeds vary depending upon whether the species are herbaceous or woody and whether they are to be transplanted after germination or will be grown at the planting site.

Procedure for Planting Seeds of Herbaceous Greenhouse Crops

1. Fill greenhouse flats with a propagating medium that has been pasteurized and is well aerated, well drained, and fertile.

2. Tamp and firm the medium to drive out air pockets. Leave about 1/2 inch of space at the top of the flat to allow for watering.

3. Plant the seeds in furrows or in holes made with a pegged board. Tiny seeds may be broadcast over the surface. The seeds may be covered with additional propagating medium. Shaking the medium through a wide screen sieve will assure that the seeds are covered evenly.

4. If the seeds are large enough so that they will not be washed away, the flats can be watered from above. For fine seeds, the flats should be immersed in water before seeding. The planted seeds should be kept moist for rapid germination, yet not so moist that damping-off disease becomes established. Moisture may be applied by hand watering or by use of a mist system. Moisture loss can be reduced by covering the flat with glass or transparent plastic for a portion of the day. Care must be taken to assure that the seedlings do not remain too wet at night.

5. A germination temperature of 70° to 75° F is suitable for most greenhouse plants. Tropical plants may require higher temperatures. Plants should not be allowed to overheat.

6. After germination, reduce the temperature to between 65° and 70° F and maintain moderate light intensities. A soil drench to guard against damping-off disease is advisable. A slight reduction of water to a point that the new roots are moist while the surface is moderately dry will also aid in control of damping-off.

7. After two to four true leaves, not the seed leaves (cotyledons), have formed, transplant the seedlings to allow healthy growth to continue. If conditions at the transplant site are more stressful than at the germination site (as when seedlings are moved from greenhouse to field conditions), a *hardening-off* period is needed. Hardening-off occurs when reduced temperature and water

cause carbohydrates to accumulate in the plant tissue, allowing it to survive transplant shock. The hardening-off process is described in more detail later in this chapter.

Procedure for Planting Seeds of Woody Crops for Container Production

1. Following any preconditioning that may be necessary, the seeds can be planted as in the previous procedure. The time required for germination may be considerably longer with woody crop seeds.

2. When the seedling roots have reached the bottom of the greenhouse flat, they should be pruned to promote branching.

3. Transplanting should follow, with each seedling moved from the germination flat to individual peat pots or other types of containers.

4. After roots extend an inch or so beyond the peat pot, they should be pruned off to encourage still more branching.

5. Set the plant into a production container for further growth, Figure 14-2.

Procedure for Planting Seeds Directly in the Ground

1. The in-ground planting site may vary from a nursery field for landscape plants to a hillside for a reforestation project. The soil should be prepared and conditioned as necessary to assure that it is aerated, well drained, fertile, reasonably weed-free, and of the correct texture. It may be necessary to work the soil with a rototiller or larger piece of agricultural equipment. There should be no large clods. Additions of organic matter, mixed well into the soil by the tiller, will help assure the uniform size of the soil aggregates. A medium, loamy texture is ideal for seeding.

2. Seeds should be preconditioned to prepare them for planting. Certain species can be planted in the fall and after-ripened naturally in the soil. They will then germinate in the spring. Other species will germinate

FIGURE 14-2.
Following its germination in the peat pot, this nursery seedling is transplanted into a container for further growth. It may be transplanted several times before it it sold.

more uniformly if preconditioned by the grower before planting.

3. Seeds can be treated before planting to protect against damping-off, insects, birds, and rodents.

4. The seedbed will benefit by application of both pre-emergence and post-emergence herbicides during production of the seedlings, which may extend over several years. Otherwise, the seedlings will suffer from competition with weeds for light, nutrients, and space to grow.

5. The seeds may be planted either by hand or by machine. The depth of planting will vary with the species. Two or three times the diameter of the seed is usually a satisfactory depth.

6. A mulch should be applied to shade the new seeding and aid water retention. The mulch also reduces the force of water on the new seeds and helps maintain their spacing. Lightweight organic mulches work best, but additional nitrogen should be applied to compensate for that tied up by microbes during the decomposition of the mulch.

7. The new seeding, if in a nursery seedbed, should be shaded. Whether in a seedbed or unshaded nursery row, it should be kept moist. Temperature extremes and drastic changes should be avoided.

8. The plants may remain in the seedbed or nursery row for a season or more before being transplanted.

PROPAGATION BY PLUGS

Plugs are seedlings that retain their undisturbed root system within a core of media. They are planted as seeds into shallow, chambered flats known as *plug sheets*, Figure 14-3, which vary depending upon the size of the plug being produced. Typical plug sizes range from 7/16″ x 3/4″ x 3/4″ deep to 13/16″ x 1¼″ x 7/8″ deep on 21″ x 11″ plug sheets that contain from 288 to 800 chambers per sheet. The plug chambers may be either square or round, resulting in a root plug of the same shape.

FIGURE 14-3.

Flat with 98 separate cells for seedlings

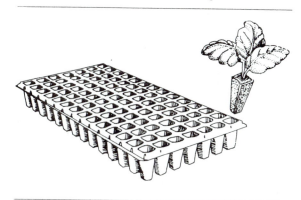

While plugs are used in many areas of horticulture, in the production of ornamentals they are most widely used by the growers of bedding plants. A majority of growers prefer to use plugs over other seed propagation techniques, for several reasons:

- Transplant shock, common to bareroot seedlings, and transplanting time are reduced.
- Plugs do not overcrowd as quickly as seedlings in germination flats, so they can be held longer awaiting transplant.
- Sowing of the seed can be automated and there is no need to thin the seedlings after germination, resulting in the need for fewer seeds and in time saved.
- Plugs can be transplanted automatically, further reducing time and labor costs.
- Shorter crop production time permits more crops to be produced in the same facility.

Although some growers do start their own plugs, the trend is toward specialization in plug production by large wholesale growers who can provide the controlled environments necessary to produce high-quality plugs at reasonable costs. They then sell the plugs to other growers for transplanting and further growth.

Planting the Seeds

As a result of the specialization in plug production, the large plug propagators sow the seed automatically unless a particular species has a seed type that does not work in automatic seeders.

After the plug sheets are filled with the proper pasteurized propagating medium, each chamber is seeded with one or two seeds. The use of two seeds per chamber ensures maximum and uniform production per sheet, but it does add some labor costs if thinning is later needed.

Germination of the seeds in the plug chambers can be done conventionally in the greenhouse benches under misting systems or in growth chambers if particular species require special humidity and/or temperature conditions. Whether in the bench or growth chamber, water application must be even and gentle to avoid washing out the seeds. Automated mist systems are preferable to hand watering, even with fogger nozzles. Due to the shallowness of the plug sheets, there is always the danger of overwatering or underwatering. The sheets cannot become watersoaked; neither can they be permitted to dry out.

The Problem of Stretching

If kept too long in a growth chamber after germination, the plugs will stretch and overgrow. Likewise, when the seeds in a plug sheet do not germinate uniformly, the early ones will stretch while the grower awaits the germination of the slower seeds.

Michigan State University researchers have published convincing evidence that in many plants, especially bedding plants, stretching can be reduced by keeping the greenhouse temperatures cool during the day and warm at night. A day temperature cooler by 5°F during production of plugs can result in a higher number of plug sheets displaying near-maximum germination, without undesirable overgrowth.

Chemical growth retardants also have proven effective against plug stretching. The retardants are somewhat species-specific though and should not be used until the grower is certain no damage will result to the plugs being produced.

Fertilizing Plugs

After water, nutrients rank as the most probable limiting factor in plug production. Fertilization is vital, since the small core of propagation medium in each chamber is unable to offer much to the new plant. Major and minor elements are both critical. Low nitrogen fertilizer (about 50 ppm) can be applied at the time of seeding, with concentrations increasing as the seeds germinate and begin to develop. Local problems such as high soluble salt levels or pH extremes (outside the optimum 5.8 to 6.0) need to be addressed as well.

PROPAGATION BY CUTTINGS

Cuttings are pieces of roots, leaves, or stems that are removed from the parent plant and placed in an environment that promotes their development into total plants. In most cases, the new plant will be genetically identical to the parent plant since propagation by cuttings is an asexual method of reproduction. As measured by sheer numbers of plants produced, propagation by cuttings is the most widely used method of reproducing ornamental plants asexually. The reasons are that the technique is quick, easy, and inexpensive. Greenhouse growers and nursery growers both propagate important commercial crops from cuttings.

Propagation by cuttings is possible because of the ability of plant cells to revert to an undifferentiated, actively growing (meristematic) condition, from which they can once again initiate the root, stem, and/or leaf tissue necessary to form a complete plant. Years of research have been directed toward understanding why and how cuttings develop into complete plants and why certain species are easier to propagate than others. Although not everything is known or understood, these points are commonly accepted:

- Adventitious roots are initiated in herbaceous plants from points just outside or between the vascular bundles. In woody plants, they originate next to and out from the center of the vascular core. The adventitious *root initials* (growing points) may develop after cuttings are taken or they may be preformed but dormant in the vegetative tissue before the cutting is taken. Individual species vary.
- Adventitious roots and adventitious shoots do not develop at the same rate within an individual plant or among different species.
- The rate of root and shoot initiation and formation in cuttings is controlled in part by growth regulators, principally auxins, gibberellins, and cytokinins. Some growth regulators occur naturally in the tissue. Others can be applied by the propagator. Those that occur naturally may promote or inhibit growth depending upon the growth regulator and plant species involved and the concentration of the growth regulator.
- Auxin is necessary for the initiation of adventitious roots on stem cuttings. IAA (indole-3-acetic acid) is the naturally occurring auxin. NAA (naphthaleneacetic acid) and IBA (indolebutyric acid) are similar materials, but synthetic. The synthetic auxins, applied by plant propagators, are generally more effective than the naturally occurring IAA at initiating the formation of new roots on stem cuttings.
- The formation of adventitious roots is very slow in certain plants. Depending upon the species, rooting may be inhibited by: (1) naturally occurring *rooting inhibitors* in the plant tissue, (2) the lack of one or more *rooting cofactors,* found by several researchers to work synergistically with auxin in root initiation, or (3) a *continuous sclerenchyma ring* between the phloem and the cortex, which creates a physical barrier to developing roots as they attempt to emerge from the center of the vascular core.
- Cuttings possess *polarity.* They form new shoots at the end nearest the tip of the plant (the distal end) and new roots at the end nearest the crown (the proximal end). In commercial propagation operations, care must be taken to orient the cuttings properly in the propagation bench to prevent delayed and unsatisfactory rooting.
- The best cuttings result from healthy stock plants that contain adequate nitrogen for good shoot formation and high carbohydrate levels for easier rooting. Excessive nitrogen in the cuttings is likely to inhibit root formation. Lateral shoots make better cuttings than terminal shoots in certain species because the former contain more carbohydrates while the latter are very high in nitrogen.
- Adventitious roots form more quickly on stem cuttings in the dark than in the light. Stock plants of certain species can have their stems, which are to be used as cuttings, wrapped in black plastic or tape where the roots are to form. Later, when the cuttings are made, new roots will form more quickly in the darkened portion. This is one reason why air-layering works as a propagative technique and why cuttings root best at their proximal end, within the darkened rooting medium.
- Stem or root cuttings taken from young plants root more quickly than cuttings taken from older plants. The juvenility factor is commonly

acknowledged but only partially understood. Many commercial propagators maintain their stock plants in hedge form to sustain the juvenile growth phase rather than permitting the plants to develop into their normal adult tree form.

- Cuttings from lateral shoots and others from vertical shoots may develop into plants with different growth habits. Generally, strongly vertical plant forms will result only if the cuttings are from vertical shoots. The more easily a species roots, the less important is the position of the cutting on the stock shoot.
- Cuttings can usually be taken from the stock plant regardless of whether it is in a flowering or vegetative stage. When a species is difficult to root, better results are usually obtained when vegetative growth is selected for the cuttings. Research suggests that a reduction in auxin levels during flowering is a probable reason.

The Environment for Rooting

Once separated from the stock plant, cuttings are vulnerable to injury from assorted environmental factors. To assure the rooting of the greatest number of cuttings, several precautions must be taken by the propagator.

Moisture Moisture is critical to the successful rooting of the cutting. The irony of the situation is that the leaves need to produce food to support root formation, but lack of roots prevents the cutting from taking up water the leaves need for photosynthesis to produce food. The more slowly a species forms roots, the more critical the situation becomes. To reduce water loss from transpiration, a mist line is widely used by propagators. Seldom applied continuously, intermittent mist under the control of a time clock maintains high humidity around the leaves of the cuttings. The film of water that settles onto the leaves also keeps the cuttings cooler than the drier air.

Temperature Temperature controls the rate of root and shoot development in cuttings. If the air temperature in the propagation bench is too high, shoots will form faster than roots, placing even greater moisture stress on the cutting. Instead, the rooting medium should be warmer than the air to promote root growth in advance of shoot development. To accomplish this, an electric heating cable and a thermostat may be installed in the propagation bench. With controlled daytime air temperatures between 70° and 80° F, night temperatures of 60° F, and the rooting medium slightly warmer at all times, the cuttings should root satisfactorily.

Nutrition Nutrition supplied during rooting may influence the quality of the cuttings produced. Since the rooting medium is not high in nutrients and the misting leaches minerals from the plant tissue, it is often beneficial to add a liquid fertilizer to the mist water.

Acidity/Alkalinity The pH of the rooting medium can affect the number and quality of adventitious roots produced. Media having a pH near neutral have generally been found to produce the best root systems. There are selected exceptions however. For example, azaleas root better in an acidic medium.

Light Quality and Intensity As noted previously, high carbohydrate levels in cuttings promote good root formation. Therefore, high light intensity from sunlight or an artificial source high in red and/or blue wavelengths (most usable by plants) is necessary for good root development. As the cuttings develop, the day length may begin to affect the shoots either positively or negatively. Research has shown the effect of photoperiod to vary depending upon the species of plant being propagated, the season of the year, and the type of cutting.

Oxygen Content As long as the propagating medium is porous and the bench well drained, sufficient oxygen will usually enter the medium to satisfy the plants' needs. If the cuttings are found to be producing adventitious roots only near the surface of the medium, it is an indication that the medium or bench is not draining well.

Types of Cuttings and Methods of Propagating Them

In Table 14-3, various types of cuttings are compared. Many species of plants can be propagated by more than one technique. Which method is most practical is a decision to be made by the commercial propagator, Figure 14-4.

Hardening-Off the Cuttings

Hardening-off means the gradual adaptation of plants to environmental conditions more stressful than their present ones. Cuttings rooted under conditions of high light, high humidity, and warm temperatures must be hardened-off soon after the roots form if normal growth is to occur. Allowing the cuttings to remain in the propagation environment longer than necessary for rooting can cause damage to both new roots and new shoots. However, the abrupt cessation of misting or other special propagating conditions can result in severe shock to the cutting.

Commercial propagators use various techniques to harden-off cuttings. Among these techniques are the following:

- *Gradually decrease the misting.* Either reduce the misting interval or the time of system operation.
- *Root cuttings in peat pots for direct transplanting or directly into their sales container.* This eliminates any chance of root disturbance.
- *Transplant during the dormant season.* The reduced metabolic rate of the cuttings reduces transplant shock.
- *Hold transplants temporarily under high humidity.* This allows roots to adjust to the new medium with no disturbance to the foliage.
- *Remove the propagating frame from around the cuttings.* The cuttings can then send roots into soil beneath the rooting medium and continue their growth undisturbed.

PROPAGATION BY GRAFTING

A *plant graft* is the union of parts from two or more plants into a single plant. The technique of grafting requires a knowledge of which plants can be grafted, which ones benefit from grafting, and how to do the numerous types of grafts. Knowledge of *how* to graft plants has existed for centuries. Knowledge of *why* only certain plants can be grafted and why they grow as they do is more recent and is still incomplete. Much of the research into grafting has been accomplished with citrus and other fruit trees because of the commercial importance of those grafted plants.

Ornamental propagators graft either when the resulting plant is superior to those provided through other methods of propagation or when no other propagative method will work. Other ornamental horticulturists such as groundskeepers turn to grafting to repair, brace, or otherwise save damaged plants. Occasionally, grafting is used to create special and unusual plant forms, such as tree-form shrubs or weeping-form trees. While grafting can be done on both woody and herbaceous plants, it is most common to woody plants.

A typical plant graft involves the following parts, Figure 14-5:

- *Stock:* the portion of the graft that will develop into the root system of the new plant
- *Scion:* The portion of the graft that will develop into the branches and foliage of the new plant. It may constitute some or all of the canopy depending upon how much of the top growth of the stock was removed. The scion may be a small branch or a bud. It should not be a sucker or water sprout.
- *Graft union:* the point where stock and scion join. It usually remains visible throughout the plant's life.
- *Interstock:* a piece of plant stem placed between the stock and scion, resulting in two graft unions. It is not a part of every grafted plant. It is used to join two species that will each unite with the interstock species but not with each other. It may also contribute additional characteristics to the new plant.

How the Graft Forms

While methods are varied, the goal of the propagator is always to place the separate cambiums of

TABLE 14-3.

A Guide to Propagating Cuttings

Type of Cutting	Typical Plants Used	When to Take Cuttings	Part of Plant to Select	Location of the Cut	Length of Cutting
Hardwood stem (deciduous)	Some deciduous trees and most deciduous shrubs	Late autumn to early spring	Central and basal portions of the stem	Basal cut: just below a node Distal cut: 1/2 inch above a node	4 to 30 inches
Hardwood stem (narrow-leaf evergreen)	Young seedling stock plants of most narrowleaf evergreen trees and shrubs	Late autumn to late winter	Mature terminal shoots of previous year's growth	4 to 8 inches below the terminal shoot	4 to 8 inches
Semihardwood stem	Woody, broadleaf evergreens	Summer	New shoots, only partially matured	Basal cut: just below a node	4 to 6 inches
Herbaceous stem	Succulent herbaceous shrubs and florist crops	Usually successful at any season	Strong stem sections, only partially matured	Basal cut: just below a node	3 to 5 inches
Leaf	Tropical foliage plants	Successful at any season	Leaf blade *or* Leaf blade and petiole *or* Pieces of the leaf blade	—	—

Preparation and Planting	Preferred Rooting Medium	Special Requirements	Special Methods of Handling
Treat cuttings with a root-promoting material; plant immediately or place in a box of peat until callus forms, then plant; insert into rooting medium so that 2 inches to one-fourth of the cutting is planted.	Loose sandy loam	Select cuttings from stock plants growing in full sun.	* See notes
Remove all leaves from lower half of cutting; treat with IBA and a fungicide; place into rooting medium up to the leaves.	sand *or* half perlite and half peat	High light intensity and high humidity are necessary but not heavy wetting. Bottom temperature of 75° to 80° F is recommended.	Cuttings may root more easily if a piece of old wood is left at the base. Also, wounding the basal end of the cutting may help promote rooting.
Remove all leaves except two or three at the distal end; if leaves are large, cut them in half to reduce transpiration; treat with a root-promoting material; place in medium up to the leaves.	half perlite and half vermiculite *or* half peat and half perlite	Use intermittent mist and bottom heat.	Take cuttings in early morning to assure turgidity, or keep wood cool until ready to take cuttings.
Remove all leaves except two or three at the distal end; if leaves are large, cut them in half to reduce transpiration; place in medium near but not up to the leaves. Use of a root-promoting material is optional.	half perlite and half vermiculite *or* half peat and half perlite	Use intermittent mist and bottom heat.	Take cuttings from stock plants kept cool to assure turgidity. Keep cuttings cool and plant as soon as possible.
It is usually necessary to use a leaf or section of leaf with one or more wounded veins included. With some species, the basal ends are inserted into the medium. Other species are pinned flat onto the medium, vein side down. Treatment with a root-promoting material may help.	Loose sandy loam *or* sand	High humidity is needed. Also, polarity must be noted.	There is great variation in the techniques used. Some can be propagated best in greenhouse flats; others in closed petri dishes.

TABLE 14-3.

A Guide to Propagating Cuttings (continued)

Type of Cutting	Typical Plants Used	When to Take Cuttings	Part of Plant to Select	Location of the Cut	Length of Cutting
Leaf-Bud	Tropical shrubs, broadleaf evergreens and greenhouse crops	When the foliage is actively growing	Leaf bud, petiole, and a piece of stem with bud attached	At a node	—
Root	Deciduous shrubs	Late winter or early spring	Root sections	—	1 to 6 inches (larger roots make larger cuttings)

FIGURE 14-4.

Cuttings being placed in the propagation bench to promote root formation

FIGURE 14-5.

Typical plant grafts with and without an interstock

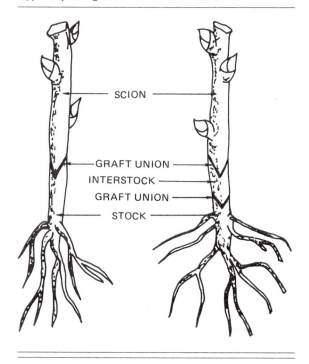

SCION

GRAFT UNION
INTERSTOCK
GRAFT UNION
STOCK

Preparation and Planting	Preferred Rooting Medium	Special Requirements	Special Methods of Handling
Treat cut surfaces with a root-promoting material. Insert into the rooting medium with the bud buried 1/2 to 1 inch beneath the surface.	half peat and half perlite	Select cuttings with well-developed axillary buds. Maintain high humidity and bottom heat.	—
Cut ends may be treated with a fungicide. Plant either with the crown end upward and at soil level, or horizontally 1 to 2 inches deep.	sand *or* sandy soil	Correct polarity must be noted.	Cuttings are frequently obtained as nursery crops are being balled-and-burlapped.

* Notes: Deciduous hardwood stem cuttings are handled differently depending upon when the cuttings are taken and when the local climate allows planting. Where winters are mild, cuttings can be made in the autumn and planted immediately. Some callus formation and rooting will often occur before winter. Cuttings taken in late fall or winter can be stored outdoors in mild areas or in cool, nonfreezing storage indoors. They should be placed upside down or horizontally in moist sand, sawdust, or shavings. Callus tissue and root initials will usually form during the storage period. Some cutting stock can be collected during the winter and kept in cool, nonfreezing moist storage until early spring. Cuttings can then be made and planted immediately.

the stock and scion into close contact in order that the two may fuse. The joining of the separate parts is made possible by the production of *callus,* or wound tissue, from the parenchyma cells of the cambiums of both stock and scion. The cells of the callus tissue intermingle until a binding results. Cell differentiation follows and parenchyma cells become cambium cells. The new cambium, a product of the two previously separated plant parts, produces new vascular tissue. The formation of a vascular connection between the stock and scion is essential for the success of the graft. When the graft union is successful and the two different plants become one, it is termed a *compatible* graft. If the two plants do not unite even though there was no error in the technique used, it is termed an *incompatible* graft. In some instances, the graft may appear successful until months or years later when the plant suddenly dies or snaps off unexpectedly at the graft union. *Delayed incompatibility* is the term used to describe such a situation.

The Uses and Limitations of Grafting

As noted earlier, propagation research is ongoing. The uses of grafting will no doubt increase as more is learned about the reasons for its current limitations. Some of the widely accepted information about grafting can be summarized as follows:

- Grafting can increase the winter hardiness of certain plants by providing them with a root system that is more tolerant of low temperatures.
- Replacing the root system with that of a more resistant species can permit the growth of plants normally susceptible to soil-borne pathogens.

- Trees girdled by animals, lawn mowers, or vandals can often be saved by bridge-grafting.
- A pistillate dioecious plant can have a scion from a staminate counterpart grafted to one of its branches to assure the pollination needed for fruiting.
- Multiple cultivars can be grown on one established tree assuming all are compatible with it. Thus, it is possible to have rose bushes with several different-colored blooms on one plant, apple trees that bear four or five different kinds of apples, and the same with citrus fruits.
- Some hybrid plants grow more vigorously if removed from their own root system as seedlings and grafted onto another rootstock.
- The rootstock can greatly influence the subsequent growth of the scion after graft union. The effect may be one of dwarfing or acceleration, altered growth habit, increased or reduced flowering and fruiting, pest resistance, or hardiness.
- Not all plants graft and unite as easily as others. It is not a problem of incompatibility but of species variation. Also, some species graft more successfully by one technique than by another.
- Not all plants can be grafted. Generally, the closer the botanical relationship between the plants involved, the greater the probability of a successful graft. For example, grafting among members of the same clone is predictably successful; among different species of the same genus, some will graft and some will not; among different genera, grafting is highly improbable.

The Materials Needed for Grafting

The tools and accessory materials required for grafting have changed little during the hundreds of years that the craft has been practiced. Today's materials have the benefit of being produced and sold commercially, so it is unnecessary for propagators to make their own. Even so, many still choose to do so. Regardless of how the materials are obtained, grafting requires the following basic items:

A Knife   The knife must be well constructed of good steel that can be sharpened repeatedly and retain a good edge. Fixed-blade knives are preferable to folding knives because they have greater strength, especially important with woody material. Budding (described later) requires a modified knife styled for the more detailed and intricate needs of the technique. *NOTE:* Severely cut fingers are common even among experienced propagators. Finger guards for the thumb and index finger are recommended, especially for beginners.

Tying Materials   Some grafts require tying. The tying material must have strength with elasticity, so that it will not girdle the plant as the graft union forms and expands. Materials range from ordinary string to commercially manufactured plastic or rubber tapes. All must either rot away or be removed soon after the graft union forms.

Grafting Wax   Wax serves as a sealant to keep moisture inside the graft region and to keep insects and inoculum out, Figure 14-6. The wax must be water repellent or it will be washed away. It must be elastic enough to permit expansion as the graft union grows. In addition, it must not melt in warm temperatures. Both hot and cold waxes are available from commercial sources. If hot wax is used, a source of heat is required to melt the wax. A brush is used to apply the wax.

Assuring the Success of a Graft

Assuming that the stock and scion have the potential for a compatible graft union, the propagator can maximize the percentage of successful unions by taking several precautions.

- Obtain scions from the current season's growth when it is matured, not succulent. Avoid terminal shoots; central to basal shoots are better.
- Select scion and stock material from plants that are free of pathogens and abnormalities.
- Use herbaceous materials immediately after collecting to prevent their drying out. Woody material may be collected at the time of grafting or prior to it. If collected and stored until use, the cut material should be kept moist (but not

FIGURE 14-6.

Materials used for grafting. Front (left to right): finger guards, string, tape, twistems; center (left to right): whet stone, fixed blade knife, budding knife, patch bud knife, wedge, rubber ties; Rear: wax and lantern heater with brush.

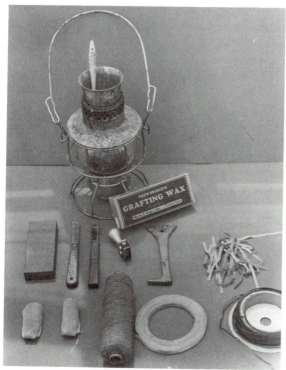

surface wet) and cool (40 to 50° F). Peat moss, shavings, or similar noncohesive materials are suitable for storing scions.

- Place the cambiums of the stock and scion in the closest contact possible. This will assure the strongest and most rapid graft union. Wrapping the stock and scion can assist a close contact.

- After the graft is assembled, keep it moist. Coating the grafted area with wax will usually suffice. Placing the grafted plant into a covered frame or moistened holding medium will also provide the proper environment around the grafted area.

- Pinch off any shoots that develop from the stock.

Types of Grafts

Methods of grafting vary with the size of the plants. Some methods are more effective than others with herbaceous materials. Still other methods were developed to accommodate specific plants. Figure 14-7 shows and describes the most common types of grafts used by ornamental horticulturists. It is not all-inclusive. A more detailed treatment can be found in texts devoted solely to plant propagation.

PROPAGATION BY BUDDING

Budding is actually a type of grafting but the techniques are sufficiently specialized to warrant separate consideration. The rootstock is similar to that used in other grafting methods; that is, seedlings, rooted cuttings, and branches of young trees and shrubs. The scion is markedly different, though. It is only a single bud with a patch of bark and sometimes a bit of wood attached. Budding is comparatively simple as a grafting method. It is the most rapid method of propagating roses for commercial sale. Fruit trees and other nursery stock are also propagated and top-worked through budding. (*Top working* is the grafting of scions onto the existing framework and root system of a large, established tree. It is done to produce a crop faster than natural growth permits.) Since one budstick can provide as many scions as it has buds, budding is often a means of maximizing limited propagative materials.

Budding involves removing a piece of bark on the stock plant and replacing it with a comparably sized piece from a compatible plant. The added piece contains a bud which develops into a new shoot. Afterwards the stock above the newly grafted shoot is cut back, along with any buds that develop on the stock below the graft.

Successful union requires that the bark of both the stock and scion separate easily (termed *slipping*) for removal. The bark slips most readily when the plant is actively growing. Budding can be done al-

FIGURE 14-7.

A guide to grafting techniques

TYPE	TYPICAL USES	SEASON TO COLLECT SCION	SEASON TO MAKE GRAFT	PREPARATION OF THE SCION	PREPARATION OF THE STOCK	ADDITIONAL PREPARATIONS REQUIRED	TIE THE GRAFT UNION?	WAX THE GRAFT UNION?	DIAGRAM OF THE TECHNIQUE
Approach Graft	Used to join plants that do not graft together easily.	Any season	Any season (best in spring and summer)	A 1- to 2-inch slice of wood is removed where the graft union is to be. The cut should be flat and smooth.	A 1- to 2-inch slice of wood is removed where the graft union is to be. The cut should be flat and smooth.	For a tight fit, a tongue can be added with an extra upward cut on the scion and a downward cut on the stock. Stock top and scion roots are not removed until after the union is formed.	Yes	Yes	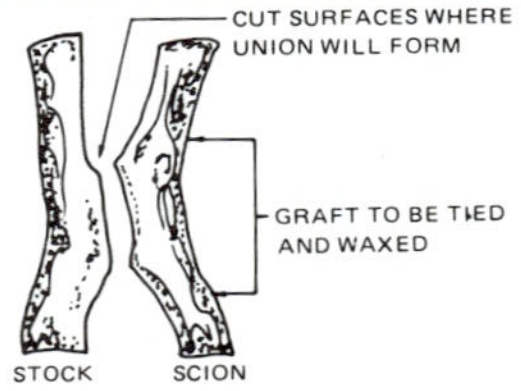
Bark Graft	Used to top work trees and branches up to 12 inches in diameter.	Late fall to early spring	Done in the spring using dormant scions and stocks	It should be about 4 inches long and contain several buds. A sloping cut should be made on one side of the base, 2 inches long and about one-third the way into the stem. On the opposite side, make a short, sloping cut to create a wedge end.	Prepare the stub as for a cleft graft. In several places around the stub, make a cut 2 inches long through the bark and into the wood. Separate the bark from the wood slightly to receive the scions.	Several scions are used around the stub. Do not make them too thin or they may break off. Flat-headed nails are often used to secure the loose flaps of bark and hold the scions in place.	Yes	Yes	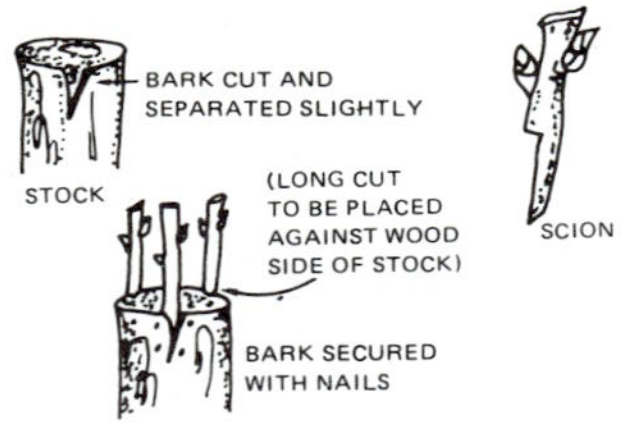
Bridge Graft	Used to repair damaged bark areas around tree trunks. It is the only hope when complete girdling has occurred.	Late fall to early spring	Early spring	Dormant, one-year-old wood should be used, 1/4 to 1/2 inch in diameter. Each scion should have a wedge end cut at its top and bottom ends.	Remove all damaged bark around the area to receive the graft. Only live healthy bark should remain. Cut slots into the healthy bark at the top and bottom to receive scions. Leave a flap of bark as described for in arching.	Scions are inserted under each flap of bark all around the damaged area. They should then be nailed into place.	No	Yes	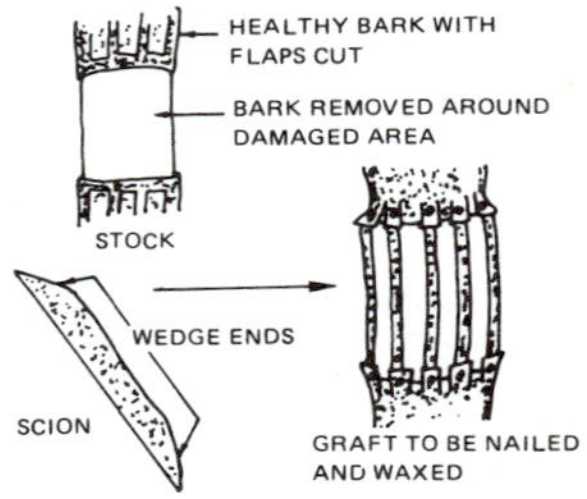

Method	Uses	Season	Season	Scion	Stock	Comments			Illustration labels
Cleft Graft	Used to top work trees, changing the cultivar of an established tree	Late fall to early spring. Use dormant wood, one year old	Any time while the tree is dormant.	It should be 3 to 4 inches in length and contain several buds. The base should be 1 1/2 inches long and shaped as a long wedge, sloped on both sides.	Cut it at a 90° angle to the direction of growth of the branch. Make a vertical split 2 to 3 inches down the center of the stub branch. (A knife and mallet can do it nicely.)	Two scions are placed into the split of the stock. Cambium of the stock and scion must touch as much as possible for the entire length of the graft. Use a wedging tool to hold the split open until scions are placed.	No	Yes	SCIONS; WEDGE CUT; SPLIT; STOCK; WAX; SIDE VIEW
In Arching	Used to repair or strengthen the root system of an established tree	The established tree is the scion.	Early spring as active growth begins	Slots 4 to 6 inches long are cut from the bark of the old tree to match the size of the stock seedling to be inserted. A flap of bark is left at the top of each slot to secure the wedge end of the stock.	Seedling plants planted around the old tree at 6-inch intervals are the stocks. Each receives a shallow 4- to 6-inch long cut along the side next to the tree trunk. Opposite the cut is another short one to create a wedge end.	Flat-headed nails are used to secure the seedlings to the tree trunk and also to hold the flap of bark over the stock inserts.	No (nails sufficient)	Yes	SCION (OLD TREE); STOCK SEEDLINGS SECURED WITH NAILS; SLOT TO RECEIVE STOCK SEEDLING
Saw-Kerf Graft	Used to top work trees, often larger than suitable for cleft grafting	Late fall to early spring	Any time while the tree is dormant.	It should be 4 to 5 inches in length and contain several buds. The base should be prepared as for a cleft graft.	Prepare the stub as above for the cleft graft. Then make three cuts about 1 inch toward the center of the stub.	The cuts can be widened as necessary to accommodate the scion. Cambiums must touch as much as possible.	No	Yes	SCIONS; WEDGE CUT; STOCK
Splice Graft	Used as above but with plants having inflexible stems	Late fall to early spring	Winter to early spring	Same as whip or tongue graft, but without the second cut.	Same as whip or tongue graft, but without the second cut.	Same as whip or tongue graft	Yes	Helpful	SCION; STOCK; TIED AND WAXED

FIGURE 14-7. (Continued)

TYPE	TYPICAL USES	SEASON TO COLLECT SCION	SEASON TO MAKE GRAFT	PREPARATION OF THE SCION	PREPARATION OF THE STOCK	ADDITIONAL PREPARATIONS REQUIRED	TIE THE GRAFT UNION	WAX THE GRAFT UNION?	DIAGRAM OF THE TECHNIQUE
Side Graft	Used for adding branches to young trees	Late fall to early spring	Early spring	Two 1-inch long sloping cuts are made at the base to create a wedge-shaped end.	With a sharp knife or small chisel, make a cut 30° up from the stem and about one-third of the width of the stem.	The stock is cut off above the graft union.	No	Yes	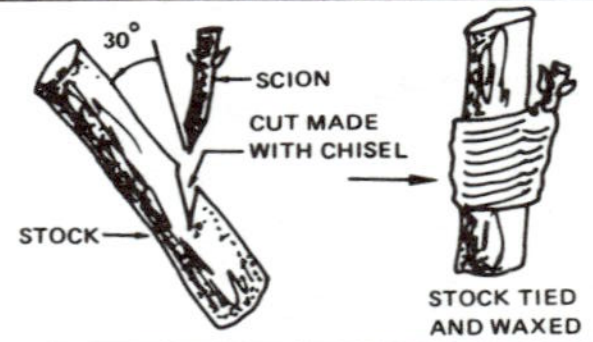
Spliced Side Graft	Used commonly for grafting small woody potted plants	Early spring	Early spring	A shallow sloped cut 1 to 1 1/2 inches long is made on one side of the base, intersecting with a short downward cut from the other side.	A matching cut is made into the stem of the stock just above the crown.	High humidity must be maintained for at least a week after the graft is made. After union, the stock is cut off above the grafted area.	Yes	Helpful	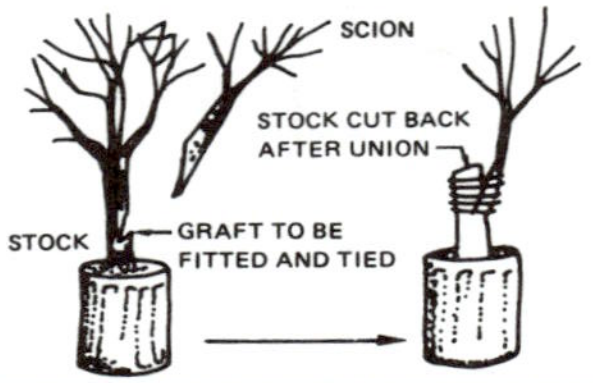
Whip or Tongue Graft	Small plant materials such as deciduous nursery stock that is 1/4 to to 1/2 inch in diameter	Late fall to early spring	Winter to early spring	It should contain several buds. The cut should be sloped and 1 to 2 1/2 inches long at the base. A second cut should be made, half the length of the first and vertical to the first.	The stock should be of the same diameter as the scion and cut at the top, but at the top. A second cut should be made as described for the scion.	The cut surfaces should be smooth. After the graft is assembled, the scion should not overlap the stock.	Yes	Yes	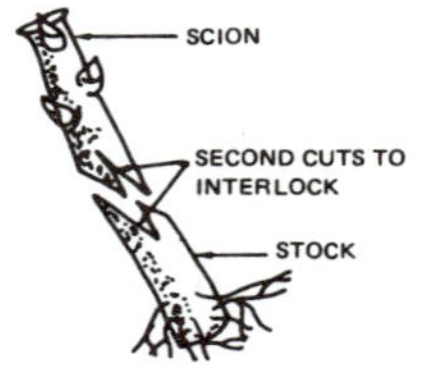

328

most any time from spring through fall as long as the bark is slipping and buds, not shoots, are available from the scion plants.

Fall and spring are the best seasons for budding, but attention must be given to the state of development of the buds. In the fall, buds may be used after they have formed prior to the onset of winter. Normally, they will remain dormant through the winter and begin growing in the spring. For spring budding, the budstick often must be collected early in the winter while still dormant, then kept in moist, cool storage until the bark begins to slip on the stock plant outdoors. Summer budding, most often done in June, requires that the leaves on the scion be sufficiently mature to have formed an axillary bud that can be used. When a leaf is attached to the bud, the blade should be cut away allowing the petiole to remain. The petiole provides a handle to lift the bud patch. It will later die and drop away.

Tools for Budding

The grafting knife described previously can be purchased with a pointed, blunt end useful in inserting the bud piece. Other tools have been devised to make the budding technique faster and easier. These include an assortment of double-bladed knives for patch budding that permit simultaneous parallel cuts on opposite sides of the bud to be made in a single stroke. The bark patch removed from the stock plant will be of a similar size to the bud patch, Figure 14-8. The result is a tighter fit and more rapid healing.

Methods of Budding

There are several different methods of budding. Some require actual removal of bark from the stock. Others require only an incision into the bark permitting the bud to be slipped in behind it. Following are some of the methods.

T-budding The most widely used method of budding, T-budding derives its name from the shape of the incision made in the bark of the stock plant, Figure 14-9. It is most adaptable to young, thin-

FIGURE 14-8.

A special cutter is used for patch budding. It removes identically sized patches from both the stock and scion plants.

barked nursery stock up to an inch in diameter. The bud is cut from a slipping budstick as a shield-shaped piece. The cut should be shallow, just beneath the bark, avoiding the wood as much as possible. The bud shield should be cut from an inch above the bud to half an inch below it. Using the thumb, a gentle sideward push separates the bud shield from the wood. The stock is prepared with a 1-inch long vertical cut through the bark on a smooth internodal section of stem. A horizontal cut across the top completes the T-incision. The blunted end of the budding knife lifts the stock's bark and the bud shield is inserted. When the upper edge of the bud shield matches the top of the T-incision, the bark flaps are closed around, but not over, the bud. Then the bud graft is tied, usually with a rubber budding strip. The strip assures a tight fit of stock and scion and prevents drying. After several weeks, the rubber strips will decompose and fall away. Waxing is not necessary. In some parts of the country, the T-incision is reversed to keep out rain that could promote rotting of the bud. This is termed *inverted T-budding.*

Patch-budding Using rootstocks and budsticks that are nearly the same size (about an inch in diam-

FIGURE 14-9.

Steps in making a T-bud

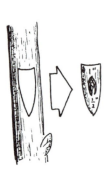   

SCION

THE BUD SHIELD
IS CUT AND
REMOVED.

STOCK

A T-CUT IS MADE
THROUGH THE
BARK.

THE SHIELD IS
INSERTED UNTIL
TOPS OF THE
SHIELD AND
T-CUT ARE EVEN.

THE BUD IS LEFT
EXPOSED WHILE
REMAINING SUR-
FACES ARE
WRAPPED TIGHT-
LY WITH RUBBER
STRIPS.

eter), a rectangular patch of bark is removed from both the stock and the scion, Figure 14-10. The scion patch contains the bud that will be the shoot of the new plant. The special tools described earlier help to keep both patches the same size. The most appropriate use of patch budding is with a stock plant that has thick bark and a scion plant whose budsticks have thin bark. With heavy bark, a T-bud graft does not fit tightly enough. The closer the fit is, the better are the chances of a successful graft union. Following insertion of the scion patch into the stock plant, the union is wrapped, leaving the bud uncovered.

I-budding This method is a combination of the techniques thus far described. The scion is prepared by cutting a patch from the budstick, Figure 14-11. The stock does not have a patch of bark removed; instead, an I-shaped cut is made into the bark. The two horizontal cuts and one vertical cut create two flaps of bark. These can be lifted up and the bud patch slipped beneath. Following insertion of the bud patch, and after assuring that stock and scion meet, the graft should be tied.

Chip Budding When either the season of the year or environmental conditions result in a nonslipping bark condition, it is still possible to propagate some plants by budding. Chip budding unites a chip of wood containing a bud from the scion with a stock plant that has had a comparably sized chip removed from an internodal area near the base, Figure 14-12. The technique works best with small wood up to an inch in diameter. The chips are removed with two downward cuts that intersect. With the scion, one cut begins 1/2 inch above the bud and proceeds downward and inward to a point about 1/4 inch beneath the bud. The other cut is beneath

FIGURE 14-10.
Steps in making a patch bud

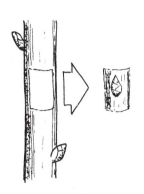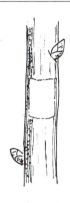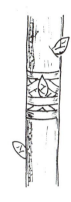

SCION   STOCK

| HORIZONTAL AND VERTICAL CUTS ARE MADE AND THE BUD PATCH IS REMOVED. | A SIMILAR PATCH IS REMOVED FROM THE BARK OF THE STOCK. | THE SCION PATCH IS MATCHED AND INSERTED INTO THE STOCK. IT IS TRIMMED TO FIT IF NECESSARY. | THE BUD IS LEFT EXPOSED WHILE REMAINING SUR-FACES ARE TAPED TIGHTLY. |

FIGURE 14-11.
Steps in making an I-bud

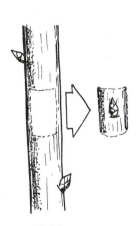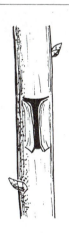

SCION   STOCK

| HORIZONTAL AND VERTICAL CUTS ARE MADE AND THE BUD PATCH IS REMOVED. | TWO HORIZONTAL CUTS AND ONE VERTICAL CUT ARE MADE TO CREATE TWO FLAPS IN THE BARK OF THE STOCK. | THE BUD PATCH IS INSERTED BENEATH THE FLAPS. | THE BUD IS LEFT EXPOSED WHILE THE REMAINING SURFACES ARE TAPED TIGHTLY. |

FIGURE 14-12.
Steps in making a chip bud

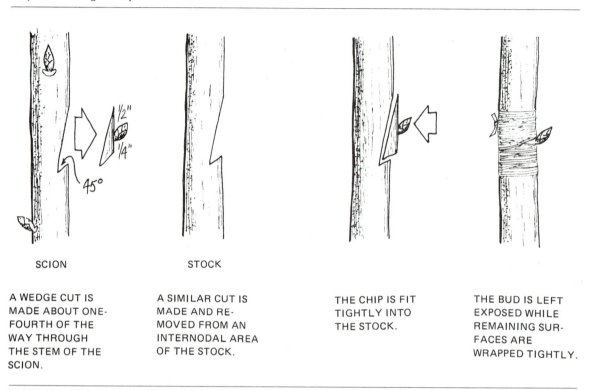

SCION   STOCK

| A WEDGE CUT IS MADE ABOUT ONE-FOURTH OF THE WAY THROUGH THE STEM OF THE SCION. | A SIMILAR CUT IS MADE AND RE-MOVED FROM AN INTERNODAL AREA OF THE STOCK. | THE CHIP IS FIT TIGHTLY INTO THE STOCK. | THE BUD IS LEFT EXPOSED WHILE REMAINING SUR-FACES ARE WRAPPED TIGHTLY. |

the bud, downward and angled sharply inward to intersect the longer cut from above. The cut into the stock is made the same way. As with all bud grafting, the closer the two chips are in size, the better is the chance of rapid healing of the union.

Follow-up Regardless of which budding technique is used, the follow-up procedures are the same. If budded in the fall, the shoots of the stock may be retained until the following spring. Then they are cut off above the graft union as soon as the buds begin to break dormancy. If budded in the spring, the stock can be cut back as soon as the graft union is established, in ten to fourteen days. Summer budding usually results in dormant scion buds that do not break until the following spring even though the union has been established.

They can be treated in the same manner as fall-budded plants.

PROPAGATION BY LAYERING

Certain species of plants do not develop roots rapidly enough to be propagated by cuttings. In these cases, the stem dies from lack of water and nutrients before the roots can be initiated. Layering permits the stem to remain partially attached, receiving water and nutrients through the still intact vascular system, while initiating roots at the point where separation from the parent plant will eventually occur.

Compared to other methods of propagation, layering is usually slower, more expensive, and produces fewer new plants per parent plant. For reproduction of large numbers of the most common nursery and greenhouse crops, layering has limited use. Amateur horticulturists use the techniques frequently when quantity is not important. Specialty nurseries use layering for unusual or rare species that do not reproduce from cuttings. Landscape gardeners whose professional efforts are directed to a single landscape such as an estate, cemetery, or school campus also use layering techniques.

Successful layering depends upon treatments applied to the stem to create an accumulation of carbohydrates, auxins, and other organic materials in the layered region. The treatments commonly applied to the stem include:

- elimination of light in the region where rooting is desired
- intentional injury or girdling where rooting is desired
- application of growth substances to promote rooting
- provision of a rooting medium that is well aerated, moist, and of a consistent and moderate temperature

Methods of Layering

Simple Layering A dormant, one-year-old branch is bent to the ground and covered with soil except for its tip, which is left exposed to the light. A short distance back from the tip, the branch is bent sharply upward. This bend may be accompanied by a cut on the underside of the branch, Figure 14-13. It is at the bend or cut that rooting will occur. The exposed section of branch is usually staked for support. Following rooting, the new plants can be severed from the parent plant.

Tip Layering A shoot from the current season of growth is bent to the ground and covered with soil. If layered by hand, a hole can be made for the tip to rest in, which is angled downward on the side closest to the parent plant and vertical on the other side. Sometime later, the shoot will turn

FIGURE 14-13.
Simple layering

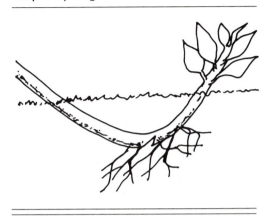

FIGURE 14-14.

Tip layering (From H. E. Reiley and C. Shry, Jr., *Introductory Horticulture,* 2nd edition, © 1983 by Delmar Publishers Inc.)

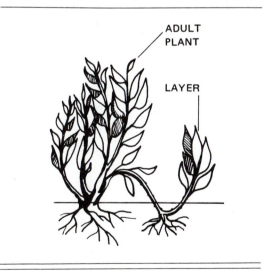

around and begin growing upward out of the soil. The layered tip will eventually form a new plant complete with roots and terminal shoot, Figure 14-14. It can be separated just prior to transplanting. Many species of plants propagate naturally by tip layering, and propagators promote the technique by pinching back the tips of plants to encourage additional shoots to form.

Mound Layering The parent plant is cut back to ground level while dormant, encouraging a proliferation of new shoots in the spring. As the shoots grow, soil is added periodically to keep each shoot covered to one-half its height, Figure 14-15. Within the mound, each shoot will develop roots. Each rooted shoot can then be cut off at the base and grown as a new plant. The parent plant can be used repeatedly to produce new plants.

FIGURE 14-15.
Mound (or stool) layering (From H. E. Reiley and C. Shry, Jr., *Introductory Horticulture,* 2nd edition, © 1983 by Delmar Publishers Inc.)

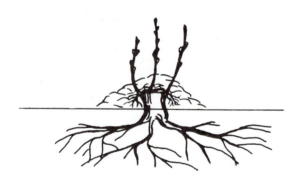

CUT PARENT PLANT TO GROUND LEVEL.

AS NEW SHOOTS BREAK, ADD SOIL TO KEEP EACH SHOOT COVERED TO ONE-HALF ITS HEIGHT.

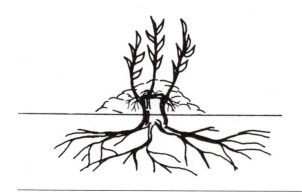

AS NEW ROOTS DEVELOP, SHOOTS CAN BE CUT OFF AT THE BASE AND GROWN AS A NEW PLANT.

Air Layering Amateur horticulturists frequently use this technique on interior plants. It results in the formation of roots on an aerial portion of the stem following a slit or girdling injury, Figure 14-16. Portions of stem used should be a year or less in age. The injury should be made 8 to 12 inches back from the tip of the branch. If girdled, a strip of bark 1/2 to 1 inch wide should be removed, with care taken to assure complete removal of the phloem and cambium. If slit, the cut should be 1 to 2 inches in length, slanted upward and toward the center. Following the injury, the stem can be treated with a root-promoting chemical, usually in a talcum powder carrier. A quantity of damp (not wet) sphagnum moss is wrapped around the injured area of the stem and fastened with polyethylene film which is then tied at each end. A split Jiffy pot® makes a perfect medium for rooting when wrapped around the cut or girdled area.

FIGURE 14-16.
Air layering: The stem is girdled to stimulate root formation. (From H. E. Reiley and C. Shry, Jr., *Introductory Horticulture,* 2nd edition, © 1983 by Delmar Publishers Inc.)

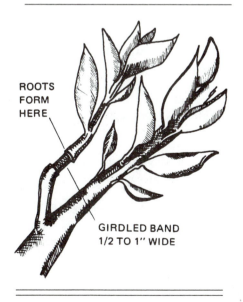

ROOTS
FORM
HERE

GIRDLED BAND
1/2 TO 1" WIDE

After the new roots are formed, the new plant can be cut from the parent plant. Air layering is usually performed in the spring and the new plant removed in the fall after becoming dormant. Following separation, the new plants are potted and kept in a cool, humid location for hardening-off. When produced commercially, the new plants may be placed under intermittent mist and gradually hardened-off.

Serpentine Layering Another layering technique that is more practical for amateur propagators than commercial growers, serpentine layering is best suited to plants that have a pendulous growth habit and flexible branches, Figure 14-17. The technique is similar to simple layering except that the branch being layered is buried in the soil at several places, resembling a serpent threading its way in and out. Each above-ground loop should possess at least one bud to develop into a new shoot. Each below-ground loop should develop roots for the new plant. Rooting will be encouraged and hastened if the undersides of the buried loops are cut before burial. If started in the spring, new rooted plants are usually obtained by fall, permitting them to be cut, separated, and transplanted.

The Time for Layering

Table 14-4 summarizes the seasonal timing required for successful layering.

PROPAGATION BY TISSUE AND ORGAN CULTURING

Most propagative techniques, whether natural or developed by horticulturists, have been recognized and practiced for centuries. The twentieth century, however, heralded the beginnings of new techniques presently termed *tissue culturing* and *organ culturing*. These techniques permit the reproduction of certain species from tiny pieces of plant organs such as embryos, pollen grains, and shoot tips or from undifferentiated plant tissue, usually callus. The piece of organ or tissue is removed from the parent

FIGURE 14-17.
Serpentine or compound layering (From H. E. Reiley and C. Shry, Jr., *Introductory Horticulture,* 2nd edition, © 1983 by Delmar Publishers Inc.)

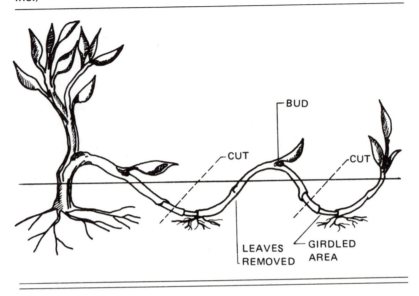

TABLE 14-4.
Timing for Layering

Method of Layering	Time of Layering	Time of Separating
Simple layering	Early spring	Autumn or next spring
Tip layering	Late summer	Late autumn
Mound layering	Spring to midsummer	Late autumn
Air layering	Spring	Late autumn
Serpentine layering	Early spring	Autumn or next spring

plant under sterile conditions, using precision scalpels, forceps, and tweezers, Figures 14-18A, 14-18B, and 14-18C. Immediately after removal, it is transferred to a sterile nutrient material (in either a liquid or gelatinous agar state), usually contained in a clear glass flask, tube, or petri dish.

Totally sterile (aseptic) conditions are essential throughout the process to prevent contamination of the culture by microorganisms such as fungi and bacteria. While methods to assure aseptic conditions vary, the following precautions are typical.

• The transfer chamber is kept free of all contamination. Whether the chamber is a special room used solely for the purpose or a small box-like compartment, it must be rid of all dust, spores, bacteria, and other contaminants, Figure 14-19. Sterilization of small chambers may be ac-

FIGURE 14-18A.
The tip of a strawberry plant runner is held by forceps as a scalpel removes a tiny piece. (Courtesy United States Department of Agriculture)

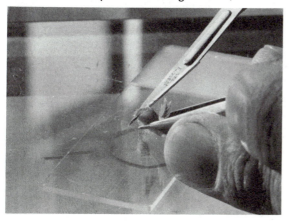

FIGURE 14-18B.
The piece of tip is placed into a container filled with sterile nutrient agar. (Courtesy United States Department of Agriculture)

FIGURE 14-18C.
Later the tip has differentiated and reproduced new, virus-free plants ready to be transplanted and grown to maturity. (Courtesy United States Department of Agriculture)

complished with ultraviolet light. (*Caution:* Ultraviolet light can be blinding and should not be looked at directly.) Larger chambers may be washed down with chlorine bleach or ethyl alcohol.

- All tools, containers, and solutions are kept sterile. They can be made that way initially by subjecting them to high temperatures under pressure in an autoclave, Figure 14-20. During the culturing operation, tools and the openings of containers can be kept sterile by passage through the flame of an alcohol lamp or bunsen burner.

FIGURE 14-19.
A transfer chamber requires an antiseptic environment, a heat source to flame instruments prior to each use and a supply of sterilized media and containers.

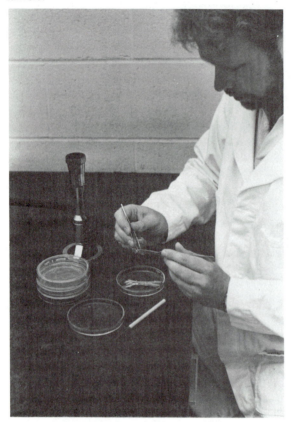

- All plant tissue is surface sterilized before use, care being taken not to damage the plant tissue. Disinfestants commonly used include Merthiolate, mercuric chloride, calcium hypochlorite, and sodium hypochlorite. Disinfectants include hot water, steam, and formaldehyde.

The nutrient media used for organ and tissue culturing vary with the species of plant being propagated. Agar media are the most common, but liquid media can also be used as long as a filter-paper wick is placed in the tube to hold the tissue or organ piece out of the solution. The media usually contain inorganic mineral elements (both macro- and micronutrients), sucrose sugar, growth regulators, and vitamins. Assorted other additives are found in the several media developed by researchers. You may wish to consult texts on plant propagation for the formulations of Knop's solution, the Murashigi and Skoog medium, White's medium, or Knudson C medium. The media take their names from the scientists who first developed or reported them.

Following a successful transfer from parent plant to the aseptic and nutritious environment of the tube, flask, or petri dish, the piece of plant tissue (termed an *explant*) enters a period of growth. New tissues and organs are differentiated and a new, complete plant is formed. Not all explants are able to differentiate; they continue to produce callus tissue, rendering the culture useless. The failure of certain species to differentiate is probably due to our failure to provide the right conditions rather than to any

FIGURE 14-20.
A laboratory autoclave provides steam heat under pressure to sterilize media, tools, and containers. Here flasks of agar are being set into the autoclave.

biological limitation, however. The full differentiation from callus or organ explant to complete plant may require several aseptic transfers of the developing plant, each time altering the size of the growth container and the composition of the nutrient medium.

Tissue and organ culture is costly, time consuming, and difficult compared to other methods of plant propagation. Its greatest application today is as a research tool for plant pathologists, geneticists, and plant breeders. However, commercial use in ornamental horticulture is increasing. Chrysanthemums and carnations are easily and commonly infected by pathogens that make propagation by cuttings risky. Organ culturing, using the shoot tips of desirable mum and carnation cultivars, permits pathogen-free material to be obtained and propagated. Orchids (as seeds and as shoot tips) and bromeliads are also cultured, shortening the time required to propagate them by conventional means.

SUMMARY

Plant propagation exemplifies the application of science to the craft of ornamental horticulture. It is a craft that changes continually as new facts are discovered, yet respects the skills and techniques that have evolved over many years of practice.

The medium for propagation must provide support, moisture retention, drainage, aeration, a pest-free environment, and sometimes nutrition. It may be made up of one or more of the following: soil, sand, peat moss, sphagnum moss, perlite, vermiculite, or fired clay. It must be pasteurized before use.

Propagation can occur in greenhouses, hotbeds, cold frames, or lathhouses. The environment must provide proper light, high humidity, warmth, and ventilation. Seed propagation may require a light or dark environment for germination, depending upon the species. It may also require scarification or stratification if the seed possesses one or more types of dormancy. Planting techniques for seeds vary depending upon whether the species are herbaceous or woody and whether they are to be transplanted after germination or will be grown at the planting site.

Propagation by cuttings is the most widely used method of asexual plant reproduction. It is quick, easy, and inexpensive. It is made possible by the cell's ability to revert to an undifferentiated condition, once again initiating the root, stem, and leaf tissues necessary to form a complete plant. There are different types of cuttings and different methods of propagating them, some capable of producing greater quantities than others. Soon after rooting occurs, the cuttings must be hardened-off to adapt them to more stressful growing conditions.

The technique of grafting necessitates a knowledge of which plants can be grafted, which ones benefit from grafting, and how to make the numerous types of grafts. Grafting is usually done either because the plant benefits from it or because no other propagative method will work. Grafting is possible on both herbaceous and woody plants but is most common on woody plants. The tools needed for grafting are a knife, finger guards, tying materials, grafting wax, and a source of heat. Methods of grafting vary with the type of plant (woody or herbaceous), its size, and sometimes the species.

Budding is a method of grafting that requires somewhat specialized techniques. It involves the removal of a piece of bark on the stock plant and its replacement with a comparably sized piece containing a bud from the scion. There are several different methods of budding.

Layering as a propagative technique permits the stem of a new plant to remain at least partially attached to the parent plant while new roots are initiated. There are different methods of layering. Simple layering, top layering, mound layering, and serpentine layering find frequent use by professionals; air layering enjoys popularity with amateurs.

Tissue and organ culturing are comparatively recent propagative techniques. They permit the reproduction of certain species from tiny pieces of plant organs such as embryos, pollen grains, and shoot tips or from undifferentiated plant tissue, usually callus. The propagative material is collected under aseptic conditions and transferred to sterile nutrient media, usually contained in flasks, tubes, or petri dishes. The piece of tissue then enters a period of growth and new tissues and organs are differentiated to form a complete new plant.

Achievement Review

A. MULTIPLE CHOICE

From the choices given, select the answer that best completes each of the following statements.

1. The propagation medium must be _____.
   a. unpasteurized   b. pasteurized
2. The medium must _____ moisture.
   a. repel   b. retain
3. Fired clay, sand, or perlite can improve the _____ of the media.
   a. drainage   b. moisture retention
4. The best sand for use in propagation media is _____ sand.
   a. fine   b. quartz
5. Propagation of seeds requires that the medium contain _____.
   a. nutrients   b. agar

B. TRUE/FALSE

Indicate if the following statements are true or false.

1. Propagation structures must exclude all light.
2. Intermittent mist systems are useful in the prevention of wilting in cuttings.
3. The application of a soil drench can reduce damping-off of seedlings.
4. Temperature controls the rate of root and shoot development in cuttings.
5. In a cutting bed, the rooting medium should be cooler than the air temperature.
6. Most cuttings form better root systems in acidic rather than neutral propagating media.
7. High light intensity is necessary for vigorous root formation on cuttings.
8. Grafting wax keeps moisture outside the graft union.
9. High humidity discourages the formation of callus tissue in grafts.
10. Once roots have formed on cuttings, the propagation chamber should be ventilated.

C. DEMONSTRATIONS

1. Collect or purchase seeds of several herbaceous and woody species common to your region of the country. Precondition as necessary. Prepare a suitable soil for germination and fill separate greenhouse flats for each species to be planted. Plant the seeds in furrows and cover. Water the flats either by immersing them in water (if the seeds are fine) or by sprinkling gently. Cover with sheets of plate glass during the day and remove at night. Monitor the day and night temperatures as closely as conditions permit.

   For added interest and learning, vary the planting techniques. Treat some seeds with fungicide before planting and leave others untreated. Apply a soil drench to some flats and not to others. Keep records of the percentage of seeds that germinate, develop damping-off disease, and develop into healthy plants.

2. Plant 100 seeds of the same species of flowering annual in each of two plug sheets. (Use regular greenhouse flats if plug sheets are not available.) Maintain the same growing conditions for each sheet (watering, low-concentration N fertilizer), but grow one sheet at 80°F day and 60°F night temperatures and the other sheet at 70°F day temperatures and 75°F night temperatures. After both sheets are showing approximately the same percentage of germinated seeds:
   a. count the number of seedlings in each sheet.
   b. measure the length of each seedling and calculate an average seedling length per sheet.
   c. compare the results and draw a conclusion if one is possible.

3. Using Table 14-3 (A Guide to Propagating Cuttings) as a reference, take several differ-

ent types of cuttings from local plants. The season of the year and the species of the plant will determine the types of cuttings taken. Propagate them as indicated in the table. Keep records of the percentage of cuttings that do and do not develop roots. You may want to use different media to compare their effects on the number and quality of roots formed and the number of days required for rooting.

4. Practice making some grafts. Use as many different techniques as the available plants and conditions allow. With Table 14-4 (A Guide to Grafting Techniques) as an aid, select and prepare stocks and scions from compatible species. For beginners who have no knowledge of species compatibility, the stock and scion may be of the same species for practice purposes. If grafts do not unite, have them checked by a teacher or commercial propagator to determine the reason.

5. Practice budding on roses or fruit trees, using either greenhouse or landscape plants. Try T-buds, I-buds, and patch buds. Evaluate which ones worked most successfully and why.

6. Using some large garden shrubs as stock plants, layer some branches using the techniques of simple and tip layering. If pendulous shrubs are available, try a serpentine layer as well. Initiate the layering in the spring and separate the plants in the fall. For comparison, wound the buried stems of some branches and leave others unwounded. Mark the branches so that a count of successfully rooted plants can be made and the techniques compared.

7. If a large stalky foliage plant is available for use, make an air layer. When the roots are visible within the plastic wrapping, sever the plant.

D. SHORT ANSWER

Answer the following questions as briefly as possible.

1. Define tissue and organ culturing.
2. Why is sterility of the transfer chamber and the growth container so essential?
3. In what physical states do nutrient media for culturing commonly occur?
4. What tools are used to excise and transfer the explants?
5. What are the two major benefits to commercial propagators from tissue or organ culturing?

SECTION III

THE PROFESSIONS OF ORNAMENTAL HORTICULTURE

15

The Floriculture Industry

Objectives

Upon completion of this unit, you will be able to

- describe the categories of employment in ornamental horticulture.
- describe the professions of floriculture.
- outline the education each one requires.

CAREERS IN ORNAMENTAL HORTICULTURE

Ornamental horticulture used to be a craft occupation, engaged in by persons who enjoyed working with their hands in fields, greenhouses, and shops. Little education was required, although college-level programs existed, and most practitioners opted for apprenticeship. Limited capital was needed to start up a small business, and many people viewed it as a second-income or postretirement occupation.

In many ways, all of this is still true. And ornamental horticulture is still a small business industry—even the largest national firms are relatively small. Yet today's industry has changed and advanced markedly in the past fifty years. Increased public awareness of the enrichment provided by ornamental crops has shown itself in an increased demand for more products, more services, innovative ideas, and an insistence upon higher quality. The old system was unable to meet the demand,

so new career opportunities have become available, attracting people with new ideas. Finally, after years of virtually ignoring ornamental crops, university and corporate researchers have begun the development of new and improved varieties; market researchers have offered new methods of measuring the public's needs; and industry trade organizations have discovered the benefits of working together to promote their products and services to consumers eager to learn and spend more. Colleges and universities have developed extensive curricula at the under-graduate and graduate levels, lending a new respectability to the industry and attracting young people from urban as well as the traditional suburban and rural areas.

Today ornamental horticulture offers employment to individuals with a diversity of backgrounds, educations, and career goals. The high school graduate and the Ph.D. can both find satisfying work, as can the scientist and the artistic craftsperson. It is not impossible to find all working in the same firm. As would be expected, they have different responsibilities, reflecting their different interests and formal preparation.

Persons working in the industries of ornamental horticulture usually fall into one of the following categories:

- unskilled laborer
- skilled laborer
- middle management
- owner/operator
- educator/researcher/specialist

The categories are general and apply to an industry whose practitioners range from floral delivery truck drivers to horticulture therapists, but they can be helpful in defining entry-level positions and career goals.

Generalizing further, the qualifications for employment within each category are commonly found to be as follows.

Unskilled laborer   The employee may work indoors or out depending upon the firm. Contact with customers may or may not be required. The typical worker hired often has no schooling beyond high school and may not have completed high school. This category includes part-time and summer workers still in school, retirees, migrant workers, and persons comfortable doing heavy labor and repetitious work. This category of worker usually is paid the federal minimum wage.

Skilled Laborers   These employees possess technical skills needed by the firm. They may have acquired the skills through previous work experience or through vocational courses taken at the high school or college level. As vocational and technical education has gained acceptance as a legitimate alternative or supplement to the traditional liberal arts and sciences curriculum, the number of skilled laborers has increased, as has the public image of the ornamental horticulture profession.

Middle Management   These employees generally have some college training and have made the decision to follow careers in the field of ornamental horticulture. This category offers great opportunities to individuals with knowledge, formal school-

ing, an interest in the business, but no money to begin their own operations. The category is comparatively new, having developed in response to the increased number of college-trained young people in the job market and the rapid expansion of the industry.

Owners/Operators   These individuals may have no college schooling, but the trend is definitely toward college education for tomorrow's entrepreneurs and away from reliance on apprenticeship. The level of college education may range from a two-year associate degree to bachelor's and graduate degrees. Certain occupations may require specialized degrees such as the B.L.A. (Bachelor of Landscape Architecture) and state licensing or certification before practice is permitted.

Educators, Researchers and Specialists   Due to the nature of their work, these professionals usually need the greatest amount of university training. The category includes teachers of both high school and college students, and handicapped and nontraditional learners, scientific researchers, Cooperative Extension agents, and consultants.

Any of the categories can include entry-level positions, depending upon the qualifications of the individual. All permit movement into other categories as an individual's skills and aspirations change. None of the positions should be regarded as more or less prestigious than the others. This lateral movement of talented people from roles as teachers or owners to roles as practitioners and managers is common and good for the industry.

THE PROFESSIONS OF FLORICULTURE

The floriculture industry is involved with the production, distribution, and utilization of floral products and related goods and services. It is a multibillion dollar industry annually in the United States. Businesses range in size from small neighborhood flower shops to corporations engaging in international trade. Products can be as old-fashioned as a prom corsage and as modern as new computer technology. The following descriptions can only provide a rudimentary idea of what floriculture workers do. Diversity is commonplace, and diverse skills, abilities, and interests are an asset to anyone who would seek employment within this complex industry.

Growers

Floriculture crops are grown in greenhouses and other growing structures (described in Chapter 20). They are also grown outdoors in climates where the winter season is short or nonexistent. Firms that grow flowering and foliage plants may emphasize cut flower production or potted crop production or both, Figures 15-1 and 15-2. Firms may also choose not to grow crops to maturity, but instead to propagate them and provide rooted cuttings or other reproductive stock to other growers.

Growers should have a knowledge of assorted crop plants, although their knowledge can easily become limited when they grow only a few crops on a repeating basis. Knowledge of proper production hygiene is essential, coupled with the ability to prevent and control the insects and diseases to which the crops are susceptible. The essence of production know-how is the ability to manipulate the environment around the plants in a way that will

benefit the crops, not hamper their development, Figure 15-3.

The grower may be the owner of the operation or an employee. In production operations, the grower is a key figure who is given great responsibility and permitted little error. Large operations may employ several chief growers, each supervising several assistant growers. Except for the grower who also owns the business, growers are regarded as skilled laborers. They may acquire their skills through university programs, on-the-job training, or both. In any case, they must constantly update their knowledge through attendance at workshops, seminars, and industry tours. Competition from

FIGURE 15-2.
Potted Easter lilies in their forcing bed (Courtesy Rodney Jackson, Photographer)

FIGURE 15-1.
A typical cut crop bench. These are chrysanthemums, necessitating the overhead support racks to hold black cloth for day length control.

FIGURE 15-3.
A grower adjusts the thermostat to control the greenhouse temperature.

other firms both inside the country and abroad requires that growers stay abreast of the latest information.

Most growing operations are wholesale firms whose clientele are retail florists. Still, there are small growers who retail their own products and produce only what they need. Rural communities often attract or perhaps necessitate grower-retailers because of their distance from suppliers.

In addition to the professionals described, a growing operation may also employ several unskilled laborers who perform valuable tasks to support the growers' efforts. These laborers may have no technical education or experience when they begin work, but are often highly motivated to learn.

After several years, they may qualify as skilled laborers within the same firm or in a similar operation.

Wholesale Suppliers

These vendors serve as centralized sources for the countless items needed by the profession's producers and retailers. Most wholesale suppliers stock materials needed by either growers or retail florists but seldom try to serve both due to the dissimilarity of their needs.

Wholesalers who supply growers provide almost everything but plants. Their inventory includes tools of all types, artificial media and soil additives, lights, containers, greenhouse materials, pesticides, and hundreds of other items needed to produce and distribute quality crops.

Wholesalers who supply the nation's flower shops and floral designers serve as sources for cut flowers from all over the world. They also provide the materials necessary to transform cut flowers into flower shop products: dried, plastic, and silk materials; decorative containers; ribbons; paints; gift items; register tapes; boxes; and so on.

Staffing the wholesale supply house are warehouse workers, who keep the extensive inventories accounted for and organized, and sales floor personnel to serve the customers. Many wholesalers also have a staff of traveling salespeople to call on customers personally, to take and deliver their orders. Numerous middle management positions are available in wholesaling. College education in horticulture with a business emphasis provides the best background.

Flower Shops and Garden Centers

The retail outlets for the floriculture industry are the flower shops and garden centers of the nation. Some are small corporate chains, and some are operated as part of larger retail stores, notably Sears Roebuck, Montgomery Ward, and K Mart. Most are small, privately owned businesses. It is through these retail shops that floral products reach the final consumer, Figure 15-4. The typical flower shop provides a full floral design service and carries

FIGURE 15-4.
Most retail flower shops operate in an atmosphere of informal cordiality. Repeat customers are common and are usually known to the florist.

potted flowers and foliage plants on a regular or seasonal basis. The shop may also sell gift items, greeting cards, or candy. Christmas decorations may be added during the holiday season. Garden centers always carry woody plant materials, but many have a floral design service as well. Garden centers are a major outlet for the retail sale of bedding plants and bulbs. Their inventory may be diversified with floral products as well as pet supplies, firewood, Christmas trees, and seed.

The retail outlet may consist of a store with nearby parking, or it may include a sales yard and/or greenhouse(s). The owner or manager must be a skilled merchant with business training as well as training in horticulture. Training of shop personnel, techniques of promotion and marketing, customer relations, purchasing practices, accounting, and computer use are only some of the business subjects that a retail operator must deal with daily. A retail center requires a general practitioner of business management, not a specialist.

Since a retail flower shop or garden center is a small business, the owners are likely to be directly involved with the services or goods offered. They may be floral designers, greenhouse workers, or salespeople. Unloading trucks, developing displays,

answering customers' questions, and running the cash register—all are as likely to be done by the owner as by an employee. There are few opportunities to play chairman of the board in a retail flower shop or garden center.

The education required to be a horticulture retailer can be obtained through college programs at the two- or four-year level. The ideal program would include technical training in both ornamental horticulture and business management. Many retailers, unable to attend a university full time, enroll in part-time programs. Others take advantage of the short courses and educational workshops offered by the Cooperative Extension Service or industry trade organizations. Still others keep up by reading texts, journals, and magazine articles that cover retailing in general and horticulture in particular.

Floral Design and Sales

Floral design is the artistic craft of the floriculture profession. In the hands of floral designers, floral crops are transformed into the products desired by consumers, Figure 15-5. The floral designer must possess a sufficient knowledge of postharvest plant physiology to prolong the life of cut flowers and foliage as long as possible. The designer must also be able to select and assemble colors, shapes, textures, and materials for floral arrangements tailored to the needs of each client. The designer must have the vision of an artist yet be comfortable repeating many of the same tasks dozens of times.

Floral designers are usually expected to be salespeople also. Waiting on customers, taking orders on the telephone, consulting with future brides, expectant fathers, and bereaved families require sensitive interpersonal skills. Knowing when to offer a suggestion and when to just listen are desirable qualities in those who would sell as well as design.

The position of floral designer is highly important in any shop or garden center that offers floral arrangements. The designer may or may not be the owner of the business and may or may not be the only designer on the staff.

Floral designers are trained in several ways. Courses are offered at two- and four-year colleges

FIGURE 15-5.
A floral designer constructs a novelty arrangement for Thanksgiving.

as part of degree programs in ornamental horticulture. A course or two may also be offered as electives in nonhorticulture programs, but they are usually directed at hobbyists and are of limited use. Floral design training schools also exist, where aspiring designers receive concentrated training. These are noncollege programs that provide the graduate with a certificate upon completion.

Teacher/Researcher

One of the most difficult positions to fill on an agricultural college faculty is that of the floriculture teacher. With graduate degrees required, the number of qualified applicants is significantly less than for any other type of agriculture teaching. Apparently, few individuals pursue education in floriculture beyond the associate or bachelor degree level and even fewer choose teaching or research as career

fields. Most college-educated floriculturists go to work for industry. The shortage of qualified floriculture teachers has existed for some time and shows no sign of disappearing.

Teachers find good opportunities at both the high school and university levels. The shortage of scientific researchers is not as critical, although it is not an overcrowded career field either. Usually a minimum of a master's degree is required for teaching, and a doctorate is needed for a university research position.

Teachers help students measure the depth of a career field that seems to hold some interest for them. The selection of a career field of study is a critical decision in everyone's life, and a good teacher can be of tremendous importance to the student. Teachers are also responsible for introducing students to many of the skills and techniques used daily by floriculture professionals, Figure 15-6. By representing the industry in the classroom, teachers provide their students with insight into the careers they are seeking and help them build confidence in their ability to succeed.

Scientific researchers are constantly seeking ways to improve crops and methods of production, Figure 15-7. Their research may include selecting and breeding to develop new and better plant varieties. It may involve searching for better control of old and recurring pest problems. It may involve searching for ways to lengthen the postharvest life of flowers or ways to produce more crops in less bench area or for less cost. It is the researchers of the profession who keep the industry looking and moving forward. It is the teachers who train young professionals to seek new ideas and methods and apply them.

SUMMARY

Traditionally, ornamental horticulture has been a craft industry requiring little formal education. While still a small business industry, it has changed and advanced markedly in the past fifty years. Today the industry offers employment to persons with a diversity of backgrounds, educations, and career goals. Predictably, they have different responsibili-

FIGURE 15-6.
These floriculture students are planting hanging baskets with the guidance of their teacher. Educators trained in floriculture are in high demand as professionals.

FIGURE 15-7.
Automated equipment is used at the University of Maryland to test simultaneously for several plant nutrients. (Courtesy United States Department of Agriculture)

ties, reflecting their different interests and formal preparation.

The general categories of ornamental horticulturists are unskilled laborer, skilled laborer, middle management professional, owner/operator, and educator/researcher/specialist.

The floriculture industry is involved with the production, distribution, and utilization of floral products and related goods and services. Its businesses range in size from small shops to international corporations. Numerous career opportunities are available.

Floriculture growers may work as propagators or produce cut flowers, potted crops, or foliage plants. They may be owner/operators or skilled laborers. They acquire their education through university programs, on-the-job training, or both.

Wholesale suppliers are the centralized source of the items needed by growers or retailers. Numerous middle management positions are available. College education in horticulture with a business emphasis is the best preparation.

Flower shops and garden centers are the retail outlets for the floriculture industry. While some are chain stores, most are small, privately owned businesses. The owner/manager of a retail outlet must have both business and horticulture training, and is likely to be involved directly as a floral designer, greenhouse worker, or salesperson. College level training in horticulture and business is ideal preparation for retail floristry.

Floral designers create the floral products desired by consumers. They must be able to arrange materials in ways that are tailored to the needs of each client. They usually need sales skills also. Training can be acquired through college and noncollege courses as well as on-the-job training.

Teachers and researchers are in demand within the industry. Their positions require the greatest educational preparation.

Achievement Review

A. SHORT ANSWER

In the table that follows, place Xs where appropriate to match job characteristics with floriculture occupations.

	Grower	Wholesale Supplier	Flower Shop or Garden Center Owner	Floral Designer	Teacher or Researcher
Centralized source for items required by growers					
Designs flower arrangements					
Propagates floral crops					
Maintains sanitary production environment					
Develops new varieties of plants					
Their customers are retail florists					
Sales skills are important					
Business management training is important					
Trains future practitioners					

	Grower	Wholesale Supplier	Flower Shop or Garden Center Owner	Floral Designer	Teacher or Researcher
May stock Christmas ornaments, gift items, and pet supplies					
Artistic skills required					
Attendance at short courses and update sessions is important					
Requires the most formal education					
Training can be obtained through noncollege private programs					
Pathology and entomology training is helpful					
Skilled laborer					
Middle-management positions are common and available					
The position *may* be filled by an owner/operator					
College training is not mandatory but is a helpful shortcut					

16

The Nursery Industry

Objectives

Upon completion of this chapter, you will be able to

- explain the differences between the floriculture and nursery industries.
- describe the different types of nurseries.
- explain the relationship of the nursery industry to the landscape industry.

THE SCOPE OF THE BUSINESS

Two features more than any others distinguish the nursery industry from floriculture: the *size of the operation* and the *type of crop produced and sold.* Nurseries deal largely with *woody* plants (trees, shrubs, vines, and groundcovers); while florists handle more *herbaceous* (nonwoody) materials. The plants produced by a nursery usually require several years to reach a size suitable for retail sales. Those grown by a florist normally reach the harvest stage within a shorter time span, often less than a year. Also, the plants grown and sold by a florist are most commonly used indoors—on tables, as corsages, to decorate for parties and weddings, and so on. Nursery crops usually grace the outdoors— around our homes, shopping centers, and public buildings, and in our parks. They are the plants that keep a touch of nature in our urbanized environment, Figures 16-1 and 16-2).

The amount of land and capital investment required for nursery production is greater than that required for a greenhouse or flower shop. A successful floriculture operation may be located on part of an acre, often in an urban or suburban location. A nursery may occupy many acres; some large ones exceed a thousand acres. Like the floriculture industry, the nursery business is intensive agriculture

FIGURE 16-1.
A typical field nursery

354

FIGURE 16-2.
Ornamental grasses, espaliered vines, perennial flowers, and shrubs are all products of the nursery craft used in the landscaping of this home.

dealing with high-value crops. The gross sales of nursery plants and related products exceed a billion dollars annually.

DIFFERENT TYPES OF NURSERIES

The nursery industry is a complex one, not easily categorized into distinctly different branches. There are plant growers, retail sellers, and associated supply and/or service businesses, but they are not always separate operations. Each of the following types of businesses is part of the nursery industry:

- propagation nursery
- wholesale nursery
- wholesale nursery supplier
- retail nursery
- privately owned garden center
- chain store garden center
- landscape nursery

Each function will be described separately, but remember that countless nurseries throughout the country combine two or more functions. As in any seasonal business, diversity is often the way to keep the cash register ringing during the off-season. With summer the peak period for nursery production,

and the sale of nursery plants occurring almost entirely in the spring and fall, nurseries may rely on the sale of flowers, Christmas trees, decorations, birdhouses, and endless other products to expand their profits. Some may combine the propagation of woody plants with their wholesale production and sale. Others may combine wholesale and retail operations.

Propagation Nurseries

A propagation nursery initiates plant production. (Propagation techniques are defined and described in Chapter 14.) Nurseries that engage totally or partially in propagation may buy or collect seeds from all parts of the world. These seeds are then germinated and grown to a size that allows the plants to be passed on to other nurseries. A propagation nursery may also reproduce plants from cuttings or similar methods of vegetative propagation, Figure 16-3.

Propagation nurseries are responsible for the introduction of new species and cultivars to the nursery industry as a whole. They are also key sources of disease- and pest-free stock material. These nurseries depend upon the sale of large quantities of plants, since the per-plant value of a young seedling or rooted cutting is very low. A propagation nursery may specialize in only a few crops or grow a wide range of material.

Wholesale Nurseries and Suppliers

A wholesale nursery is one that grows plants to a salable size in the field or in containers. What a salable size is cannot be defined precisely except to say that the plant is not at full maturity but is ready to install in the landscape.

The wholesaler may propagate the seedlings or cuttings, but large operations usually buy stock from a propagating nursery. Rooted cuttings ready for transplanting into containers or the field are called *liners* or *lining out stock*, Figure 16-4.

A plant may spend from one to five years or more in the wholesale nursery depending upon its rate of growth and the sale size desired. While in

FIGURE 16-3.
A typical mist propagation system for rooting

FIGURE 16-4.
A nursery liner is containerized for further growth.

the nursery, each plant is provided with space, fertilizer, weed control, shearing, shaping, and pest control to bring it to market as soon as possible. The longer it remains in the nursery the more it costs the grower, a cost that is passed on to the buyer and eventually to the retail customer.

To hold down production costs, many wholesale nurseries limit the number of different species they grow to those requiring similar production techniques. In so doing, the nurseries are able to maximize the use of expensive equipment while minimizing the number of different fertilizers, chemical sprays, and growing environments necessary to produce high-quality plants. Unfortunately, this eco-

nomic good sense leads to a limited number of species available for use in landscapes.

Wholesale nurseries sell their plants to other businesses that resell them at a higher price to the final consumer. Since the customers of a wholesale nursery are widespread and distant (perhaps as much as several states away), it is critical that the nursery be near transportation routes. Nearby truck or rail lines keep down the costs of handling and shipping. For the convenience of retail operators and landscapers who come directly to the wholesale nursery to pick up their plants, it must be near a major highway system.

Retail Nurseries

Retail nurseries and garden centers are two major customers of the wholesale nursery; they are also the final link in the marketing chain. Through the retail nurseries, the privately owned garden centers, and the garden centers operated by regional or national chain stores, nursery products reach the consumer. It is to these businesses that the homeowner comes when wishing to purchase trees and

FIGURE 16-5.
Chain garden centers seek to sell large volumes of popular plant materials in a short period of time. These geraniums are being promoted in the same manner as dry goods inside the store.

FIGURE 16-6.
Private garden centers often try to project an image of greater plant quality and concern for the customer.

shrubs, sprays, fertilizers, grass seed, garden tools, and similar items for the home property, Figures 16-5 and 16-6.

Retail nurseries and garden centers seldom have growing areas of any significance. Instead, they have sales yards, where the materials purchased from the wholesalers can be displayed and maintained, Figure 16-7. The method of display will vary with the place and with the type of material involved. Plants growing in pots or other containers may be set on an asphalt or crushed stone surface. Trees and shrubs sold in balled-and-burlapped root forms or as bare-root stock may require holding in deep bins of peat moss or sawdust to prevent their drying out.

Some retail nurseries and garden centers also have a greenhouse from which they sell tropical houseplants and potted flowers. Like the outdoor salesyard, the greenhouse is usually open for customer browsing and promotes sales by providing

FIGURE 16-7.
Customers are encouraged to enter and select their own plants during this nursery's special rose promotion.

FIGURE 16-8.
Plants are retailed inside and outside of this nursery greenhouse. Wagons are provided for the customers to assist in carrying their purchases.

a realistic setting in which to display the plants, Figure 16-8.

The location of the retail operation is important to its success. It must be near developing residential areas and easily reached by car. If located on a major thoroughfare, it has a better chance of attracting customers from the passersby. A remote location for a retail outlet increases the amount of money that must be spent on advertising.

In addition to supplying plants and other materials to retail customers, the garden center or nursery is usually the local center for garden advice. "When should I plant it?" "What colors will the flowers be?" "What kind of fertilizer does my lawn need?" "How can I attract bluebirds to my backyard?" Questions of this type besiege the retailer and present an opportunity to promote good customer relations. Some retail operators write garden columns for local newspapers. Others are frequent speakers at garden clubs and on radio and television programs. Successful retailers of nursery materials have a thorough understanding of ornamental plants, their culture and care; but equally important, they have good communication skills and an interest in working with the general public.

It is primarily in the degree of customer service that privately owned garden centers and nurseries differ from those operated by chain stores. The chain store garden center may not be in operation for the entire year. Instead it may concentrate its sales effort on the two periods of heavy customer demand, the spring and the fall. The chain garden center may be staffed by personnel with limited horticultural knowledge since they spend most of their time in some other sales area such as hardware or clothing. There is seldom much interpersonal contact between sales personnel and customers in a chain store operation. Still, the quality of the merchandise may be as good as in a private garden center, and often the prices are lower since the larger operation can take advantage of discount buying, Figures 16-9, 16-10, and 16-11.

Landscape Nurseries

The landscape nursery may grow plants, buy them from a wholesale nursery, or both. In addition to providing a retail outlet for plants, the landscape nursery provides an installation service for homeowners who do not wish to develop their landscapes

FIGURE 16-9.
The privately owned garden center can often convey a sense of permanence that a chain store garden center cannot.

FIGURE 16-10.
A chain store garden center is often set up for spring and fall sales and the space is used for other purposes in other seasons.

themselves. Such a nursery must employ several types of workers: retail salespeople, landscape crews, and often a separate crew for nursery production.

THE LINK BETWEEN THE NURSERY AND LANDSCAPE INDUSTRIES

The landscape industry is composed of a large group of professionals involved with the design, installation and maintenance of public and private gardens. In their work, landscapers require great quantities of plant materials, construction materials, chemical sprays, fertilizers, mulches, tools, and similar items. *Wholesale nurseries* supply the landscapers with the necessary plants. *Wholesale supply dealers* provide the important accessory materials.

One of the frequent difficulties of a landscaper is finding plants, especially trees, of the type and size specified by a designer. Large trees and unusual species are often not carried by wholesale nurseries because of limited demand. There is a definite need in today's nursery industry for operations willing to carry large plants and uncommon species (*specimen plants*) for use by landscapers. The market area and operating costs for such a nursery are much greater than for a more conventional wholesale nursery, though, which explains their scarcity.

FIGURE 16-11.
Merchandising in the chain garden center may be directed to selling large numbers of selected high-profit items quickly.

SUMMARY

Two features distinguish the nursery industry from floriculture: the size of the operation and the type of crop produced. Like floriculture, the nursery industry deals with high-value crops under intensive production, but its complexity makes it more difficult to categorize into distinctly different branches. There are propagation nurseries, where plant production is initiated and new species and cultivars are introduced.

Wholesale nurseries grow plants to salable sizes in the field or in containers and sell them to retailers and to landscapers. It is important that they carry an inventory of popular species in a range of sizes.

A need exists for more nurseries willing to carry large plants and uncommon species. Retail nurseries and garden centers are the major consumer outlets for nursery products. They market the products through sales yards and greenhouses. Landscape nurseries may grow and/or purchase plants for retailing or their own use. They also provide installation services for homeowners.

Achievement Review

A. SHORT ANSWER

Answer each of the following questions as briefly as possible.

1. Indicate which of the following are characteristics of florists (F), nurseries (N), or both (B).
   a. Their crops are most likely to be used outdoors.
   b. Their crops are usually herbaceous.
   c. Their crops usually reach maturity in a shorter time.
   d. There are both wholesale and retail operations.
   e. More land is usually required for the business.
   f. They deal with high-value, intensively cultivated crops.
   g. They supply the landscape industry.

2. Match the characteristics on the left with the type of function on the right.

   a. supplies the landscape industry with plants
   b. may operate only seasonally
   c. major customer of a wholesale nursery along with garden centers and landscapers
   d. sells rooted cuttings in large quantities
   e. differs from chain store operations in the quality of customer service
   f. may be hundreds of acres in size
   g. provides installation services to homeowners

   1. propagation nurseries
   2. wholesale nurseries
   3. wholesale nursery supplier
   4. retail nurseries
   5. privately owned garden centers
   6. chain store garden centers
   7. landscape nurseries

3. Arrange the following groups of three in the chronological order of their involvement with plant materials.
   a. retail nursery
      propagation nursery
      wholesale nursery
   b. propagation nursery
      landscape nursery
      wholesale nursery
   c. homeowner
      wholesale nursery
      retail nursery
   d. wholesale nursery
      homeowner
      landscaper

17

The Landscape Industry

Objectives

Upon completion of this chapter, you will be able to

- describe the different types of landscapers.
- outline the education each one requires.

THE LANDSCAPE

Broadly defined, all the spaces outside our buildings constitute the landscape. Beautiful or hideous, natural or scarred by human blundering, the landscape usually reflects some degree of human control. A *landscaper* is someone whose profession it is to control and develop the landscape.

DIFFERENT TYPES OF LANDSCAPERS

Landscapes can range in size from a small backyard to a vast parkland; and control and development can range in scope from the pruning of a single plant to the design of entire gardens. It follows that landscape careers will vary in their day-to-day operations and in the academic preparation they require. Each of the following has a role in the landscape industry:

- landscape architect or designer
- landscape contractor
- landscape gardener or maintenance supervisor
- landscape nurseryman

Landscape Architects and Designers

Landscape architects and designers conceive the ideas that later find form in the landscape. They see the gardens in their minds before the gardens exist in reality. It is the landscape architect who seeks out and studies sites that are to be developed to meet specific human needs. The landscape architect identifies all the capabilities and limitations of the site to determine its suitability for the client's purposes, Figures 17-1 and 17-2.

In addition to being a competent horticulturist, the landscape architect must possess enough knowledge of geology, hydrology, agronomy, engineering, and architecture (see Glossary) to work closely with professionals in these fields. Such affiliations become necessary in large or complex site studies.

Once the site has been selected and the needs of the client determined, the landscape architect formulates the design that will satisfy those needs. Working through a series of stages (rough drafts, concept designs, final designs, planting plans, and working drawings), the landscape architect turns ideas into graphic plans. These enable the client

FIGURE 17-1.
A landscape architect completing work on a presentation plan

FIGURE 17-2.
The landscape architect uses surveying skills to study and record the topography of a site.

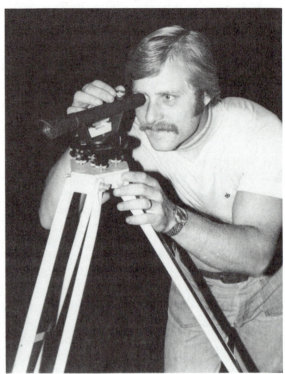

to react to the ideas and approve them or suggest changes. Later the ideas become detailed explanations to the contractor of how the design is to be installed, Figures 17-3, 17-4, 17-5, and 17-6.

Landscape architects seldom actually install the landscape but usually supervise those who do. The landscape architect is the client's representative from beginning to end and normally has the right of final approval when the installation is complete.

Landscape architect and *landscape designer* are terms that are often used interchangeably. In most states they are licensed professions and can be practiced only by those holding state certification. In other states, landscape architects may require certification while landscape designers do not. In a few other states, neither term is restricted by such rules. The general trend across the country is to license landscape architects and designers. Entry into this area of landscaping usually requires extensive schooling and work experience.

Landscape Contractors

Landscape contractors construct and install the landscapes designed by the landscape architects or designers. Where legally permissible, landscape contractors may also offer a design service to their clients, usually at the residential or other small-scale level. In general, though, a landscape contractor is an ornamental horticulturist and a builder rolled into one.

The procurement and installation of the plants specified by the landscape architect/designer is a major responsibility of the contractor. Proper installation may require laying out the planting beds, conditioning or replacing the soil, placing the plants,

FIGURE 17-3.

A concept drawing for the entry road to a college campus (Courtesy James Glavin, Landscape Architect, Syracuse, NY)

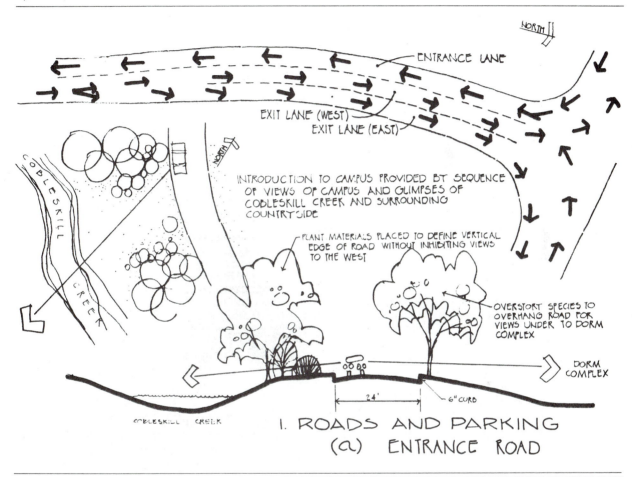

NORTH

ENTRANCE LANE

EXIT LANE (WEST)
EXIT LANE (EAST)

INTRODUCTION TO CAMPUS PROVIDED BY SEQUENCE OF VIEWS OF CAMPUS AND GLIMPSES OF COBLESKILL CREEK AND SURROUNDING COUNTRYSIDE

PLANT MATERIALS PLACED TO DEFINE VERTICAL EDGE OF ROAD WITHOUT INHIBITING VIEWS TO THE WEST

OVERSTORY SPECIES TO OVERHANG ROAD FOR VIEWS UNDER TO DORM COMPLEX

DORM COMPLEX

24'

6" CURB

COBLESKILL CREEK

COBLESKILL CREEK

I. ROADS AND PARKING
(a) ENTRANCE ROAD

staking, mulching, and other necessities of successful transplanting. If lawn installation is required, the landscape contractor may need knowledge of cultivator operation, seed bed preparation, sodding, watering, and other techniques.

Many landscape plans also specify the construction of walls, walks, fences, patios, pools, lighting systems, fountains, and other nonplant items. The landscape contractor must be prepared to offer these services or to hire, or subcontract the job to, some other specialist who can do the work, Figures 17-7 and 17-8.

Landscape contracting is active, outdoor work. Those who pursue it as a career should enjoy physically demanding labor. The financial investment required to operate successfully can be considerable. Large and expensive machinery is needed if land grading, back-hoeing, tree-moving, or similar services are offered. The labor, transportation, and equipment costs of several crews are great. Cost of land is relatively minor, though, since the work usually requires only an office and storage space for equipment and material.

FIGURE 17-4.
A complete presentation drawing

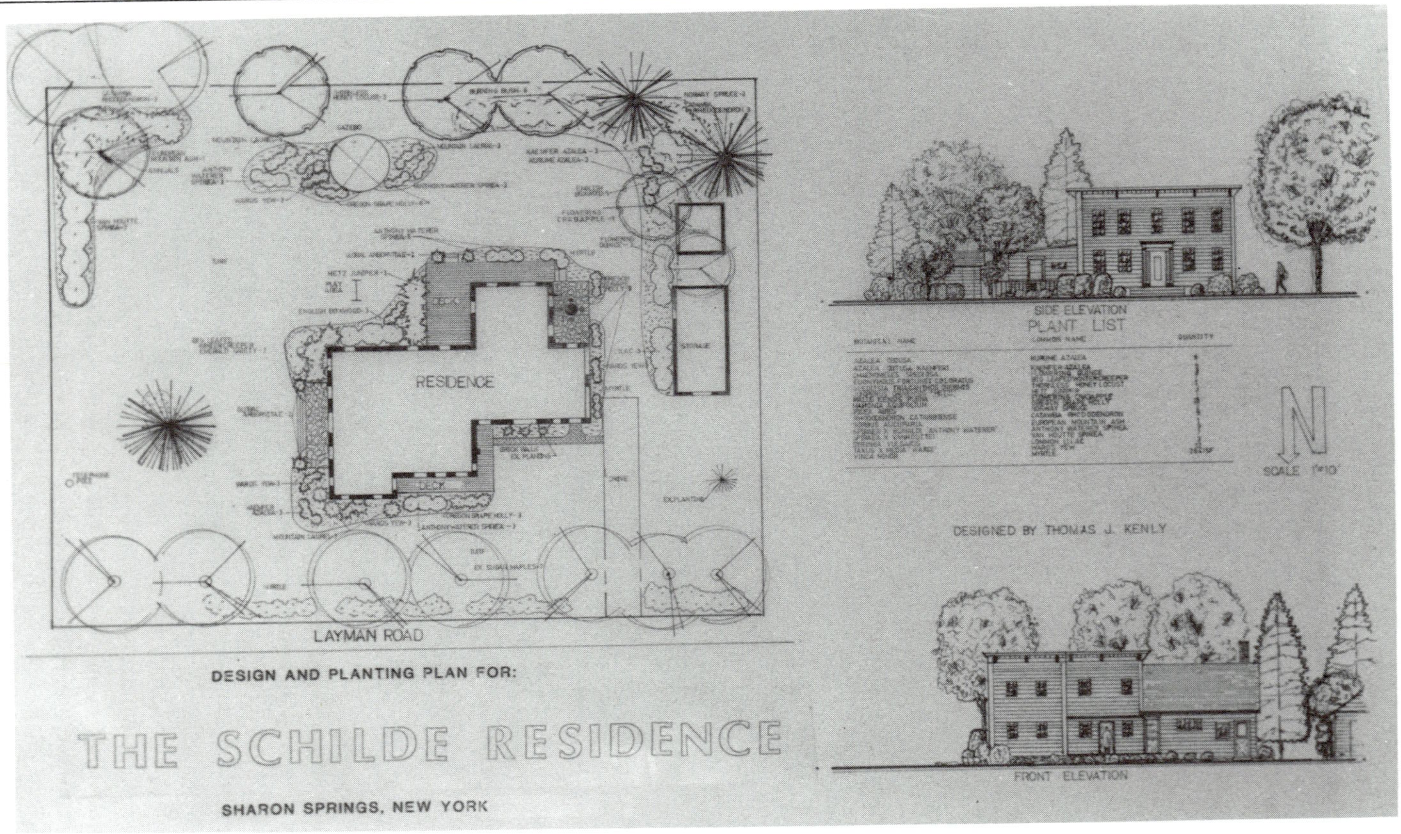

FIGURE 17-5.

A perspective view can be time-consuming to prepare if done in this detail, but provides a client with the best understanding of what the designer is proposing.

PERSPECTIVE DRAWING
SCOTT RYAN   MARCH 1977

Landscape Gardeners and Maintenance Supervisors

Landscape gardeners and maintenance supervisors are responsible for the care of landscapes after they have been installed by the contractors and approved by the landscape architects/designers.

Landscape maintenance involves such tasks as mowing, edging beds, pruning shrubs and trees, planting flower beds, weeding, fertilizing, watering, repairing surfacing and walls or fencing, replacing dead plants, and directing the growth of plants over a span of years. Because good landscapes take many years to mature, the role of the maintenance landscaper is of vital importance. Once mature, the land-

FIGURE 17-6.

A detail drawing for walk lights on a college campus (Courtesy James Glavin, Landscape Architect, Syracuse, NY)

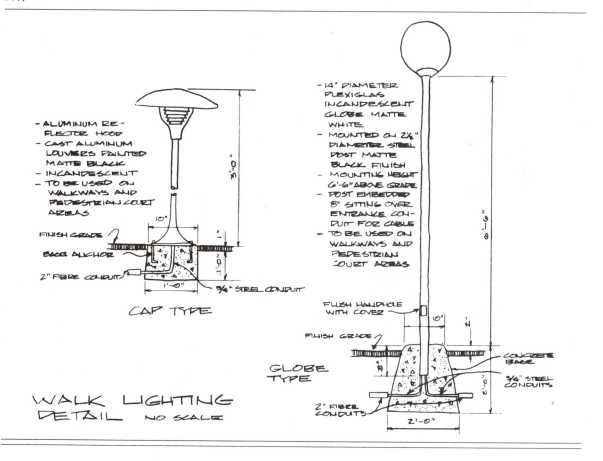

FIGURE 17-7.
A landscaper using a tile saw to cut pavers for a patio

FIGURE 17-8.
The construction of lawns, as in this sod installation, is an important part of a landscape contractor's work.

FIGURE 17-9.
Maintaining just what can be seen in this photo requires knowledge of tree and shrub care, turf care, plumbing, electricity, edging, brickwork, and flower care.

scape requires care and attention to assure its continuing success, Figures 17-9 through 17-13.

Maintenance landscapers may be resident employees (as on large estates) or they may be independent businesspeople who work under long- or short-term contracts for clients. A knowledge of horticulture, appreciation of plants, and understanding of the many factors that can affect plant growth are essential to success in this career field. Many opportunities exist, and the investment capital needed to begin is considerably less than for landscape contracting. A young person just out of school can begin by offering a few maintenance services (those not requiring expensive equipment) and build the business as profits allow.

The best landscapes result from close cooperation between the landscape architect, the landscape contractor, and the landscape maintenance gardener. Many problems of installation and maintenance can be avoided and the intent of a design best achieved if the designer talks with the persons responsible

FIGURE 17-10.
This foundation planting relies heavily upon flowers and topiary pruning for its effectiveness. As such, it requires higher maintenance than other entries may require.

FIGURE 17-12.
Weeds are controlled around this rhododendron by careful application of a contact herbicide. Landscape maintenance often requires attention to each individual plant.

FIGURE 17-11.
Changing flower plantings from spring to summer and summer to fall is a common task in landscape maintenance.

FiGURE 17-13.
Mowing is one of the most frequent tasks performed by landscape maintenance specialists. Here an attached grass catcher eliminates the need for follow-up raking or sweeping.

for implementing the design and caring for it afterwards.

Landscape Nurserymen

The sale of plant materials is the primary objective of the landscape nurseryman's business. An installation service is also offered as a means of promoting and supplementing the sale of the plants. Serving homeowners more than other types of clients, the landscape nurseryman engages in small-scale projects such as foundation plantings, patios and installing trees and shrubs. The business seldom expands into areas such as construction or follow-up maintenance. As suburban living spreads, the future for landscape nurserymen looks bright. There is a good market of homeowners who want professional assistance in getting their residential properties planted but are then willing to care for the landscape themselves.

EDUCATIONAL REQUIREMENTS

As the skill levels vary in different landscaping occupations, so do the amounts of formal schooling required. Some tasks can best be learned by doing, although it is often as important to know *why* something is done as *how* to do it. The best education preparation for success in the landscape fields is a combination of college and work experience.

Landscape contractors, landscape nurserymen, and maintenance landscapers should seek two to four years of college in strong ornamental horticulture programs. All types of plant science courses should be taken and, where offered, business and mechanical equipment courses should be taken as electives.

Landscape architects, where states mandate, must complete a four- or five-year university program accredited by the American Society of Landscape Architects. Several years of apprentice training follow in the office of a licensed landscape architect and finally a state certifying examination. Quite obviously, the landscape architect's position is the most difficult one to attain among the landscaping professions. Many schools of landscape architecture give greater emphasis to graphic art skills, engineering training, and the behavioral sciences than to basic horticulture. Therefore, aspiring landscape architects would be wise to seek elective courses that will add to their appreciation and knowledge of plants. Such training will produce landscape architects and designers who can better anticipate the problems their designs may create for those who must install or maintain them.

SUMMARY

A landscaper deals with the control and development of our outdoor spaces. Due to the range of project sizes and needs, the types of careers and training vary.

Landscape architects and designers conceptualize the landscapes. They seek out and analyze the sites that are to be developed. They represent the clients from the beginning of a project to the end. Landscape architecture requires extensive schooling and work experience and certification to practice in most states.

Landscape contractors construct and install the landscapes designed by the landscape architects and designers. Some landscape contractors also offer a design service to their clients. It is the landscape contractor's responsibility to acquire and install the plants and to build the nonplant elements of the design.

Landscape gardeners and landscape maintenance supervisors are responsible for the care of landscapes after their installation. They may work as resident employees or as independent professionals.

Landscape nurserymen have the sale of plant materials as their primary objective. In addition, they offer an installation service to promote and supplement the plant sales.

Achievement Review

A. SHORT ANSWER

Match one or more of the professions on the right to each of the characteristics on the left.

a. responsible for the care of the landscape after it has been installed.

b. conceives how the site should be developed

c. responsible for weekly mowing

d. responsible for installing the landscape

e. does the original site survey

f. primarily interested in the sale of plants

g. lays out new planting beds and prepares the soil for planting

1. landscape architect/ designer
2. landscape contractor
3. landscape gardener/ maintenance supervisor
4. landscape nurseryman

h. determines the needs of the customer

i. responsible for pruning back overgrown plantings

j. most involved in construction of a new stone wall

k. weeds flower beds

l. needs the greatest understanding of plant growth requirements

m. needs an understanding of geology, hydrology, agronomy, engineering, and architecture as well as ornamental horticulture

n. entrance into the field most regulated by law

o. requires the greatest initial capital investment

p. may be a resident employee

q. requires the greatest amount of schooling

r. may do some designing in addition to landscape installation and construction

18

Specialized and Nontraditional Careers

Objective

Upon completion of this chapter, you will be able to

- describe how interest, training, and skills in ornamental horticulture can be adapted to a variety of career fields.

CHOOSING OR CHANGING A CAREER

There are two truly difficult tasks in almost everyone's life: choosing a spouse and choosing a career. This text can offer no assistance with the first task, but it can with the second, at least to those with a yet untested interest in ornamental horticulture.

So far, this section has described the traditional careers in ornamental horticulture. Those careers have been attracting people to the profession for many years and will continue to do so. Many persons have entered the profession after preparation at colleges and trade schools; others have entered as the sons and daughters of industry personnel; still others have had few prior credentials but a sincere interest and willingness to work hard. The industry is sufficiently diverse and healthy to accommodate a work force that ranges from the wage-earning laborer to the salaried scientist.

Still, there are individuals whose interest in ornamental horticulture is just as genuine, but who seek different ways to apply that interest. They include, but are not limited to, people who wish to:

- supplement their income with part-time work in ornamental horticulture
- be involved with both the applied and the scientific branches of the profession
- combine their interest in ornamental horticulture with their desire to serve others
- specialize in the care of a single type of plant
- combine their interest in ornamental horticulture with an interest in writing or other modes of communication

PART-TIME WORK

For many, ornamental horticulture is an avocation, a hobby and leisure-time pursuit. They enjoy arranging flowers from their garden and filling their homes with plants. They may have a small backyard greenhouse. Others may have studied the profession in high school or college but later stepped out of the business world to raise children. Such persons often seek to supplement the family income by working on weekends or during peak holiday periods in flower shops, greenhouses, or garden centers. For many, it is a satisfying way to turn avocation to vocation. Florists are often eager to employ the temporary services of persons able to make bows and puffs, add greenery to arrangements, and wait on customers during holidays or as replacements for vacationing staff members. Greenhouse growers also need vacation replacements and workers to water and maintain the crop during weekends. Landscapers and nurseries often need a large work force during the spring and fall but not year round, so part-time and temporary workers fit their needs ideally.

Some part-time workers are self-employed. Operating from their homes, they often specialize in dried or silk arrangements, terrariums, and novelty items. Others may sell spring bedding plants or cemetery arrangements for use on Memorial Day, then close down and return to their regular jobs. It is impossible to speculate on how much income is generated by part-time, self-employed practitioners because little data is available.

Other workers have full-time careers that involve them with the ornamental horticulture industry during part of the working day. Examples include the small engine repairer who works on the lawn mowers, sprayers, chain saws, and other power equipment of landscapers and groundskeepers. General maintenance staffers often are required to care for interior or exterior landscape installations in office complexes, shopping centers, hospitals, and schools. Restaurant and resort managers may have flower arrangements on tables and in lobbies and need a regular floral supplier just as they need a food supplier.

COOPERATIVE EXTENSION SPECIALISTS

In nearly every state, the Cooperative Extension specialist is considered a member of the professional staff of the state college of agriculture. The specialist's constituency is off-campus and includes industry practitioners, homeowners, and 4-H youth.

The purpose of the Cooperative Extension Service is to process the latest information from universities and private research firms worldwide into a form that can be understood and used by the Service's constituency; then to disseminate that information. Cooperative Extension is a university outreach service, and its specialists need skills appropriate to the task.

Specialists must be university educated and have industry experience. They must read well and be able to organize and simplify complex material. Cooperative Extension specialists must enjoy organizing and leading workshops and tours. They must be comfortable with public speaking, and they must write well. Like teachers, Extension agents must have an above-average mastery of all communication skills.

People are at the center of the Cooperative Extension effort: people with problems and questions, people of all ages and intellects. Predictably, many of their questions are the same, so agents must also have patience and the desire to help.

Some Cooperative Extension specialists work from offices, responding to homeowners' questions by telephone. Others are traveling troubleshooters responsible to growers in several counties, Figure 18-1. Still others work through radio, television, and the news media. The particular location and constituency usually define how the agent must work to reach the greatest number of people.

Most Cooperative Extension specialists have bachelor's degrees in their fields of specialization and many have master's degrees.

HORTICULTURE THERAPISTS

Both teaching and Cooperative Extension offer an opportunity to work in ornamental horticulture while also working with people on a personal level. In both situations, the clientele are usually healthy and typical of average Americans.

In addition, special populations of citizens can be served by ornamental horticulturists. The elderly, the physically handicapped, the mentally retarded,

FIGURE 18-1.
The Cooperative Extension agent (right) advises a greenhouse grower having a bedding plant problem.

FIGURE 18-2.
A horticulture therapist uses plants to enrich the lives of these elderly citizens. (Courtesy United States Department of Agriculture)

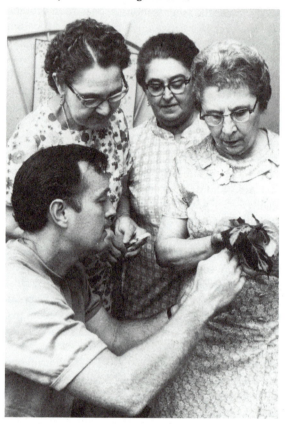

and the imprisoned have all been found responsive to therapists with horticultural skills.

Certain crafts of ornamental horticulture are eagerly learned by elderly citizens at club meetings, church groups, and even in nursing homes, Figure 18-2. Flower arranging, terrarium construction, houseplant care, seed propagation, and patio gardening are all activities that can be taught as part of a senior citizens activity program. Many of the elderly bring a rich background of gardening to their senior years and are knowledgeable students. Even the more infirm enjoy working with plants. Often their hands are still agile and talented even though their bodies are confined to chairs. Working with living, growing materials is satisfying for nearly everyone; the elderly are no exception. It also offers them an opportunity for role-reversal, in which they can direct the growth and well-being of a living thing instead of being the recipient of direction.

Although job opportunities exist in horticulture for handicapped workers, some may not yet be confident or skilled enough to enter the work force. For the physically handicapped, training in ornamental horticulture skills can increase manual dexterity, build self-confidence, alleviate frustration, and perhaps offer hope of a meaningful career.

Some of the mentally handicapped are trainable and employable; others are marginal, at best. Many also have physical handicaps or limited coordination. Still, many possess the patience and desire to learn that are necessary to accomplish some of the craft skills of ornamental horticulture. Therapists use skill training as one means of developing a sense of self and accomplishment in their clients.

Prisoners have also proven responsive to programs that allow them to work with the materials of nature while developing skills that offer hope of employment after their release. As a rehabilitation tool, ornamental horticulture training is proving its value, yet the number of programs nationally is limited due to a lack of instructors willing to work within the penal system.

While the clients of therapists can vary greatly, the education and personal qualities required of the therapist are similar. An increasing number of colleges are offering courses, even majors, in the use of horticulture as a tool of therapy. Training in the techniques of horticulture is combined with courses in the behavioral sciences, physical education, and the needs of special populations. Compassion, patience, and a call to human service are important personal characteristics of professionals in the field of horticulture therapy. Also necessary is the ability to remain objective toward the often tragic circumstances of the clients. Horticulture therapy is not a career that should be pursued by overly empathetic people, harsh though that advice may sound.

ARBORISTS AND LAWN CARE SPECIALISTS

Two specialized career fields have developed quite recently compared to other fields in ornamental horticulture.

Arborists

Arborists are tree maintenance specialists. Their services include insect and disease control, trimming, fertilizing, cabling, cavity treatment, woodlot thinning, and tree and stump removal, Figure 18-3. Some would dispute their classification as an occupation of ornamental horticulture, preferring the term *urban forestry*. Indisputably, arborists work with trees of all types: forest trees, street trees, lawn trees, ornamental trees, and fruit trees. It is vigorous outdoor work and potentially dangerous. The use of power tools, the climbing involved, the possibility

FIGURE 18-3.
Careful use of ropes permits this arborist to remove a dead tree next to a busy street with minimal danger or disturbance to passing people or traffic. Tree removal is only one of the many tasks performed by professional arborists.

of unexpected weak limbs and the difficulty of seeing power lines all call for care and skill.

Training is available through university programs in ornamental horticulture, forestry, or both. There is some conflict over which academic discipline has the closest affinity to the profession. Interested individuals should take a discerning look at the curriculum offerings of the college they are considering, not the program titles, to determine the quality and suitability of the training. College-level programs are offered at the two- and four-year levels. Laborer positions are available for nongraduates.

The industry is becoming increasingly professional in its methods and its expectations. This is due in part to the efforts of the National Arborist Association and its state chapters and the International Society of Arboriculture. Frequent local, national, and international meetings, in addition to newsletters and audio-visual programs keep the industry's professionals abreast of new techniques. An attempt to certify all arborists nationwide is now underway, confirming the profession's commitment to both education and skill development.

Lawn Care Specialists

Lawn care specialists are concerned with the maintenance of turfgrass installations, both residential and commercial. Clients contract for their services, which include a full season of insect and disease control, fertilization, grassy and broadleaf weed control, aeration, and, in small firms, may also include installation and mowing.

If there is a genuine American horticulture success story, it is the lawn care industry. Essentially, it created itself in 1969 when a novice company began convincing homeowners and commercial firms that it could maintain lawns better than the clients could and at an affordable price. That first year, annual sales totaled $218 thousand. By 1979, annual sales for the same firm were nearly $87 million. Inspired by the success of the industry leader, which now operates nationwide, smaller operations have sprung up in every metropolitan area and expanded the industry still further, Figures 18-4 and 18-5.

The services are primarily chemical application services. Lawn care specialists have a route of customers who are all to receive a particular treatment program. Each day the specialist will mix a tank truck full of the fertilizer or pesticide to be applied and make the rounds. Lawn care specialists may also be responsible for preparing bids to acquire new customers. During the off-season, the repair of trucks, sprayers, and other equipment and attendance at trade shows and short courses assure that

FIGURE 18-4.

A lawn care specialist applies a mixture of herbicide and fertilizer as part of a scheduled maintenance program. (Courtesy ChemLawn Corporation)

the specialist stays current in the field and is ready to begin again when the growing season resumes.

If the career has a limitation, it is the very long hours required during the spring, summer, and fall. A route may include several hundred customers, each property requiring four to six treatments each season. Generally, off-season hours are shorter, and vacations are taken then.

Preparation for the position of lawn care specialist should include at least a two-year college education, since the hiring trend is definitely in that direction. Majors in ornamental horticulture, turfgrass management, or agronomy can all provide the background needed for successful placement in the field. Additional training or personal skills in power

FIGURE 18-5.
This large lawn care firm has diversified its operations and now provides spraying of tree and shrub plantings. (Courtesy ChemLawn Corporation)

FIGURE 18-6.
Technical texts and journals are one means of staying abreast of new developments in industry. (Courtesy Rodney Jackson, Photographer)

equipment care and maintenance can prove helpful, as will good communication skills. Special operator's licenses may be needed for some equipment, and pesticide applicator certification is also required.

COMMUNICATIONS

Men and women who enjoy ornamental horticulture and have the ability to write, speak, photograph, or prepare instructional materials are in short supply. The demand far exceeds the number of people entering the profession.

Most newspapers carry a garden column, either prepared locally or syndicated. Papers in large ur-
ban areas often have a garden editor who reports on topics of interest to area gardeners. Texts, such as this one, serve students and other professionals in training, while other books are written for home gardeners and hobbyists. Some are children's books for beginning gardeners and some are highly detailed for the most experienced horticulturists. All have a service to provide, Figure 18-6.

Magazines for the home gardener and the horticulture industry publish thousands of articles annually. Some are prepared by full-time staff writers and others are free-lance writers. Equipment and supply manufacturers need technical writers with knowledge of horticulture to prepare instruction manuals and promotional materials for their products. Persons who can write usable technical material are needed throughout the professions of ornamental horticulture.

Photographers and technical illustrators are also needed to provide the graphics for books, magazine articles, and promotional and technical publications. Despite the many commercial artists being graduated each year, it is rare to discover someone who has the biological and horticultural training necessary to illustrate a complex article or text. Effective horticulture photography also necessitates training and interest in the natural world. Understanding the mechanics of picture taking is necessary, but it is not enough without the interest and insight of a horticulturist.

Another affiliated career is the educational media specialist. He or she is responsible for much of the software that is used in classrooms nationwide. The educational media specialist is also responsible for selecting visual and audio learning aids for teachers, Cooperative Extension agents, and garden club speakers. Some of these learning aids include slide series, overhead transparencies, plastic models, educational games, charts, computer software, motion pictures, filmstrips, microfiche, and cassettes.

Preparation in addition to university training in ornamental horticulture should include courses in journalism, technical writing, public speaking, audiovisual materials, graphic arts, data processing, and/or photography.

WHAT ELSE?

The objective of this chapter was to assure you that it is all right to be uncertain about the precise direction your career will take. If one of the traditional industry jobs described earlier sounds interesting,

give it a try. If you like the work but do not like the employer, change employers. If the traditional positions do not meet your career needs, consider some of the nontraditional alternatives described and illustrated here. And look beyond what has been described. Ornamental horticulturists find work in unpredictable locations; for example, military bases, resorts, ski centers, hospitals, restaurants, department stores, and supermarkets. If horticulture is but one of several major interests in your life, it can probably contribute at least partially to a satisfying career.

SUMMARY

It is possible to pursue a career in ornamental horticulture outside the traditional professions. Whether part-time or full-time, horticultural work can be combined with other interests or concerns.

Particular career possibilities discussed in this chapter were part-time self-employment, Cooperative Extension specialist, horticulture therapist, arborist, lawn care specialist, and communication specialist.

Achievement Review

A. TRUE/FALSE

Indicate if the following statements are true or false.

1. Ornamental horticulture can be a vocation or an avocation.
2. Ornamental horticulture offers full-time and part-time, temporary and permanent jobs.
3. The Cooperative Extension Service is an outreach program of private industry.
4. Horticulture therapy can be studied as a college major.
5. Horticulture therapy deals only with a client's skill development.
6. Both arborists and lawn care specialists need training in entomology, plant pathology, soil science, and botany.

7. Lawn care specialists usually work with turf maintenance only, not with other elements of the landscape.
8. Lawn care specialists need good communication skills since they deal directly with the customers.
9. Technical writers direct their writing more to the scientific audience than to the industry practitioner or hobbyist.
10. Nontraditional careers in ornamental horticulture should be considered only as a last resort if traditional career opportunities are not available at the time of the job search.

SECTION IV

THE PRODUCTION TECHNIQUES OF ORNAMENTAL HORTICULTURE

19

Greenhouses and Other Growing Structures

Objectives

- list and compare five types of growing structures.
- list the characteristics of various greenhouse and shade house coverings.
- list advantages and disadvantages of steam heat, hot water heating, and unit heaters in greenhouses.
- describe five methods of ventilating or cooling greenhouses.
- describe the latest methods employed to conserve energy in greenhouses.
- diagram three common methods of arranging greenhouse benches.

THE PURPOSE OF GROWING STRUCTURES

If the climate throughout the world were consistent, having sunny, temperate days and dry, cool nights; with sufficient rainfall to keep soil moist but not overly wet; with no hailstorms or damaging winds; and with day lengths always suitable to promote flowering; then there would be no need for greenhouses or other growing structures. The commercial production of high-quality, intensively cultivated plant material requires just such conditions. Thus the purposes of a greenhouse reveal themselves:

1. to provide a controlled growing environment for plants whose economic value justifies the expense
2. to permit the growth of plants in regions where survival outdoors is not possible
3. to extend the season of growth for plants at times when they would normally go dormant

381

TYPES OF GROWING STRUCTURES

Following are some of the many different types of growing structures that exist for the production of ornamental horticulture crops.

Superstructure: aluminum, iron, steel, or wood (Figure 19-1)
Glazing materials: glass or plastic
Advantages:
- best environmental control
- strongest structure
- maximum light allowance

Disadvantages:
- most expensive to build and heat
- wastes land between houses

FIGURE 19-1.
A detached A-frame truss greenhouse (Courtesy Rodney Jackson, Photographer)

FIGURE 19-2.
A typical ridge and furrow, A-frame, truss greenhouse range (Courtesy Lord & Burnham, Division of Burnham Corporation)

Superstructure: aluminum, iron, steel, or wood (Figure 19-2)
Glazing materials: glass or plastic
Advantages:
- less expensive to build due to lack of interior walls
- good environmental control if a single crop is in production
- less expensive to heat
- wastes no land between houses

Disadvantages:
- environmental control difficult when different crops are grown
- structurally weak under heavy snow buildup

FIGURE 19-3.
A quonset-style greenhouse

Superstructure: pipe arches (Figure 19-3)
Glazing materials: hard or soft plastic
Advantages:
- less expensive to build; needs no extensive foundation or roof support
- ideal for production of seasonal crops (such as bedding plants); afterwards uncovered or changed to a lath house
- may be either freestanding or grouped and interlocked in a ridge or furrow style

Disadvantages: • requires new covering nearly every year
• more difficult to ventilate

FIGURE 19-4.

A lath or shade house (Courtesy United States Department of Agriculture)

FIGURE 19-5.

A brick cold frame or hotbed

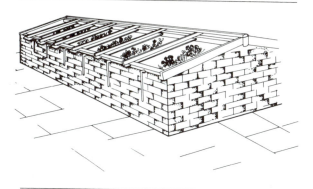

Covering materials: snow fencing or shade fabric, Figure 19-4

Purposes: • to provide a shaded area for production of heat-sensitive plants
• to provide a cool holding area for plants on sale to retail customers or awaiting wholesale shipment

Side materials: concrete blocks, wood, or similar materials, Figure 19-5

Glazing materials: glass or plastic

Purpose: • to provide supplemental growing space for a greenhouse operation; used for propagation, starting, and hardening-off

Differences: • cold frames use only sunlight passing through the glass or plastic for warmth. Hotbeds supplement solar energy with decomposing manure, electric cables, or heating pipes.

MATERIALS USED TO COVER THE STRUCTURES

The material selected to cover the superstructure of the greenhouse, shade house, or cold frame depends upon the amount of light needed, the length of service the growing structure is expected to offer, and the time and cost of maintenance and materials that the grower is willing to absorb. These variables are compared in Table 19-1. A special comment is warranted concerning cost of materials. Traditionally, the initial cost of glass greenhouses has been much greater than that of plastic houses. Since the rise in cost of all petroleum-based products, plastic and fiberglass now rival glass in initial cost. Only the simpler superstructure needed to support the lighter-weight plastics is keeping the cost of plastic houses below that of comparably styled glass houses.

TABLE 19-1.
A Comparison of Materials Used to Cover Growing Structures

Material	Length of Service	Amount of Sunlight Passing Through	Maintenance Required
Glass	Ten to twenty years depending on material used as glazing bars	Full sunlight less that blocked by settled dust and shading of the superstructure; glass gives greatest amount of light	Wood glazing bars require glass to be removed, new caulking applied, and the bars repainted every ten years. Aluminum glazing bars need little maintenance except periodic resetting of glass to correct slippage.
Fiberglass	Five to ten years depending on quality of fiberglass, maintenance, and intensity of sunlight causing clouding	75 to 85 percent of that transmitted by glass	The fiberglass must be scrubbed down every other year to remove dirt that collects on exposed glass fibers.
Soft plastic (polyethylene)	Six months to two years depending on quality of plastic and intensity of sunlight causing clouding	50 to 80 percent of that transmitted by glass depending upon type of plastic, number of layers, age, and degree of clouding caused by ultraviolet rays	More frequent replacement needed than with other coverings; plastic clouds and is broken down by ultraviolet rays in sunlight.
Shade fabric	Normally used for part of a year, then stored until next season	30 to 50 percent of that transmitted by glass; multiple layers of fabric used for heavier shading	None of importance
Soft plastic (vinyl)	Four to five years depending upon thickness	85 to 90 percent of that transmitted by glass	The material attracts dust, which must be washed off.
Soft plastic (polyvinyl fluoride)	Ten years or more	92 percent of that transmitted by glass	Less than the other soft plastics
Acrylic rigid panels	Ten years or more	83 percent of that transmitted by glass	Periodic washing to remove dust and dirt
Polycarbonate rigid panels	Ten years or more	75 to 80 percent of that transmitted by glass (less as it ages)	Periodic washing to remove dust and dirt

TYPICAL STRUCTURAL PLANS

Figures 19-6 through 19-9 show typical construction details for the major types of growing structures previously described and illustrated (Figures 9-1 through 9-5).

HEATING THE GREENHOUSE

Anyone who has spent even a moment in a greenhouse on a bright, sunny day can attest to its ability to magnify the sun's energy. The ability to admit light and use the energy for warmth and plant

FIGURE 19-6.

A typical aluminum gable greenhouse

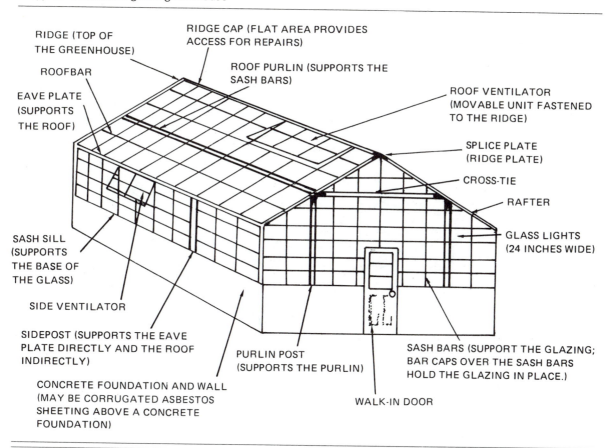

RIDGE (TOP OF THE GREENHOUSE)

RIDGE CAP (FLAT AREA PROVIDES ACCESS FOR REPAIRS)

ROOFBAR

ROOF PURLIN (SUPPORTS THE SASH BARS)

EAVE PLATE (SUPPORTS THE ROOF)

ROOF VENTILATOR (MOVABLE UNIT FASTENED TO THE RIDGE)

SPLICE PLATE (RIDGE PLATE)

CROSS-TIE

RAFTER

GLASS LIGHTS (24 INCHES WIDE)

SASH SILL (SUPPORTS THE BASE OF THE GLASS)

SIDE VENTILATOR

SIDEPOST (SUPPORTS THE EAVE PLATE DIRECTLY AND THE ROOF INDIRECTLY)

PURLIN POST (SUPPORTS THE PURLIN)

SASH BARS (SUPPORT THE GLAZING; BAR CAPS OVER THE SASH BARS HOLD THE GLAZING IN PLACE.)

CONCRETE FOUNDATION AND WALL (MAY BE CORRUGATED ASBESTOS SHEETING ABOVE A CONCRETE FOUNDATION)

WALK-IN DOOR

FIGURE 19-7.
A typical hotbed (Courtesy United States Department of Agriculture)

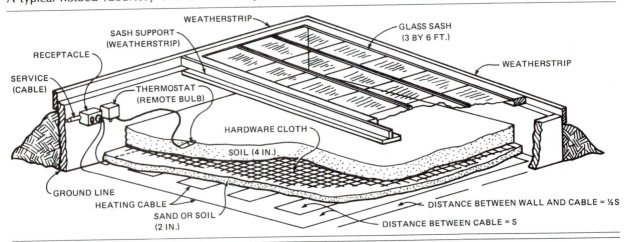

FIGURE 19-8.
A typical lath house

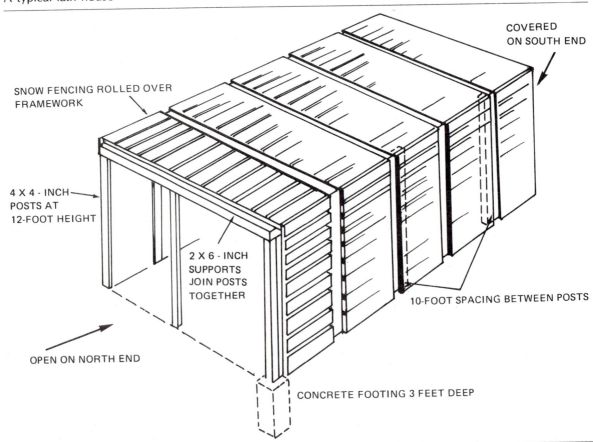

FIGURE 19-9.
Structural plan for greenhouse glazing

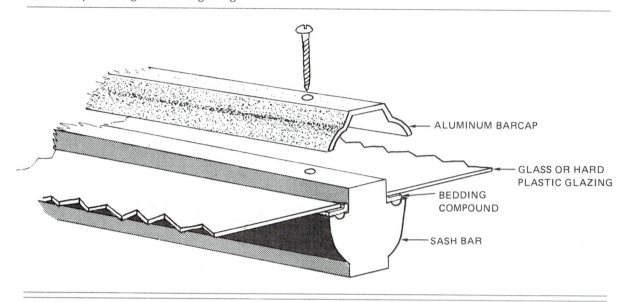

ALUMINUM BARCAP

GLASS OR HARD
PLASTIC GLAZING

BEDDING
COMPOUND

SASH BAR

growth is one side of the story. The other side is the inability of the greenhouse to retain any significant portion of that heat energy after the sun sets. Any attempt to insulate the greenhouse in a conventional manner would severely reduce the admission of light that is critical to good plant growth. Thus the need for supplemental heating is apparent, Figure 19-10.

With rising energy costs and long-term fuel shortages confronting every greenhouse professional, the choice of heating systems deserves some thought where the opportunity for a choice exists. Nearly all systems are automated now, and many involve both primary and back-up (emergency) systems. The choices are compared in Table 19-2.

VENTILATING AND COOLING THE GREENHOUSE

The even distribution of air throughout a greenhouse is essential for consistent temperatures and uniform plant growth. Also, greenhouse air should move slightly as often as possible to minimize mildew and other diseases of plants that proliferate under stagnant air conditions.

FIGURE 19-10.
A typical greenhouse unit heater (Courtesy Rodney Jackson, Photographer)

TABLE 19-2.

A Comparison of Greenhouse Heating Systems

Type of System	Advantages	Possible Disadvantages
Steam	• Can transport heat throughout large ranges (greenhouse physical plants) without cooling. • Requires smaller piping than hot water systems, reducing installation costs • Makes steam available for heat-treating greenhouse soil	• Control of greenhouse temperatures not as subtle with steam as with hot water and house may temporarily overheat. • Cost of maintenance greater with steam due to damaging effects that high pressure has on pipes.
Hot water	• Permits more accurate and responsive thermostatic control of greenhouse temperatures • Distributes heat more evenly with fewer hot spots to injure plants • Less potential for danger to workers if a hot water line ruptures than if a pressurized steam line breaks	• Larger piping required, increasing cost of installation • Steam still required for soil pasteurization
Unit heater	• Adaptable to small greenhouse areas • Excellent back-up systems in the event of boiler breakdown or power failure of larger system	• Fuel cost (gas or oil) may be greater than cost of expanding existing steam or hot water system. • Heat distribution uneven but can be improved by attaching plastic sleeves that extend length of house. Bench level temperatures may still be cooler than nearer roof.

Ventilation of greenhouses is accomplished by using:

• roof and side ventilators
• exhaust fans
• fan and convection tube systems

Use of roof and side vents for air exchange is based on the principle that heated air rises. Hot air exits through the roof vent, while fresh, cooler air enters through the side vents. When ventilators run the length of the roof and sides, air exchange

FIGURE 19-11.
Side vents permit cool air to enter the greenhouse, forcing the hot air out through the top vents.

is consistent throughout the house. Use of ventilators alone is the oldest method of ventilating greenhouses, Figure 19-11.

Exhaust fans can be used in warm weather to draw old air out of the house and pull fresh air in. Fans can supplement roof and side ventilators or be used in place of them as in small, plastic-covered quonset houses, Figure 19-12.

When the outside air is so cold that its direct entry into the greenhouse would cause plant injury, a plastic convection tube can be used to permit warming and even distribution of the air before it enters the greenhouse atmosphere. A polyethylene sleeve with holes, the tube inflates and pulls outside air into the house when fan jets operate. The convection tube system may be paired with exhaust fans as well as heaters or cooling systems to provide temperature-adjusted fresh air for the greenhouse, Figure 19-13.

Cooling the greenhouse is necessary many times during the year and is vital if the greenhouse is to be used during the summer months. Three basic approaches are taken depending on the time of year and geographic location:

1. shading the glass to reduce light intensity
2. ventilating to allow cooler outside air to replace warm inside air
3. promoting heat exchange through water evaporation

FIGURE 19-12.
A greenhouse exhaust fan

FIGURE 19-13.
Use of a convection tube assures even distribution of heated air through the greenhouse. (Courtesy Rodney Jackson, Photographer)

Shading

Greenhouse shading can be obtained by spraying a compound onto the glass externally. It is a temporary shading that gradually washes away with the rain. In the South a more permanent shading compound may be used. The compounds are available in both liquid and powdered forms. Those containing binders persist longer, Figure 19-14.

Shading can also be applied internally as cotton or Saran cloth suspended above a crop to reduce the light intensity, Figure 19-15. The saran is available in different weaves which determine the amount of light reduction.

Ventilation

As described earlier, greenhouse ventilation systems bring fresh air into the greenhouse to replace the warm internal air. As long as the outside air is cooler than the temperature desired for the greenhouse, ventilation will cool the greenhouse. During especially hot weather, ventilation alone may not be enough.

Heat Exchange and Water Evaporation

When water evaporates or changes from a liquid to a vapor it absorbs heat from the atmosphere and cooler air results. That is the principle behind two greenhouse cooling systems: fan and pad and fog evaporative.

Fan and pad cooling uses exhaust fans and continuously wet pads of excelsior (a fibrous porous material), cross-fluted cellulose, aluminum fibers, or glass fibers. Through a recirculating water system, the pads are kept wet at one end of the greenhouse, while fans at the opposite end of the house pull outside air through the pads into the greenhouse. The system goes by other names as well: wet-pad cooling or washed air cooling. Regardless of name, the principle is the same. Figure 19-16A and 19-16B.

FIGURE 19-14.
This shading compound (Varishade-2®) can be applied by rolling, brushing, or spraying. It will provide shade on hot, dry days, yet go clear on wet days. (Courtesy Solar Sunstill, Inc.)

FIGURE 19-15.
Saran shading aids in light control for production of these tropical foliage plants. (Courtesy Rodney Jackson, Photographer)

FIGURE 19-16A.
In this pad and fan cooling system, continuously wet pads of excelsior cool the air drawn through them into the greenhouses.

FIGURE 19-16B.
These fans at the opposite end of the greenhouses pull the outside air through the pads.

Fog evaporative cooling uses a high-pressure pump to create a fine mist whose water droplets are so tiny that they remain in suspension in the air, then evaporate. As the fog disperses through the greenhouse, the evaporation of the tiny droplets causes the desired cooling without getting the plants wet.

Both fan and pad cooling and fog evaporative cooling systems can reduce greenhouse temperatures from 10° to 30°F below the outside air.

CONSERVING ENERGY IN GREENHOUSES

As the era of cheap fuel passes into history, a new era of energy conservation has begun. Greenhouse growers have felt the impact of increased fuel costs directly in their cost of production. Since most greenhouse crops are discretionary rather than essential consumer purchases, the cost of production cannot be allowed to rise as much as the cost of fuel or sales may decline.

The greenhouse industry has responded in several ways to the need for energy conservation to curb rising production costs. Current trends in crop production show an orientation of crop types to geo-

graphic regions. Crops needing high temperatures for optimum production are being grown less commonly in the northern states and in greater numbers in the South. Cool-temperature crops such as carnations are replacing warm-temperature crops such as roses in northern greenhouses. It must be noted, however, that the technology for energy conservation in greenhouses is still in its infancy. No doubt some of the techniques to be described will prove their merit and gain acceptance. Others will probably prove impractical and be replaced.

Double-Layered Plastic Covering

Covering the plastic greenhouse with two layers of plastic instead of one and inflating the air space in between using a small fan makes the covering more airtight and more heat-retentive. Even some glass houses are being covered with a plastic outer shell to aid in heat retention. The principle disadvantage of the extra plastic layer is the reduction in light intensity.

Foundation Insulation

Large amounts of heat are lost through the concrete or wooden sides of the greenhouse. The siding material may be supportive or merely provide a skirt between the glass and the ground. In either case, the addition of insulation can plug an important area of heat loss.

The insulation can be an earthen bank against the outside of the wall. It can also be a layer of insulating material such as fiberboard or foil-covered styrofoam against the interior wall. With pierced concrete block foundations, the holes can be filled with insulating material at the time of construction.

North Wall Insulation

With no direct sunlight entering through the north wall of the greenhouse and maximum heat loss occurring there, the north wall is an energy liability. Some commercial and university greenhouse ranges are experimenting with full ground-to-roof insulation of the north wall, using concrete or wood and appropriate insulating material. The energy saved is proving to be significant and the light reduction minimal.

Thermal Blanket Over Crops

Most of the heat loss in a greenhouse occurs after dark; hence, that is when fuel consumption is greatest. Heating the air between the crop and the greenhouse roof at night has been a necessary evil in the past although admittedly of no value to the plants. Now a thin thermal blanket supported on wires can be pulled over the plants at night to hold warmth around the benches where it is needed.

The principal drawback to this technique is the awkwardness of its operation. If hand-pulled across the crop, it results in a loss of some daylight, because the greenhouse staff does not report for work at sunrise each day nor work until dark. To date, automated blanket systems have not worked flawlessly either.

Other Methods

Other energy-saving techniques are being devised by researchers worldwide. Lining greenhouse benches with black plastic tubes of water is one such technique. In theory the water tubes will absorb sunlight energy during the day and release it slowly at night. Tighter sealing of glass glazing panels is another technique for slowing heat escape from the glass greenhouse. The installation of a plastic false ceiling to cut off some of the roof space that otherwise requires heating can also reduce energy needs.

Many methods are being tried; the next decade will begin to see their results. There is no question that the greenhouse industry has entered into a period of change.

GREENHOUSE BENCHES

The benches of a greenhouse may be raised or at ground level. They may actually contain the crop or merely support it. The style of benches, materials used, and arrangement within the greenhouse all depend upon the crop being grown. However, regardless of the crop grown or style of bench selected, greenhouse benches must fulfill three functions:

1. They must drain quickly.
2. They must be of a width that allows workers to reach into their center.
3. They must maximize the crop's exposure to light.

The choice of bench styles depends upon the crop being grown. For example, greenhouse vegetable production (principally of tomatoes and lettuce) requires *ground beds*. Production methods are similar to those used in field plantings, but the ground beds need special preparation for proper drainage. Such preparation includes approximately 6 inches of crushed stone beneath 6 to 8 inches of porous soil and a properly installed drainage tile system to provide for subsoil drainage. Similar bed construction is required if the ground benches are used for cut flowers such as roses. The benches are usually limited to 3 feet in width to allow access to the center of the bed by workers who must tie the crop, cut the blossoms, and so forth. Sides may be built as an edging for the ground beds to define walkways more clearly and reduce the possibility of plant injury, Figure 19-17.

Raised benches are used in the production of

FIGURE 19-17.
These roses are being grown in ground beds as a cut flower crop. (Courtesy Rodney Jackson, Photographer)

FIGURE 19-18.
A bench for potted plant production

pot crops as well as cut flower crops. In the case of potted plants, the bench serves much like a table to bring the crop up to a level where it can be worked with easily. With cut flower crops, the bench not only elevates the crop but contains the growing medium for the plants. As such, it must be from 6 to 12 inches deep to allow for proper root development and drainage, Figures 19-18 and 19-19. Raising the bench above ground allows air to flow more freely around the crop and permits a warmer root-zone temperature to be maintained. Often heating lines are placed beneath the benches to attain maximum benefit from the warmth.

Bench Materials

Greenhouse benches may be purchased as prefabricated units or improvised by the grower to fit particular needs. The most important quality of a bench is capacity for rapid drainage. A number of materials are commonly used:

- corrugated transite (very durable and expensive)
- aluminum (very durable and expensive)
- welded wire fabric
- wooden slats (less durable and less expensive)
- snow fencing (lath) (less durable and least expensive)

FIGURE 19-19.
A cut flower bench must be from 6 to 12 inches deep to permit root development and proper drainage.

Wooden supports, concrete blocks, and wire fabric or lath can be combined easily to make an affordable bench that can be adapted to all types of containerized crops. Care should be given to preserving any wood used in bench construction. Either naturally rot-resistant woods such as redwood or preservative-treated woods (stained with copper naphthanate) can be used.

Bench Arrangements

Since crops cannot be grown and money cannot be made in space devoted to aisles, greenhouse benches are arranged carefully to provide the greatest area for growing space. The number and width of aisles are determined by the width of the carts needed in the aisles and whether the crop can be worked from one side or needs to be accessible from both.

Three methods of arranging benches in greenhouses are commonly used, Figures 19-20, 19-21, and 19-22.

To increase the efficient use of space in greenhouses employing raised benches, some growers have placed benches atop rolling pipes, thereby permitting the movement of benches back and forth, Figure 19-23. This system eliminates the need for permanent aisles and reduces their numbers. Essentially the aisle can be shifted to where it is needed.

FIGURE 19-20.

A cross-bench arrangement

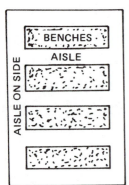

BENCH ARRANGEMENT

CHARACTERISTICS
AISLES ALONG THE SIDES AND BETWEEN THE BENCHES; LEAST EFFICIENT USE OF SPACE FOR GROWING.

CROSS-BENCHING

FIGURE 19-21.

With peninsular benching, each bench is accessible from both sides and the aisle end.

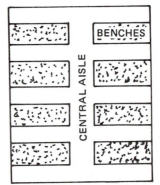

BENCH ARRANGEMENT

CHARACTERISTICS
FEATURES A WIDE CENTRAL AISLE TO ACCOMMODATE CARTS AND OTHER EQUIPMENT; NARROW AISLES BETWEEN BENCHES ALLOW ACCESS BY WORKERS AND EFFICIENT USE OF SPACE FOR GROWING.

PENINSULAR BENCHING

FIGURE 19-22.
Length-of-house benching. The central bench features a trickle irrigation system.

BENCH ARRANGEMENT	CHARACTERISTICS
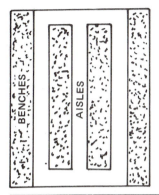	UTILIZES GROWING SPACE EFFICIENTLY, BUT WORKERS MUST WALK LENGTH OF HOUSE TO GET TO OTHER SIDE OF THE BENCH.

LENGTH-OF-HOUSE BENCHING

FIGURE 19-23A.
Benches atop rolling pipes permit aisles to be changed as needed.

FIGURE 19-23B.
The rolling pipe system beneath a bench

SUMMARY

Plant growing structures exist to provide consistent, controlled environmental conditions for the growth of plants. There are several different types of structure. Greenhouses exist in many forms depending upon their age, location, and use. They can be built of aluminum, iron, steel, or wood and glazed with glass, hard plastic, or soft plastic. Their styles include: (1) detached A-frame truss, (2) ridge-and-furrow A-frame truss, and (3) quonset. Other growing structures include shade or lath houses, cold frames, and hotbeds. Typical structural plans for these structures are included in the chapter.

The material selected to cover the growing structure depends upon the amount of light needed, the length of service required, and the time and cost of materials and maintenance that the grower can accept.

Greenhouses require heating systems to supplement natural solar heat by day and at night. The choices include steam systems, hot water systems, and unit heaters. Each has advantages and disadvantages.

Greenhouses are ventilated by using roof and side ventilators, exhaust fans, or fan-and-convection-tube systems. Ventilation is needed to distribute air evenly throughout the greenhouse, maintain consistent temperatures, and reduce the risk of disease.

The need to cool the greenhouse is greatest during the summer months. Three approaches to cooling are used depending upon the time of year and geographic location of the greenhouse: shading the glass, ventilation, or heat exchange and water evaporation.

Energy conservation has become increasingly important in recent years as expensive fuels have sent greenhouse production costs climbing. Techniques such as double-layered plastic coverings, foundation insulation, north wall insulation, and thermal blankets are being tried now. The future will see additional techniques developed.

The style of benches, the materials used in their construction, and their arrangement within the greenhouse all depend upon the crops being grown. Benches may be raised or at ground level. They may contain the crops or merely support them. They may be commercially prefabricated or improvised. They may be arranged as cross-benching, peninsular benching, or length-of-house benching. Regardless of these differences, all benches must drain quickly, be of a width that allows workers to reach into their center, and allow crops maximum exposure to light.

Achievement Review

A. SHORT ANSWER

Answer each of the following questions as briefly as possible.

1. List the three purposes of a greenhouse. Then list four crops that are grown commercially in your region of the country for each purpose given.

2. Match the characteristics on the left with the growing structures on the right.

   a. most efficient land use of all greenhouse styles
   b. rounded style that may be covered with hard or soft plastic
   c. low structures used for holding plants, propagation, or hardening-off bedding plants
   d. strongest greenhouse style
   e. greenhouse style offering the best environmental control
   f. another use for a soft plastic greenhouse during the summer
   g. not well suited for the production of crops requiring different environmental conditions
   h. lacks interior walls between houses
   i. may be heated by solar energy or electric cables
   j. greenhouse style not requiring a deep foundation

   1. Detached, A-frame truss greenhouse
   2. Ridge and furrow greenhouse
   3. Quonset greenhouse
   4. Shade house
   5. Cold frames or hot beds

3. Match the characteristics on the left with the covering materials on the right.

   a. requires scrubbing with a wire brush every other year to remove collected dust
   b. a summer covering used over quonset frame greenhouses
   c. allows the highest percentage of sunlight to pass through
   d. allows the least percentage of sunlight to pass through
   e. has the shortest time of service of the soft plastics

   1. clean glass
   2. polyethylene
   3. shade fabric
   4. vinyl
   5. polyvinyl fluoride
   6. fiberglass

4. Label the parts of the following structural plans.

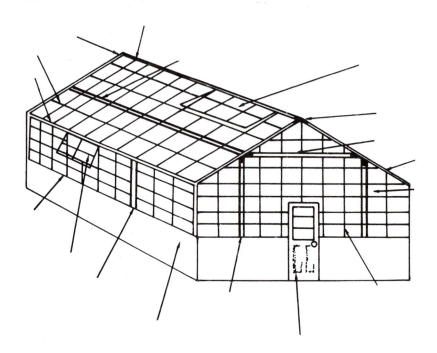

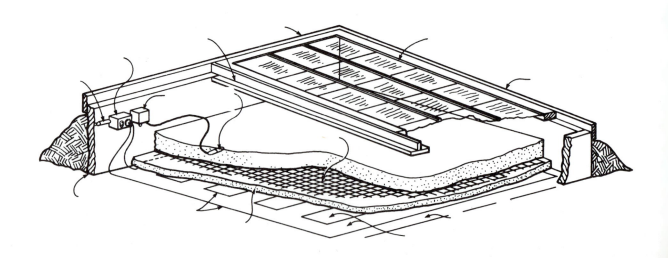

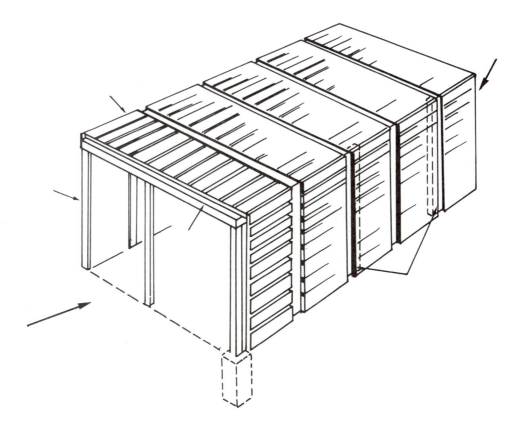

5. Match the characteristics on the left with the types of heating systems on the right.

a. adaptable to small greenhouse areas and often used as emergency back-up

b. requires large piping, thereby increasing installation costs

c. uneven heat distribution

d. heat can be transported through a large range of greenhouses without cooling

1. steam
2. hot water
3. unit heater

e. permits the most accurate thermostatic control of temperatures

f. requires smaller piping, keeping installation costs lower

g. may be used with a plastic convection tube to permit more even distribution of heat

h. a safer system with less potential danger in the event a line ruptures

i. most adaptive to heat treatment of soil

j. cost of maintenance is greatest

6. Several bench styles are described below. For each, list the attributes that are characteristic of it.
   a. A ground bed edged with concrete to separate it from the walk contains a field of drainage tile and 6 inches of crushed stone.
   b. A raised bench of corrugated transite contains no soil.
   c. A raised bench of concrete blocks, wooden supports, and welded wire fabric.
   d. A raised bench of wood has sides and contains soil.
   e. A raised bench of concrete blocks, wooden supports, and lath snow fencing.
7. Diagram and label three common methods of arranging benches in a greenhouse.

B. ESSAY

1. Write a short paragraph explaining how shading, ventilation, and water evaporation each serve to cool a greenhouse differently.

C. TRUE/FALSE

Indicate if the following statements are true or false.

1. Heat can be lost through both the foundation and the glazing material of a greenhouse.
2. Energy conservation methods employed in greenhouses are aimed at retaining as much heat as possible and plugging all avenues of escape.
3. Present energy conservation technology is highly advanced.
4. We can anticipate no advances in energy conservation technology in the future.
5. The use of a small fan to inflate a double layer of plastic covering a greenhouse is a fairly effective method of conserving heat energy.
6. Industry trends are toward the production of warm-temperature crops in northern greenhouses.
7. Thermal blankets are used to maintain warmer night temperatures near crops while allowing the rest of the greenhouse to cool down.
8. Since maximum heat loss occurs through the west wall of a greenhouse, full ground-to-roof insulation is proving to be effective in cutting heating bills.
9. Since warm air rises, heating the roof area of a greenhouse is an inefficient use of energy.
10. Placing heating lines beneath benches is more efficient than running the lines overhead.

20

Greenhouse Production Techniques

Objectives

Upon completion of this chapter, you will be able to

- explain the need for a crop production schedule.
- explain how and why greenhouse root media vary.
- describe four methods of pasteurizing growing media.
- list three reasons for frequent testing of greenhouse soil.
- list and describe the containers used in greenhouse production.
- list the methods of reproduction used for greenhouse crops.
- label crops on a greenhouse bench.
- describe methods of spacing, watering, and fertilizing greenhouse crops.
- describe techniques of integrated pest management in greenhouses.
- plan production schedules for a representative sampling of greenhouse crops.

CROP PRODUCTION SCHEDULING

The production of greenhouse crops can be compared to the manufacture of nonplant items in one way: scheduling is important. No one buys a poinsettia on December twenty-sixth or an Easter lily on the Monday after Easter. Retail florists need large quantities of red roses for Valentine's Day, and they sell more mums during the autumn than the spring. In northern states, bedding plants are of interest to consumers for about a six-week period in the spring, and after that they can barely be given away. Unlike other manufacturers, who can store

excess inventories, greenhouse growers lose their perishable products and the money invested in them if they don't sell. Timing is everything in the flower production business.

To serve the market when consumer demand is greatest, and minimize losses due to overproduction or underproduction, greenhouse growers plant, care for, and harvest their crops in accordance with production schedules. In theory, plant production schedules are logical and not difficult to follow. A particular crop can be expected to require a certain number of weeks at a given temperature to go from seed or cutting to harvest. Thus by counting backwards from the desired harvest date, the date of planting can be determined. By keeping accurate yearly records, a grower can determine whether to increase or decrease the number of plants produced. Dates of pinching, fertilizing, shading, application of growth retardants, or repotting can be determined from previous years' data or provided by seed or cutting suppliers.

Theory meets reality, though, when unexpected warm weather sends greenhouse temperatures soaring above those specified in the crop's production schedule, and crop development advances by a week in only a day or two. Equally troublesome is an extended period of overcast skies, or an outbreak of disease, or a malfunctioning heating system, or a disrupted photoperiod at a critical stage of plant development. That is when growers must apply their education and past experience to compensate for the unexpected and get the crop back on schedule.

Complete crop records are essential for effective production scheduling. In addition, growers must stay attuned to the economic forecast to anticipate changes in consumer buying. If last year was a record year for sales but the current year's forecasts are for restrained consumer spending, the crop should be reduced. On the other hand, when a downward trend in the economy is followed by a steady increase, more plants can be grown for sale.

GREENHOUSE ROOT MEDIA

Greenhouse plants are grown in a wide variety of root media, including those that contain actual soil as well as those that are entirely soil-less. The media can be formulated by the grower, adapting and modifying one of the traditional mixes such as those described in Chapter 14, or the media can be purchased from commercial sources.

Whether soil-containing or soil-less, root media must all do the same things: provide nutrients to the plant, retain water for use by the plant, allow aeration for the roots, and anchor the plant firmly in the container or bench.

The research comparisons between greenhouse crops grown in soil versus those grown in artificial media are extensive. However, the grower's decision to use one medium rather than another is likely to be based upon economic factors more than research findings. Is there a close source of good field soil that can fill ground beds or benches? Is the potted crop going to be shipped over a long distance, making weight a cost and handling factor? Does the grower have the facilities and desire to mix and pasteurize large quantities of soil, or is a ready-mixed product more practical?

When field soils are used in greenhouse crop production, they must be conditioned and pasteurized. Drainage is essential, so sand, peat, vermiculite, fired clay, and even gravel may be added. Even a loam soil, ideal for nursery and landscape use, is not satisfactory for greenhouse use without additives. Pasteurization with steam or chemicals is needed because of the presence of weed seeds, pathogens, and insects in the soil. The specifics of pasteurization are described later in this chapter. Once in the greenhouse bench, the soil may serve for many years if properly reconditioned after each crop is harvested.

While good field soil in a bench offers the cultural predictability that is essential to high-quality crop production, the unpredictability of field soil taken from different sources each time is what led to the development and widespread use of soil-less mixes for potted crops.

The commercially prepared soil-less mixes differ in their formulations but not in their functions. Regardless of brand names, the artificial mixes are either *bark-based* or *peat moss–based* mixes. Brand name examples of each formulation include:

Bark-based growing media

Ball Growing Mixes I and II
Choice Container Mix
Fafard Mix, Number 3 and 4
Metro Mixes 300, 350, and 500
Pro-Mix Peat-Bark Mix
Strong-Lite Bark Mix
VJ #1 Mix and #3 Mix

Peat-based growing media

Ball Germinating Mix

Fafard Peat-Lite Mix
Jiffy Mix
Ogilvie Professional Mixes 2 and 5
Premier Germinating Mix
Pro-Mix A
Redi-earth Peat Lite Mix

The principal advantage of artificial root media is their uniformity of composition and the predictable response of plants to fertilization, watering, and other production techniques when grown in them. By using soil-less media, a grower eliminates one of the variables that can make production difficult. An added advantage is that these media are lightweight, making their handling less burdensome.

PASTEURIZING THE MEDIA

Greenhouse growing media that contain field soil or artificial media that have been exposed or previously used can contain undesirable microorganisms, insects, and weeds. The elimination of these undesirables is termed *pasteurization*. Unlike sterilization, which kills all life in the soil, pasteurization kills just the harmful elements. Soil pasteurization is preferred to sterilization and, fortunately, is easier to accomplish.

Pasteurization is usually accomplished either with steam or with chemical fumigants. The fumigants are more expensive than steam, which is usually available from the greenhouse heating system. However, some weeds are not killed by the 180° F soil temperature of steam treatment and require fumigation. Small quantities of soil may be pasteurized with electric heat, but it is not a common method for commercial use.

Steam is applied to the bench, pots, or ground beds in different ways:

- from the surface
- through pipes buried in the soil
- in a closed container

For bench soil pasteurization, benches must have evenly spaced openings in the bottom to permit air to escape as steam enters.

Surface Steaming

Surface steaming, like all steaming, is done after the soil is conditioned and of uniform consistency. All nutrients and pH adjustors should be mixed evenly into the soil before steaming except those that would be harmed by the treatment, such as time-release fertilizers formulated with plastic coatings.

The soil is then leveled in the bench and a perforated pipe or hose is placed on top to distribute the steam evenly down the bench. One pipe or hose is used per 3 feet or less of bench width. Next, at intervals of 4 to 5 feet down the center of the bench, inverted flats or concrete blocks are set into place. Finally, a solid cover, such as a canvas tarpaulin or sheet of heavy plastic, is placed over the bench and weighted down around the outside edge. The flats or blocks prevent the cover from adhering to the soil, Figure 20-1.

FIGURE 20-1.
This greenhouse ground bed is being steam pasteurized. A tarpaulin weighted with sandbags prevents the steam from escaping.

Steam is then introduced into the perforated pipe or hose from the boiler. Probe thermometers placed into the medium at intervals around the bench record the soil temperature as the steam permeates it. When the most distant parts of the bench attain a soil temperature of 180° F (or 140° F if air is being injected into the steam) for thirty minutes or longer, pasteurization is complete. The tarp, blocks, and pipes can be removed and the soil permitted to cool for planting.

Buried Pipes

Buried pipes are a preferred method of steaming if ground beds are being pasteurized, although the technique is suitable for benches also. In this method, perforated pipes are placed in the center of the bench and then buried at one-half or more of the soil's depth. The bench is covered and the temperature measured as in surface steaming.

Closed Container Steaming

Closed container steaming is used for pots and other containers, with or without soil in them. It may also be used for tools and other production materials that could be contaminated. A thermometer must be outside the container to permit temperature reading.

Chemical Fumigants

Where growers do not have the facilities for steaming soil, they may use chemical fumigants. The fumigation should be done outside the greenhouse to help avoid injury to greenhouse crops or workers, since many of the products are toxic to plants or humans. Benches can be disinfested with a dilute antiseptic. The fumigants are in two forms: pressurized gas canisters and liquids that turn to gas after application. The pile of soil being fumigated must be covered with a gas-impermeable canopy to hold in the gas and allow it to permeate the soil. The toxicity of some of the products is one of the disadvantages of fumigation. Another

disadvantage is the waiting time required before the soil can be planted without injury to the crop. Up to three weeks may be needed for safety.

The fumigants currently used are chloropicrin (tear gas), methyl bromide, Vapam, and Formalin. Because methyl bromide is odorless and highly toxic, 2 percent chloropicrin is added to provide an odor in the event of leakage. Other combinations of chloropicrin and methyl bromide include both 55–45-percent and 66–34-percent mixtures.

SOIL TESTS

One important item of the greenhouse production schedule is regular soil testing for both bench crops and potted plants. The grower should not wait until plants display symptoms of difficulty before soil testing. Tests should be run monthly on benches with crops. The frequency of testing needed for potted plants will depend upon how long the particular crop takes to reach salable size. For example, potted azaleas require more production time than most herbaceous plants. They also require an acidic pH. Several soil tests are needed during their production to assure optimal growth.

Growers need soil tests for three main reasons:

1. to check the pH
2. to check for nutrient deficiencies
3. to measure the soluble salt content

Soil acidity is important in the production of certain crops, such as the azaleas noted above, and needs to be measured whenever field soil is used in the growing medium or in regions where irrigation water may alter the pH. Testing for the level of nitrates, phosphates, and potash in the growing media is usually sufficient unless another deficiency or excess is suspected. Any of several commercially available soil test kits can provide the grower with same-day results. The do-it-yourself kits are affordable, easy to use, and generally accurate enough to be of use to the grower. More accurate tests can be performed by the state's college of agriculture or commercial labs, but there is a time lag of several days or longer.

Soil samples need to be selected from several

parts of the bench or from random pots of a crop, then mixed, dried, and tested, Figure 20-2. The most reliable results will be obtained if:

1. the same person does the testing each time
2. all glassware is kept clean and rinsed each time with distilled water
3. the reagents are fresh and kept in a cool, dark storage space when not in use. Regularly scheduled reordering of the reagents, rather than waiting until the old ones are gone, will assure their freshness.

Knowledge of the soil's soluble salt levels is considered by many growers and educators to be even more important than pH information for good greenhouse production. The soluble salts of a soil can be increased by steam pasteurization, excessive fertilization, and failure of the irrigation water to drain sufficiently to leach the fertilizers from the soil. Soluble salt level can be measured easily using a Solu Bridge®, Figure 20-3.

Most college courses in soil science or greenhouse production include training in soil testing. All the techniques required for greenhouse production are easily learned. Obtaining a truly representative soil sample and maintaining consistency in the testing are the most difficult aspects.

FIGURE 20-2.
Testing the soil of a greenhouse crop

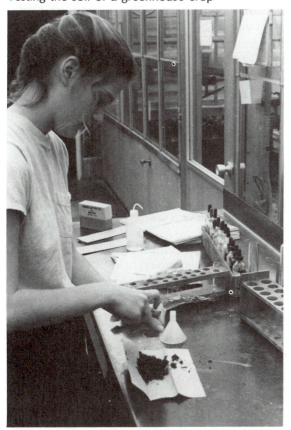

FIGURE 20-3.
A Solu Bridge® (Model SD-B15) is used to test the total soluble salt levels of solutions and media. (Courtesy Beckman Industrial Corporation, Cedar Grove Division)

GREENHOUSE CONTAINERS

The containers selected for use in greenhouse potted plant production vary depending upon the crop and the time of year.

Pots are round containers whose height and diameter are equal. Made of clay or plastic, they are used for production and sale. Made in small sizes of pressed peat moss, they are used for rooting or seeding and transplanting. *Azalea pots* are round containers whose height is three-fourths their diameter. They are made of clay or plastic and are preferred by some growers for the production of poinsettias and chrysanthemums as well as azaleas. *Pans* are clay or plastic containers whose height is one-half their diameter. They are preferred by some growers for bulbous flowers such as tulips and hyacinths.

Clay containers are porous, which plastic is not. Thus gas and air can permeate clay containers, and soil dries more rapidly. This is advantageous during the winter season, when plants can often be too wet in the nonporous plastic containers. The situation reverses in the warmer months, when clay may permit soil to dry out too quickly.

Plastic containers may be either hard or of styrofoam. Either is much lighter than clay, and that is often a reason why plastic is preferred by a grower. Plastic containers cannot be heat pasteurized, though, while clay containers can be heated without fear of their melting. Also, the combination of lightweight container and artificial soil can result in topheavy plants; so clay containers will probably always have their proponents, Figure 20-4.

Peat pots are manufactured either individually or in strips, similar to egg cartons. They range in diameter from 1 1/2 to 4 inches. They are a great convenience for transplanting because there is no need to remove the pot. It is set directly into the new container or garden, where it decomposes. The peat container is fragile, especially after being wet, and will not tolerate much movement. Growers usually place them in flats (which will be described shortly) to provide the support necessary for moving and handling. Peat pots and strips, Figure 20-5, are widely used for the production of bedding plants.

FIGURE 20-4.
Typical greenhouse pot crop containers (left to right): Azalea pots, pots, and pans. All are shown in (left to right) clay, styrofoam, and hard plastic.

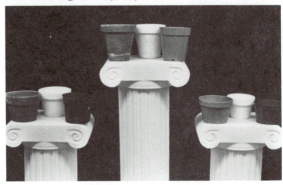

FIGURE 20-5.
Square and round peat pots and production strips

They are also used for the production of lining-out stock (which will be described in Chapter 21).

Molded plastic packs are also used for greenhouse production, especially for bedding plants. They resemble ice-cube trays and are available in different sizes. Since they do not decompose in the soil, plants must be removed for transplanting. They are firmer than peat strips and are salable as they are. They also retain moisture better than peat strips do.

Hanging baskets are specialized production containers made of wire or plastic, Figure 20-6. Both may be solid or meshed. The mesh baskets require a liner to hold the soil and they drip after being watered. The solid baskets require no liner nor do they drip.

Flats are shallow, rectangular containers that may be used to start seedlings, root cuttings, or hold less sturdy peat pots and strips. Traditionally constructed of inexpensive wood, often as an off-season job for greenhouse workers, flats may also be made of plastic. The plastic flats are manufactured to hold a number of the molded plastic packs neatly, Figure 20-7.

Understandably, the size of the containers determines the amount of pasteurized growing medium needed at the time of planting and transplanting. Table 20-1, developed by Dr. P. A. Hammer of Purdue University, can help growers determine their needs in advance.

REPRODUCTION METHODS

Chapter 14 details the many methods used by ornamental horticulturists for reproducing plants. Greenhouse production uses some reproduction methods more than others for particular purposes:

- *seed* for production of bedding plants and geraniums
- *runners* for foliage plants like the spider plant
- *bulbs* for flowering perennials like tulips, daffodils, crocuses, and gladioli
- *layering* for foliage plants like the fig and rubber plant
- *cuttings* for geraniums, chrysanthemums, and many foliage plants
- *grafting* for some azaleas
- *budding* for roses
- *division of the crown* for foliage plants and African violets
- *tissue and organ culturing* for orchids and bromeliads and for production of disease-free stock for chrysanthemums, carnations, and other flowering plants

The greenhouse grower may propagate the crops

FIGURE 20-6.
Examples of greenhouse hanging baskets. The papier-mâché baskets are given added strength by wire. Others are molded plastic. Half baskets are used against the walls.

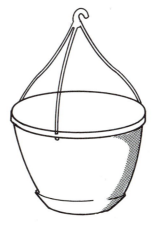
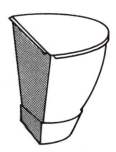

FIGURE 20-7.
Examples of greenhouse flats. The plastic flats are lightweight but strong. The traditional wood flat is sturdy but heavy by comparison.

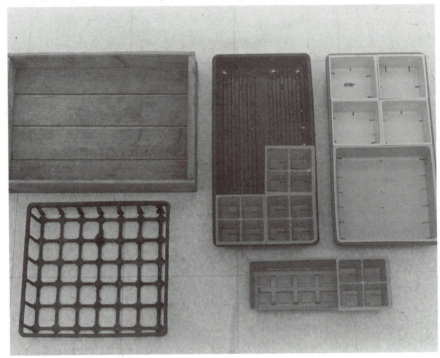

and grow them to maturity or buy cuttings and other propagative stock from firms that specialize in propagation. Of the numerous methods of reproduction used by greenhouse growers, seeds and cuttings are the most common, Figure 20-8. Despite the comparative ease of propagation by cuttings, many species are protected by plant patents, which require new cuttings to be purchased each time from the supplier.

SPACING AND LABELING A CROP

The spacing of a greenhouse crop has a direct impact on the costs of production and the quality of the crop. Closer spacing permits more plants to be grown, but if it is too close it reduces the quality of the product. Spacing that is too great produces excellent plants but at the expense of crop numbers and profit.

Different methods of growing require different techniques of spacing. With bench crop production, the spacing decision made at the time of planting remains unchanged throughout the life of the crop. Potted crop production can be the same, with no movement of the pots after planting (termed *fixed spacing*); or the crop can be started in small pots with close spacing, and both container size and spacing can be increased as the crop matures (termed *expanding spacing*). For example, the crop may be started in 3-inch pots and advance through several transplantings to mature size in 6-inch pots. Each transplanting requires labor and more space. The space can only be provided by moving out some other crop. When certain holidays fall close together, such as Valentine's Day and Easter or Easter and Mother's Day, there may not be enough time to space out the next crop before damage is done due to close spacing. Expanding spacing is com-

TABLE 20-1.
Number of Pots Filled with One Cubic Yard of Medium (Compiled by Dr. P. A. Hammer, Purdue University)

Container (Diameter in Inches)	Number Filled Per Cubic Yard
Standard round pot	
2 1/4	6,952
2 1/2	4,840
3	2,948
4	1,276
5	704
6	396
7	242
8	176
12	44
Azalea pot	
4	1,540
5	770
5 1/2	648
6	440
Bulb pan	
5	1,100
6	858
Hanging baskets	
8	324
10	135
Flats	
11 1/2 x 21 1/4 x 2 1/4	85

FIGURE 20-8.
A greenhouse worker unpacks rooted mum cuttings for continued growth as a potted crop.

monly practiced, though, and probably will continue to be. Expanding and fixed spacing are compared and summarized in Table 20-2.

Whether spacing bench crops or potted plants, the objectives are the same. The plants must receive adequate sunlight and air to permit photosynthesis to proceed unimpaired. Crowded spacing can reduce photosynthesis and increase the possibility of disease due to shaded, overly moist foliage and reduced air circulation. Where expanding spacing is used, the dates of repotting and spacing must be part of the production schedule for each crop. Consider-

ation must also be given to future crops. As the current crop reaches maturity, its time of sale must be anticipated to permit the wider spacing of another crop.

Labeling the crop is the means of identifying the plants on the bench. It is necessary where different varieties of similar plants are grown, since they often cannot be identified otherwise until they flower. Chrysanthemums and geraniums are two common examples, but exotic orchids are just as easily confused when only the foliage is visible.

Bench crops are often labeled on the end of the

TABLE 20-2.
A Comparison of Expanding and Fixed Spacing

Expanding Spacing	Fixed Spacing
• Plants are transplanted one or more times.	• Plants are not transplanted.
• Several production containers are required per plant.	• A single production container is required per plant.
• More labor is required.	• Less labor is required
• Less bench space is required at the beginning. Other crops in later stages of maturity can also be grown.	• Greater bench space is required at the beginning, which may limit the number of other plants grown annually.
• Roots often develop better when pots are not larger than the plant needs.	• Early root development may be limited with small plants in large pots.
• Small pots dry out more quickly.	• Large pots do not dry out as quickly.
• Plants can be damaged if repotting or increased spacing are delayed.	• Certain crops, such as bulbous flowers, do not transplant well and require fixed spacing.

FIGURE 20-9.
Two methods of labeling crops on a greenhouse bench: A bench crop is identified on the outside end. A pot crop is identified on the pot closest to the front and left side.

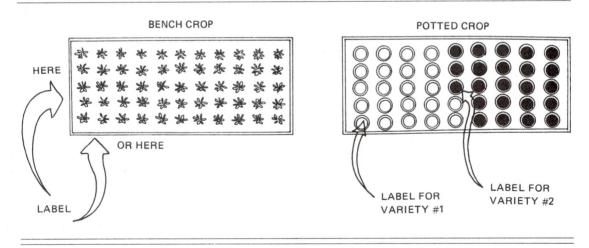

bench when a single variety is being grown. Potted crops may be labeled on each container, especially if the crop is protected by a patent. These labels are furnished by the propagator supplier. Other potted crops are not individually labeled. Instead the pot closest to the front and the left side of the bench is labeled. All plants spaced behind and to the right are the same variety until the next label marks the next variety, Figure 20-9. In addition to the variety name, the label may contain information such as the flower color, date of planting, and date of harvest.

WATERING

Greenhouse crops are watered either by hand with a hose or with semiautomatic watering systems. In large greenhouse ranges, hand watering is too time-consuming, and automated systems are necessary.

Watering done correctly becomes *irrigation,* the maintenance of a proper balance of both moisture and air in the soil of the crops. Too much water leaves too little air space, and the root system may rot or at least be dwarfed. Too little water does not sufficiently drive old air out through the bottom and draw fresh air in from the top. Too little water can also cause shallow root development, soluble salt buildup, and reduced plant and blossom size.

The amount of water applied to greenhouse plants is totally under the control of the grower. Nature can be blamed for a lot that goes wrong in a greenhouse but not a water excess or deficiency. Correct irrigation techniques are not the same for all crops, nor are they uniform for a single crop throughout the year. The first and perhaps most difficult thing that a new greenhouse worker must be taught is how to water plants correctly. Proper irrigation can be encapsulated as follows:

- Sufficient water should be applied each time to drain from the bench or pot slightly but not excessively.
- Potted plants must have room at the top of the container for water. The soil level in each pot must be the same.
- Surface drying of the soil between irrigations is desirable. It permits new air to enter the soil.
- Clay containers require more frequent watering than plastic containers.
- Larger plants require more water than smaller plants.
- Artificial media require more frequent watering than all-peat or field soil media.
- Less irrigation is needed during the winter and on cloudy days than in the summer and on sunny days.
- The need for water should not be determined by sticking a finger into the soil if a systemic pesticide has been used on the crop. The pesticide may be toxic and absorbed through the skin.
- Potted plants and benches do not dry out uniformly. Usually the soil at the edge of a bench or the pots along the edges dry sooner than those at the center of the bench. Also, plants nearest the heating system or ventilators can be expected to need water sooner than other plants. The determination to water must be based upon the needs of plants away from these early drying locations. Even semiautomated irrigation systems may require hand watering of crops in the early drying locations.
- Several kinds of moisture meters are available to help determine when to water and how much to apply. They are not foolproof and should be regarded as guides only.
- Fertilizers and pesticides can be applied as liquids through the irrigation system. To do so saves time.
- Water should be applied at the base of the plants. Wetting the foliage should be avoided.
- Watering should be completed early enough so that surface water can dry before nightfall.

Hand Watering

Greenhouse hoses are usually 3/4 of an inch in diameter and flexible for easy handling. They must either extend the full length of the benches they are to water or at least reach to the center if the water supply is at both ends of the greenhouse. An assortment of nozzles is manufactured for greenhouse use. Some are *breakers,* which diminish the force of the water but not the rate of flow. Others are *foggers,* which permit the water to be applied as a fine mist to seedlings and other tender plants. Water should never be applied at a rate of flow that furrows the soil in the bench or the container. To do so creates a condition favoring uneven water absorption thereafter.

When watering, a hose should first be stretched down the aisle to its full length. The crop should then be irrigated while moving with the hose back toward the faucet. Two benches may be watered from one aisle. A pattern of watering should be established and followed so that the technique becomes almost habitual. For example, the nozzle and hose can be extended between rows of pots and both rows irrigated from the rear of the bench forward. Regardless of the pattern, each pot should be filled to the top with water as a means of assuring that each container receives the same amount each time. The technique is imprecise but the best possible with hand application, Figure 20-10. Hand watering is more effective with potted crops than bench crops. The rate of application to a bench can be very inconsistent. *NOTE:* To avoid possible contamination by pathogens, the hose end should rest in a clip between uses. It should not be allowed to fall on the ground.

Semiautomatic Watering

Semiautomatic irrigation systems can be time-saving for large greenhouse operations. Most are designed for bench production, but systems for potted plant production also exist. The systems are designed to apply a standard amount of water to a crop. By linking the system to appropriate controls, several benches or crops can be watered in succession. However, the grower must still determine the frequency of irrigation and the amount of water to be applied. The system must be checked frequently to assure that no nozzles are clogged and that no leaks have developed in the pipes or tubes.

The semiautomatic systems use plastic (PVC) pipes, tubing, joints, and nozzles. They may be either directed spray systems or trickle systems, termed so by the manner in which the water is delivered. *Spray systems* deliver the water through nozzles spaced at intervals along the pipes, which run around the perimeter of the greenhouse bench. *Trickle systems* deliver the water through holes in inflatable plastic tubes stretched down the bench, Figure 20-11. A form of trickle system is used for potted plants. Nicknamed the spaghetti system, it delivers water through thin plastic tubes attached by pins or weights to the pot on one end and to a

FIGURE 20-10.
Potted gerbera are being watered by hand. A breaker is used to reduce the intensity of the water pressure. The grower holds the hose away from the plants to prevent damage.

FIGURE 20-11.
A semiautomatic irrigation system used in a ground bed for watering roses. Note the control valve, the pipes along the side of the bed, and the ooze tubes crossing the bed at intervals.

3/4-inch PVC pipe on the other end. The PVC pipe runs the length of the bench to transport the water, Figure 20-12.

Not quite fitting into the category of trickle systems is the *capillary mat,* which is most suitable for the production of potted plants that should not have wet foliage. This mat is made of fibrous material and is placed on a bench that is first lined with plastic, Figure 20-13. Water and nutrient solutions periodically flood the mat and are then absorbed by the plant through capillary action. The mat system is also useful for the production of plants in pots of various sizes. The system has the disadvantages of becoming unsightly due to algal growth and harboring insects and pathogens within the matting. Periodic sterilization of the mat and bench is needed.

FIGURE 20-12.
A trickle system delivers water to each hanging basket on this production rack. The 3/4-inch PVC pipe is attached to the top of the rack. As seen here, the nickname "spaghetti system" is understandable.

FIGURE 20-13.
Capillary mat showing the bench and the plastic liner.

FERTILIZING

Fertilizers can be applied to greenhouse crops in dry or liquid form. If applied in dry form, the fertilizer may be incorporated as the soil is mixed and before it is pasteurized unless a slow-release fertilizer is used. In that case the fertilizer can be incorporated after pasteurization. Nutrients can be replenished later by supplemental applications of dry fertilizers to the surface of the benches.

Dry fertilizers require time to apply, especially if slow-release forms are *not* used. For potted crops, repeated application of dry fertilizer adds to labor costs. In addition, slow-release dry fertilizer is believed to be used more efficiently by the plant with less loss through leaching than fertilizer applied in liquid form.

The popular alternative to dry fertilizer is liquid fertilizer, dissolved in water and applied as the crop is irrigated. Such methods save on labor but use more fertilizer overall.

The greenhouse watering system is the most common means of applying liquid fertilizer to crops. Three types of systems are used: the venturi proportioner, the positive displacement pump, and the eductor. In each case, concentrated liquid fertilizer is held in a container and injected into the greenhouse irrigation system when water is used or when fertilization is desired.

- The *venturi proportioner* draws the fertilizer into the water system through a small tube that connects the water hose to the fertilizer concentrate. As the water pressure and flow rate vary, so does the rate at which the concentrate mixes with the water. The venturi proportioner is not precise but is serviceable for many crops.
- The *positive displacement pump* may be water driven or use an electric motor to power a fertilizer concentrate pump. It maintains a uniform proportion of fertilizer to water regardless of the water pressure or rate of flow. It is usually not adjustable to permit a change of the injection ratio.
- The *eductor* places the fertilizer concentrate in a pliable bag surrounded by water under pressure. As the pressure increases, it forces the fertilizer through a metering system and into the water line. The eductor system is a flexible one that does permit the injection ratio to be varied. Under normal operating conditions, it maintains a constant proportion of concentrate to water.

These fertilization systems must be checked for accuracy at least once each year and should be one of the first things checked if a crop begins to show signs of trouble. The choice of a system is usually based upon the size of the operation, whether the injection ratios will be standard or must be varied, whether the system will be mobile or permanently installed, and how easily a particular system can be serviced when necessary.

Fertilizers for the injectors can be mixed by the grower or purchased premixed from suppliers. Either way, the fertilizers must be easily soluble in water and contain a dye that will be visible in the water supply even when highly diluted to provide a visual check that the system is working.

The amount of fertilizer applied per application depends on the strength of the concentrate, the dilution ratio of the injector, and the amount of water applied. Most commonly, a ratio of 1:100 (one part concentrate to ninety-nine parts water) is used, but 1:200 is not uncommon. If fertilizer is only applied every few weeks, the concentrate must be stronger. If applied with each watering, the concentrate must be weaker.

Measuring Greenhouse Fertilizers

Greenhouse fertilizers are measured by the number of parts of nutrients to a million parts of water, or *ppm* (parts per million). Most soil test reports are expressed in parts per million. So too are most fertilizer recommendations for greenhouse crops.

To determine ppm, multiply the percent of any element in a fertilizer by 75. The answer will be the ppm of 1 ounce of the fertilizer in 100 gallons of water. For example, a 20-5-30 fertilizer is 20 percent nitrogen. In a solution of 1 ounce of 20-5-30 fertilizer in 100 gallons of water, the ppm of nitrogen is .20 × 75, or 15.

To make a 100 ppm solution, the number of ounces of fertilizer required per 100 gallons of water must be calculated. To do so, divide 100 by the ppm in 1 ounce. Continuing the example above, 100 ÷ 15 = 6 2/3 ounces.

GROWTH RETARDANTS

Some greenhouse plants get too tall to make desirable potted plants. Poinsettias, Easter lilies, chrysanthemums, and some hydrangeas can all get leggy if started too early or subjected to overly warm temperatures during their periods of most rapid vegetative growth. The application of growth retardants to the plants will help to regulate their size. For a listing of the most commonly used florist crop growth retardants, their mode of action, and the crops they are effective with, refer to Chapter 4. The use of growth retardants and the time of their application must be included in the production schedule for each crop.

PEST CONTROL

Greenhouses provide an ideal environment for plants and pests alike. The conditions of high temperature and high humidity coupled with close spacing of a monoculture make pest control a necessary element in the greenhouse production schedule. Also necessary is employees' cooperation and understanding of how they can personally assist in the control effort. Pest control should be included in the job training for all greenhouse workers.

Chapter 6 describes the pests of ornamental plants: insects, infectious parasitic pathogens, weeds, and other injurious agents. It also describes the profitable control level sought by growers and the four principles of control. Efforts to prevent pests from becoming established in the greenhouse apply the principle of *exclusion*. For example, sources for plant materials that guarantee disease-free and insect-free stock are essential. *Eradication*, another principle of control, seeks to remove or eliminate pests that are already present in the greenhouse. Destruction of infected or infested plants, elimination of alternate host weeds beneath benches, soil pasteurization, and use of chemical sprays all apply the principle of eradication. *Protection* is the third principle of control. It seeks to place a barrier between the crop and the pests. Screens on greenhouse vents and doors and growth conditions that favor the hosts more than the pests are methods of protection. Spraying plants with chemicals before insects or pathogens arrive also applies the principle of protection. The principle of *resistance* is applied each time a grower selects a resistant variety for production.

Pathogenic inoculum, insects, and weed seeds enter the greenhouse in numerous ways. Insect eggs, viruses, fungal spores, or bacteria can be on cuttings or stock plants purchased from another grower. *Oxalis*, other weed seeds, and nematodes can be in the soil of plants purchased for forcing. Pests may be carried in by the wind, on equipment, or on the shoes and clothing of greenhouse workers. In retail operations, every customer who comes through the door is a potential source of inoculum of many common greenhouse pests.

Once inside the greenhouse, pests are disseminated by irrigation water, air currents, workers, hoses, and other contaminated tools.

Pest control in greenhouses is complicated by two factors:

1. the complex life cycle of many insects and some fungi that allows them to be unaffected by many chemical pesticides throughout much of their lives
2. the rapid development of resistance to particular pesticides that characterizes some pests, especially insects

For example, many insecticides are only effective against the adult stages of an insect. The larval, nymph, and egg stages may not be affected at all by a fumigant or spray. Hence, two or three days after spraying, the crop can again be overrun. With insects such as whiteflies, an insecticide may have to be applied every two or three days for a month to control a succession of adult populations before they are able to lay eggs.

The problem of natural resistance develops when the same pesticide is used repeatedly. No spray, dust, or fumigant kills 100 percent of a pest population. A few are always unaffected. As they survive and breed, a pest population builds within the greenhouse, immune to the pesticide that once controlled effectively. To avoid the buildup of resistant pest populations, a grower must rotate the choice of pesticides frequently. This should be part of the production schedule for each crop.

Common Forms of Greenhouse Pesticides

Depending upon the crop, the size of the operation, the pesticide being used, and the equipment available, greenhouse growers apply chemical pesticides in one or more of the following forms:

• *Systemics:* These may be applied as soil

drenches or as foliar sprays. They remain in the plant and kill the pest when it arrives.

- *Sprays:* The material is mixed with water in a hydraulic sprayer and applied to the foliage of the crop. A spreader/sticker agent may be needed to provide uniform coverage of the foliage. All parts of the plant must be covered or the untouched insects will remain alive. Careless or incomplete spraying accomplishes little more than temporary population reduction.
- *Dusts:* Coverage must be thorough as with sprays. A duster can cover a large area of crops quickly and often at less expense than sprays. Many greenhouse pesticides are not available in dust form, however.
- *Aerosol bombs:* The pesticide material is packaged in a container with a propellant liquid or gas that disperses the pesticide uniformly through the greenhouse. The grower must calculate the cubic area of the greenhouse before use to assure that the proper amount of pesticide is applied. Aerosols are quick and easy to apply but expensive.
- *Smoke fumigants:* The pesticide is packaged with a flammable, smoke-producing material. When the container is lit, the smoke carries the pesticide throughout the greenhouse and to all parts of the crop. As with aerosols, the cubic area of the greenhouse must be calculated for correct application.
- *Steam line vaporizers:* The pesticide is painted onto cold steam lines, then all vents of the greenhouse are closed and the steam turned on. The chemical vaporizes and fills the house. Cubic areas must be calculated.
- *Foggers:* The grower must mix the pesticide in an oil solvent and fill the fogger. The fogger then heats the mixture, causing it to disperse uniformly. The oil can plug the machines, so thorough cleaning is important.
- *Misters:* The pesticide is mixed with a solvent that evaporates quickly once out of the mister. An air blast disperses the mixture in a fine mist.

Sprays and dusts often leave a residue of pesticide on the plant leaves that may prove objectionable to consumers. Certain systemics can remain invisible yet toxic in the soil for many weeks. Their application must be discontinued far enough in advance of their sale to assure that the plant product is safe for release to consumers. Growers who produce vegetables must take special care that no pesticides will remain active in the plant, since it will eventually be consumed by humans.

Specific recommendations for greenhouse pest control are available from state colleges of agriculture and Cooperative Extension specialists. Many states have active florist trade associations also that share common problems and solutions through newsletters and statewide meetings. It is important for growers to take part in these associations.

Pesticide Safety in the Greenhouse

The safe application of pesticides in the greenhouse or elsewhere requires an understanding that:

- The health of the plant is more important than killing the pest.
- The health of the greenhouse staff and the consumer are more important than either the crops or the pests.

Sometimes priorities can get confused, and risks are taken to produce a marketable product regardless of the dangers. Other times, safety procedures are bypassed for reasons of expediency or personal comfort. But *people are more important than plants or pests* is the first rule of safe pesticide use.

For a review of pesticide safety, see Chapter 6. Safety rules should be posted where all staff and supervisors can see them frequently. The rules should be enforced without exception. Where the level of toxicity is high, as indicated by the skull and crossbones on the label, workers should wear protective equipment during all contact with the material (mixing, application, and cleanup). This equipment includes rubber boots, gloves, pants, hat, raincoat, face shield, and gas mask for the application of a highly toxic spray, Figure 20-14. Too often,

FIGURE 20-14.

Safe application of pesticides in the greenhouse requires the use of protective clothing, gloves, boots, and respirators.

in hot weather, workers want to spray without donning the rubber suits. Such a practice should not be condoned by the supervisor. Either a less toxic pesticide should be selected, or a fumigant or similar form used that eliminates the danger to the applicator.

Another danger of pesticide use is that people will reenter the greenhouse too soon after the pesticide has been applied. The reentry time, in hours or days, is specified on the pesticide label and should be adhered to strictly. After spraying, a warning sign should be posted on the greenhouse doors and the reentry time noted. The greenhouse should be aired as much as the weather permits following pesticide application.

If systemic pesticides are used for potted plants, workers must be careful not to touch the soil in the pots. When expanding spacing or disbudding or staking are required, it is nearly impossible to avoid coming in contact with the soil. In such cases workers should wear plastic or rubber gloves. This is another situation where workers will object. The

gloves are hot and hands perspire quickly. Safety, not personal comfort, must be given priority and the rule enforced.

Integrated Pest Management

As noted in Chapter 6, the most current approach to pest control in greenhouses is integrated pest management. IPM is a step back from total reliance upon pesticides to play the lead role in profitable control. As the number of approved pesticides diminishes and the resistance of insects and pathogens increases, the chemical control of pests is declining in its effectiveness. Also, as concern for the natural environment grows, pesticides are no longer blindly accepted by everyone. Many growers share the concern of non-horticulturists about the excessive, indiscriminate use of pesticides in the past.

Integrated pest management in greenhouses uses temperature and humidity control, resistant varieties, screens and other entry barriers, monitoring devices, predatory insects, and pesticides to keep pest problems below an economic damage level. Pesticides are applied only when the pest is most vulnerable and then only where needed. They are not necessarily broadcast through the entire greenhouse.

IPM is costly in terms of the time required to monitor insect and disease development within a crop, as well as the expense of the predatory inoculum. Both costs are ongoing. Greenhouse crops may require careful, methodical inspection twice each week. That can include looking under leaves and/or lifting the pots of sampled plants to find insects in their hiding places. Yellow sticky cards, placed at regular intervals throughout the crop, can help the grower survey the type of insects present in the greenhouse and their stages of development, Figure 20-15. The cards are changed weekly. All data concerning the type and number of insects are recorded, along with observations of disease symptoms and the areas where the pests are located within the crop. Such data permit the grower to select the best and most cost-effective means of control, while targeting the area(s) of the crop where it should be applied.

FIGURE 20-15.
Checking the yellow sticky card regularly helps a grower monitor insect numbers and stages of development.

To incorporate fully a program of IPM in a greenhouse necessitates the following:

- elimination of weeds throughout the greenhouse
- a program of sanitation throughout the structure to include bench and/or container pasteurization
- thorough inspection of all plant materials before bringing them into the greenhouse
- screening of all vents, doors, and other openings to the outdoors
- regular, ongoing scouting to identify, quantify, and record the presence and location of pests within the crop
- control of the greenhouse environment to favor the crop rather than the pests
- application of the principles of control to include biological control methods that can reduce the amount and frequency of chemical pesticide application

Integrated pest management requires a heavy commitment of time in monitoring the crop and the recurring expense of purchasing and releasing beneficial predators into the greenhouse environment. However, the savings in pesticide costs and the reduced occurrence of pesticide resistance and chemical injury to crops can offset that expense.

SELECTING CROPS TO GROW

The high cost of producing certain plant species in northern regions during the winter has driven some greenhouse operations out of business and drastically altered the selection of crops grown by others. In addition, increasing competition from growers in other countries and the desire by retail florists to use more uncommon flower species in their arrangements have caused an overhaul of greenhouse crop choices in all parts of this country. The industry is definitely in a state of transition. Growers who once raised extra quantities of bedding plants, chrysanthemums, and poinsettias on the chance that they might sell and because the space was available, are now growing for contracts only and shutting down all or part of their operation until time to start the next contracted crop. Other growers are changing from warm crops (mums, roses, poinsettias) to cold crops (carnations, cyclamen, kalanchoe) to reduce their energy costs.

Perhaps motivated by increased energy costs, yet still to the long-term benefit of both grower and the consumer, growers are trying new plant products. Crops formerly grown only as bench crops, such as carnations and snapdragons, are now being produced as potted plants because they are cold crops. Lilies once grown only in the garden are finding their way into greenhouse production and flower shops.

The future will probably see more varieties in the northern greenhouse, not fewer, and they will replace some old favorites, now too costly to produce except in warmer regions. The new varieties can be predicted to require lower temperatures for production, less light, and perhaps greater manipulation of photoperiodic response by the grower.

Table 20-3 summarizes the production requirements of some common greenhouse crops. It is neither a complete listing of all greenhouse crops nor a comprehensive guide to the culture of the plants. It is intended merely as an introduction.

SUMMARY

To serve the consumer market when demand is greatest and minimize losses due to over- or under-production, greenhouse growers plant, care for, and harvest their crops in accordance with production schedules. Accurate records aid in determining whether to increase or decrease the production quantities of previous years. Production schedules are frequently altered by changes in the weather and require the immediate attention of the growers to get the crops back on schedule.

Greenhouse soils differ greatly from field soils. They often are totally artificial. To be most useful to the greenhouse grower, the growing media must be predictable. Predictability is the major advantage of the artificial media.

The growing media are pasteurized to eliminate undesirable microorganisms, insects, or weeds. Pasteurization is usually accomplished with steam applied from the surface, from pipes buried in the soil, or in closed containers. Chemical pasteurization is also possible using chemical fumigants.

Regular soil testing is an important part of a greenhouse production schedule. Soil tests are needed for three reasons: to check the pH, to check for nutrient deficiencies, and to measure the soluble salt content. Soil test kits are available for personal use, or tests can be performed by the state's college of agriculture or commercial laboratories.

The containers used for greenhouse potted crop production vary depending upon the crop being grown and the time of year. They include pots, pans, and azalea pots in clay, hard plastic, or styrofoam; peat pots and strips; molded plastic packs; hanging baskets; and flats.

Greenhouse crops are reproduced from seeds, runners, bulbs, and cuttings; and by layering, grafting, budding, division of the crown, and tissue/organ culturing. Propagation of the stock may be done by specialized propagation firms or by the greenhouse growers who will grow it to maturity.

Different methods of growing require different techniques of spacing. Both expanding spacing and fixed spacing have their uses depending upon the crop and the production schedule. Labeling methods vary, too, and are important for the identification of cultivars, flower color, date of planting, and date of harvest.

Irrigation is the maintenance of a proper balance of moisture and air in the growing media. Irrigation techniques are not the same for all crops, nor are they uniform for a single crop throughout the year. Both hand watering and semiautomatic watering systems have their place. The watering system can also be used to apply fertilizer to the crops via injection devices. An alternative is the use of fertilizer in dry form.

Since greenhouses provide a near-optimum growing environment for plants, it is predictable that plant pests can flourish as well. Thus pest control must be a regular part of the greenhouse production schedule. The principles of control (exclusion, eradication, protection, and resistance) must be applied continuously. Specific recommendations for greenhouse pest control are available from state colleges of agriculture and the local Cooperative Extension Service. All chemical pesticides must be applied in accordance with strict safety standards.

Greenhouse crops are being grown more cautiously than in past years. High fuel costs, stronger competition from sources worldwide, and an increasing demand for more unusual flowers have caused more growers to seek contractual agreements before planting and to change the crops they grow to reduce energy costs and satisfy consumer preferences.

TABLE 20-3.

A Comparison of Selected Floriculture Crops

Crop	Pot or Bench Crop	Day Temperature	Night Temperature	Method of Propagation	Production Time Required
Azalea	Benched for propagation; potted for sale	70°–75° F before refrigeration; 60° F after refrigeration	60°–65° F	Cuttings to propagate; most purchased as rooted liners or precooled and partially grown for forcing	Up to two years; refrigeration needed for four to eight weeks at 50° F to mature flower buds; also need twelve hours of light each day; four to six weeks needed in greenhouse to force the flowers
Bedding plants	Molded plastic packs, clay or peat pots	60°–65° F	70°F to reduce stretching; then 40°–45°F to harden-off for sale	Seed	Approximately three months depending upon the species
Begonia, wax	Molded plastic packs, clay or peat pots	60° F	60° F	Seed	Four to six months
Calceolaria	Potted	50°–65° F	50°–60° F	Seed	Twenty-five weeks total; five weeks for seedling production at 75° F; five weeks in flats at 70° F; fifteen weeks in pots at 50° F
Carnation	Traditionally a benched crop; recently tried as a pot crop	55°–60° F	50° F	Cuttings to propagate; often purchased as rooted cuttings to avoid virus disease problems	Eight to nine months from planting to flowering
Chrysanthemum	Benched	60°–65° F	60° F	Cuttings to propagate; often purchased as rooted cuttings to avoid virus disease problems	Twelve to seventeen weeks depending on the season

Spacing	Shading	Growth Retardant	Pinching	Disbudding	Special Requirements
Fixed at 10x10 inches for forcing	Some cultivars initiate buds best under short-day conditions.	B-Nine or Cycocel for flower bud initiation	Used to increase flowering shoots during first year; shearing needed to shape plants for sale during second year.	Buds are not removed, but wild growth around flower buds should be removed.	An acidic soil is needed with a pH of 5.0 to 5.5. Azaleas are grown in pure sphagnum moss or a mix of sphagnum and sand. The soil should be low in fertility, but iron and manganese must be added.
One plant per pack cell or pot.	Unnecessary	Varies with species	Sometimes done to encourage branching and to prevent legginess in plants before sale.	Unnecessary	New seedings must be kept moist and not allowed to overheat. No drying is allowable after germination. If not direct seeded, they should be transplanted as soon as possible. The transplant hole is made with a dibble for pots and a dibble board for molded plastic packs. Also can be automated. High light intensity improves plant quality.
Expanding	Unnecessary	Unnecessary	Unnecessary	Unnecessary	
Expanding to 4x4 inches	Heavy shade needed in summer; none in winter	Cycocel needed for some cultivars	Unnecessary	Unnecessary	Seedlings are transplanted to flats for further growth and finally to pots. Calcelaria is a long-day plant. Extended days in the winter can promote flowering.
Fixed at 6x8 inches	Unnecessary	Unnecessary	Once or twice; second pinch not as extensive	Side buds are removed as soon as they form.	The plants must be supported by wires as they grow. The wires are raised as the plants grow. Pinching should occur at the first nodes to elongate. When harvesting, cut stems at the first internode to elongate. Flowers should be opening before harvest.
Fixed at 7x8 inches (standards); 6x7 inches (pompons)	Mums set flower buds when given twelve hours or less light each day. Artificial light used to prevent flowering at certain seasons and stages of growth; shade cloth used to induce flowering at other stages and seasons.	Unnecessary	Multiple-stem plants are pinched one to three weeks after planting. Single-stem plants are not pinched.	Spray mums are not disbudded. Single-blossom mums are disbudded as soon as the side buds form.	The plants must be supported by wires as they grow. The wires are raised as the plants grow. Lights must be used from August to May to provide long days. Black cloth must be used from mid-March to August to provide short days. During the summer, high temperatures beneath cloth should be vented to prevent flower abortion.

TABLE 20-3.

A Comparison of Selected Floriculture Crops (continued)

Crop	Pot or Bench Crop	Day Temperature	Night Temperature	Method of Propagation	Production Time Required
Chrysanthemum	Potted	62°–63° F	62°–63° F	Same as benched chrysanthemum	Same as benched chrysanthemum
Cyclamen	Potted	55°–65° F	50°–55° F	Seeds to propagate; often purchased as partially grown plants to reduce length of time in greenhouse	Twelve to fifteen months if started from seed
Daffodil	Potted	55°–63° F to force	55°–60° F to force	Bulbs	Up to twenty weeks required; precooling for four to five weeks; eight to twelve weeks for rooting; ten days to three weeks for flowering after rooting
Foliage plants	Potted	80° F	70° F for production; 60° F prior to sale	Seed and assorted asexual methods depending on species; stem and leaf cuttings common	Four weeks or more after rooting and transplanting
Geranium	Potted	65°–70° F	60° F	Seeds and cuttings; often purchased as rooted cuttings	Eight to ten weeks from cuttings; sixteen to eighteen weeks from seeds
Gloxinia	Potted	70°–75° F	65°–70° F	Seed	Five to seven months

Spacing	Shading	Growth Retardant	Pinching	Disbudding	Special Requirements
Expanding to 15x15 inches	Same as benched chrysanthemums	B-Nine or Phosfon or A-Rest	Plants are given a soft pinch to promote at least three breaks from each cutting.	Remove side buds to allow one bud per stem as soon as buds form.	Cuttings are planted at an outward-directed angle (about 45°) to create a fuller and more shrub-like plant. There are usually five cuttings per 6-inch pot.
Expanding	Unnecessary	Unnecessary	Unnecessary	Unnecessary	To promote flowering, the plants must be a bit pot-bound and the soil kept on the dry side. They will flower about ten weeks after their final potting. Avoid high fertility.
Fixed with pots touching	Unnecessary	Unnecessary	Unnecessary	Unnecessary	Bulbs may be precooled at 48° F. Storage may be outdoors or in controlled refrigeration. If not precooled before planting, time must be provided during the rooting period. Temperatures are gradually reduced during the rooting period from 48° F to 32° F. Bulb suppliers can provide specific schedules.
Expanding	Most foliage plants require partial shading.	Unnecessary	Usually unnecessary	Unnecessary	Accelerated growth is possible with high humidity, heavy watering, and frequent fertilization. The plants must be hardened-off before sale.
Expanding to allow good ventilation between plants; crowding promotes disease	Light shade needed to prevent blossom burn	Unnecessary	Stock plants pinched to promote shoots for cuttings; production plants given a hard pinch about four weeks after potting and a soft pinch four weeks later	Unnecessary	Fans in the greenhouse will keep disease down. Flowers should be picked off while the plants are in production.
Expanding to 12x12 inches	Saran or light greenhouse shading compound	B-Nine or Cycocel	Unnecessary	When the plant becomes well budded, the first two buds to show color should be removed for a more uniform plant.	Plants should be watered carefully to avoid wetting the crown. Flower formation is diminished if the crown remains wet overnight.

TABLE 20-3.

A Comparison of Selected Floriculture Crops (continued)

Crop	Pot or Bench Crop	Day Temperature	Night Temperature	Method of Propagation	Production Time Required
Hydrangea	Potted	60°–65° F (greenhouse)	60° F (greenhouse)	Cuttings	One year; cuttings need three weeks to root; grown outdoors during the summer; placed into cold storage until forcing in midwinter (50° F); fifteen weeks needed to force
Lily, Easter	Potted	60°–65° F	60° F	Bulbs	Twenty weeks; six weeks of cooling (45° F) either before or after potting; fourteen weeks needed for forcing
Poinsettia	Potted	80°–85° F	60°–64° F	Cuttings; often purchased as rooted, unrooted, or callused cuttings	Three weeks to root cuttings; about eleven weeks from potting to sales
Rose	Bench	65°–70° F	60° F	Budding; grafting; cuttings	Five to eight weeks for a flower to develop after a stem is pinched
Snapdragon	Bench	60°–65° F	50° F	Seeds	Fifteen weeks for single-stem crops; nineteen weeks for pinched crops
Tulip	Potted	68° F	60°–63° F	Bulbs	Fourteen to eighteen weeks for precooling and root formation; about three weeks for flowering

Spacing	Shading	Growth Retardant	Pinching	Disbudding	Special Requirements
Expanding to 14x14 inches	Outdoor growth best under partial shade	B-Nine or Gibberellic Acid or Ancymidol	A hard pinch in mid-summer	Unnecessary	Heavy irrigation and high humidity are needed for best outdoor growth. Aluminum sulfate is used to turn the pink flower color to blue. Soil should be slightly acidic.
Fixed with 7 inches between pot centers	Unnecessary	A-Rest	Unnecessary	Unnecessary	Greenhouse temperatures may need to be altered as the time of sale approaches to accelerate or slow the time of bloom. Tallest plants must be kept in the center of the bench and all plants must be rotated to prevent stem curvature.
Expanding to 15x15 inches	Black shade cloth may be necessary from October 1 to 21 to assure a fourteen-hour dark period for bud set. Before that, a two-hour lighting in midnight will keep the plants vegetative.	Cycocel or A-Rest	Required in second week after potting if multi-flowered plants desired; four or five nodes should be left below the pinch.	Unnecessary	Molybdenum deficiency is common and must be anticipated and offset by the fertilizer formulation.
Fixed at 12 inches between centers	Greenhouse shading compound needed during summer	Unnecessary	Young plants pinched to promote shoot development; mature plants pinched to promote flower development; all pinches above a five-leaflet leaf	Unnecessary	Plants must be supported in the bench. Mulch must be applied during the summer. Roses must be cut twice a day before they open. Flowers are kept in a cooler (45° F) in water until grading and sales.
Fixed at 4x5 inches for single-stem crops and 7x8 inches for pinched crops	Greenhouse shading compound needed during summer	Unnecessary	Only pinched if multistemmed plants desired	Unnecessary	Plants must be supported in the bench.
Fixed with pots touching	Cover with newspaper during first few days of forcing to stretch flower stems.	A-Rest for certain cultivars	Unnecessary	Unnecessary	Bulbs are placed into pots or pans with tips even with the rim. A reservoir for water should be left in each container.

Achievement Review

A. TRUE/FALSE

Indicate if the following statements are true or false.

1. Floral crops are seasonal in their appeal to consumers.
2. Certain consumer demands can be anticipated by a grower.
3. Crop production scheduling is a total guessing game each year.
4. The environment outside the greenhouse has no appreciable influence on the crop production schedule.
5. Economic trends in the local market area influence the crop production schedule.
6. Irrigation should be sufficient each time to allow some water to drain from the pot or bench.
7. Plants should never be permitted to surface-dry.
8. Pots should be filled to the rim with soil.
9. Bench crops do not dry out uniformly, but potted crops do.
10. Fertilizers and pesticides can be applied through the irrigation system.
11. Plants should be watered in the evening for greatest benefit.
12. Less liquid fertilizer than dry fertilizer must be applied for the same effect.
13. The amount of fertilizer applied through an injection system is determined by the strength of the concentrate and the dilution ratio of the injector.
14. Injector fertilizers contain dyes to provide a check of the dilution accuracy.
15. Growth retardants are applied through the irrigation system.

B. MULTIPLE CHOICE

From the choices given, select the answer that best completes each of the following statements.

1. The elimination of all life in the soil is termed _____.
   a. protection
   b. sterilization
   c. pasteurization
   d. steam treatment
2. The elimination of undesirable microbes, insects, and weeds from the soil is termed

   _____.
   a. protection
   b. sterilization
   c. pasteurization
   d. steam treatment
3. Pasteurization is accomplished with

   _____.
   a. steam
   b. chemicals
   c. steam or chemicals
   d. solar radiation
4. Steam can be applied to a bench of soil either _____ or _____.
   a. in canisters/as hot water
   b. from the surface/through buried pipes
   c. through vented pipes/in aerosol cans
   d. by hand/by machine
5. A disadvantage of chemical fumigants is their _____.
   a. toxicity
   b. waiting time
   c. both of these
   d. none of these

C. SHORT ANSWER

Answer each of the following questions as briefly as possible.

1. List two advantages and one disadvantage of using artificial soils for greenhouse crop production.
2. List three reasons why greenhouse growers need soil tests.
3. List the options available to a greenhouse grower when a soil test is needed.

4. Match the types of containers on the left with the description on the right.

a. clay pots	1. specialized production containers made of wire or plastic and often containing a liner to retain the soil
b. plastic pots	2. porous containers whose height and diameter are equal
c. peat pots	3. compartmented trays used for both production and sale
d. pans	4. shallow, rectangular trays without compartments, constructed of plastic or wood
e. molded plastic packs	5. biodegradable containers that permit direct planting without removal of the container
f. hanging baskets	6. clay or plastic containers whose height is one-half their diameter
g. flats	7. lightweight nonporous containers whose height and diameter are equal
h. azalea pots	8. round containers whose height is three-fourths their diameter

5. Give some examples of greenhouse crops commonly reproduced by the following methods.
   a. tissue and organ culturing
   b. division of the crown
   c. budding
   d. grafting
   e. cuttings
   f. layering
   g. bulbs
   h. runners
   i. seed

6. Indicate whether the following apply to expanding spacing (E) or fixed spacing (F).
   a. less bench space required for immature crops than for mature ones
   b. more transplanting required per plant
   c. early root development limited by placement of small plants in large pots
   d. used for bench crops
   e. more handling and labor required

7. List the items of information that may be included on a bench label.

8. List the four principles of pest control.

9. List the protective clothing that should be worn when spraying highly toxic pesticides in a greenhouse.

10. List at least four alternatives to pesticides that can be used in a program of integrated pest management.

D. DEMONSTRATION

From Table 20-3 (A Comparison of Selected Floriculture Crops), select a group of crops that can be grown together in a single greenhouse. For that group, prepare production schedules that will keep the greenhouse in full production for an entire year. Provide for special sales at Christmas, Easter (April 5th), and Valentine's Day. The production schedule for each crop should include the date of planting, transplanting, potting, staking, pinching, fertilizing, shading, wrapping, disbudding, and growth regulator application whenever appropriate. It should also include the greenhouse temperatures, method of fertilizing and formulations of fertilizers, the spacing, and dates of expansion. Staking, wrapping, and dates of sales should also be included. Consult other references for necessary details.

21

Nursery Production Techniques

Objectives

Upon completion of this chapter, you will be able to

- explain the differences between container and field nurseries.
- explain how plant species are selected for production and how quantities are determined.
- describe the concerns for media preparation.
- list the factors that determine how field nurseries are laid out.
- describe the methods of irrigating nursery crops and disposing of excess water.
- outline methods of fertilization, irrigation, and pest control for nursery production.
- label nursery plantings correctly.
- explain the techniques of pruning and field harvest.
- explain the use of cover crops and crop rotation in nursery production.

CONTAINER VERSUS FIELD NURSERIES

Nursery plants are produced in specialized production facilities under controlled, monitored conditions to maximize their rate of growth and standardize the quality of the harvested products. The product may be young plants that will be grown on by another nursery. Propagative materials such as grafts or liners (rooted, unbranched plants that will be grown to a larger size before sale) are examples of such young plants. The harvested product may also be plants ready for placement into landscape settings by landscape contractors or homeowners. Sizes of such plants can range from partially grown to fully mature. It is in the methods used to produce this great variety of plant products that container nurseries and field nurseries differ.

Field nurseries are similar to crop farms. Plants are grown directly in the soil. These nurseries require sites with rich, well-drained soil, few rocks, a reliable and adequate fresh water supply, and proximity to transportation, usually a highway. Field nurseries traditionally produce trees and shrubs that require from one to ten years to reach the intended size for harvest. Once installed in the field, plants are seldom moved prior to harvest. Therefore field nursery growers must make careful decisions about spacing between plants and between rows in order to make efficient use of the land and allow field equipment to reach the plants during their production years.

Field nurseries are vulnerable to the uncertainties of nature. Care or harvesting of the crop can be brought to a costly halt when excessive rainfall renders the fields too wet for working. Also, due to the large acreages involved, winter injury and rodent, deer, and rabbit damage can be greater than in closed production facilities.

Container nurseries do not require fertile field sites. They do need level locations and may be surfaced with concrete, asphalt, or crushed stone. Like field nurseries, container production facilities need a good water supply and access to their markets.

Many of the same plants grown in field nurseries are adaptable to container production. However, some evergreens grow better under field conditions, and many deciduous shrubs can be produced faster and less expensively in the field. Usually plants produced in containers are kept in production for only one or two years and seldom longer than four years. Because container-grown plants have their root systems intact and suffer less from transplant shock, their acceptance in the marketplace is good. Current trends suggest that even higher percentages of nursery plants will be produced in containers in the future.

CROP SELECTION

The choice of plants for nursery production is governed by several factors: market demand, production requirements, and production capabilities. *Market demand* is the driving force that dictates what plant species will sell, by whom they are sought, what sizes are desired by the purchasers, and what quantities of each species and size will be required. Plants headed for garden centers and other retail outlets for homeowner purchase are often smaller than plants used by landscape contractors. Nurseries frequently specialize in the production of plants for one type of market rather than trying to serve them all.

Production requirements are simplified when the grower limits the choice of plant species to those having similar propagation methods, nutrient and water needs, pest control requirements, light needs and tolerances, spacing, and harvest methods. Such simplicity and specialization saves time and avoids the costs required to produce a more diverse, less homogeneous array of species.

Production capabilities can be developed to parallel the requirements for crop production. Only facilities, tools, and equipment required for the crops being produced need to be purchased. Capital is not tied up in expensive machinery or buildings that are required only infrequently or for the production of a species marketed in small quantities or that return small profits.

There are limits to the simplification of crop choices, however. One reason is the public interest in new varieties of plants, which is increasing the demand for a wider array of plant species in the marketplace. Another limit is the danger of great financial loss to the nursery grower who grows only a few species. A severe insect or disease infection or environmental calamity can wipe out years of effort and investment almost overnight. A change in market demand can be almost as disastrous. In nursery production, as in all areas of agricultural production, monocultures should be avoided.

DETERMINING THE NUMBER OF PLANTS TO GROW

Nursery plants are costly to produce. Each plant represents an investment of labor and materials as well as heat, water, and other cultural expenses. Growers must determine the total number of plants needed in the size required at the time of harvest. For some growers the harvested crop is ready to be planted in homeowners' yards. For other growers the harvested crop is young grafts or liners that will be further grown by another nursery that has chosen not to do its own propagating.

Profits are lost when an excess of plants is produced and cannot be sold. They are also lost when too few plants are ready when needed. Growers attempt to anticipate and allow for losses due to mortality during production, failure to meet grading standards, and replacements required for buyer rejections.

The number of plants grown is calculated by working backward from the desired harvest quota. The technique is adaptable to all phases of nursery production regardless of whether the nursery specializes in propagation, wholesale field production, or container production. The formula is a simple one:

Number of plants necessary to start = The production quota ÷ The estimated return percentage

where the estimated return percentage = (100% - the percent mortality)

Example 1: If a production quota (representing 100 percent of the crop) required 5,000 trees for a landscape installation on a certain date, and the landscape contractor traditionally rejected 10 percent of the plants on first inspection, insisting on replacements, the nursery grower would calculate the number of plants to be grown to harvest size as follows.

Number of trees to be grown = 5,000 trees ÷ (100% − 10%)
= 5,000 trees ÷ .90
= 5,556 trees

Example 2: To calculate the number of young trees to plant in the field several years in advance of the harvest, the grower first estimates at 15 percent the percentage of loss due to insects, diseases, rodents, equipment injury, or environmental damage. The formula is then applied again, using 5,556 trees as the production quota.

Number of saplings to install in the field = 5,556 trees ÷ (100%−15%)
= 5,556 trees ÷ .85
= 6,537 trees

In a similar manner, the grower could work backward to calculate the number of field liners needed to produce the saplings, then the number of seedlings or grafts required to produce the liners.

MEDIA CONSIDERATION FOR NURSERY CROPS

A review of Chapter 2 will highlight basic information about soil, which is central to the production of nursery stock. As in other areas of agricultural plant production, the composition of the soil or even the decision to use soil-less media in place of soil, is determined by criteria such as the type of crop being grown, the stage of growth at the time of harvest, the cultural techniques being used, the methods of harvest, and whether the harvested plant will be bare rooted or balled and burlapped.

Nursery soils need a structure that promotes nutrient availability to the plants and a proper air and water relationship. They must drain well, while simultaneously retaining the moisture needed for growth. The soil pH should be between 5.00 and 7.2 for most nursery crops. Ericaceous plants are sometimes an exception, requiring a lower pH for optimum growth.

Soils for field production of woody nursery plants should have these characteristics and also be free of noxious (difficult to eradicate) weeds, insect pests, pathogenic inoculum, or nematodes. The herbicide history of a field should also be known before planting with nursery stock to assure that a prior crop's herbicide is not latent in the soil, waiting to damage the new nursery crop. Also, field soil used for nursery crops should be high in organic material. Good soil fertility is also desirable as long as it is not too acidic or basic, which can tie up micronutrients.

Container-grown nursery stock also needs media with the properties already described. In addition, due to space limitations of containers, the media must be heavy enough to anchor the plant and prevent container tipping, have a high cation exchange capacity, provide desired levels of nitrogen throughout the growing season, and retain the moisture needed for optimal plant growth.

Porosity of the medium is of great importance in

container plant production. Porosity is the size of spaces between the particles that comprise the medium. It determines how much water and air remain in the root environment after the container has been watered and allowed to drain. Large pores hold the air within the medium. Smaller pores hold the water within the medium. Shallow containers tend to hold a greater percentage of water following irrigation and drainage than do deeper containers; whereas deeper containers usually have good aeration but less water retention capacity.

These differences can be partially compensated for by incorporating materials into the media of shallow containers to improve the aeration porosity. Materials such as sand, perlite, turface (calcined clay), and various wood barks are commonly mixed with fine-textured media for that purpose.

Container media must be readily available in the amounts needed when they are needed. Accordingly, the media must be either commercial mixes or locally prepared easily. They must also be inexpensive, easy to handle and affordable to transport, pest free, blendable with other additives, and stable over time, allowing them to be stored without concern for changes in their chemical or physical characteristics. The artifical media described in Chapter 14 are commonly used for nursery crops. The University of California and the Cornell mixes may be used in unaltered forms or with additives to improve aeration porosity. Other container media used include aged and composted hardwood or softwood barks. Hardwood barks combined with sand and with ammonium nitrate (for added nitrogen) and elemental sulfur and iron sulfate (to stabilize the pH) is a common mix.

LAYOUTS FOR FIELD PRODUCTION

Nursery production fields are not all planted in exactly the same way. The grower usually considers a number of factors that determine how each field will be laid out:

1. eliminating opportunities for erosion
2. using the land as efficiently as possible

3. grouping together plants with similar requirements for cultivation but, possibly, different harvesting schedules
4. making cultivation and harvest as easy as possible
5. knowing the length of time the crops will be in the field and the size they will attain
6. knowing whether the field will be harvested all at one time or over a period of several seasons

The simplest layout for a nursery is a single species planted in side-by-side rows on level land, Figure 21-1. Erosion is minimal, and cultivation equipment can pass easily down the rows and across perpendicular to the rows between the plants. Harvesting is just as simple when this layout is used, making it ideal for bare-rooting trees and shrubs. Partial or selective harvesting can also be accomplished by digging plants in a diagonal pattern across the field. The open space remaining allows further growth by the plants that remain.

Side-by-side rows can also be used on rolling land, with the rows oriented to follow the contours of the slope and thereby minimize erosion. Cross-cultivation is seldom possible on slopes because of the soil erosion it would encourage. Also, total harvesting is not wise; some plants should always be left to hold the soil until replacement plants can become established.

Another variation of row planting is the staggered or offset layout, Figure 21-2. Plants in alternate rows are not set directly opposite those in adjacent rows but in a staggered pattern similar to the fives on dice. Cross-cultivation is not possible with this arrangement. Harvesting is by rows, not diagonally. Every other row can be harvested the first year and either all remaining rows or every other remaining row harvested in the second year. All plants would be cleared by the end of the third year. As rows are cleared, the soil can be conditioned and planted again or a cover crop seeded to hold the soil. (Cover crops are discussed later in the chapter.)

When space is at a premium, narrower aisles

FIGURE 21-1.
An example of a single species field planting in rows that are evenly spaced. Arrows indicate the two directions possible for cultivation. The broken diagonal lines indicate a possible harvest pattern that allows additional growing room for the plants that remain.

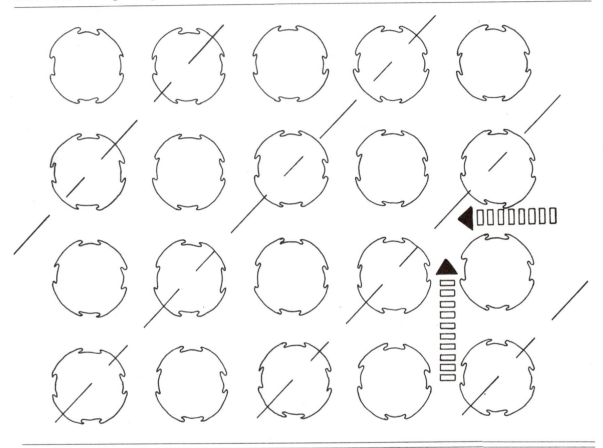

can be left between alternate rows, Figure 21-3. This permits large equipment to be moved into the field while only slightly increasing the difficulty of cultivating and harvesting.

Mixing species within a single production field adds some complications but is not uncommon.

First, the species must have the same soil and moisture requirements. Second, if shading or winter protection is required, it must benefit all the plants. Finally, if the plants are of different sizes or on differing harvest schedules, cross cultivation may be impossible. Species are often mixed between rows

FIGURE 21-2.
An example of a single species field planting in rows that are offset. Arrows indicate the single direction of cultivation possible. The broken lines indicate a first-year harvest pattern, leaving remaining plants to grow another season or two.

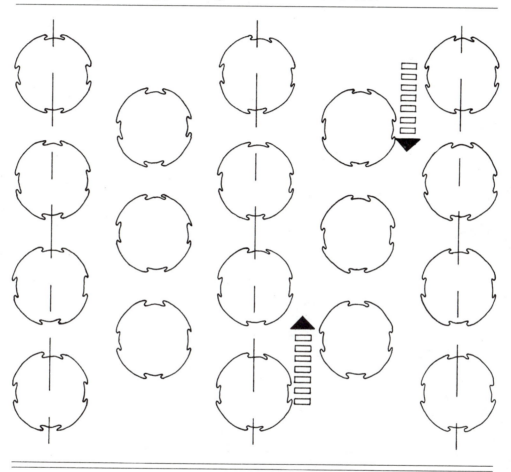

to combine fast-growing and slow-growing species, the first for quick sale and cash turnover, the latter for future sale, Figure 21-4.

MAINTAINING NURSERY CROPS

In some ways, maintaining nursery crops is similar to maintaining floriculture crops. The major distinc-

tions are in the magnitude of the operations and the degree of grower control. Watering, fertilizing, and controlling pests in the greenhouse are almost totally under the grower's control and have little impact outside the operation. With nursery production, the leaching and runoff of surface and irrigation water, fertilizers, and pesticides have the potential to affect the environment of people far beyond the nursery's

FIGURE 21-3.
To conserve space, alternate rows of a nursery planting can be placed closer together. Arrows indicate the directions of cultivation. The broken lines show first-year harvest pattern.

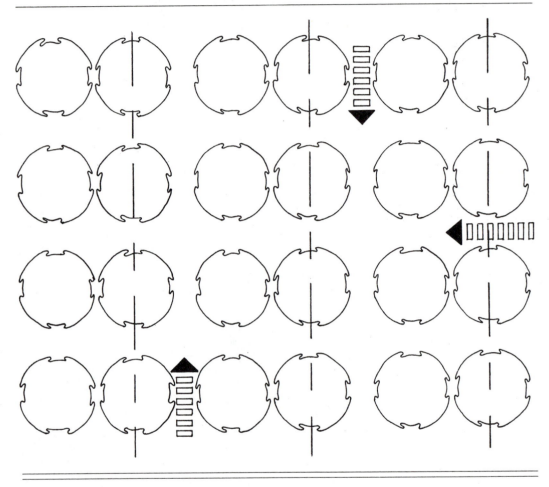

property lines. Consequently, full consideration must be given not only to what is applied but to where it goes after application.

Irrigation

Both field and container nurseries receive the natural irrigation of rainfall and supplemental irrigation controlled by the grower. Provision must be made for a reliable source of water, such as an irrigation pond, deep well, or municipal water supply. (Use of a municipal water supply can add greatly to the cost of production.) Following application, the runoff water must be absorbed on site or diverted into municipal storm sewers or nearby streams. Water that is not absorbed on site must be free of toxic

FIGURE 21-4.

Mixing fast-growing and slow-growing species in alternating rows allows the fast-growing species to be harvested while giving the slower one more time and space to develop. Arrows indicate the single direction of cultivation.

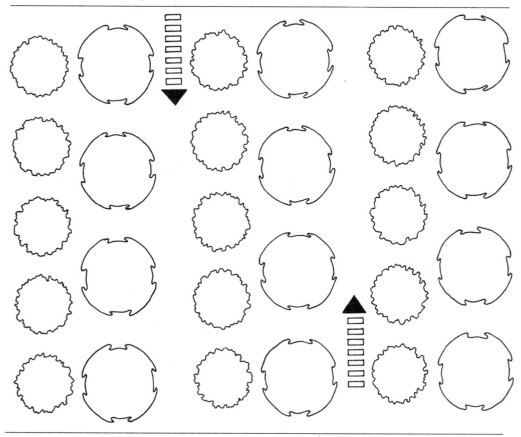

waste, excessive fertilizer salts, and other nonbiodegradable materials. Excess irrigation water is often collected in a holding pond and used again to water container crops, thereby minimizing the amount of water that leaves the nursery. Where soluble salt levels are too high, the runoff water can be diluted with other water until a safe salt level is attained. A holding pond will permit some heavy particulate materials to settle out before the water is reused. In other situations, leach fields may be necessary to remove undesirable materials before the water is reused or permitted to escape into streams or public systems. Regulations differ among communities, and the prospective nursery grower should consult local water authorities while the nursery is still in the planning stages. The authorities can offer suggestions on how the local water quality standards can be met most efficiently. This assistance should be sought, not avoided, by nursery growers.

In a few regions of the United States, the local water is too alkaline for direct use on nursery crops. The soluble salts in the water would quickly harm the plants, and irrigation equipment would plug up almost as quickly. Therefore, settling ponds are often needed to permit precipitation of the salts before the water is used on the plants. In extreme situations, the water must be distilled before it is suitable for use.

Supplemental Irrigation Systems

The delivery system for supplemental irrigation of nursery crops may be permanent, semiportable, or portable. The choice of system is usually determined by the frequency and duration of its use. Where the same species are grown repeatedly and the amount of rainfall is predictable, a choice is easily made. Arid regions are likely to require permanent systems due to the need for regular irrigation. Nurseries in temperate or subtropical areas may receive enough rainfall to maintain healthy field production and only need supplemental water for container areas or newly transplanted field crops.

The three degrees of permanency can be compared as follows.

1. *Permanent irrigation system*
   - The power source is either an electric or a diesel engine. It is permanently installed at the site.
   - The pump is also permanently installed and is usually either a deep well turbine or a horizontal centrifugal one. The deep well turbine is used when water must be lifted to a height of twenty feet or more.
   - The system must be near the water supply and on a solid and level base.
   - The main and lateral water lines are sta-

tionary and below ground (deep enough to permit crop cultivation above them).
   - The lines are constructed of concrete or PVC piping.
   - The system is the most expensive to install but least time-consuming to operate.

2. *Semiportable irrigation system*
   - The power source may be an electric or a diesel engine. It is permanently installed at the site.
   - The pump is also permanently installed and is usually either a deep well turbine or a horizontal centrifugal one.
   - The system must be near the water supply and on a solid, level base.
   - The main water lines are stationary and below ground with surface couplers for the attachment of portable lateral lines.
   - The main lines are constructed of concrete or PVC piping. The lateral lines are usually of PVC.
   - The system is less expensive to install but more time-consuming to operate than a permanent system.

3. *Portable irrigation system*
   - The power source is usually a gasoline engine, transported to the site on a flatbed truck, or a PTO (power takeoff) shaft on a tractor.
   - The pump is transported on a truck or cart with the power source.
   - All water lines are portable and lightweight. They are not buried and are constructed of either PVC or aluminum, Figure 21-5.
   - The system is the least expensive to install but most time-consuming to operate.

As the irrigation water leaves the pump and water lines, it is distributed to the crop through sprinkler heads. Container nurseries use overhead sprinklers or a PVC tube spaghetti system (described in Chapter 20). The sprinkler systems are more wasteful of water than the tube system, which applies water directly to each container and not in between. Either method requires occasional hand-

FIGURE 21-5.
This supplemental irrigation system delivers water to each pot. It requires exact spacing of each container.

watering. In one case, some plants are not reached fully by the sprinklers. In the other case, the thin PVC tubes occasionally become plugged. The use of filters in the water lines can reduce the frequency of plugging. Some growers also insert a fiber mulch ring around each containerized plant to retain moisture and discourage weeds.

Field production areas may apply supplemental irrigation from rotating sprinklers coupled to the lateral water lines, Figure 21-6. Portable systems may make use of a traveling sprinkler that carries its own water supply and dispenses it rapidly as it is pulled down the nursery rows. Neither system should apply the water faster than the field soil can absorb it.

In production fields where crops are widely spaced, as with large trees or high-value specimen plants, supplemental water may be applied through trickle irrigation, Figure 21-7. The low-pressure tube system reduces the waste and expense of watering expanses of field where the moisture is not needed. Instead, water is applied slowly to the root zone of each plant,

FIGURE 21-6.
A rotating sprinkler system on two different levels keeps this container production area irrigated.

which allows for both optimum growth and cost efficiency. Trickle systems require a timer to turn on the irrigation water each day when the plants are most in need of it. They also require a flow regulator and

FIGURE 21-7.
Trickle irrigation systems permit some movement of the containers. They sometimes plug up, requiring the grower to clear the tube.

pressure-compensating emitters to control the rate of water delivery to each plant.

Fertilization

Nursery plants usually require fertilization to promote healthy growth and bring the plant to the harvest stage as rapidly as possible. Due to the variety of sizes and production methods, it is predictable that the methods of fertilization will vary. As is the case with other cultural procedures, grouping together plant species that have similar fertilizer requirements will make the fertilizer application process easier for the grower.

Small plants growing in ground beds usually are fertilized once a year. Typical rates of application are:

- needled evergreens   4 lb per 1,000 sq. ft.
- broadleaved evergreens   3 lb per 1,000 sq. ft.
- deciduous shrubs   5 lb per 1,000 sq. ft.

Small plants growing in containers need more frequent fertilization during the growing season because of the limited amount of media around the roots. These plants lack the nutrient reserve that is available in ground beds. Container-grown crops may be fertilized with dry, slow-release fertilizers or during irrigation with the fertilizer injector systems described in Chapter 20.

Plants growing in the field usually have dry, low-analysis fertilizer or manure incorporated into the soil before planting. After planting, nitrogen fertilizers are usually added annually. Typical rates of nitrogen application for field production include:

- needled evergreens   175 lb per acre
- broadleaved evergreens   125 lb per acre
- deciduous trees and shrubs   225 lb per acre

Actual rates are dependent upon the crop, the soil type, and the length of growing season. Rather than applying all of the fertilizer at once, it is usually applied in two or three applications. A spring and early fall or a spring, early summer, and early fall application schedule allows the most efficient use of the fertilizer by minimizing nutrient loss due to leaching. Further cost efficiency is made possible by applying the fertilizer as *side dressing* (applied along the side of the rows) rather than broadcasting it over the entire field. Side dress applications are also safer for the off-site environment, since the potential for chemical runoff is lessened.

Regardless of how the fertilizer is applied, all applications of nutrients should be discontinued with the approach of fall to allow the stock to harden-off before winter sets in.

Weed and Pest Control

Weeds, insects, and diseases are as inevitable with nursery crops as with any other plants. Production of a monoculture with close spacing, high fertility, and ample irrigation assures an assortment of prolific pests. All the principles of pest control discussed in Chapters 6 and 20 are applicable to nursery production, but there are some differences in the ways they are applied. The differences are due to the openness of the nursery production area and the length of time the crop is under production. For example, soil pasteurization is helpful in nursery container production but far less effective than in greenhouse production. The soil, being outdoors, becomes contaminated much more quickly. Still, the quantity of weed seeds, insect larvae, and pathogenic inoculum is reduced initially, giving the crop a better chance of becoming established, so pasteurization can be helpful. Weeds are also much more troublesome for the nursery grower than for the greenhouse grower. Their seeds can persist in the nursery field for many years, with a fresh crop waiting to germinate each time the soil is turned over. In addition to their competition with the desired plants, which has been described, weeds can reduce plants' sales appeal when they appear in the same container.

Insects and diseases are controlled by using pest-free propagative stock, by growing plants in regions of the country where their pests are not as common, by selecting resistant varieties for growth, by preventing the growth of weeds that can harbor the pests, by cleaning up crop debris that can serve as sources of inoculum, and by spraying or dusting with chemical protectants and eradicants. Where chemical pesticides are required, the state's college of agriculture can provide listings of effective products approved for use on specified crops. As noted elsewhere, growers should take care to mix only the quantity of pesticide that can be applied in a single spraying, and avoid spraying or dusting on windy days. If the field is sloped, run-off can also be a hazard, so the product should not be applied to the point of dripping.

Weeds can be controlled with pre-emergence herbicides applied in the autumn and early spring. Later in the spring, post-emergence products can keep the nursery free of broadleaved weeds. The herbicides are usually applied in granular form as a side dressing. Many of the products are specific against certain plants. They cannot be applied universally since certain crops could be harmed. The product labels list the plants that can be treated safely as well as the weeds controlled by the chemical. Some of the herbicides require mixing into the soil to become active. Others are dissolved by rainwater or artificial irrigation and do not require cultivation. Some are reduced in effectiveness if the soil is disturbed, so mechanical cultivation must be avoided during crop production.

Some pests bother nursery crops more than floriculture crops. These include rodents, rabbits, and deer. Rodents can be controlled, though seldom eliminated, with poisoned bait placed throughout the nursery. Rabbits can be discouraged with repellants sprayed onto the base of plants or with plastic coils wrapped around the trunks of young trees. Deer are much more difficult to control, although some benefit is claimed for repellants. Fencing the nursery can prevent deer from winter browsing in the production fields.

Integrated pest management is possible with nursery crop production, but it is effective only if every employee who works with the crops understands and implements it in a timely manner. *Scouting* of the crops on a regular, frequent basis is essential to detect the presence of insects, pathogens, noxious weeds, or animal injury. *Immediate response* to the scouting report is then mandatory if the pest is to be contained. Control procedures must be applied as soon as the scouting report is filed. In so doing, the pest can often be controlled without the need for more expensive control practices that would be required if the problem were permitted to spread.

LABELING NURSERY CROPS

Growers must label their nursery crops in a standardized manner that is easily understood by all employees. Production scheduling, inventory control, and the filling of orders depend upon accurate knowledge of where each species is growing and how many are available for sale or in a particular stage of production. Should a label be destroyed, the very identity of certain look-alike cultivars could be lost permanently. Therefore, labels of nursery stock must be durable as well as correctly placed, Figure 21-8.

FIGURE 21-8.

A label imprinter such as this one can print permanent, weather-resistant plant labels in colors, giving each label a professional appearance. (Courtesy Economy Label Sales Company, Inc.)

Two methods of labeling are commonly used. The method used for container stock and equally effective for many field nurseries places the label in front of the first plant of that cultivar and pertains to all plants to the rear and right of that label until the next label is reached, Figure 21-9. The other method is used almost solely for large field nurseries and follows the direction of planting, with a new label inserted into the ground each time the cultivar changes. When mechanical planters are used and more than one row is planted simultaneously, the label changeover becomes a bit more complex, Figures 21-10 and 21-11. Mixing the species in alternate rows requires that each row be labeled.

FIELD PRUNING NURSERY STOCK

There are few similarities between landscape pruning (described in Chapters 11 and 12) and the pruning of nursery plants in production. The objectives of landscape pruning are to improve the health or

appearance of the plant or to limit its growth; the objective of nursery field pruning is to promote and direct branching and create a fuller plant as quickly as possible. Fullness adds measurably to the sales appeal of the plant. Shade and specimen trees and certain specialized plants, such as fruit trees, require individual attention and hand-pruning during production. Plants grown in mass quantities for the seasonal market cannot be given such time-consuming and costly attention.

Trees require more time to shape and develop than shrubs do. Trunk development and crown shaping are both of importance. To develop straight trunks with smooth bark and strong scaffold limbs with wide crotch junctions requires individual attention at least once during each year of field growth.

Evergreen trees must have their candles (the new growth) pinched in the spring before growth fully expands if the developing trees are to benefit by the pruning. If the trees are intended to be cut for Christmas sales, the only objective is to add fullness. If the trees are intended for landscape use, the pruning can direct growth to fill in holes in the plant, remove conflicting branches, and add fullness.

Evergreen shrubs are sheared in the early spring before their annual flush of growth begins. The shearing promotes lateral branching of the shoots and creates a fuller plant. If the shrubs have a pyramid or columnar form, care must be taken to respect their natural shape and not cut off the lead shoot(s). If the shrubs are rounded or spreading forms, the shearing is directed to heading back the lateral branches.

Deciduous shrubs, particularly those that are to be bare-rooted at harvest, often get the least individual attention, because they bring the least return at the market. They are commonly pruned with a sharp cutting bar driven down the nursery row, removing about a third of the previous season's growth in the process.

Root Pruning

Using an underground blade in a manner similar to that of the cutting bar just described, nursery growers prune back the root systems of shrubs and

FIGURE 21-9.
Labels may be placed in the first container to identify all plants of that variety behind and to the right of the label. In this example, four different varieties are labeled.

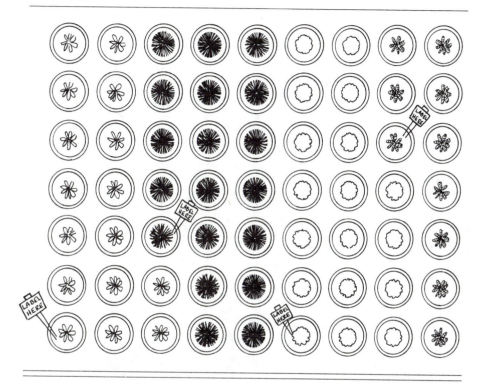

young trees. Done in the fall after the plants have ceased their active growth for the year, the pruning technique promotes a compact root system. Such a root system gives the plants a better chance of surviving their eventual harvesting, handling, and transplanting. Without root pruning, most of the fibrous nutrient- and water-absorbing roots would be lost at the time of harvest, and nursery plants would have little more chance of survival then plants collected from the woods.

For larger trees and shrubs, the sickle bar technique is inadequate. Instead, several months to a year in advance of the harvest, a trench must be dug around each plant just outside of what will be the soil ball at the time of harvest. The trench should be at least the width of a spade and as deep as the soil ball will be. The trench is then back filled with sand. This process cuts the lateral roots of the plant and encourages new ones to develop within the soil ball. The sand discourages root growth out of the ball.

Containerized nursery plants do not require root pruning, since the entire system is intact within the container. If kept too long in the container, the plants may send roots out through the drainage holes and into the soil of their holding area, making it hard to lift them. This can be discouraged by spreading a layer of heavy black plastic beneath the containers. The plastic also discourages weeds from growing.

FIGURE 21-10.
An example of how a nursery field can be labeled if two rows of the same variety are planted at the same time. In this example, four different varieties are labeled. The labels follow the direction of planting.

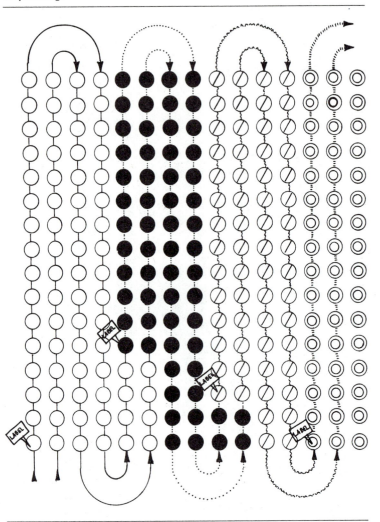

HARVESTING

Nursery crops, like any other product, must be made ready for sale before the time they are sought by the customer. Since the greatest sale of nursery plants occurs in the spring, most plants are dug, bare-rooted or balled-and-burlapped, graded, and prepared for sale in the fall. In regions of the country where spring arrives late and the ground freezes deep, it is impossible to dig enough material in the spring. Fall digging also allows the nursery workers

FIGURE 21-11.

An example of how a nursery field can be labeled if four rows of the same variety are planted simultaneously. In this example, four different varieties are labeled. The labels follow the direction of planting.

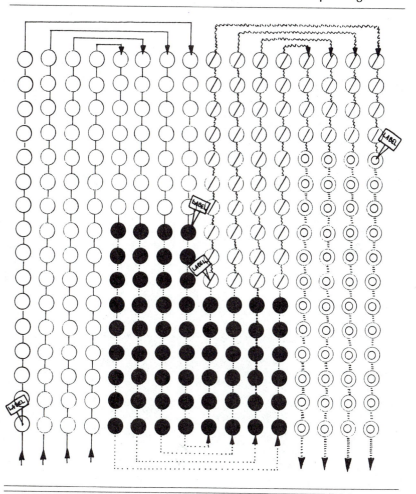

to grade, prune, bundle, and package the stock throughout the winter months.

If an entire row of shrubs is to be dug and bare-rooted, a forked, U-shaped blade pulled by a tractor may be used. As it lifts the plants from the ground, nursery workers shake off the soil and place the plants on a truck. The plants are taken to shaded storage areas where the roots are covered with soil or other materials to prevent drying until they can be graded and bundled.

FIGURE 21-12.
This conifer has been dug with a large soil ball and now must be burlapped and tied.

FIGURE 21-13.
An example of drum-lacing. The technique binds the burlap and soil ball tightly. It is helpful with large, heavy plants to prevent breakage of the soil ball.

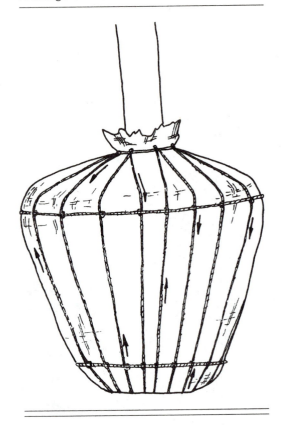

Plants that are to be balled-and-burlapped may be dug by rows or by a pattern that harvests some plants while leaving others for a later time. Mechanized tree and shrub spades are used by some nurseries to dig plants for burlapping but the majority are dug by hand. Squares of burlap large enough to cover the ball are available precut or can be cut off a roll. Burlap is also available in a treated (rot-resistant) or untreated form. Treated burlap is more expensive but is needed if the plants are to be stored unplanted for a while. Untreated burlap can be used when the plants are going to be transplanted soon after harvest.

For best sales appeal, the soil ball should be nicely rounded, not flat or misshapen. It should contain enough roots to assure survival of the plant, yet not be unnecessarily heavy, Figure 21-12. The burlap can be secured with twine or with pinning nails. Larger plant balls can be secured using the technique of *drum-lacing*, in which twine is wound around the ball and then laced in a zig-zag pattern, Figure 21-13. All balled-and-burlapped plants should be lifted and carried by the soil ball, not by the branches, Figure 21-14. Large plants can be moved with a forklift, tractor, crane, or ball cart. The ball should be handled carefully to prevent it from breaking apart.

After burlapping, the plants can be bedded (or *heeled in*) with sawdust, peat moss, or shavings to keep the roots and soil ball moist until grading and sale. In warm climates, some harvested plants may require special care. For example, cacti need a week for the roots to air-dry and heal. In the southeast, tropical plants grown for interior use require carefully controlled light reduction to prepare them for transplanting.

FIGURE 21-14.
A balled-and-burlapped plant should be lifted and supported from beneath. The soil ball should never be permitted to pull on the plant. (Courtesy Rodney Jackson, Photographer)

FIGURE 21-15.
This tree ball has been placed directly from the field into a burlap lined wire basket, then tied at the top with twine.

In recent years, nursery growers have introduced several variations on the traditional balled and burlapped plant. *Wire baskets* lined with nontreated burlap accept the harvested plants' soil balls directly from the field. The baskets are then tightened around the balls, with soil added if necessary to create the roundness desired for marketing appeal, Figure 21-15. *Containerized plants* harvested from the field are also commonly available today. Immediately following digging, the plant balls are set into containers then held for several weeks while the root system regenerates. This method of harvest can be done almost any time the weather permits digging.

Nursery stock is graded in accordance with the American Standards for Nursery Stock developed and published by the American Association of Nurserymen (AAN) in Washington, D.C.

Containerized plants require little preparation for sale other than cleaning weeds from the containers, grading, and perhaps tagging them with names, plant patent numbers, or other information. If a fiber mulch ring has been used for moisture retention or weed control, it should be removed before sale.

COVER CROPS AND CROP ROTATION

Since a nursery crop is only as good as the soil it is grown in, it is vital that nursery fields be stabilized and reconditioned after each crop is harvested. Root balls remove significant quantities of soil from fields, including valuable top soil. Where fields are completely cleared by a harvest, erosion and further soil loss become dangers. The first priority of a nursery operator must be to prevent soil erosion. The second priority is to rebuild the soil before the next crop of nursery plants is set out.

A *cover crop* is a rapidly growing, nonnursery crop used to stabilize the soil and prevent erosion. It may cover an entire field or only an area where insufficient nursery crops remain to hold the soil. *Green manure* is a crop that is rotated in the field with nursery crops. It too deters erosion but also improves the structure of the soil, fixes nitrogen, and adds valuable organic matter. The deeper the roots of the cover and green manure crops, the greater their benefit to the soil. It is possible for one species or mix of species to function as both a cover crop and a green manure.

Cover crops are frequently planted in the fall after the harvest. Sometimes they are planted between the rows of nursery plants when harvesting is done over a period of several years. Cover crops are rototilled or plowed under each spring, returning organic matter to the soil. The seeding of another cover crop may then follow if the field is not planted right away.

Green manure crops are usually seeded in the spring and grown through the summer, then turned under before they develop seeds that could create weed problems for the nursery crop that is to follow. A winter cover crop may then be seeded to protect the soil from erosion until planting in the spring. Plants commonly used as cover crops or green manure include buckwheat, sudan grass, oats, rye, field corn, annual ryegrass, soybeans, alfalfa, clover, and hybrid sudan-sorghum grass.

Animal manures are also valuable soil conditioners and can be plowed under at the same time as cover and green manure crops. They can be obtained in quantity from nearby farmers and slaughter houses.

SUMMARY

Although the crops are often the same, field nurseries and container nurseries have different production objectives, different requirements of the production site, and different growing techniques. Field nurseries require productive natural soil, while container nurseries can totally control their soil, often using purely artificial media as all or part of the mix. The containers used are similar to those of the greenhouse industry but larger. They may also include wooden baskets and soft plastic bags. Nursery stock may be started from seed or as liners. Nearly all propagative techniques are used in some way within the industry.

The layout of production fields is determined by the contour of the land, the amount of space available, the cultivation techniques used and the methods or schedules of harvest.

Nursery maintenance has the potential to affect people beyond the boundaries of the nursery. There must be a reliable and plentiful source of good water. Its runoff must be controlled to assure cleanliness before it enters a public waterway. The delivery system for supplemental irrigation of nursery crops may be permanent, semiportable, or portable.

Fertilization may be done with dry fertilizer or with liquid fertilizer in the irrigation system. Use of pest-free propagative stock and resistant varieties, good field hygiene, and application of chemical pesticides are all methods employed in pest control.

Two methods of labeling are commonly used. One method identifies all plants to the right and rear of the label until the next label is reached. The other method follows the direction of planting, changing the label as the cultivar changes.

The objective of nursery field pruning is to direct branching and create a fuller plant as quickly as possible. Certain plants warrant individual attention because of their potential value; others are given minimal attention since their profit return is less. Both top growth and root growth are pruned. Containerized plants do not require root pruning.

Use of pest-free propagative stock and resistant varieties, good field hygiene, directed use of pesticides, and an integrated pest management system that employs frequent scouting and rapid response are all modern methods employed in pest control.

The nursery is usually harvested in the fall to assure an ample supply of plants in the spring when sales are greatest. Plants may be dug by hand or by machine. Containerized plants require little preparation for sale except to clean weeds from the containers, grade, and label.

After the harvest of a field nursery, the soil must be stabilized and reconditioned. Cover crops, green manure, and crop rotation are techniques utilized to rebuild the soil.

Achievement Review

A. SHORT ANSWER

Answer each of the following questions as briefly as possible.

1. Indicate whether the following are characteristic of container nurseries (C), field nurseries (F), or both (B).
   a. Larger trees and shrubs are produced.
   b. Soil mix may include artificial media.
   c. Good drainage, adequate water, and light are needed.
   d. Nearby transportation is needed.
   e. Most crops are grown for two years or less.
   f. Production soils are similar to those of a crop farm.
2. List three factors that determine the choice of plants to grow in a nursery.
3. List the three most common methods of plant propagation used by nursery growers.
4. If the production quota for a nursery plant is 8,000 plants and the grower estimates that 12 percent are likely to be unfit for sale due to the traditionally harsh winter weather, how many plants should be planted in the field?
5. Indicate whether the following statements apply best to permanent (P), semiportable (S), or portable (I) irrigation systems.
   a. The main water lines are stationary and the lateral lines are portable.
   b. The main water lines and the lateral lines are stationary and below ground.
   c. The power source can be a PTO shaft.
   d. The power source and pump are both mounted on a flatbed truck or cart.
   e. The system is the most expensive to install.
   f. The system is the least expensive to install.
   g. The system is the most time-consuming to operate.
6. Describe the two ways that nursery crops are usually fertilized after planting.
7. Describe how weeds are controlled in field nurseries.
8. List the two methods most often used to label crops in field nurseries.
9. How do the pruning objectives for nursery plants differ from those for landscape plants?
10. Insert one or two words to complete each of the following sentences.
    a. Evergreen trees are pinched back in the _____ of the year.
    b. Evergreen shrubs in the nursery are sheared in the _____ of the year.
    c. Root pruning of nursery plants is done in the _____ of the year.
    d. _____ nursery plants do not require root pruning.
    e. Most nurseries harvest their field crops in the _____ of the year.
    f. Large soil balls are burlapped using the technique of _____.
    g. Balled-and-burlapped plants should be lifted and carried at the _____ to prevent injury to the plant.
    h. _____ nursery plants require less time to prepare for sale than field grown plants.
    i. A rapidly growing nonnursery crop used to stabilize the soil of a nursery and prevent erosion is termed a _____ crop.
    j. An alternate crop used to improve soil

structure and nutrient content and increase organic matter is termed a _____crop.

B. TRUE/FALSE

Indicate if the following statements are true or false.

1. All nurseries have the same planting layout.
2. Species cannot be randomly mixed in a field nursery.
3. Space efficiency is of concern to both container and field nursery growers.
4. Contour planting is a means of controlling soil erosion in field nurseries.
5. A side-by-side row planting on flat land does not permit cross-cultivation between plants.
6. All plants in a nursery planting must be harvested at the same time.
7. Different crops can be grown within the same nursery field.
8. Not all nursery aisles must be the same width.
9. Irrigation water runoff is of more concern in field nurseries than in container nurseries.
10. Where natural rainfall is frequent, supplemental irrigation of field nurseries is usually unnecessary.

22
Beginning and Promoting an Ornamental Horticulture Business

Objectives

Upon completion of this chapter, you will be able to

- describe and compare three legal forms of business operation.
- explain the value of a market survey.
- list sources of capital for new and established businesses.
- list factors to be considered in choosing a business site.
- draw up a physical plant layout plan and a staff organization chart.
- outline the major laws and regulations that affect horticulturists.
- describe the methods, values, and limitations of advertising.
- list characteristics of effective advertising.
- list characteristics of effective displays.

WHY HAVE YOUR OWN BUSINESS?

Let us begin by saying that most people do not own their own businesses; they never will; nor do they wish to do so. Still they have interesting and rewarding careers.

For those who do operate or seek to operate their own business, a desire for one or more of the following may be the motive:

- the opportunity to exercise leadership and make all major decisions that impact upon the operation of the business (in other words, the chance to be your own boss)
- the opportunity to build a business around your own interests and personal strengths
- job security, insofar as this ever exists
- certain tax advantages
- money (however, the perception that the owner

of a business makes the most money is not always correct)

Conversely, the advantages of nonownership include:

- lack of major responsibility for decisions that affect the future of the business
- the opportunity to pursue a career interest without the problems of ownership
- freedom to change jobs and relocate as the desire or opportunities develop

If there are reasons why self-employment appeals to certain people, there are also personal qualities that make business ownership satisfying. Leadership skills and a fondness for both the industry and people are vital to successful operation. Liking plants is not enough. Not everyone is equally comfortable with workers and consumers or tolerant of their behavior under a myriad of conditions. The owner sets the tone for the way customers are treated and the spirit that exists among the staff. Someone who prefers working alone or is unusually shy might be better advised not to seek a career as an owner.

FORMS OF BUSINESS ORGANIZATION

Once the decision is made to begin a business or to reorganize an existing firm, another decision must be made: how the business will be legally organized. Three types of organization are common: the sole proprietorship, the partnership, and the corporation. Each type has certain advantages.

Sole Proprietorship

The sole proprietorship is the simplest form of doing business. It is the easiest to begin and the most private. After assuring that no zoning laws or local ordinances are being violated, the horticulturist can simply obtain the necessary licenses and insurance, set up shop, hang out a sign, plant a field or load a lawnmower on a truck, and be in business. The sole proprietor puts up the capital, hires and directs all employees, reaps all profits, absorbs all losses, and is personally and totally re-

sponsible for all debts. There is no separation between the sole proprietor's personal finances and those of the business. A sole proprietorship may be advantageous if you have substantial personal assets. However, should the new venture fail, your personal assets may be lost.

Sole proprietorships have other disadvantages. One is that the business can be limited by the weaknesses of the owner. The business usually grows at the outset, then reaches a plateau due to the proprietor's inability to direct its growth further. Another disadvantage is that the business ends with the death of the owner. If the death is premature and unexpected and no provision has been made for the sale or continuation of the business, liquidation may be the only recourse.

In the ornamental horticulture professions, sole proprietorships are common. Retail flower shops, garden centers, and small landscape maintenance firms are frequently organized in this form, at least at the beginning.

Partnership

A partnership is a business engaged in by two or more persons. It is also a common form of business in ornamental horticulture. It works best when the partners have interests and abilities that complement rather than duplicate each other. For example, a flower shop partnership might involve one partner with an interest in design and another who prefers sales and customer contact. A nursery could be owned by one partner who looks after promotion, sales, and administration and another who oversees outdoor field production. Should the partners decide to add a greenhouse range for propagation, they might take in another partner to provide capital and/or take charge of the greenhouse operation.

In modern partnerships, all agreements between partners are in writing before the business begins. The terms of agreement should include the percentage of original capital to be provided by each partner, the responsibilities assigned to each partner, the business objectives of the partnership, and provision for the dissolution of the partnership in the event a partner dies or wishes to leave the business.

While such a contract, prepared in advance, will not ensure that partners work together compatibly, it will help to expose potential areas of misunderstanding or disagreement before the pressures of the work arena complicate them.

In the classic partnership, the two partners contribute the same amounts of capital and divide the work equally. They each draw the same weekly salary, and at the end of the year they divide the profits equally. In such a situation, each partner is an agent of the business, empowered in the eyes of the law to act on behalf of the business, to contract with other firms, receive credit, and obligate the company. Thus both partners become liable for the actions of either partner.

There are variations on the classic partnership. In one, a single partner provides most or all of the capital, and the other(s) provide the technical and administrative skills to operate the business. Another variation is the limited partnership, where one or more partners perform some function, such as partial capitalization, but limit their claims on the profits or liabilities to a share agreed to in advance. Where liability is limited for some of the partners, at least one partner must agree to be legally responsible for all the liabilities of the company. Limited partnerships usually restrict the right of the limited partner(s) to obligate the firm.

Corporation

Corporations are the third form of business operation. A corporation is the most public method of doing business and more governed by state and federal laws than are sole proprietorships or partnerships. More expense is involved in establishing a corporation, but it is not excessive. The corporation is given its legal life by the state, much as cities and villages are. Therefore, to the state, the corporation becomes a legal entity, separate from its owner and managers. The profits of the corporation are taxed independently of the incomes of the company's owners. The liabilities of the corporation are also legally separate from the personal debts of its owners. A corporation can file or be the defendant in a lawsuit without affecting the personal finances of the owners.

Since the corporation is a separate entity, it has greater flexibility than noncorporate businesses. Management decisions may be easier to make if the business is not tied to the personalities of one or two owners. Some corporations, are completely managed by nonowners. Thus, the ownership of a corporation can change while the business continues without interruption.

The ownership of a corporation is obtained through the purchase of shares, each share having a particular monetary value. When major decisions must be made, requiring a vote of the owners, each shareholder is entitled to a vote proportionate to the number of shares owned. At the end of the fiscal year, the profits of the corporation are paid to the shareholders on the basis of the number of shares owned. The profits are termed *dividends*. If the corporation does not make a profit, no dividends are paid, and the value of the stock may be reduced.

There are several types of corporations, two of which are common in the ornamental horticulture industry. A *private corporation* sells its stock publicly and can be owned by many people, often as an investment. A *closed corporation* does not sell its stock publicly. It is often owned by the members of a single family. If additional capital is needed to finance expansion of the business, either corporation may sell more stock.

Operating decisions are made by a board of directors, elected by the stockholders. Stockholders cannot obligate the company. Since most ornamental horticulture corporations are closed corporations with few stockholders, the election of directors, their titles, and the conduct of meetings are usually informal. The incentive to incorporate is usually the desire to limit the liability of the owners.

MARKET SURVEYS

The desire to start a business is not enough to assure its success even when the owner has enough capital to get the business started properly. Hard work, great talent, and the support of family and friends will not guarantee the survival of a business. Simply stated, the success of a business depends upon enough people wanting the products or services of the business and having the money to pay for them.

The geographic area from which a business attracts most of its customers is termed its *market*. For garden centers, flower shops, retail nurseries, lawn maintenance firms, and landscape contractors, the market area may be limited to a city or county. For wholesale nurseries, mail-order nurseries, landscape architects, and propagators, the market area can often extend beyond state lines. The safest way to begin a business is to study the market beforehand. An analysis known as a *market survey* can answer the following questions:

- *What is the extent of the market area?* If deliveries are to be made, how far can the truck travel before the cost of delivery becomes prohibitive? How far must customers travel to reach the business? Are they more likely to use the telephone to call in orders or to come in personally?
- *Is there an established desire for the products or services of the business or must one be created?*
- *How strong is the economy of the market area?* In the opinion of some people, ornamental horticulture products and services are not essentials. Do the local people have sufficient funds to support their desire for what the business is selling?
- *How many similar businesses are already established in the market area?* A pie can only be cut into so many pieces before they become too small to satisfy. A market area will only support a certain number of similar firms profitably. A saturated market area is not the place to begin a new operation.
- *How good is the existing competition?* It is difficult to gain an edge against a competitor that enjoys a good reputation in the community and backs it with competent performance and fair prices.
- *Can qualified employees be attracted and retained?* In upper income market areas, there may be a scarcity of workers, and workers may

be unable to move into the area at the wages a new business can pay.

If possible, the would-be owner should let someone else do the market study. Wishful thinking and ego can get in the way of an objective analysis. A hopeful newcomer may be sure that customers will beat a path to the door of a new business when someone finally does it right. The facts may indicate that the business would be better located in a different area. In a saturated market area, someone must eventually close each time a new business opens. The chances are good that it will be the new business.

SOURCES OF CAPITAL

For someone considering a new business, the initial excitement of plans and dreams can be quickly replaced by the realities of financing. Where does the money come from to start a business? How much is necessary to start and support the business until it begins to show a profit?

Since there is no predetermined period of time after which a business is guaranteed to survive, capital must be discussed in terms of initial funding and subsequent funding necessary to sustain the business. Capital is needed for two reasons. One is to purchase *fixed assets* such as real estate, coolers, furniture, equipment, and similar items. The other is to pay wages, purchase supplies, and pay utility bills. Such funds are termed *working capital*.

There are three ways to obtain capital for a business:

1. Use your own money or that of other people willing to invest their money in your business.
2. Borrow from lending institutions such as banks, insurance companies, and loan companies; or from persons seeking to buy into the business as limited partners; or from other companies that will market your future crop and are willing to pay you in advance (termed *contract growing*).
3. Return the profits of the firm to the business for use as capital.

Each method of raising capital has features that may make it the appropriate or necessary choice.

Using Your Own Money

You should use your personal savings as much as possible in starting a new business. There are two main reasons for this. First, there is no interest charge when you use your own money. If money is also invested by family or friends, it can be with the written understanding that no interest will be paid on the initial investment but that they will share in the profits for a predetermined period of time. If interest is to be paid on the money of friends or family, it may be less than that charged by a bank. The second reason for using personal savings is that it is evidence of your commitment to the business. Such evidence is important when credit is sought later from other sources. Certainly the most healthy financial situation is to have enough capital available from personal sources that the borrowing of working capital is only necessary during peak seasons, and the loans are of short duration.

Borrowing the Money

For the novice as well as the established horticulturist, borrowing may be the only way to obtain capital; but it is also the most expensive way. In addition, it requires you to reveal a good deal of information about your background and personal finances, precise details of how the money is to be used, and in-depth records of the financial history of the business. Lending institutions tend to be conservative and cautious when making loans, especially to new businesses with no financial history; hence their inquiring attitude. Regardless of the institution, it will require *collateral,* or something of value to be guaranteed as security for the loan. Big ideas, enthusiasm, and a college degree are not collateral, although the personal credentials of the borrower can lend credibility and promote confidence. Usually accepted as collateral are such things as land, buildings, vehicles, equipment, crops, or accounts receivable.

Several kinds of loans are available.

Character or Signature Loans   These are short-term loans granted to businesspeople with good credit ratings. To obtain working capital for thirty to ninety days requires only the borrower's signature if the bank approves a line of credit in advance. This kind of loan is reserved for the well established and successful. A young person starting out will seldom qualify for a character loan, but it is the kind of loan all should aspire to attain.

Term Loans   These long-term loans are repaid over a period of years, usually by the month. This is an expensive loan due to the interest rate charged. Term loans may be made entirely by a private lending institution or by the federal Small Business Administration (SBA) in association with a bank. The SBA supports small businesses by granting a percentage of loans to borrowers that in its opinion deserve the chance for success. Through the agency's support, the borrower obtains capital at a rate of interest lower than what the bank alone would charge.

Accounts Receivable Loans   These are obtained by established businesspersons by pledging money owed to the business from customers. This is a short-term loan that is repaid as the customers pay their bills.

Loans on Life Insurance   These can be obtained using the cash value of the borrower's life insurance as collateral.

Limited Partnership Loans   Loans made by individuals who provide capital in return for a share of the profit, this type of capital is termed *equity funds.* Most owners do not want strings attached to the loan, particularly attempts by the lender to influence management decisions, so the terms of the loan need to be spelled out in writing.

Contract Growing Loans   Made to growers by firms that will eventually market their crops, these loans provide the grower with the money necessary to produce a particular crop while assuring the lender of a guaranteed price. Recently, contract growing has been used by growers to assure that they will have a market for their crops and thus avoid expensive overproduction.

Reinvesting Profits into the Business

Retaining the profits that a business makes rather than paying them out to the owners is an important method of raising interest-free capital. While it is not enough for a developing and expanding business, it can be of major significance for a successful, established firm that needs capital to replace inventories and depreciated equipment or to pay taxes and other operational costs. In order to gain the greatest benefit from this source of capital, the administrator of a business must be attentive to money owed the company. The longer customers withhold payment, the more it costs the business, since an unpaid account represents money tied up in expended materials and services that is unable to be used.

SELECTING THE BUSINESS SITE

One of the initial uses of capital is to obtain a site for the new business. The opportunity to select from among several sites is not always available. When it is, however, numerous factors must be considered.

Size   Size of the site must be evaluated in terms of how much area the business needs at the beginning and how much expansion is planned if the business is successful. A field nursery requires more area than a garden center, and both need more space than a retail flower shop. A landscape designer does not require much space to operate but would need a lot more to expand into a contracting business. Small sites that offer no room for expansion should be avoided if the horticulturist foresees the business eventually needing greenhouses, lath houses, or other production or storage facilities. Conversely, large sites should be avoided by small retail businesses. Taxes and other costs of maintaining a large site can be good reasons for not purchasing a piece of property that is larger than will be needed now or in the future.

Natural Features   Natural features of the site and adjacent areas are of special concern to production firms. Fertile, rock-free soil of good texture and drainage is necessary for a field nursery. Good quality air, available water, and ample sunshine benefit all production nurseries as well as greenhouses. Level terrain that is easily accessible and areas that are protected from harsh extremes of weather are also of importance for any production firm.

Garden centers, landscape nurseries, flower shops, and maintenance firms may benefit from the presence of shade trees, bodies of water, rock outcroppings, or other natural features that contribute to the aesthetics of a site.

Zoning Regulations   Zoning regulations determine what use can be made of a site. Some areas of a community may be zoned only for residential development and perhaps for churches and schools. Other areas may be zoned for heavy industry and still others for light industry or commercial purposes. Zoning ordinances may restrict expansion, regulate architectural style, or even specify landscape features. The zoning of an area can also affect the rate of taxation on the land and buildings. Careful investigation of the zoning patterns of a community and the frequency of zoning changes can prevent a site that seems otherwise perfect from becoming a liability.

Utilities   Utilities must be on the site or easily installed to avoid high costs of hookups. Nurseries and greenhouses need adequate sources of fresh water. Everyone requires electricity, and special wiring or special forms of electricity may sometimes be needed. Oil, gas, or other fuel suppliers must be nearby and reliable.

Access   Access to the site is of prime concern. The type of access depends upon the type of business. A retail operation needs to be close to main streets and highways so that suppliers can make deliveries. Growers who wholesale their plants must be near highways and railways if the plants are to reach their markets. Equally important is access to the business for customers. A landscape nursery needs to be on a major road near the community of clients. A flower shop or garden center has the best chance of attracting impulse buyers if it is located along a major road where traffic is heavy and access is convenient. However, if traffic is too hectic, as at certain intersections, motorists may avoid stopping. One-way streets can also be bad sites if homeward bound traffic is passing in the opposite direction on another street. Where walk-in traffic is an important source of customers, the store could benefit from a location along a pedestrian mall, in a shopping center, or even in a major hotel lobby.

Compatibility   Compatibility of the new business with other businesses around it will frequently determine whether customers are attracted to the operation. Other businesses generate customer traffic. Since most retail flower shops, garden centers, and nurseries benefit from people willing to linger and shop, it is advantageous to be located among other businesses that cater to shoppers. Convenience stores that cater to a dash-in and dash-out trade do not make compatible business neighbors. Stores that serve customers who remain in their cars, such as banks and fast-food restaurants, make especially poor nearby neighbors.

Merchants' Associations   Merchants' associations are common in retail areas and are often very helpful to new and established businesses alike. The members of active associations often share the expenses of group advertising, holiday decorations, security systems, remodeling, group insurance plans, and other items of collective interest. At times, as necessary, they can even work as an effective political lobby. In evaluating a potential retail site, it is worthwhile to ask nearby merchants about the quality and effectiveness of the local merchants' association. Lack of such an association may indicate an area on the decline.

Lease Agreements   Lease agreements are usually required when land or buildings are rented rather than purchased. A lease should not be signed until the business operator has thought carefully about the future. If a long-term lease is sought, the agreement may need to permit expansion when needed. If another site is likely to be preferable when capital availability permits it, a short-term lease may be preferable. Any responsibilities of the landlord should be written into the lease agreement. Maintenance responsibilities are especially important, and the period of time for their accomplishment should be specified. If possible, an agreement that vacant properties nearby will not be leased to competing businesses should also be written into the lease.

Cost of Acquisition   Cost of acquisition is the final factor in the evaluation of a site. For some would-be business owners it may be the most important factor of all. The cost of the site should be weighed carefully against its advantages and disadvantages to determine if the price is fair or if another site is a better value.

Buying an Established Business

An alternative to beginning a business from the ground up is to purchase an established operation. In many cases, the cost of entry into the profession can be reduced if an established business can be purchased at a sacrifice price.

A horticulture operation will usually be put up for sale for one of three reasons:

1. The owner has reached the age of retirement and has no one to carry on the business.
2. The owner has become ill and is unable to carry on the business.
3. The business has been unsuccessful.

The first two situations can provide excellent opportunities to acquire a successful business at a reasonable price. The third situation does not provide such an opportunity. An unsuccessful operation may have a bad location, obsolete or improper equipment, or a history of poor products or service that leaves it with a bad reputation. It is difficult to overcome such obstacles.

When a successful business is purchased, a certain following of established customers (termed *goodwill*) comes with the purchase. It is an important asset but not one that can have a value (and price) assigned to it easily. Sellers usually try to inflate its value; buyers should resist paying much for something that is so difficult to appraise.

When considering the purchase of an established business, the future owner should look at its past performance as an indication of how it will do in the future and also assess the potential of the business under new ownership. The seller should be asked to supply profit-and-loss statements for the preceding five years. These should be taken to an accountant and studied carefully. The earnings, the increase or decline of sales, the salaries of employees, and the budget for advertising during the five years can all provide a picture of the past performance of the company. In attempting to determine how the business will do under new ownership, there must be a strong indication of steady growth over the next five- to ten-year period. The business must promise not only a good salary for the owners, but a return on the invested capital. If the projection by the accountant does not favor both salaries and profits, it would be best to keep looking for a more suitable business.

If and when a good business is found for purchase, the selling price must be evaluated in terms of the factors that it represents:

- *Equipment* will seldom be brand new. Its value at the time of purchase will be greater if it is newer. If it is old and has been fully depreciated, it may be a liability.
- *Existing inventory* is not always usable. It may not be what the buyer would have stocked and may contain outdated products that have been rejected repeatedly by customers. The inventory values on the books of the seller should not be accepted without investigation. Items may be carried at full value when in reality they are unsalable and nearly worthless.
- *Accounts receivable* are monies owed to the seller that will be owed to the buyer after the sale is finalized. If a large percentage of those receivables are long-standing bad debts, they should be discounted in all or in part.
- *Goodwill,* as already mentioned, has a vague value and should not represent a large percentage of the sale price.
- *Use of the existing name* may be part of the sale price or may have no value assigned to it. It may be advantageous to use the established name for a while, especially if the firm has a good reputation, but the advantage should be carefully assessed. It is difficult to assign a value to reputation.

USING SPACE EFFICIENTLY

Once acquired, the site must be put to use in a manner that will maximize the return on the capital.

Space must be assigned to each of the major areas of the business. For example, a retail business such as a garden center usually requires areas for:

1. customer sales and product display (indoors and outdoors)
2. inventory storage
3. plant holding
4. parking
5. employee use
6. climate control

In addition, production operations such as greenhouses or nurseries require areas for:

7. plant growth
8. equipment and supply storage
9. plant grading and packaging
10. shipping and receiving

Each area can be further divided into subareas. Examples of these subareas would include:

1. *customer sales and product display*
   - space for shelves of products
   - display cooler for cut flowers
   - display greenhouse for foliage plants and potted flowers
   - outside sales yard for woody plants, bedding plants, lawn ornaments, and bulk materials
   - rest rooms
   - check-out area
   - landscaped setting for the business

2. *inventory storage*
   - warehouse for bulk materials and equipment
   - coolers
   - storeroom inside the retail center

3. *plant holding*
   - heeling-in/overwintering
   - acclimatization
   - overwintering, bare-root storage

4. *parking*
   - customer parking
   - employee parking
   - access roads
   - space for snow to be piled

5. *employee use*
   - offices
   - workrooms
   - lockers and rest rooms

6. *climate control*
   - central heating
   - air conditioning
   - windbreaks

7. *plant growth*
   - greenhouses
   - nursery fields
   - lath houses
   - propagation beds

8. *equipment and supply storage*
   - garage
   - fuel tanks
   - small engine repair shop
   - tool shed

9. *plant grading and packaging*
   - grading and packaging room
   - overwintering, bare-root storage

10. *shipping and receiving*
    - maneuvering space for trucks
    - loading dock
    - access roads
    - space for snow to be piled

The use of space in a business must be analyzed at several levels. First there is the allotment of space to selected subareas. For example, parking lots must be large enough to hold the number of vehicles expected to be there at one time on a busy day at peak season. Assigning too little space to the lot will result in lost sales and annoyed customers. However, assigning too much space to the lot may deprive other areas of the space they need. Similarly, customer sales areas must allow enough room for comfortable browsing through a diversified array of products. Allowing too much space between products will limit the number of items that can be displayed, and crowding too much in will make the shopping area look like a bargain basement. Some florists and landscape design firms assign space to customer consultation areas. Although

such areas are nice to have, the space might be used more profitably if the retail area is small.

Use of space must also be analyzed in terms of inventory control. Quantity buying is generally regarded as wise because it results in lower unit costs. However, it also requires more space to store the inventory and ties up capital until the inventory is sold. In small shops where the storage area is small, quantity buying can result in limited product offerings.

Space efficiency is also important in the production of plants. Production schedules must be

FIGURE 22-1.
An example of proper layout for a flower shop. Comparative sizes of areas are shown and circulation patterns are predicted.

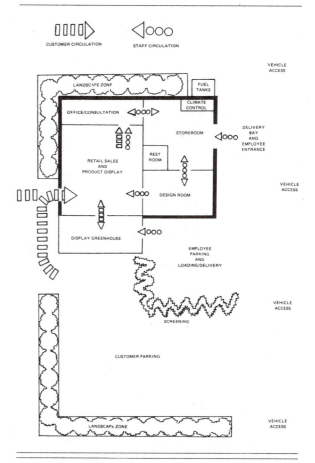

FIGURE 22-2.
An example of proper layout for a production nursery. Predictable circulation patterns are shown. Conflicting intersections are minimal.

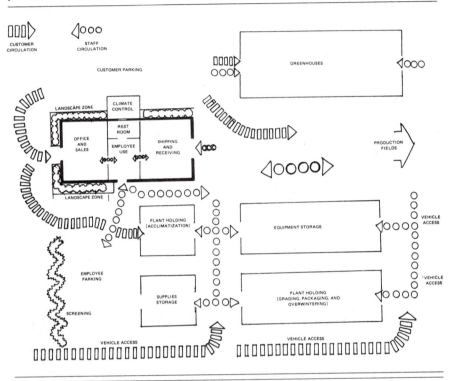

planned so that bench space is available when it is time to expand the spacing of a crop. Planning must also foresee when a bench will be empty so that another crop can be started as soon as possible. Heating a partially empty greenhouse drives up the grower's cost of operation. Equally costly and inefficient is a partially harvested nursery field if no new crop is planted. With land as expensive as it is, few growers can afford inefficient use.

ORGANIZATION PLANS

It is a good idea to plan the layout of the business on paper—how big each area and subarea are to be and how each is to relate to the other. This *physical plant layout plan* should be made while the business is still in the developmental stage. It allows the owner to think about the business in an organized way and also provides a beginning point for rethinking the layout if expansion is needed later.

Proper sizing of each area and *ease of circulation* are the keys to a successful layout, Figures 22-1 and 22-2. Examples of sizing have already been given. Circulation patterns are equally important. Both the customers and the employees must be considered. Customers will seek the most direct route from their cars to the products they came to buy. They do not want to pass through work areas or go past the loading dock in the process. They are generally receptive to passing other products on the way to the product they want, however. Meanwhile, employees do not want to be routed through the customer sales area in order to use the rest room. Neither do employees want to take their coffee break in the sales yard.

FIGURE 22-3.
An example of the staff organization in a medium-sized garden center

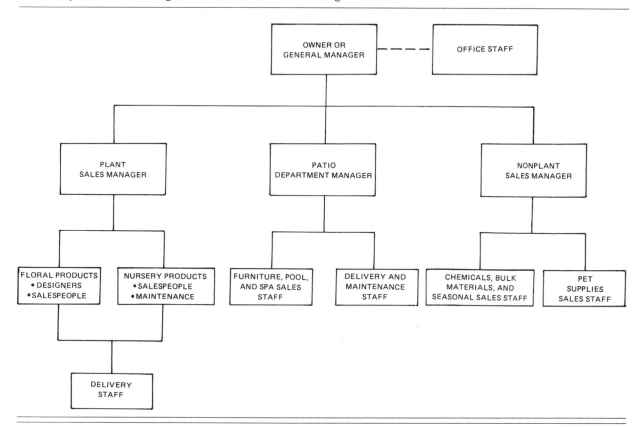

A staff organization plan can also be useful when a business is starting out and each time that expansion or reduction in the staff causes duties to be reassigned, Figures 22-3 and 22-4. A staff organization plan helps employees understand their roles within the company. This reduces conflict over who has responsibilities in a particular area. No business is too small to benefit from an organization plan.

BUILDING THE INVENTORY

Once a business is established and has earned a good credit rating with suppliers, it will have little difficulty obtaining materials and supplies as needed. A new business may encounter some problems at the outset, however. First, suppliers are wary of extending credit to someone who has no credit his-

tory. Second, if shortages develop, suppliers are likely to service their established customers first, leaving newcomers with unfilled orders.

The best way to begin the purchase of materials and supplies is to order small quantities that can be turned over quickly and paid for promptly. Large purchases tie up capital too long and can create credit problems. Also, many items are highly seasonal (such as holiday items) or highly perishable (such as flowers and bedding plants), and these should only be purchased in quantities that can be turned over quickly.

When budgeting for inventory items, it is important to anticipate all of the costs that will be encountered. Cost of the stock is obvious, but there are others, such as the staff time spent purchasing, receiving, and storing the stock, and the cost of keeping stock control records.

FIGURE 22-4.
An example of staff organization in a landscape contracting firm

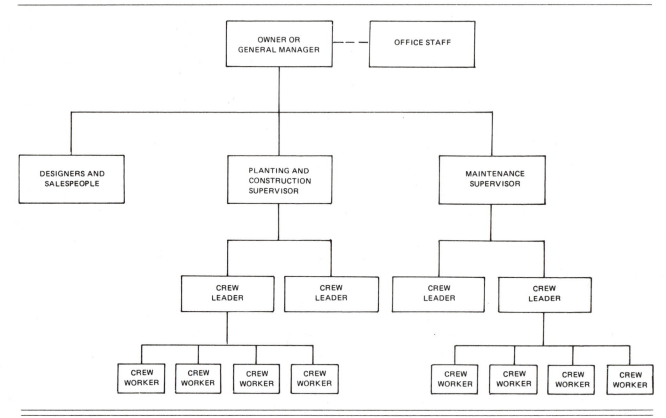

To assure an uninterrupted supply of materials, a reliable source (*vendor*) should be found for each item that will be needed regularly. Records should be kept regarding the vendors' terms, the credit lines they have extended for the business, who pays the shipping charges, how delivery is made, and the time usually required to fill orders. It is also desirable to establish cordial relationships with the salespeople who represent the vendors in your area.

LAWS AND REGULATIONS

Like all businesses, horticultural businesses are subject to regulation by all levels of government. Horticulturists may have an above-average number of laws and regulations to comply with, especially if they are involved in plant production. Some of the laws and regulations are described in Table 22-1.

ADVERTISING THE BUSINESS

The purpose of advertising is to focus public attention on items or services that are for sale. Advertising is related to, but different from, promotion, publicity, and public relations. All are needed for the successful growth of any business.

A modern horticultural business must advertise to retain old customers and add new ones. To be productive, advertising must generate both *sales* and *profits* and generate them in quantities greater than could be expected without advertising. Customer traffic and sales volume are not enough to make advertising successful. It must make profits for the business.

FIGURE 22-5.
This example of immediate response advertising was mailed to all regular customers of a flower shop prior to Mother's Day. It promotes the shop and features the holiday specials of its wire service affiliation. (Courtesy Speedling Florist)

the telephone or come into the store soon after reading or hearing the advertisement. Such advertising must offer merchandise or services at a time when customers are receptive to them. For example, flowers and special arrangements are of predictable interest to consumers at holiday time and should be advertised by florists just prior to the holidays, Figure 22-5. Home landscaping is on the minds of property owners in the spring more than in the winter. The timing of the advertising is important for maximum return on the advertising dollar. The purpose of *attitude advertising* is to build the reputation of the business. Done on a regular basis, attitude advertising seeks to implant in customers' minds an understanding of the services or products available from the business along with a sense of its status or reliability, Figure 22-6. Successful attitude advertising will cause customers to seek out the business in the future without comparative shopping.

The Advertising Media

The media through which a business can advertise are known to nearly everyone: newspapers, magazines, billboards, telephone directories, radio, and television. In addition, small businesses can utilize direct mail, handbills, display windows, county fairs, flower and garden shows, and a variety of other means.

Advertising is a cost of doing business and must be budgeted in advance. The business owner must know which media will result in the greatest return on the money spent. The point to remember is that *advertising must result in measurable profits*, then certain futile expenditures of advertising capital will be avoided. Examples of wasted advertising dollars include high school yearbooks, lodge and church publications, local athletic team programs, pencils, and matchbooks. The population reached by such media is predictably small, yet the costs can be high compared to other methods of advertising. While some business owners find it difficult to say "no" when asked to advertise in their own lodge or church publications, it is necessary if the money assigned to advertising the business is to be used pro-

There are two types of advertising, each with a different objective. The purpose of *immediate response advertising* is to make customers reach for

TABLE 22-1.
Laws and Regulations Affecting Horticulturists

Law or Regulation	Purpose	Cost		Comment
		Employer	Employee	
Deceptive Practice or Product Regulations	To protect the consumer from unrealistic descriptions and claims for products and services	No costs		Regulations such as these may be consumer protection laws or industry-initiated trade practices.
Fair Labor Standards Act	To establish a minimum wage for all workers; to establish standards for payment of overtime work; to prevent abuse of child labor	Cost borne by employer		Equal pay must be given for equal work on jobs that require equal skills and responsibilities.
Farm Labor Contractors Registration Act	To assure fair treatment of migrant workers. They must be accounted for through regular payroll accounts and must be provided by each employer with information about their work area, housing, insurance, transportation, wage rates, and charges.	Cost borne by employer and paid to a farm labor contractor		Nurseries are most likely to be affected.

Act	Purpose	Compliance	Cost	Notes
Occupational Safety and Health Act	To assure safe and healthful working conditions for employees	No cost, but required to comply with the law by using all safety devices properly	Cost borne by employer	All employers with at least eleven employees are covered by the law. The law is enforced by the states.
Pesticide Application Certification (Environmental Pesticide Control Act)	To limit the use of toxic pesticides and to assure that they are used only by persons knowledgeable about their safe and proper use	Same	Cost (minimal) borne by applicator	Once certified, the applicator must stay up to date by acquiring a number of recertification points by a certain date. Points are awarded for participation in various professional meetings.
Plant Patent Act	To protect the right of the inventor or discoverer of a plant to be the sole asexual propagator of the plant		May increase cost of production for a grower	Plant patents are valid for seventeen years.
Plant Pest Act	To regulate by quarantine the entry and distribution of plants and plant products into and across the United States in order to protect against foreign plant pests		Fees charged for import/export permits and nursery inspections	Quarantines may control the movement of plants between countries, between states, or within individual states.

TABLE 22-1.
Laws and Regulations Affecting Horticulturists (continued)

Law or Regulation	Purpose	Cost		Comment
		Employer	Employee	
Price and Services Act	To assure that all commercial customers are treated equally by a seller by forbidding selective prices or services	No cost, but employer must assure compliance		The use of printed wholesale price lists is a common means of proving compliance with the law.
Social Security	To provide financial assistance to persons after their retirement; to provide income for surviving dependents after a worker's death	A percentage of a base amount of employee's wage (subject to change over time)	An equal amount is withheld from paycheck	Current trends are for both the percentage and the base amount to increase regularly to meet the expenses of the system.
Unemployment Compensation	To provide financial protection for workers who lose their jobs for reasons other than firing	3.4 percent of the first $6,000 of wages (subject to change over time)		
Warranties Act	To define the terms "full warranty" and "limited warranty" as used by tradespeople	Replacement cost borne by employer		Horticulturists are not required to offer warranties, but if they do they must either replace a product completely or in part in compliance with the terms of their stated warranties.
Workmen's Compensation Insurance	To provide compensation to employees injured on the job	Cost borne by employer		Coverage is required for all businesses having at least one employee in a hazardous occupation or at least three employees.

FIGURE 22-6.
This example of attitude advertising was run in the newspaper in the early spring. It states clearly what the firm does and suggests quality and reliability. (Courtesy Albert Glowacki Landscape, Inc.)

> Nantucket's largest professionally trained staff in design build and landscape maintenance is currently quoting for summer delivery of landscape construction and installation.
>
> We will be pleased to discuss your requirements for decks, fencing, terraces, or any specialized construction. We feature custom design services, including AUTOMATED IRRIGATION SYSTEMS.
>
> Weekly deliveries from New England's best known growers of high quality plant material are scheduled throughout the season.
> Thorough maintenance services have always been our specialty.
>
> ALBERT GLOWACKI LANDSCAPE, INC.
> P.O. BOX 928, NANTUCKET, MASSACHUSETTS 02554 . TELEPHONE: (617) 228-1867

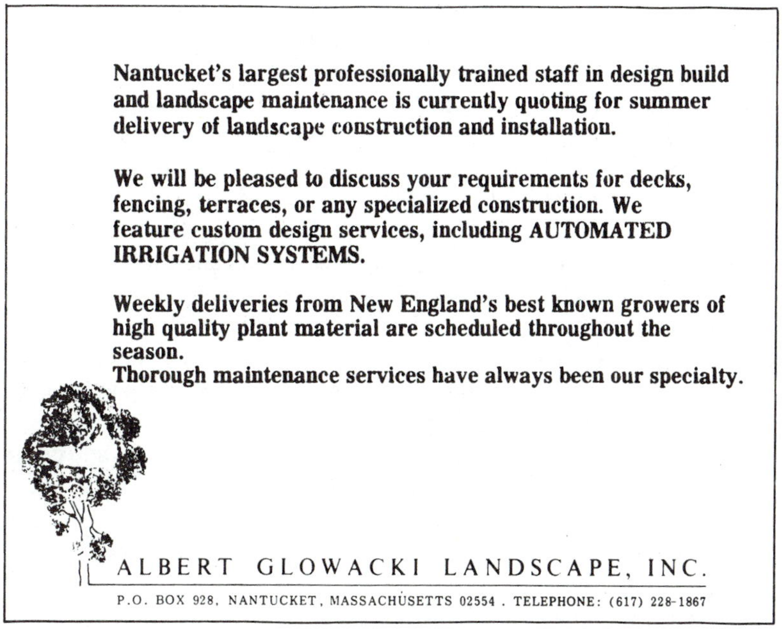

ductively. Money given in support of local church, school, and civic organizations should be regarded as public relations or charitable donations, not as advertising.

Of the established advertising media, some work better than others for different businesses. A retail florist may find that a weekly advertisement in the local newspaper builds public awareness of the shop (attitude advertising) and also increases the sale of weekly specials (immediate response advertising). Garden centers, nurseries, and landscapers may use the local newspaper, television and radio stations, and even billboards during the spring and fall when customers are thinking about their home landscapes. Such advertising is usually attitudinal, yet it can evoke immediate response if it features specific products or services at promotional prices.

Wholesale nurseries, large landscape contractors, and landscape architects gain little benefit from newspaper or broadcast advertising. They must reach a customer population that often extends across state lines. Magazines, particularly trade journals, are one means of reaching their clientele. Exhibition at industry conventions is another means of putting the business before the proper audience. Direct mail advertising is also used. As much as possible, journal advertisements, exhibitions, and direct mail should induce an immediate response, although attitudinal advertising is valuable and commonplace in this market also.

Telephone directory advertising can vary in its value to the business. In a small community, most citizens are familiar with the few horticultural operations in the calling area. Their choice of one over the other is not likely to be influenced by the presence or size of the advertisement in the yellow pages.

In a larger city, customers new to the area or infrequent users of horticultural products or services may use the directory to select a flower shop, garden center, retail nursery, landscaper, or lawn care firm. How satisfactorily they are served will usually determine whether they return to the firm or to the directory. Many established retail operations and most wholesale businesses find limited value in telephone directory advertising, especially since it is almost totally attitudinal in approach.

Immediate response advertising has a short-term value. It must motivate the customer to seek out the advertised product or service the same day, the same week, or the next at the latest. The value of such advertising is both direct and indirect. Directly, the items advertised must sell in excess of what would have been sold without advertising and generate enough revenue to pay for the advertising while leaving a profit on the advertised items. Indirectly, the customers advertising attracts to the business may make additional impromptu purchases. A new business operator should study the response of consumers to determine which medium or combination of media is most effective for immediate response advertising.

The value of a particular medium may depend upon the items being advertised. For example, a flower shop might invest in a series of thirty-second radio commercials. If prom corsages are advertised, a classical FM station will not reach the teenage market as well as an AM rock station. Advertising prom corsages could generate a profit if done at the right time and through the right medium. Still, it would be unlikely to result in impromptu sales due to the generally limited interest and finances of a teenage clientele. If the same advertising dollars were spent on immediate response advertising to an older and wealthier clientele, the profit return would probably be greater.

The single greatest value of any advertising is to let customers know that you have a good supply of what they want at the time they want it. Advertising to create a demand for a new product is often beyond the capabilities of a small business and must be left to manufacturers and trade organizations.

Advertising has definite limitations. It can never fully compensate for a poor business location or inadequate parking facilities. Also, it can never sell a shoddy item or service more than once, and it may actually lose customers for the horticulturist who tries to unload poor-quality merchandise through special sales. The quality of the merchandise or service must be high even when the price is reduced.

Since advertising can be attitudinal, it follows that another limitation is the danger of a negative attitude being created through a poorly produced advertisement. For example, businesses that seek an upper-income clientele should avoid advertising like high-pressure used-car lots. Businesses that serve a middle-income clientele should avoid advertising that suggests high prices and Park Avenue attitudes.

Promotion, Publicity, and Public Relations

A complete advertising program must include promotion, publicity, and public relations. Often, no clear distinctions can be drawn among them. In general, promotion is the range of activities whose purpose is to establish goodwill or further the growth of a firm. For example:

- sending greeting cards, complimentary plants, or other gifts to important customers at holiday time
- sending clipped newspaper announcements of engagements to brides
- sponsoring free lawn clinics, floral design shows, pruning clinics, or similar events just before major sales periods

Publicity and *public relations* activities are intended to bring the business and its staff to the attention of the public (and potential customers) in order to generate favorable public opinion. For example:

- writing articles about plants, flowers, or landscaping for the local newspaper
- holding an open house before major holidays or planting seasons
- providing shirts for the local Little League
- joining and speaking before civic groups on horticultural topics

The return on the dollars spent in these areas is not easily measurable, and so the budget should not lump them together with advertising. To do so would complicate and distort the interpretation of the effectiveness of the advertising program.

Budgeting for Advertising

At the outset, the business owner must decide what percentage of the anticipated sales revenue will be spent on advertising. The percentage will vary with the financial condition of the business, the amount of competition in the market area, and the nature of the products or services sold. Multiplying the total anticipated sales for a year of operation by the percentage allotted for advertising, a dollar amount can be assigned to the advertising program for the year.

Some business owners determine how much to spend each month by correlating the percentage of advertising dollars with the percentage of total sales that each month contributes. For example, if June contributes 12 percent of the annual sales, then 12 percent of the advertising budget is spent for June sales. Some months are naturally high-volume sales months, however. Christmas sales in a flower shop are an example. The sales for the month of December may be 35 percent of the annual sales, yet only 20 percent of the advertising budget may be needed to generate the sales.

An advertising budget should also allocate funds to special, nonrecurring public relations or promotional events such as the store's grand opening, historical commemorations, or the sale of an entire truckload of plants obtained at a special price. Special-event advertising should be added to the regular advertising budget to arrive at the total allocation.

The United States Small Business Administration, in its *Advertising Guidelines for Small Retail Firms,* recommends that a monthly record of advertising expenditures be kept to determine if the advertising budget is being spent as planned. The chart in Table 22-2 can be used to keep such a record.

Writing Descriptive Copy

The best way to obtain advertising copy that will make customers call or stop in is to hire a professional advertising agency. A small business often cannot afford professional assistance, however, so the owner or manager must either prepare the copy or provide the necessary information to the advertising staff of the newspaper or radio or television station. Therefore, a new business owner should have a basic understanding of how to construct a good advertisement.

A lot of money is spent on nonproductive, even counterproductive, advertising. Humor that falls flat or wears thin after repeated exposure does not make good advertising. Advertisements that do not contain information about specific products or services are unlikely to result in increased sales. Omitting prices or allowing customers to suppose them higher than they really are is not productive. Advertisements that criticize a competitor or make unbelievable claims for a product or service may harm the reputation of the business.

To prepare honest and descriptive advertising copy that sells, remember the following points.

1. *Study the buying patterns of the customers and advertise to appeal to them.* For example, bronze mums are properly advertised in the fall but are not what customers are seeking during the Christmas season. Hardy bulbs for fall planting should be advertised then, not during the winter, when they are obvious leftovers.

2. *Advertise representative products and services.* For example, avoid featuring low-

TABLE 22-2.
Advertising Expense and Budget Record

Account	Month		Year to Date	
	Budget	Actual	Budget	Actual

Media
  Newspapers
  Radio
  Television
  Literature
  Other				

Promotions
  Exhibits
  Displays
  Contests				

Advertising expense
  Salaries
  Supplies
  Stationery
  Travel
  Postage
  Subscriptions
  Entertainment
  Dues				

Totals				

cost items if the business normally carries a better line of merchandise.

3. *Give customers good value for their dollars.* Promoting broadleaved evergreens in an area that is too cold to assure their survival will not create goodwill. It is better to feature plants of predictable hardiness.

4. *Be definite about the price of what is advertised.* Avoid giving price ranges that make it hard to determine what an item will cost. Featuring flowering shrubs for $5 to $30 is too vague. Pricing them at $5 and up is even worse. Tie the price to a size or quantity so that customers know what they will get for the money.

5. *Describe the benefits to the customers of purchasing the product or service.* The size or length of the advertisement will determine how many benefits can be described, but the most obvious benefit as well as the one most likely to motivate the customer should be mentioned. For example, a lawn service firm can advertise benefits such as "rich, green, weed-free turf" (an obvious benefit) and "more free time to spend enjoying the summer" (a strong motivator).

6. *Suggest a need for the item.* This can often be incorporated into the description of benefits.

7. *Encourage the customer to act immediately.* Close the advertisement with a statement that motivates customers to pick up the telephone or stop by the store while it is still fresh in their minds.

8. *Keep the layout of printed advertisements simple.* Use a heading in large type and an illustration to catch the reader's eye; follow with the copy; close with the price and store name, address, and hours.

9. *Develop a recognizable format for a series of advertisements.* Although specific items may change, the look or sound of the advertisement should strike a familiar note. The use of standardized graphics can help to create the desired identification.

10. *Keep broadcast advertisements conversational.* Avoid an impression of aloofness.

11. *Devise a way to measure public response to the advertisement.* Use of a coupon in a printed advertisement allows a count of the number of persons responding. Radio and television advertisements can offer some special benefit to customers who use a key word when placing their orders. Figure 22-7 is an example of a good written advertisement for ornamental horticulture.

FIGURE 22-7.
Advertisements such as this are placed in trade publications by wholesalers to introduce new products to the industry. It is not directed to retail consumers. (Courtesy Lake County Nursery Exchange)

Using Displays to Advertise

The business itself may be its own best advertisement. A customer who comes to buy one item may be encouraged to buy others if they are displayed freshly and attractively, Figure 22-8. In the jargon of the professionals, such displays are known as *point-of-purchase advertising.* Florists, garden centers, and retail nurseries may use or feature live materials in their displays. Landscapers and lawn care services may use color photographs of completed properties. Other displays may feature bottles, boxes, or bulky bags of chemicals, seeds, and fertilizers, Figure 22-9. The following techniques will help make in-store displays more successful.

1. *Keep the display simple.* Feature either a single product or service or a limited choice of related products or services. Complexity in the display can create indecision in the customer's mind.

2. *Suggest a use for the product.* This may establish a need for the item in the customer's mind.

3. *Keep the display small.* Display a few of the items, but avoid a mass display. Create a qualitative image rather than a quantitative one.

4. *Keep the display complete.* If customers are to lift items out of the display for purchase, replace the items quickly.

FIGURE 22-8.
A display of silk flower basket arrangements in a retail flower shop

5. *Coordinate the display with other advertising.* The display may feature merchandise advertised in the newspaper or on the radio or television.
6. *Keep all plant material in peak condition.* Do not let wilted blossoms, yellow leaves, wilted foliage, or broken branches suggest merchandise of inferior quality.
7. *Neatness counts.* All signs should be neatly lettered, unfaded, and free of spattered water, soil stains, and dust.
8. *Light the displays.* Show windows offer opportunities for effectively lit displays even after the store closes for the day.

9. *Change the displays frequently.* Regular customers will cease to see them if they remain unaltered for too long.

SUMMARY

Ornamental horticulture businesses may be structured as sole proprietorships, partnerships, or corporations. Each has features that make it appealing, but the corporation offers the greatest legal protection to the owners. Also, a corporation may offer the greatest opportunities for diversification and growth of the business.

No one should begin in business without a careful market survey to determine the probability of success in the location under consideration. The study should analyze the range of the market area, the need for the business, the economy, the competition, and employee availability.

A business requires capital to purchase fixed assets and to keep the business operating (working capital). Capital can be obtained by using personal savings, borrowing from lending institutions, or by reinvesting profits into the business.

In selecting a site for the business, the buyer should consider factors such as size, natural features, zoning regulations, utilities, access, compatibility of the new business with other businesses nearby, strength of merchants' associations, lease agreements, and the cost of acquisition. If purchasing an established business, the buyer should consider factors such as the equipment included and its condition, the existing inventory and its market value, the accounts receivable, goodwill, and use of the existing name.

Once acquired, the business site must be allocated to different uses. Proper sizing and ease of circulation are the keys to a successful layout of the physical plant. A staff organization chart can help employees understand their roles within the company.

When starting out, materials and supplies should be ordered in small quantities that can be turned over quickly and paid for promptly. The less capital that is tied up in inventory, the better.

FIGURE 22-9.
A prominent display of spreaders and wheelbarrows is placed where customers will pass them as they enter and leave this garden center.

Advertising seeks to focus public attention on items or services that are for sale. To be productive, advertising must generate sales and profits in quantities greater than would be expected without advertising. There are two types: immediate response advertising and attitude advertising. A complete advertising program must also include promotion, publicity, and public relations.

This chapter also includes a summary of the laws and regulations that affect horticulturists.

Achievement Review

A. ESSAY

1. Prepare a brief essay on why you would or would not like to own your own business. Discuss the factors that you think will contribute most to your job satisfaction and how ownership or nonownership would affect them.

B. SHORT ANSWER

Answer each of the following questions as briefly as possible.

1. Indicate if the following apply to sole proprietorships (S), partnerships (P), or corporations (C).
   a. the most public method of doing business
   b. the easiest form of business to begin
   c. the most private method of doing business
   d. a business owned by two or more persons
   e. a legal entity independent of its owners

f. liabilities legally separate from the owners

2. Explain the value of a market survey undertaken before investing money and time in a new business.

3. List three ways to obtain capital for a business.

4. Explain briefly how each of the following factors influence the choice of a business site:
   a. size of the site
   b. natural features
   c. zoning regulations
   d. utilities
   e. access to the site
   f. compatibility
   g. merchants' associations
   h. cost of acquisition
   i. lease agreements

5. Identify each law or regulation described below.
   a. provides financial assistance to persons after their retirement; supported by payments from employees and employers during the working years
   b. assures safe and healthful working conditions for employees
   c. provides compensation to employees for injuries incurred on the job
   d. assures that all commercial customers are treated equally by a seller
   e. limits the use of toxic pesticides to those knowledgeable about their safe and proper application

6. Complete the following statements.
   a. _____ advertising has a short-term value.
   b. The main purpose of _____ advertising is to create a positive image of the business in the minds of customers.
   c. _____ activities are intended to bring the business, its owner, or its personnel to the attention of potential customers and to create good will.
   d. The purpose of all _____ is to focus public attention on items or services that are for sale.
   e. Advertising must result in _____ profits.

C. TRUE/FALSE

Indicate if the following statements are true or false.

1. Parking should be assigned to whatever space is left on the site after all other uses are given their necessary space.
2. Quantity buying may be unwise if storage space is limited.
3. A staff organization plan helps employees understand their roles within the company.
4. Most horticultural businesses are too small to need a staff organization plan.
5. Unsold inventory represents capital that cannot be used for other purposes.

D. DEMONSTRATION

Select an item or service that would be sold by the type of ornamental horticulture business that interests you. Prepare a radio advertisement for it. Specify the time of year the advertisement should be featured.

23

Human Relations: Personnel Management and Customer Sales

Objectives

Upon completion of this chapter, you will be able to

- define and state the objectives of personnel management.
  - list the characteristics of a good personnel manager.
    - list the characteristics of a good salesperson.
      - describe an effective sales procedure.

THE IMPORTANCE OF HUMAN RELATIONS

The owner or manager of a horticulture business is at the top of a human relations triangle that includes the employees of the business and the customers of the business, Figure 23-1. No business can operate for long with a serious problem between any members of the triangle. It is the responsibility of the manager to see that such problems do not arise and, if they do, to resolve them. Employees must be kept happy with their work and productive on behalf of the company. Customers must be pleased with the products and services they obtain

from the company and equally pleased with the people who wait on them, answer the telephone, deliver to their homes, install or maintain their landscapes, or respond to their complaints. The two groups will be discussed separately, but the dependency of the business upon a compatible relationship between employees and customers is obvious.

PERSONNEL MANAGEMENT

The direction of workers in a manner that brings out their best efforts and attitudes on behalf of the business is termed *personnel management*. While the owner of the business is the foremost personnel

FIGURE 23-1.

The triangular relationships within a business

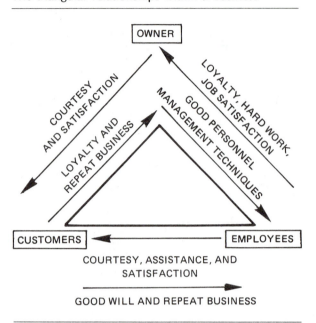

manager, other persons in supervisory or leadership positions must also develop good human management skills.

Much of what constitutes good personnel management seems to be somewhere between common sense and the Golden Rule. In addition to being an application of courtesy and friendliness in the work place, personnel management requires good leadership skills.

Setting Realistic Expectations

The supervisor of any group of employees needs to understand what those employees can be expected to accomplish. Expectations may change as employees gain experience on the job.

Certain job skills are more physical; others are more intellectual. Some require charm and personality; others require the ability to be decisive and inspire customer confidence. A supervisor usually has different expectations of a college-trained permanent employee than of a part-time high school youngster.

In any case, a supervisor's ability to get along with all employees begins with the ability to set realistic expectations.

Depending upon its type and size, a horticulture operation may have up to three basic categories of employees. A supervisor will probably have different expectations of each.

Career-Directed These employees can be expected to have personal goals for advancement. They may want to stay with the company for many years or they may intend to start their own business after a few years of on-the-job training. They may have technical training or other education that is useful to the business. They can be taught new skills or given increasing responsibilities and be expected to learn quickly. Career-directed employees need frequent and varied rewards, including periodic salary or wage increases. They usually have opinions and suggestions for improvements, which should be listened to and respected.

Part-Time These employees' career goals usually do not involve advancement in the business. The job may be a way to supplement family income, a way to fill empty time, or a way to earn money for a short-term objective such as schooling, a new car, or a vacation. Part-time employees may have education or practical experience to bring to the business. They may possess the maturity and skills needed to staff the business over weekends or during vacation periods. They often feel as much loyalty to the company as career employees but do not seek or expect major leadership responsibilities.

Temporary Seasonal These employees are hired full-time when there is more business than the permanent staff can handle. Examples are temporary sales help hired in a flower shop at Christmas time, field laborers hired at harvest time in a nursery, and planting crewmembers hired in the spring by a landscaper. These employees are hired for their ability to perform a few skills competently and quickly, such as balling-and-burlapping, operating heavy equipment, or duplicating wire-service arrangements. The employees may be students on vacation, migrant field workers, laid-off factory workers, or others who have no long-term employment goals with the company. Money and work experience are their major motivations. Loyalty to the company and an interest in its growth and success are not common in seasonal employees. They may require closer supervision than other employees.

Other Qualities of Good Leadership

In addition to realistic expectations for each employee, effective supervision requires an understanding of what motivates workers to respond positively to their work and their supervisors. Most people respond in a positive manner to a supervisor whose competence they respect, whose expectations are fair and impartial, and whose attitude is upbeat. Consider the following qualities of good leadership.

Decisiveness   The supervisor must be able to make good decisions quickly. The knowledge necessary to make the decisions may come from earlier preparation or experience. Employees need to believe in the ability of their supervisor to confront and resolve a problem correctly without a long period of indecision.

Thorough Directions   Employees may need a lot or a little direction to accomplish a task correctly the first time. Supervisors must know the abilities of their staff and provide clear instructions as work assignments are made. If the staff member has done a particular job many times before, directions can be abbreviated. If the employee is new or a new task or a new technique is being introduced, the supervisor must take longer to explain and demonstrate what is to be accomplished and the quality that is expected.

Competence   Employees seldom perform at a level that they believe is beyond the capability of their supervisor to recognize and appreciate. Thus, supervisors must perform the tasks that they assign frequently enough to retain and refresh their own skills.

Fairness   Supervisors must allocate work among their staff equitably. Two workers with the same competencies and receiving the same wage or salary should be expected to perform equally. Otherwise, feelings of resentment develop. Fairness implies equal treatment for all, whether in the choice of vacation weeks, the assignment of new tools, or the requirement to work over holidays. If benefits accrue with seniority, this should be understood by all employees at the time of their hiring. Partiality, real or perceived, can be demoralizing to a staff or crew.

Understanding   A good supervisor is predictable. Employees must be able to anticipate the consequences of performing in a way contrary to established company policies. A certain inflexibility is necessary or all discipline will quickly disappear. However, the enforcement of company policies must be tempered by understanding when circumstances warrant it. If a trusted and proven employee suddenly begins performing carelessly or erratically on the job, something is probably happening outside of work to explain it. A supervisor who points out the problem to the employee privately and without a tone of reprimand is likely to be effective and appreciated. It is seldom necessary to become involved with the employees' personal problems. Pointing out how the symptoms of the problem are being manifested on the job is often enough.

Respect   Supervisors who expect respect from their staff must demonstrate their own respect. Many employees have insights and abilities developed through years of experience on the job. Others have knowledge attained from recent classroom experience. The employee's point of view should be sought by the supervisor whenever possible. At times it may be offered without solicitation. If it has merit, it should be used. If it is inappropriate, it should at least be listened to courteously and an explanation given as to why it is inappropriate. It is not a sign of weakness or incompetence in a supervisor to use an employee's suggestion or ask an employee for an opinion. It is merely good sense.

Sincerity   All supervisor-employee relationships benefit if the staff perceives that the supervisor is sincerely interested in their well-being. Concern for the employee as well as for the company can be exhibited in numerous ways, such as asking about employees' children or vacations or whether they are satisfied with their work. Equally important is the employee's understanding that the supervisor cares about the business, the satisfaction of the customers, and the quality of work. Any sense of phoniness or indifference perceived in the supervisor will negatively affect the attitudes of the staff.

Praise Everyone appreciates recognition. When an employee or group of employees does a job well, the supervisor should recognize the accomplishment. It may be as simple as telling the employee privately that the work was well done; for example, "Nice job, Fred. You got that done in record time and it looks great." It may be more public, such as complimenting a crew or crew member in the presence of peers. Then it is best to praise the action or the results to avoid embarrassing individuals or implying criticism of other staff members. For example, a supervisor might say, "We've had a productive week. Everything went well and that job at the Johnson home really made some money for us because it went faster than we expected. They've already called to say how pleased they were." The crew responsible has been praised, and the other workers have not been compared or criticized.

Reward In return for their continuing good performance, employees expect their status within the company to improve. The most expected reward is an increase in salary or wage or a bonus. Surprisingly, monetary reward is usually short-lived in terms of employee satisfaction. We all tend rapidly to adjust our way of living to the new income level and look for our next reward. When a business's financial circumstances do not permit monetary reward of deserving employees, or when the supervisor has no authority to reward an employee with money, other forms of reward can be offered. These may even be more long-lasting sources of worker satisfaction than financial rewards. Deserving employees may be the subject of public relations news articles. Others may be presented with service awards or prizes. Some companies recognize the "employee of the month" by posting the employee's photo in a prominent location. Other companies recognize employee excellence through titles, such as assistant grower or crew leader. Something as simple as a handwritten note from the supervisor to the employee acknowledging the latter's performance can be an effective reward.

Qualities to Avoid

Certain qualities in a supervisor may cause negative feelings to develop in an employee or group of employees. The probable result of negative feelings between workers and their supervisor is low worker morale, reduced productivity, and even theft or vandalism. While no supervisor can expect to perform flawlessly, an effort should be made to avoid the following:

Inflexibility Company rules must be applied fairly to all employees, but little is gained by a policy of "no exceptions for anyone at any time." If the rules penalize a worker for tardiness, most employees will see the need for it and the logic behind it. Employees who oversleep may expect to have their paycheck adjusted to reflect the lost time. However, if an employee arrives late once because of a flat tire, enforcement of the rule will probably be seen as needless rigidity. "They don't care about me" is a difficult attitude to reverse once implanted in an employee's mind.

Partiality In the close working environment of many horticultural businesses, friendships form easily. If personal associations between supervisors and employees outside of work cause partiality at work, however, resentment is bound to result. Joking with some employees and not others or buying some of the crew a beer after work but not all members can be interpreted as partiality. Such an interpretation can be as harmful to worker morale as if favoritism actually exists. It must be guarded against.

Condescension Training a new employee or assigning a new task to an old employee usually requires the supervisor to give instructions. They must be delivered in a manner that respects the employee's intelligence, skills, and insights. Talking down to employees by oversimplifying the directions or giving shallow answers to serious questions insults their intelligence.

Subjective Criticism No one likes being criticized or reprimanded, yet every supervisor must do so on occasion. The objective of properly applied criticism must be to prevent the employee from feeling personally offended or embarrassed while assuring that the mistake in behavior or performance is corrected. If the criticism is directed to an individual, the individual should be spoken to in privacy so that no other workers overhear. Whether the criticism is directed toward an individual or a group, the supervisor should remain objective. Criticizing the actions or the outcome rather than specific persons allows employees to retain their own sense of worth and avoids a long-term negative reaction to the supervisor. All criticism should be delivered in a positive manner without anger or profanity. Emotional reprimands are much more likely to be taken personally.

Double Standards   Employees have the right to expect that policies set for them will be followed by those in supervisory positions as well. It is the responsibility of supervisors to exemplify the positive qualities and attitudes that they desire or require of their workers.

Indifference   The small-business atmosphere of most horticultural enterprises requires that supervisors give frequent positive reinforcement to their workers. When a supervisor inspects the work done by an employee or crew, the supervisor is expected to react, not merely walk on. The supervisor may silently approve, but failure to say so may cause employees to think their work is not worth mentioning. Equally damaging to working relationships can be a supervisor who displays no concern when births, deaths, illnesses, or other personal events occur in the lives of employees. "They don't care about me" is an attitude that can be avoided by a few outward expressions of concern from a supervisor.

Sarcasm   Neither humor nor criticism should be directed toward employees in sarcastic tones. Getting a laugh from one employee at the expense of another employee is certain to cause resentment. To criticize an employee with sarcasm makes the criticism much more hurtful and is totally unprofessional in a supervisor.

Complicating Factors

Even in businesses where the owner and supervisors are sincerely interested in their workers and make a concerted effort to be good managers, prob-

lems still occur. Assuming that an employee does not dislike the supervisor, but that they are still not communicating effectively, it could be for one of the following reasons:

- *Age differences.* When the employee and supervisor are of widely different ages, either may be guilty of believing that they have nothing in common. Young people may tend to believe that the growing pains of life are known only to them. Older employees can be intimidated by the strength or impetuousness of younger ones or may have no patience with the inexperience of young workers, forgetting that they were once that young and inexperienced themselves.
- *Education differences.* Less educated employees may feel intimidated by a better educated, articulate supervisor. Young supervisors recently graduated from college can expect to be regarded suspiciously by the employees they direct. If the workers are appreciably older, more experienced, and have seniority with the firm, the young supervisor may be on the receiving end of a reverse snobbery.
- *Sex differences.* Certain types of ornamental horticulture businesses have been the domain of men for many years, due mainly to the need for physical strength to perform the work. Nurseries, landscape contractors, and lawn care firms have been dominated by male workers and have only recently started to accept women as both laborers and supervisors. A female supervisor may experience resentment or balking by some workers for no reason other than her sex and the questions it raises in their minds about her knowledge and competence.
- *Ethnic and background differences.* The often illogical problems that develop between people from different cultures and economic backgrounds are known to most people of intelligence. For example, communications can be awkward between an employee and a supervisor merely because one is from the city and the other from the country. It is not a case of disliking each other but of believing that they have nothing in common.

- *Position or status differences.* Some employees are not comfortable around authority figures, no matter how hard the supervisor tries to develop a good relationship. Supervisors who are promoted from the ranks may find that their former friends on the staff suddenly turn cool for this reason alone.

If one or more of these factors leads to complications of the human relations within a firm, it may or may not affect the workers' productivity, but it will affect their job satisfaction. As much as possible, a good supervisor or manager will seek to break down barriers that stand in the way of communication. Seldom will workers initiate the action.

Reconciliation of differences requires that supervisors seek out and emphasize the many areas that they and the employees have in common. Stressing areas of common interest and similarity, while tolerating differences that do not affect the business, can eliminate or at least minimize the areas of difference.

CUSTOMER RELATIONS

The other major group of people in the human relations triangle are the customers. They are the people who keep the business in existence. Their goodwill must be cultivated as carefully as the horticulture crops they purchase. *Goodwill* and *sales* are intimately related business objectives, and the sales staff must be trained to promote them simultaneously.

Characteristics of Customers

The type of customer to be served will depend first upon whether the business is retail or wholesale. Retail sales are likely to be directed to homeowners and other nonhorticulturists who will purchase items in comparatively small quantities. Wholesaling usually means that the customers will be other industry personnel.

Although there are differences in the ways products and services are marketed to the retail and wholesale markets, wholesale and retail buyers are not as different as might be expected. Some know exactly what they want and only need assistance obtaining it. Others browse, surveying the entire inventory in order to generate an idea of what, if anything, to purchase. In between is the customer who definitely plans to purchase an item or two but is willing to purchase more if something looks interesting.

The second factor determining the type of customer to be served is the economic class toward which the business is directed. If the business is known for costly products and services, certain types of customers will shy away and others will seek it out. If low-priced merchandise is promoted, the same is true. A business located in a middle-income neighborhood can expect to serve homeowners and amateur gardeners who seek good quality and familiar merchandise at affordable prices. A more affluent clientele will expect top quality and often unusual merchandise with attentive service and expect to pay a premium price in return.

In any case, two motivations bring a customer to a horticulture business: *need* and *desire*. Need is the most direct motivation; it has an urgency about it. As soon as the customers believe they can afford the item or service, they move to satisfy the need by going to the shop or calling on the telephone. New homeowners might regard the establishment of a new lawn and a foundation planting as a need for their property. They may have the *desire* for a maintenance firm to care for the landscape, but they may not be sure they can afford it. Still they may call for an estimate.

Characteristics of Good Salespeople

Customers deserve assistance from a salesperson who has the following characteristics:

Friendliness Customers should be greeted soon after walking through the door or into the sales yard. If the customer's name is known, it should be used. Not every customer warrants a handshake, but it can be appropriate with a regular customer. Salespeople need to smile easily and frequently. Personal problems, boredom, or business problems must not be transferred to the customer in the form of a gloomy personality, a short temper, or a short attention span.

Helpfulness   Customers should be given a short period of time to get their bearings in the store or sales area and to look around. Browsers are easily spotted usually by their slow pace and meandering ways. Customers who know what they want usually seek out a salesperson soon after entering.

When approaching a customer, too many salespeople ask, "Can I help you?" Such a question is not only unnecessary, it has also become a cliche. A better approach to a customer is "How may I help you?" The customer's response can then determine further action or inaction by the salesperson: "I'm looking for a good, inexpensive lawn mower" or "What would you recommend for a new baby whose mother works in our office?"

Knowledge   Salespeople should be familiar with the plants and products that are being offered for sale. They should know the answers to questions that customers are likely to ask; for example, "How big does this plant get?" "What are the light requirements?" "Is it safe to spray this around children and pets?" In addition to responding to customer questions, the sales staff should be able to suggest appropriate products and services to supplement those initially sought by the customer. This should increase sales volume along with customer satisfaction.

Honesty   The salesperson must be honest in describing how a product or service will fulfill the client's need or desire. Delivery of the product when promised and starting the service on time are also reflections of the salesperson's honesty in the mind of the customer.

Good Grooming and Articulate Speech   Salespeople must be dressed to meet customers and create a good impression on behalf of the firm. Overdressing can make the customer uncomfortable; a sloppy or soiled appearance can be an affront. Conservative dress is usually the wisest approach in business. Equally important is the salesperson's ability to speak clearly, accurately, with good grammar, and without profanity, slang, or verbal cliches.

Courtesy   A salesperson must be polite and positive with customers, even if they do not return the courtesy. It takes little effort to say *please* and *thank you* but it leaves an impression of caring and civility with the customer. Addressing customers by name, if known, or as *sir* or *ma'am* tells them that their importance to the success of the business is recognized.

Making the Sale

A greeting and an offer of assistance should be made to customers soon after they enter the store, greenhouse, or sales yard. When the first contact is by telephone, the offer of assistance is immediate. The salesperson must then determine the needs of the customer and set about to satisfy those needs. The approach must be positive so that the customer feels good about making the purchase. The features of the product or service must be described at whatever level the customer requires for complete understanding. Obviously the description used for a retail customer is different from that used for a wholesale horticultural buyer. The salesperson should also attempt to interest the customer in products or services that support or are related to the initially needed item. The attempt should not be high pressured or appear mercenary. Emphasis should be placed upon how the customer will benefit from the purchase, not on how the company or salesperson will profit. The customer should be encouraged to ask questions, and the salesperson should answer them directly. Where the sale is potentially sizeable, the salesperson should be prepared with information regarding financing possibilities. If a service is being sold rather than a product, the salesperson must be skilled at helping the customer visualize the service. For example, photographs of typical lawns or landscapes or illustrations of plants or flowers can be shown, or quick sketches can be done as the client watches.

Successful sales result in satisfied customers who: (1) get what they came for and perhaps even more, but (2) do not spend more than they should have spent, and (3) will return again for future purchases.

SUMMARY

Human relations in a horticultural business involves the management, the employees, and the customers in an interdependent association.

Personnel management is the direction of workers in a manner that brings out their best efforts and attitudes on behalf of the business. Good supervisors are able to set realistic expectations of their employees, taking into account each worker's individual abilities and limitations, educational and work experience, and potential. Employees may be career-directed, part-time, or temporary seasonal.

Most people respond in a positive manner to a supervisor who exemplifies decisiveness, competency, fairness, sincerity, understanding, and respect for the employees, offering praise and reward when warranted and delivering directions thoroughly. Employees can also be expected to react negatively to a supervisor who frequently displays inflexibility, partiality, a condescending attitude, indifference, or sarcasm or one who gives subjective criticism or observes double standards in adherence to company policies.

Communication can be complicated by differences of age, education, sex, ethnic background, or status within the company. The supervisor must usually seek to open the channels of communication when problems arise rather than waiting for the problems to resolve themselves.

Because sales are intimately related to customer goodwill, the sales staff must be trained to promote both simultaneously. Good salespeople offer a customer friendliness, helpful assistance, a knowledge of the materials and services sold, honesty, and courtesy, and they exhibit good grooming and correct speech. Successful sales are those that result in satisfied customers who got what they came for and perhaps even more, but did not spend more than they could afford, and will return again for future purchases.

Achievement Review

A. SHORT ANSWER

Answer each of the following questions as briefly as possible.

1. Define personnel management.
2. Indicate if the following apply to career-directed employees (C), part-time employees (P), and/or temporary seasonal employees (T).
   a. The employee may be loyal but not expect major leadership responsibilities.
   b. The employee is hired during busy periods and released after the rush is over.
   c. They are expected only to do a few tasks competently.
   d. The employee hopes to spend several years with the company and to advance within the organization.
   e. These are usually the most personally ambitious employees.
3. List ways customers may differ depending upon the type of horticulture business.
4. List the qualities of a good supervisor.
5. List the qualities of a poor supervisor.
6. List the characteristics of good salespeople.
7. Describe the procedures for effective selling.

24

Business Communications

Objectives

Upon completion of this chapter, you will be able to

- compose a correct and effective business letter.
- place a business telephone call correctly.
- answer a business telephone call correctly.

THE PURPOSE OF BUSINESS COMMUNICATIONS

Although much of the business of an ornamental horticulture firm is done face to face, there remain many occasions when a letter or telephone call is the means of communication. Customers may call to place orders, ask questions, or register complaints. Staff members may receive customer orders and place their own orders for supplies over the telephone. Business letters are used to place orders, promote customer sales, and bill customers. Thus, letters and telephone conversations are extensions of the business and must be handled professionally.

When customers come in person to the business, the owner or employee is able to smile, shake hands, and in other ways let the customers know how important they are to the business. As much as possible, that same personal tone must be incorporated into business communications by letter and phone.

WRITTEN COMMUNICATION

Business letters are permanent records, unlike conversations. When a request, an offer, or an order is made in writing, it becomes much like a contract. Prepared with care, a letter reduces the possibility of error, omission, or misinterpretation.

As a tool of business, a letter must possess the following characteristics:

1. correct mechanics
2. a conventional format
3. a polite and friendly tone
4. thorough coverage of its subject(s)
5. accuracy
6. brevity

Correct Mechanics

All business correspondence should be typed, on good quality paper with a printed letterhead. The letterhead should contain the name, address, and telephone number of the firm and may also incorporate a logo or piece of line art. Since the stationery is for communication, not advertising, it should be white or only lightly color toned. The letterhead should be restrained, not gaudy, and should not occupy more than a fifth of the page, Figure 24-1.

Proper form requires that each letter have the following parts, Figure 24-2.

Date   The date line varies in its location depending upon the format and length of the letter. It is typed beneath the letterhead leaving enough space to balance the letter without crowding at the top or bottom.

Inside Address   The full name and address of the person(s) and firm you are writing to should be entered against the left margin at least four lines under the date line. The inside address should not extend beyond the center of the page.

The name and title or position of the person you are writing to should be included if you know it. Customers are usually addressed as:

Mr.   Miss
Mrs.   Ms.
Mr. and Mrs.   Messrs.

A title follows the name in the inside address. For example:

Mrs. Irene Clark, Branch Manager
Green-Gro Suppliers
213 Carriage Road
Chicago, IL

Salutation   If a business letter had a smile and a handshake, it would be the salutation. It is spaced two lines below the inside address and punctuated with a colon. It should be cordial or formal depending upon how well acquainted the writer and reader are and the purpose of the letter. The following examples include good and bad salutations.

- Dear Mr. Johnson: (The most common business salutation. It is both cordial and formal.)
- Dear Neil: (Used only if there is a close ac-

FIGURE 24-1.
The design of a business letterhead should provide essential information while suggesting dignity and professional capability.

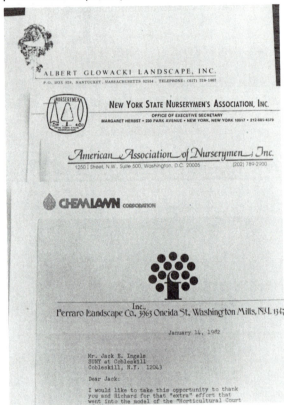

quaintance between the writer and reader. The full name is still used as part of the inside address.)
- Mr. Johnson: (Too terse. It lacks cordiality.)
- Dear Mr. Neil Johnson: (Stiff and unnatural. When the customer is addressed with a title such as Mr., the first name is not included in the salutation.)
- Neil Johnson: (Terse and stiff. It lacks both courtesy and cordiality.)
- Dear Neil Johnson: (This is sometimes a good solution when people's names give you no clue to their sex and you do not know them personally.)

Body The message to be conveyed is in the body of the letter. It begins two spaces beneath the salutation. To fill the page most attractively, the top and bottom margins should be the same, as should the side margins. Brief letters of only two or three sentences usually look best if lines are double spaced, but most letters are single spaced. A double space may be left between each paragraph.

Complimentary Close Two spaces beneath the body of the letter is the courteous ending of the communication. Its placement to the right or left of center depends upon the format being used. It never extends outside the margin. The first word is capitalized, and a comma is used at the end. Its degree of formality should correspond with the rest of the letter. If the salutation has been formal, such as Dear Mr. and Mrs. Andrews, then a similar closing is appropriate. If the letter concerns an overdue account, the closing should not be chummy. Consider the following examples:

- Yours truly,
  Very truly yours,
  (Used in most regular business correspondence.)

- Sincerely,
  Sincerely yours,
  Yours sincerely,
  (Used when greater cordiality is desired, as when the reader and writer are acquainted.)

- With best regards,
  Cordially,
  (Used when a tone of informal friendliness is desired.)

FIGURE 24-2.

A sample business letter with correct spacing and major parts identified

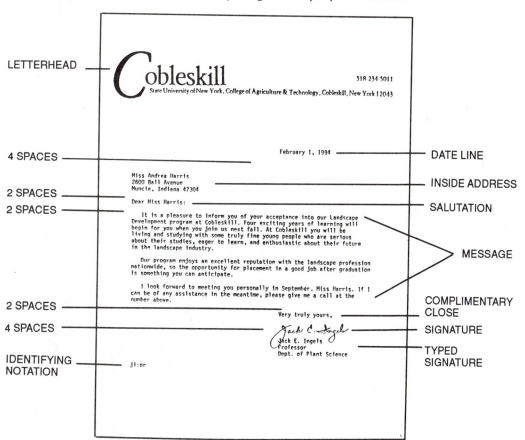

One correspondence gimmick seen too frequently is running the last sentence of the letter into the complimentary close. For example:

In the hope that we can be of service to you in the future, I remain

Sincerely yours,

The final sentence of the body should be independent of the complimentary close.

Signature   The signature is handwritten. It should be legible, not a scrawl, despite what you may have seen. Four spaces are allowed for the signature. If the name of the firm is included, it should be typed in capitals two spaces beneath the complimentary close and above the signature. The name of the writer is typed below the signature. If the writer has a title and chooses to use it, it may be added beneath the typed name. If the name of the firm precedes the signature, then the firm assumes responsibility for agreements or statements made in the letter. If only the signature follows the complimentary close, then the writer alone assumes responsibility for the contents of the letter. Obviously, not everyone corresponding on business stationery is empowered to speak for the company, so the form of the signature is important. For example:

- Yours truly,
  CHEMGREEN LAWN SERVICE

  Todd R. Martinson
  Todd R. Martinson
- Sincerely yours,
  STACEY LANDSCAPE CONTRACTING

  Linda K. Stacey
  Linda K. Stacey
  President
- Very truly yours,

  Terry A. Forsyth
  Terry A. Forsyth
  General Manager

Notations   Business letters usually carry one or more abbreviations to identify the person who composed the letter, the secretary who typed it, and other persons who will receive copies of it and as a reminder that materials were enclosed with the original letter. For example:

- JEI: or (Identifying initials placed in the lower left corner of the last page of the letter. They are either aligned opposite the typed signature or title or placed two spaces lower. The first set of initials are those of the person who composed the letter. The second set are those of the secretary or typist.)
- Enc. *or* Enc. (3) (An abbreviation of the word *enclosure* is used to indicate that something has been included with the letter. If more than one item is enclosed, the number of items may follow in parentheses. This notation is made two lines beneath the identifying initials and against the left margin.
- cc: (The initials mean *carbon copy* and are followed by the name(s) of the person(s) receiving copies of the original letter. The notation is spaced two lines beneath the enclosure notation or, if there are no enclosures, two lines beneath the identifying initials.
- P.S. (Postscripts are sometimes used in sales letters to focus additional attention on a point made in the body of the letter. As such, they are a gimmick. They should not be used in formal correspondence to add a point forgotten in the original letter. In such a case, the letter should be rewritten and typed again. Postscripts follow two lines beneath any of the above notations.)

The Envelope   Two envelope sizes are commonly used in business correspondence: 3 5/8 × 6 1/2 inches and 4 1/8 × 9 1/2 inches. The smaller size is frequently used for billing and the larger for letters. Other sizes exist for other purposes such as when folding is undesirable or when oversized enclosures are included. The usual business envelopes should be of the same color and quality of paper as the stationery they contain. Some have a cellophane window to allow the inside address on the letter to show through, thereby eliminating the need for typing on the envelope. If a windowed envelope is not used, the full name and address of the individual or firm you are writing to must be typed, single spaced, on the envelope, Figure 24-3. Business envelopes should also have your firm's return address printed in the upper left corner.

The smaller envelopes require that 8 1/2 × 11-inch stationery be folded three times before insertion. Larger envelopes usually require only two folds in the letter. In the two-fold technique, the paper is folded in thirds lengthwise, Figure 24-4. In the three-fold technique, the paper is first folded in half lengthwise and then in thirds perpendicular to the fold.

Letter Formats

An assortment of formats are acceptable in business correspondence. Most secretaries have a style that they prefer, often because their typewriters adapt more easily to it than to another style. You should give the typist specific instructions if you prefer a particular style.

Occasionally a nontraditional style is used for its eye-catching appeal. Such letters are almost always advertisements, and the unusual style is a gimmick to help assure that the message will be read. For the majority of business correspondence, one of the following styles is best:

- *Full block style:* All parts of the letter begin at the left margin. It is the easiest style for the typist but can look heavily weighted on the left side, Figure 24-5.
- *Modified block style:* All parts of the letter begin at the left margin *except* the date line, complimentary close, and signature. They may be centered or placed to the right of center, Figure 24-6.
- *Semiblock style:* The inside address, salutation, and notations begin at the left margin. The first sentence of each paragraph is indented five spaces. The date line, complimentary close, and signature are centered or placed to the right of center, Figure 24-7. If further indentations for enumerations are required, as in an order

FIGURE 24-3.
A correctly addressed envelope. The typing is centered between top and bottom and on the right side of center.

STATE UNIVERSITY OF NEW YORK
AGRICULTURAL & TECHNICAL COLLEGE
COBLESKILL, NEW YORK 12043

Miss Andrea Harris
2600 Ball Avenue
Muncie, Indiana 47304

FIGURE 24-4.

Correct preparation of a two-fold letter and placement into an envelope

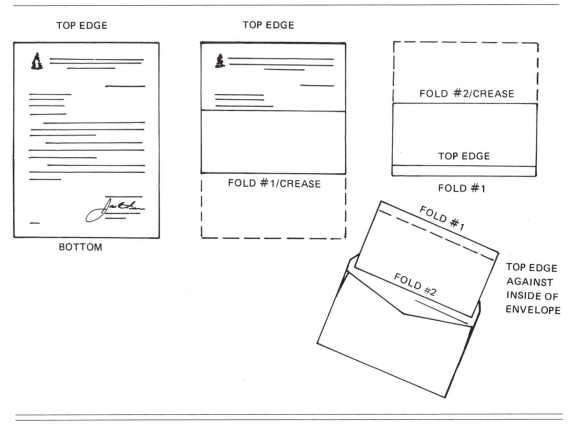

letter, the semiblock style adapts easily. Figure 24-8.

Setting the Tone of the Letter

Nearly all correspondence sent by horticulturists goes to suppliers, customers, potential customers, or members of the profession and requires the same courtesy and concern that would be extended in a face-to-face meeting. Since there is no opportunity for facial expression or voice inflexion to support or explain the words, the letter must be phrased especially carefully.

Courtesy is a quality of every good letter and can be incorporated even when firmness is required,

as with a collection letter for an overdue account.

Conversational phrasing will allow the letter to be read easily and with understanding. Business letters need not sound like insurance policies or tax forms even when legal matters are their subject.

Concern for the reader rather than the writer will encourage a better reception. For example, the words "Due to a backlog of work, we will be unable to install your new lawn until some time in May" create certain disappointment in the customer, who is not concerned about the backlog of work but is concerned about the lawn. This customer may seek another firm to provide the satisfaction desired. A better way to say the same thing is, "We appreciate your contacting us to install your new lawn and

FIGURE 24-5.

An example of a business letter written in full block style (Courtesy Albert Glowacki Landscape, Inc.)

ALBERT GLOWACKI LANDSCAPE, INC.

P.O. BOX 928, NANTUCKET, MASSACHUSETTS 02554 . TELEPHONE: (617) 228-1867

August 18, 1994

Mr. Ryan S. Bell
203 Lafayette Road
Bradford, South Carolina 12043

Dear Mr. Bell:

I received your letter and the good news that you have purchased a new summer home on Nantucket. Welcome to our little
corner of the world! It is a special place, as any of the
resident islanders will confirm.

I am delighted that you have contacted us about the landscaping for your home. Our firm will be pleased to work
with you to create a beautiful setting for the house while
retaining that Nantucket charm of which you spoke. Coinci-
dentally, we have the landscape contracts for the homes on
each side of you, so the entire setting can be customized
to your individual needs, yet pleasantly harmonious.

When you arrive on the island in September I shall be
available to meet with you and we can begin discussing
the landscape. I have already started jotting down some
ideas. I look forward to seeing you next month, Mr. Bell.

Sincerely yours,

ALBERT GLOWACKI LANDSCAPING

Eric J. Blamphin
Eric J. Blamphin
Landscape Architect

ejb:rs

will be there during the first week of May to do the job." Disappointment is not built into the phrasing, and the importance of the customer is stressed. Similarly, the words "Please drop in any time from 9 A.M. to 5 P.M. Tuesday through Saturday so that we can be of service" are more inviting than "Closed Sunday and Monday. Open Tuesday through Saturday."

Thoroughness and Accuracy

In a conversation, the listener can ask the speaker to clarify or complete statements that are not fully explanatory. In a business letter, no such opportunity exists. Therefore, the writer must be certain that the topic of the letter is covered thoroughly and accurately. Failing to state a price or stating

FIGURE 24-6.
A typical letter written in modified block style (Courtesy A. T. Bianco
Landscaping and Nursery)

A. T. BIANCO LANDSCAPING & NURSERY
198 WELSHBUSH RD.
FRANKFORT, NY 13340

Phone 315/ 724-4644

GROWERS • DESIGNERS • CONSULTANTS of DISTINCTIVE PLANTINGS

April 25, 1994

Mrs. Mary L. Fultz
1714 North Ruckle
Frankfort, New York 13345

Dear Mrs. Fultz:

I am writing to confirm our recent telephone conversations about
your new lawn. I am grateful that you contacted us about the
project. You have been one of our most valued customers in the
past and I appreciate your continuing confidence in our firm.

As we agreed, our crew will schedule your lawn installation for
the week of May 18th, while you are away on vacation. The lawn
will be seeded with a bluegrass blend following careful conditioning of the soil. We will take care of the watering until
you return on the 25th of May, after which time you will assume
responsibility for the watering. A more complete outline of
the project, your responsibilities, our guarantee and the total
cost is contained on the enclosed contract.

Thank you for contacting us, Mrs. Fultz. We look forward again
to being of assistance to you.

Sincerely yours,

A.T. BIANCO LANDSCAPING & NURSERY

Anthony T. Bianco, Jr.
Vice President

atb:ls

encl.

WHOLESALE GROWERS OF   QUALITY SHADE TREES

an incorrect price may lose a customer or obligate the firm to a loss.

The opening paragraph of the letter should state the purpose of the letter and define the subject(s) to be covered. If a review of the background is needed, it should be included in the first paragraph and kept as brief as possible. For example:

- "I am writing because we have had no response to our earlier requests for payment of your overdue account."
- "Last fall you requested an estimate of the cost for design and installation of your patio. If you are ready to proceed with the project this spring, we are eager to be of service."

FIGURE 24-7.

An example of a business letter written in semiblock style (Courtesy A. T. Bianco Landscaping and Nursery)

A. T. BIANCO LANDSCAPING & NURSERY
198 WELSHBUSH RD.
FRANKFORT, NY 13340

Phone 315/ 724-4644

GROWERS • DESIGNERS • CONSULTANTS of DISTINCTIVE PLANTINGS

December 15, 1994

Mr. And Mrs. Eric Chamberlain
1908 N. Bosart
Park Ridge, New York 47999

Dear Mr. and Mrs. Chamberlain:

Congratulations! You have been given a gift of two hundred dollars worth of nursery stock as a Christmas gift. The gift is from Mr. and Mrs. Tim McCarty of Fort Edward, New York and includes their wishes for happy holidays.

We will have a wide selection of trees and shrubs available beginning in March and will have bedding plants after the first of May. You are certain to find many new plants and some familiar old ones from which to choose.

Your gift certificate is enclosed. It is redeemable anytime during 1995. We look forward to meeting you during the coming year and join Mr. and Mrs. McCarty in wishing you the best of the holiday season.

Sincerely yours,

A. T. BIANCO NURSERY

Janine R. Bianco
General Manager

jrb:pv
enc.

WHOLESALE GROWERS OF   QUALITY SHADE TREES

Each paragraph in the letter should aid the reader's understanding of what the writer is requesting or proposing and what their mutual obligations or expectations will be. The writer should try to anticipate the reader's questions in advance and address them in the letter. All correspondence should be proofread for accuracy before it is signed and sent out. Where orders are being sent or checks enclosed, the exact amount that is to be paid or that is being sent should be stated in the body of the letter.

Brevity and Directness

The rambling dissertations that characterize personal correspondence between friends are inappropriate in business letters. Even though friendly and

FIGURE 24-8.
A business letter written in semiblock style with indentation for special enumerations (Courtesy Daniel L. Pierro, Inc.)

(201) 825-1959

DANIEL L. PIERRO, INC.
LANDSCAPE DESIGNER AND CONTRACTOR
227 SHADYSIDE RD. — RAMSEY, N.J. 07446

March 15, 1994

Prof. Jack E. Ingels
Department of Plant Science
State University of New York
Cobleskill, New York 12043

Dear Jack:

It was good to receive your letter. I appreciate your invitation to speak with your senior class in Landscape Development and am delighted to accept. As an alumnus of Cobleskill, I remember the value of informal, candid conversations between students and industry practitioners as one means of providing the young people with insights into the profession.

If you are agreeable, here is the schedule I hope to follow on April 20th:

1. Arrival on campus by 8:30 a.m. We can meet at the snack bar for coffee.

2. Speak with your class from 9 to 10 a.m.

3. Interview those students interested in summer work with our firm from 10 a.m. to noon.

If there is any particular topic that you wish me to cover, Jack, let me know. I look forward to being back at Cobleskill and to speaking with your class.

Sincerely,

Daniel L. Pierro

dlp:pb

conversational in tone, the letters should never become chatty or deal with topics unrelated to the purpose of the letter. Inquiries about the reader's health, family members, the weather, or other irrelevant pleasantries only clutter the letter and divert the reader's attention from the main topic.

Each paragraph should relate to the purpose described in the opening sentences of the letter. If it does not, it is probably unnecessary. For example:

• "I am interested in purchasing a new Ajax Tree Digger, Model R17. I saw it demonstrated recently at a trade show in Wichita and need additional information about its capabilities.

The Tree Digger should enable our firm to double our digging operation without any increase in personnel. As you know, every dollar saved these days is important.

Please have a sales representative call on me here at the nursery. Because I need the machine in time for spring digging, I shall appreciate your prompt attention to my request."

The second paragraph is of no importance to the purpose of the letter, which is to obtain more information about the tree digger. It should be eliminated.

Using Word Processors

Recent technological advances in office equipment have eliminated some of the repetitive typing required in offices where form letters that appear original are desired, yet only the inside addresses and salutations actually change. The word processor permits a typist to store the letter in the memory of the machine, insert a new inside address and salutation, then let the machine type out the letter automatically. In a similar fashion, the word processor is able to change the figures in a bid, estimate, or contract for a particular customer while retaining the basic wording common to all similar forms used by the business.

There are two limitations to the use of a word processing system by a horticultural business. One is that the amount of repetitive correspondence may be minimal, making the cost of the system prohibitive. The other is that over-reliance on the system may cause business letters to lose their personal quality and become stiff and formal in tone.

TELEPHONE COMMUNICATION

Use of the telephone for business transactions has advantages and disadvantages. Like business letters, the telephone does not permit eye-to-eye contact or a smile and handshake between the horticulturist and the customer or business associate. The telephone allows questions to be asked when clarification is needed, which the business letter does not. However, a telephone call cannot be proofread as a letter can. What is said, in haste or error, is heard immediately. Also, a letter does not have a stammer or irritating voice pitch that can create a negative impression in the listener's ear. Neither does it proceed too rapidly or slowly, requiring time-consuming repetition for total understanding.

Certain types of horticultural operations depend upon the telephone as a sales tool more than other operations do. For example, retail florists do a large percentage of their selling by telephone. A garden center or lawn care firm is likely to use the telephone more often for answering customer questions and receiving requests for estimates. In certain situations, horticulturists and their employees are more likely to initiate the phone calls. In other situations, they tend to receive them. In either case, the voice on the telephone becomes the personality of the firm. It must be pleasant and interesting to hear, and the person behind the voice must be knowledgeable, interested in the person at the other end of the line, professional in manner, friendly, attentive, and courteous. The voice should be directed into the mouthpiece of the telephone, and the receiver should not drop beneath the chin, Figure 24-9. Persons who are naturally monotonic should practice developing a more expressive voice, possibly through the use of a tape recorder, if they want to make effective use of the business telephone.

Placing a Business Call Correctly

To place a business call correctly requires preparation. The caller must have a clear understanding of what the call seeks to accomplish, what points are to be covered, the sequence in which they will be covered, what information is to be obtained or released, and what commitment is to be sought or offered on behalf of the firm. Irrelevant chatter should be avoided on the assumption that the receiver of the call is busy and will not want to waste time.

Following is a checklist for making a business telephone call:

FIGURE 24-9.
This florist is speaking directly into the telephone receiver. She keeps an order pad and other materials nearby.

- Be certain of the number (including area code) before dialing.
- Have the name of the person and firm you are calling in mind.
- Allow ample time for answering. Ten rings are usually recommended by most telephone companies.
- Identify yourself and your firm immediately. For example: "Good afternoon. This is Brian Miller of Miller's Greenhouses."
- Ask to speak to a specific person if possible. Otherwise *briefly* identify the reason for the call and ask to speak to someone who can assist you. For example: "I'm calling about the possibility of carrying the Ajax line of giftware in our shop. May I speak with someone on your sales staff, please?"
- Provide information and ask questions in an organized, sequential manner and at a rate of speed that permits the listener to comprehend and record as necessary.
- In closing, summarize the major items discussed during the conversation to assure mutual agreement and understanding of what the call accomplished.
- Close the conversation in a polite and friendly manner as briefly as possible after the purpose of the call has been accomplished.

A business call should not be made when other things require the caller's attention. The call should wait until it can proceed without distraction or interruption. If some unexpected event makes it necessary for you to leave the line for a moment, arrange to call back at another time, since no one enjoys being put on "hold." Hold lines are too often overused and may alienate more business than they serve.

Answering a Business Call

An incoming call is much less predictable than an outgoing call. It may be a customer placing an order, making an inquiry, or registering a complaint; a supplier offering information about an order; or a garden club seeking a guest speaker. Equally unpredictable are the telephone skills of the callers. They may offer their questions or information in a disorganized or emotional manner. Since the telephone provides anyone who is inclined to be abusive with a great opportunity to be so, it requires thick skin and a long fuse to deal professionally with some callers. Dealing professionally with all callers requires the ability to listen and comprehend and a voice that can convey friendliness and concern.

The ability to listen, hear, and understand is not as common as might be assumed. Today's society bombards us with almost continuous sound, and many of us develop the ability to shut it out as a defense. We can listen, nod our heads as if understanding, smile in the appropriate places, and yet be unable to repeat most of what we "heard." Employees assigned to answer business calls must often learn to listen more attentively than they have for years. Otherwise, the telephone conversation can become a series of needless requests for repetition.

In preparation for receiving business calls and making them as productive as possible, the following points are worth noting:

- Be knowledgeable about the business, its inventories and services, its pricing structure, the amount of time required to accomplish customer requests, and the dates by which they can be accomplished.
- Have a pad of order forms, a blank pad of paper, a price list, and sharpened pencils by the phone.
- Answer the phone as quickly as possible and no later than by the third ring.
- Identify the business first, then give your own name. If not too cumbersome, add an appropriate salutation. For example: "Good afternoon. Larry's Lawn and Garden. This is Janice." "Fountain View Florist. Rod Johnson speaking." "Good morning and thanks for calling Gateway Nursery. This is Barry Byrne."
- Determine the name of the caller at the outset of the conversation. If callers do not identify themselves, ask. If a name or firm is not given distinctly, ask for a repetition.
- Listen intently, shutting out other distractions, to determine the *purpose* of the call, the *needs or concerns* of the caller, and the *disposition* of the caller (for example, friendly, upset, angry, or confused).
- Avoid interrupting the caller. Resist the urge to complete sentences for a slow talker or direct the thinking of an uncertain customer before you fully understand the situation.
- Ask questions to assure that you are understanding the caller's message clearly. Do not make assumptions.
- Take notes to assure that you remember important points and that commitments made are honored.
- Be courteous and positive, even if the caller is rude, profane, or irritating. Use the caller's name occasionally to personalize the conversa-

tion. Even if the conversation does not lead to an agreement, it can be conducted courteously.
- Price must be discussed if a request for merchandise or services is the purpose of the call. Never assume that a customer will pay a price unless it has been discussed and agreed to.
- Review the major points of the conversation (for example, the item(s) ordered, specific preferences for colors or style, date needed, when workers or sales representatives will call, and the price agreed upon).
- Close the conversation courteously and allow the customer to hang up first. Nothing is gained from an extended conversation. It merely ties up the phone lines and delays the next business call.

THE LANGUAGE OF BUSINESS COMMUNICATIONS

As stated earlier, the horticulturist who would communicate most effectively incorporates a friendly and conversational tone into letters and calls and emphasizes the positive rather than the negative. In addition, horticulturists should learn to describe their products and services in ways that evoke enthusiastic and receptive responses. For example, a customer may call a flower shop before Christmas to order "something for the front door." A salesperson can suggest "an evergreen door swag" or "a cluster of fresh pine boughs with accents of red holly berries and pine cones, tied with a red velvet bow." Both descriptions are truthful and refer to the same product, but the second enables the customer to visualize the product better. In a similar fashion, a landscaper can describe a proposed patio as "800 square feet in area, made of brown brick, with a built-in grill and surrounding plantings" or as "800 square feet of patio living area done in brick of a warm earth tone and incorporating cooking facilities and cooling shade." The second description has the important quality of evoking the anticipation of use.

SUMMARY

Letters and telephone conversations are extensions of the ornamental horticulture business and must be handled professionally.

Business letters are the most permanent form of communication. Their permanency is their greatest advantage. A good business letter is written in correct form and conventional format, in a polite and friendly tone, with thorough and accurate coverage, and as briefly as possible.

Proper form requires a letterhead that is restrained and that contains the firm's name, address, and telephone number; a date line; inside address; salutation; body (message); complimentary close; signature; and notations. Letter formats include the full block, modified block, and semiblock.

Use of the telephone for business communication permits questions to be asked when clarification is needed. The development of an expressive voice is important to good telephone communication. The voice must be pleasant and interesting to hear, and the person behind the voice must be knowledgeable, interested in the caller, and professional in manner.

To place a business call correctly, the caller should know in advance what the call seeks to accomplish, what points are to be covered, the sequence of coverage, the information sought or to be released, and the commitment to be sought or offered on behalf of the firm.

Answering a business call requires good listening skills coupled with the ability to comprehend and organize information or questions as offered by a caller.

Achievement Review

A. SHORT ANSWER

Answer each of the following questions as briefly as possible.

1. List six qualities of a good business letter.
2. List the eight parts of a business letter.

B. DEMONSTRATIONS

1. Rewrite the letter in Figure 24-5 in semiblock style leaving the top fifth blank where the letterhead would be.
2. Rewrite the letter in Figure 24-8 in full block style leaving the top fifth blank where the letterhead would be.
3. Assume that you are the general manager of Dakota Gardens, a retail nursery at 1504 W. Purdue Road in Bismarck, North Dakota. Write a letter to a customer, Mr. Tracy Kostecki, at 2600 Ball Avenue in Bismarck. The letter concerns his overdue account of $187.56, which has gone unpaid for ninety days. You have written to him twice already and tried to reach him by telephone, but he has not been available and has not returned your calls. Prior to this indebtedness he has been a good and valued customer who has purchased and paid for much larger amounts of merchandise, always promptly. Your normal procedures are to turn bad accounts over to a collection agency after ninety days.
4. Practice your telephone skills by working with another person in the simulation of a business call. Before beginning, make up a realistic premise. Select a firm name and type of business. Select a customer or vendor name and problem. Make several calls, varying the problem (for example, an order, a complaint, or a request for a bid). Sit facing in opposite directions or on opposite sides of a screen that prevents any contact except by voice. The local telephone company may be able to provide a telephone simulator training kit for greater realism. Record the conversations on tape and play them back for critiques.

25

Record Keeping, Pricing, and Determining Profit

Objectives

Upon completion of this chapter, you will be able to

- list four characteristics of a good record system.
- list the four basic types of business records kept.
- explain the benefits of properly kept records.
- define profit-and-loss statements and balance sheets.
- describe different approaches to pricing merchandise and services.

KEEPING BUSINESS RECORDS

Each year, many talented ornamental horticulturists go out of business. A large number of them do not understand where they went wrong. They worked as hard as they could, their customers approved their work, there seemed to be a lot of work, yet they suddenly found themselves without sufficient working capital to continue. They often accuse assorted culprits, such as unfair competition, unreliable suppliers, fickle clientele, and, of course, government regulations and taxation. Although there are many reasons why small businesses fail, a large number fail simply because their owners or managers do not keep or properly utilize business records.

Talent and hard work, while essential, are not enough to assure survival in business. A rudimentary understanding of money management is also necessary. Too often, a person interested in growing plants, designing with flowers, building landscapes, or maintaining a property attractively is not interested in financial statements, preferring to leave those matters to accountants. Every business needs a skilled and attentive accountant, but it is the responsibility of the owner to collect the data needed by the accountant and to interpret the statements that the accountant supplies. Just reading financial statements is not enough either. They must be used to keep the business on track and moving forward.

Records give direction to the operation of a business. Current and complete records show where the business is succeeding and where it needs help or change. They permit a comparison of the present situation with past achievements. Properly established, a good record system is:

1. simple to use and structured to fit the needs of the particular business it serves
2. easily understood by the person(s) who make decisions on behalf of the business
3. accurate and consistent
4. able to provide current information whenever needed

At the time a recordkeeping system is established, the business owner and accountant must discuss the information *required* and *desired* from the system. The accountant can then explain what raw data must be collected to generate that information.

There are great differences in the ways ornamental horticulturists engage in business, but essentially four types of records are kept:

1. sales (including services performed)
2. cash receipts (monies obtained each day from cash sales and clients paying the bills that have been classed under accounts receivable)
3. cash expenditures
4. accounts receivable

In retail and some wholesale operations, sales and cash receipts may be recorded simultaneously on a cash register. The register may also be able to record the data by department within the store (a refinement that is more important to some operations than others). In service operations, such as landscape contracting or maintenance firms, sales records need to include the client name, site serviced, task performed, materials, equipment, and labor time, Figure 25-1.

At the close of each business day, all monies received that day (cash or checks) should be deposited in the bank. The exact amount received should be deposited so that the bank statement can be used to balance the cash receipts record. Some business owners take small disbursements out of the cash register drawer, or even use cash receipts to buy lunch when they run short. Such practices lead to erroneous cash receipts records. Small amounts of cash should be taken from a petty cash fund established specifically for that purpose. Personal expenses should remain that—personal. The small business owner should go to the bank or write a personal check payable to the business if cash is needed.

All cash expenditures should be recorded and a receipt or cancelled check obtained for each payment. Records should indicate the date, name of the person or firm paid, reason for the expenditure, and item(s) obtained.

Accounts receivable should be kept current. Bills must be prepared when goods are shipped or delivered, and when services are rendered. At the end of every month, each account should be listed along with the amount due and an indication of whether it is current or overdue by thirty or sixty days or longer.

Complete records of the business activities of a firm can give the owner-manager insights into many areas of the business that determine its financial health, such as:

- monthly sales
- cost of sales
- payroll
- cost efficiency
- time requirements for various services
- depreciation of fixed assets
- operational overhead expenses
- profitability of specific promotions or services

From this recorded information, the required tax and other governmental forms can be completed and filed. Thorough records, correctly interpreted, can also catch employee fraud and error, warn of excessive waste and spoilage of perishable materials, and offer guidance on which areas of business need to be promoted more or less as a result of their high or low profit return.

FIGURE 25-1.
A typical work record for a landscape firm. It provides a comprehensive yet concise summary of all costs for a particular job.

QTY.				UNIT	TOTAL
	TOTAL	A	$	MATL.	$
	SUPPLIES				
	HUMUS				
	PEAT				
	FERTILIZER				
	TOPSOIL				
	TOTAL	B	$	MATL.	$

NAME DATE

QTY.	CONST. MATL.		LABOR	UNIT	AMOUNT
	EDGING				
	AGGREGATE				
	TIMBERS				
	PAVING				
	STONE				
	TOTAL	C	LABOR $	MATL.	$
	EQUIPMENT—SUB-CONTRACT		LABOR	COST	PRICE
	DUMP				
		D	TOTAL $	TOTAL	$
	DESIGN/EST		HRS @		
	SUPERVISION		HRS @		
	OTHER		HRS @		
		E	TOTAL $	TOTAL	$
			LABOR	TOTAL	$
A	PLANTS				
B	SUP				
C	CONST				
D	SUB				
E	D/S				
TOTAL					

INTERPRETING FINANCIAL STATEMENTS

The horticulturist should have an outside accountant prepare all financial statements for the business. Few horticulturists have the time to stay abreast of ongoing changes in the tax laws or the rush to computer technology that is altering the way accountants conduct their operations. Still, the horticulturist must be able to understand and utilize the reports of the accountant once they are prepared. The horticulturist should also request reports as often as needed to gain maximum benefit. Thus, a quarterly financial statement may be all that is needed to complete required governmental forms but not sufficient to benefit the business. Monthly statements are common to all horticulture enterprises, and weekly statements are not uncommon.

There are two parts to a financial statement: a balance sheet and a profit-and-loss statement. *Balance sheets* are prepared periodically by the firm's accountant *to show the financial status of the business on the date of issue.* The company's assets, liabilities, and net worth as of that date are shown. By comparing items in the balance sheet with those at the same time last year, trends can be determined. For example, if the balance sheet indicates increasing amounts of assets tied up in inventories and accounts receivable, the owner may wish to examine and alter purchasing practices, credit extension, and/or collection methods.

The *profit-and-loss statement* records *how much money the business earned or lost over the period of time measured* (monthly, quarterly, or annually). In other words, the balance sheet shows the amount of net income at a certain point in time; the profit-and-loss statement can explain how the money was earned (or lost) over a period of time. When a profit-and-loss statement is compared with one for the same period a year ago or with the planned budget for the current period, accomplishments and failures become quickly apparent. Consider the following simplified example.

Analysis of Table 25-1 shows that increased net sales do not necessarily mean increased net profits. Although sales increased by $8,000 between 1994 and 1995, the net operating profit declined by $3,600. Further analysis offers some explanations. The cost of goods sold was unexpectedly high, with an increase of 30 percent in 1995. Inflation might be the cause, or it could be that vendor prices were not compared before purchasing. Perhaps salaries and wages should not have been raised as much as the budget called for, and the additional expendi-

ture of $500 beyond what the budget called for did not help the situation. Perhaps prices were too low.

A similar comparison, made monthly, can permit the business manager to curtail unprofitable expenditures or types of work in time to avoid losses for the year. If necessary, prices can be raised when the profit-and-loss statement indicates that excessive expenses are not to blame for the decline in profits.

FINANCIAL RATIOS

A horticulture firm may measure its financial and earning potential by calculating different ratios that serve as quantitative guides. Ratios can be applied to the analysis of balance sheets and income statements to compare relationships among the different values. Typical financial ratios used by horticulture businesses include:

- *Profitability ratios:* These ratios relate profits to total assets, or gross income, or other quantities.
- *Liquidity ratios:* Liquidity is the ability of a company to pay its bills when they are due. This ratio reveals that ability.
- *Leverage ratios:* A comparison of the company's debt with parameters such as total assets or net worth.

TABLE 25-1.
Comparative Profit-and-Loss Statement

Item	1995 Budget	1995 Actual	1994 Actual	Percent Increase or Decrease	Percent of Total Budget	1995	1994
Net sales	$95,000	100,000	92,000	+ 8.7	100.0	100.0	100.0
Cost of goods sold	32,000	39,000	30,000	+30.0	33.7	39.0	32.6
Gross profit	63,000	61,000	62,000	− 1.6	66.3	61.0	67.4
Expenses							
Salaries and wages	50,000	50,500	48,000	+ 5.2	52.6	50.5	52.2
Operational costs	4,500	4,700	4,500	+ 4.4	4.7	4.7	4.9
Interest expense	800	800	900	−11.1	0.8	0.8	1.0
	55,300	56,000	53,400		58.1	56.0	58.1
Net operating profit	$ 7,700	5,000	8,600	−41.9	8.2	5.0	9.3

Some examples of ratios are given below, based upon the data taken from Tables 25-2 and 25-3.

Current assets	$106,250.00
Inventory	50,000.00
Quick assets = current assets minus inventory	56,250.00
Current liabilities	11,630.00
Total assets	346,450.00
Total liabilities	28,630.00
Fixed assets	240,200.00
Net worth = total assets minus total liabilities	317,820.00
Net working capital = current assets minus current liabilities	94,620.00

Ratios

Quick assets to current liabilities: $\dfrac{\$56,250}{11,630} = 4.84$

Current assets to current liabilities: $\dfrac{\$106,250}{11,630} = 9.14$

Total liabilities to net worth: $\dfrac{\$28,630}{317,820} = 0.09$

Fixed assets to net worth: $\dfrac{\$240,200}{317,820} = 0.76$

These ratios are only a sampling of the financial ratios that a company may use, depending upon the nature of the business and the information sought. The figures used in the examples should not be regarded as anything more than examples for the reader. However, in a real business, the ratio of quick assets to current liabilities tells the horticulture businessperson whether there is enough money available to cover the current indebtedness. A business has liquidity and can pay its bills when the ratio is at least 1.0.

The ratio of current assets to current liabilities tells the business owner whether the business can meet its current obligations in the short term. The higher the ratio is, the better off is the business.

Sometimes called the debt-to-equity ratio, the total-liabilities-to-net-worth ratio indicates the level of investment in the business held by creditors and owners. A low ratio is desirable here.

The fixed-assets-to-net-worth ratio is a measure of how much of the company's assets are tied up in fixed assets and not available as working capital.

PRICING MERCHANDISE AND SERVICES

Determining the correct price to charge is an ongoing challenge in any business, and ornamental horticulturists are famous for doing it badly year after year. They have underpriced their products and services for so long that consumers have come to expect a great deal for comparatively little. Too many horticulturists do not know why they charge a certain price except that their competition charges that price and they match it.

If horticulturists made a single product whose material costs were constant, whose time of manufacturing was as predictable as an automobile, and whose workers were all paid the same wage, it would be easier to determine the cost of the product, and a fair profit could be calculated and added on rather simply. However, one of the great difficulties of pricing horticultural products and services is that there is little quantitative repetition. When a retail flower shop fills an order for a large number of identical wire service specialties, it comes the closest to assembly line production that ornamental horticulture ever does. Much of what ornamental horticulturists do is custom work. Most floral arrangements that leave a shop have a different combination and number of flowers, greens, and other materials. No two landscapes are the same. A greenhouse crop produced in one season is subject to an entirely different combination of environmental variables than a similar crop produced immediately before or after it. A nursery may suffer no rabbit or deer damage one winter, yet be devastated the next year. Thus, the pricing of horticultural materials, like nearly all agricultural commodities, is imprecise and subject to rapid changes. Nevertheless, it must be done knowledgeably, not with fingers crossed and a rabbit's foot in the back pocket.

Some horticultural operations sell products (flower arrangements, nursery plants, potted flowers, sod, chemicals). Others, such as landscape designers and contractors, sell mostly services. Often, these services are rendered only after successful competitive bidding. Just as there are diverse ways of doing business, so are there different methods of pricing.

Price Calculations

The price of an item or service is the sum of material costs, labor costs, overhead costs, and profit allowance.

$$Price = Materials + Labor + Overhead + Profit$$

Failure to measure any component of the equation accurately will result in a price that is unfair to either the business or the client.

The carefully kept records described earlier are the key to accurate pricing. Without them, price calculation is impossible. Even with records, pricing requires continual monitoring.

Materials Horticultural materials are both perishable and nonperishable. Nonperishable items range from floral wire to wire fencing. Their costs to the horticulturist can be found by consulting a current catalogue or by calling the distributor. Some nonperishable materials are reworked into new forms by the horticulturist (for example, wire and tape in floral arrangements and bricks and stone in a patio). Other nonperishables, such as bags of fertilizer or bottles of chemical spray, are resold without change. They are merely purchased at a wholesale price and sold at a retail price. Many of the latter types of nonperishable carry a manufacturer's suggested retail price on the packaging. The retailer is usually not obligated to sell the merchandise at that price, but usually does so for at least two reasons: competitors are using that price, and the suggested price has been carefully researched by the manufacturer to cover the average costs of handling, sales labor, overhead, and profit allowance of retailers nationally.

Perishable merchandise is much more subject to cost variation than nonperishable merchandise. This is due, in part, to seasonal supply and demand. For example, wholesale prices for red roses rise before Valentine's Day and Christmas; grass seed is most costly in the early spring; and hardy bulbs cost most in the fall. In turn, retail prices vary seasonally, requiring frequent recalculation to maintain fairness and accuracy. The pricing of perishable merchandise must include costs common to nonperishables such as handling, sales labor, overhead, and profit allowance, as well as allowance for losses from the aging and death of plants (known as *inventory shrinkage*).

The difference between the wholesale cost of materials and their selling price is termed the *markup*. It is expressed as a percentage of the retail selling price. For example, if an item costs the horticulturist $10.00 and a 35 percent markup is needed to cover associated business costs and allow for a profit, the retail selling price would be $15.38.

$$Retail\ price = \frac{Wholesale\ price}{(100\% - markup)}$$

$$Retail\ price = \frac{\$10.00}{(100\% - 35\%)} = \frac{\$10}{.65} = \$15.38$$

Once the necessary markup for materials is determined, pricing is easy. The big question is how to determine the necessary markup. The answer begins with determination of the break-even point.

Breaking Even The *break-even point* is the point in the operation of a business where total revenue equals total cost, and there is neither a profit nor a loss. Due in large part to the seasonal nature of ornamental horticulture businesses and the variable prices that result, the break-even point changes over time and may need to be determined quarterly or even monthly.

The horticulturist's total costs are usually the cost of materials, labor, and overhead. *Overhead expenses* include administrative salaries, advertising costs, rent or mortgage payments, telephone and utility costs, equipment depreciation, accountant and attorney fees, subscriptions, memberships, insurance premiums, wire service expenses, repair and maintenance costs, postage, stationery, and other costs incurred whether there is a customer that day or not. Overhead expenses are generally fixed unless there is a dramatic change in the volume of business done, salaries paid, or other costs of keeping the doors open. Using the records from the previous year, overhead costs can be determined easily.

The past year's records will also provide valuable data about the cost of materials. Material costs are variable (not fixed) and change in direct proportion to the amount of sales, market conditions, and wholesale prices. The closest estimate of material costs for the current year should be projected and a *cost of materials percentage* determined. This is expressed as a percentage of total sales. For example, if Andrew's Landscaping had $40,000 in sales for the month, and the cost of materials needed to do that business was $10,000, the cost of materials percentage would be 25 percent.

$$\frac{\text{Cost of materials}}{\text{percentage}} = \frac{\text{Cost of materials for the period}}{\text{Total sales for the period}}$$

$$\frac{\text{Cost of materials}}{\text{percentage}} = \frac{\$10,000}{\$40,000} = 25\%$$

In small retail operations, the staff may all be salaried and their labor costs included in overhead. In larger operations, particularly nurseries, landscape firms, arborists, and lawn service firms, much of the labor is provided by temporary seasonal workers who are paid an hourly wage and may receive additional benefits. From properly kept work records and time cards, the cost of labor (including wages and benefits) during a time period can be determined. Like the cost of materials, labor costs are variable and must be estimated as closely and realistically as possible using current data and the past year's records for guidance. From that cost data, adjusted to reflect current conditions, *a cost of labor percentage* can be calculated, also as a percentage of total sales. Using the same example, if Andrew's Landscaping did $40,000 in sales for the month, and the cost of labor was $7,200 (wages plus benefits), the cost of labor percentage would be 18 percent.

$$\text{Cost of labor percentage} = \frac{\text{Cost of labor for the period}}{\text{Total sales for the period}}$$

$$\text{Cost of labor percentage} = \frac{\$7,200}{\$40,000} = 18\%$$

If the year's fixed operational overhead costs for Andrew's Landscaping total $132,000, a cost of $11,000 can be assigned to a single month of operation ($132,000 ÷ 12 months = $11,000). Thus, for the month illustrated in the examples, the firm's total cost (materials, labor, and overhead) is $28,200 ($10,000 + 7,200 + 11,000). The company needs to sell that much in merchandise and/or services in order to break even.

Making a Profit   As the volume of business increases or decreases, the fixed overhead costs generally remain the same. If the volume of business increases above the break-even point and fixed costs do not, then a profit begins to be shown. If volume declines or fixed costs rise, a loss will be registered, Figure 25-2. As the chart illustrates, profits increase rapidly as the sales volume increases, assuming no rise in fixed costs. The only costs to Andrew's firm are materials and labor, which total 43 percent of sales (the cost of materials percentage and the cost of labor percentage, or 25 percent plus 18 percent). Thus for every dollar grossed in sales beyond $28,200 (the break-even point), there will be a 57¢ profit.

Break-even point analysis illustrates a basic principle of business: *Profits result from* (1) *increasing sales volume without increasing overhead costs* and (2) *raising prices of the product or service sold.* There are two points of diminishing returns that must be avoided. These are the points where (1) new investments must be made in personnel or facilities to handle the increased volume of business or (2) the price increases cause sales to decline.

Unit Pricing   Until the break-even point is reached, each unit of material or service sold must regain not only all of the material and labor costs needed to provide it, but also a portion of the fixed overhead costs. The 57¢ profit on every dollar of gross sales only begins after the break-even point is reached and the overhead costs are covered. In

FIGURE 25-2.

Relationship between business volume, fixed costs, and profit

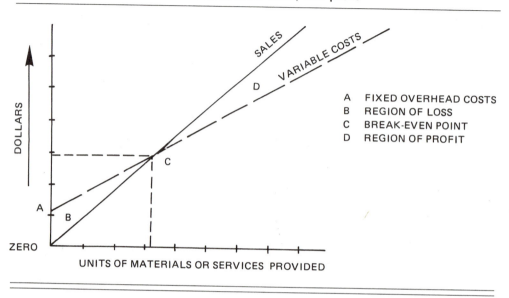

A FIXED OVERHEAD COSTS
B REGION OF LOSS
C BREAK-EVEN POINT
D REGION OF PROFIT

theory, the assignment of a portion of the overhead costs to each unit sold is simple. The overhead costs for a period (month, quarter, or year) can simply be divided by the total number of units produced in that period. The difficulty lies in deciding what constitutes a unit and in the many different units dealt with by most horticulturists. A sales unit might be a retail item such as a bag of fertilizer, a bottle of chemical spray, a dozen roses, a container, or a birdbath. It might also be a measure of time such as an hour of labor. It might be a measure of area or volume such as a square foot or a cubic yard. The selection of appropriate pricing units is determined by the type of business and by its methods of record keeping and accounting. There is no single correct way to set up the accounting program, and the accountant employed will usually have valuable suggestions about unit pricing for the firm.

In the example of Andrew's Landscaping, the $11,000 monthly overhead might be assigned to hours of labor worked during the month, from which the $40,000 in gross sales resulted. If the company's labor force worked a total of 1280 hours that month, each hour of labor charged to the cus-

tomer could include a cost of $8.59 to cover fixed overhead expenses. For example:

$$\frac{\text{Per unit overhead}}{\text{expense}} = \frac{\text{Total overhead for the period}}{\text{Total number of measurement units}}$$

$$= \frac{\$11,000}{1280 \text{ hours}} = \$8.59 \text{ per hour}$$

If, however, Andrew's Landscaping also retailed bedding plants in six-plant market packs, there would be an additional unit of measurement: a market pack. Had the $40,000 gross sales for the month been attained by $35,000 in landscape services requiring 1,000 hours of labor and $5,000 in sales of 2,000 market packs, the assignment of overhead costs would be divided among the two units in proportion to the volume of sales generated. The landscaping accounted for 88 percent of the sales ($35,000 ÷ $40,000) and must accept 88 percent of the overhead costs for the month.

$$\$11,000 \times .88 = \$9680 \text{ overhead costs}$$

The market pack sales accounted for 12 percent of the sales ($5,000 ÷ $40,000) and must accept 12 percent of the overhead costs for the month.

$$\$11,000 \times .12 = \$1320 \text{ overhead costs}$$

The per unit overhead costs for each area of operation can be then calculated as before.

$$\text{Per hour landscaping overhead expense} = \frac{\$9,680}{1,000 \text{ hours}} = \$9.68 \text{ per hour}$$

$$\text{Per market pack overhead expense} = \frac{\$1,320}{2,000 \text{ market packs}} = \$0.66 \text{ per market pack}$$

As the number of types of merchandise and services increases, the calculation of costs becomes more complex. Some firms consolidate a variety of different items under a few categories for the sake of record-keeping convenience. For example, a flower shop may classify its design work as wedding work, funeral work, holiday work, and other. Landscape firms may classify their sales as plant sales, installation projects, and maintenance projects. A nursery may use categories such as plant sales and nonplant sales or plant sales—retail, plant sales—wholesale, and so on. Grouping merchandise and services into categories for calculation of overhead costs or determination of markup can reduce the precision of pricing, but is commonly done for expediency. Each unit or category of units must have records to show what percentage of total gross sales it accounted for and how many items, hours, or other units were sold in the process.

Setting the Percent Markup

Two major influences help determine the price charged for an item or service. The price competitors are charging is one. Customers may accept a small price differential between businesses if one has a better reputation for quality, honesty, or service, but getting the most for their money is always the first consumer concern. The other major influence is how much return on equity the business owner wants or needs to have to maintain a profitable business. *Equity* is the dollar amount of assets owned by the business that is not offset by indebtedness.

$$\text{Equity} = \text{Total assets} - \text{Total liabilities}$$

The rate of return on equity is the percentage of the owner's equity that is returned as net profit during the business year. Every owner needs to know if the money invested in the business is making enough money to equal or surpass the interest it would earn if it had been deposited into an ordinary savings account or invested in bonds, money market certificates, or even a different business. The rate of return on equity provides the information needed to make that comparison. To calculate it requires two figures: the net profit for the past year and the net worth or equity of the business for the past year. The sample profit-and-loss statement in Table 25-2 illustrates how to determine net profit. The sample balance sheet in Table 25-3 illustrates how to determine net worth or equity.

From the sample financial statements, the net profit of Garland's Nursery and Greenhouses is seen to be $80,880, and the total equity at year's end is $317,820. The rate of return on equity can be calculated using the following formula:

$$\text{Percent rate of return on equity} = \frac{\text{Net profit}}{\text{Owner's equity}}$$

$$\text{Percent rate of return on equity} = \frac{\$80,880}{\$317,820} = 25\%$$

In the example, 25 percent of the owner's cash investment has been repaid in net profits for the year past. Should this highly successful rate of return on equity continue, the owner's entire cash investment will be recovered in four years. While a 25 percent rate of return is high compared with what many owners recover, it must be higher than the rate of return that other uses of the money could provide or the owner will have suffered a loss on the money invested, not to mention the time and hard work spent on the business during that period.

Using the percent rate of return and the past year's price figures as guides, pricing for the current year can be developed. If the percent rate of return was too low, prices for the next year would need to be raised, particularly if increasing the volume of business did not seem like a realistic solution. If matching the prices charged by competitors does not permit the owner to make a profit, problems will need to be recognized. Overhead costs may have to be reduced; the staff may have to be reduced or productivity increased; salaries may be too high; or the desired profit margin may be unrealistic. It is also possible that not all components of the cost

TABLE 25-2.
Sample Profit-and-Loss Statement

GARLAND'S NURSERY AND GREENHOUSES
Profit-and-Loss Statement / Final for Year
December 31, 1994

Sales		
Nonperishable	$ 87,000	
Labor	60,000	
Plant materials	153,000	
Total sales		**$300,000**
Cost of goods sold		
Cost of goods sold— nonperishable	$ 30,000	
Cost of goods sold— plant materials	8,500	
Total cost of goods sold		**$ 38,500**
Gross profit		**$261,500**
Expenses		
Wages	$ 98,500	
Payroll taxes	8,700	
Seeds and propagative stock	47,300	
Gasoline and oil	6,000	
Utilities	2,300	
Postage metering	800	
Office supplies	600	
Interest on debts	3,600	
Taxes	1,500	
Advertising	7,300	
Legal fees	1,800	
Accounting fees	2,220	
Total expenses		**$180,620**
Net profit		**$ 80,880**

of an item or service justify the same profit margin. The profit markup on material costs may differ from the profit markup on direct labor or overhead costs.

Alternative Approaches

There are several alternative approaches to pricing.

Direct Cost Pricing   The selling price is based on the direct costs of materials, labor, and overhead. For example, the direct costs of a maintenance job might be $50 for materials, $200 for labor, and $100 for overhead. To make a 20 percent profit on materials, a 40 percent profit on labor, and a 10 percent profit on overhead, the price would be determined as follows:

TABLE 25-3.
Sample Balance Sheet

GARLAND'S NURSERY AND GREENHOUSES
Balance Sheet / Year Ending
December 31, 1994

Assets

  Current Assets

Cash in bank	$ 18,500	
Cash on hand	500	
Petty cash	250	
Accounts receivable	37,000	
Inventory	50,000	
Total current assets		**$106,250**

  Fixed Assets

Land	$ 80,000	
Retail store	30,000	
Garage	25,000	
Greenhouses	30,000	
Equipment	102,500	
Less depreciation	−27,300	
Total fixed assets		**$240,200**
Total assets		**$346,450**

Liabilities

  Current Liabilities

Accounts payable	$ 7,530	
FICA payable	1,800	
Withholding payable	2,300	
Total current liabilities		**$ 11,630**

  Long-Term Liabilities

Notes payable	$ 17,000	
Total long-term liabilities		**$ 17,000**
Total liabilities		**$ 28,630**

Capital

Garland's Nursery and Greenhouses, capital	$317,820
Total liabilities and capital	**$346,450**

	Cost + Profit = Price	Profit
Material	$50 + $ 50 × .20 = $ 60	$ 10
Labor	$200 + $200 × .40 = $280	$ 80
Overhead	$100 + $100 × .10 = $110	$ 10
Totals	$350 $450	$100

Profit Margin Pricing   The selling price is calculated as a percentage of total costs. For example, if a retail florist had job costs totaling $700 and wanted a profit margin of 35 percent, the price would be calculated as follows:

Total cost × Percentage of profit margin = Profit   Price
  $700   ×   .35   = $245 $945

Time and Materials Pricing   Popular with service operations, this requires that all materials be priced at retail and include related overhead costs and profit margin. It further requires that hourly wage rates include related overhead costs and profit margin. To determine the final price requires only that accurate work records be kept of all materials used on the job and total hours of labor expended. The time and materials method, with profit already built in, makes it impossible for the horticulturist to lose money on a job unless everything dies, requiring replacement.

Direct Markup Pricing   Material and/or labor costs are multiplied by a constant figure to arrive at a price that will cover all associated costs and allow for a profit. For example, a florist may mark up all flowers and other perishable materials from three to five times depending upon the location of the shop, season of the year, and volume of business. The same shop may mark up nonperishable items only two times over cost, since spoilage is not a problem. Direct markup pricing is common in retail operations but can be imprecise if all costs are not covered. It can also result in overcharging for some items and undercharging for others.

SUMMARY

There are many reasons why small businesses fail, but one major reason is that their owners or managers do not keep or properly utilize business records. Records give direction to the operation of a business.

Four basic types of records are kept: sales records, cash receipts, cash expenditures, and accounts receivable. Properly kept, they can provide the owner with insights into monthly sales, cost of sales, payroll, cost efficiency, time requirements for various services, depreciation of fixed assets, operational overhead expenses, and the profitability of specific promotions or services.

Although horticulturists should have an outside accountant prepare all financial statements for the business, they must be able to understand and utilize the reports once they are prepared. There are two parts to a financial statement: a balance sheet and a profit-and-loss statement. Balance sheets show the financial status of the business on the date they are issued. Profit-and-loss statements record how much money the business earned or lost over a period of time (monthly, quarterly, or annually).

Pricing requires careful analysis by the horticulturist to assure that all costs are accounted for and that the profit allowance is appropriate.

$$Price = Materials + Labor + Overhead + Profit$$

Failure to measure any component of the equation accurately will result in a price that is unfair to either the business or the client. Carefully kept records are the key to accurate pricing.

The past year's records can provide valuable data about the cost of materials, although material costs are variable and change with the amount of sales, market conditions, and wholesale prices. Labor costs are also variable but can be closely estimated if the past year's records were kept well. Fixed overhead costs should be fully calculable from the records of the year past.

Profits increase as the sales volume increases, assuming no rise in fixed overhead costs. Profits also increase when the price of the product or service sold increases.

As the number of categories of merchandise, products, and services increases, the calculation of costs becomes more complex. To cope with the problem, some businesses consolidate a variety of different items under a few categories. Each category must have records to show what percentage of total gross sales it accounted for and how many items, hours, or other units were sold in the process.

The final price determination is influenced by competitors' prices and the return on equity desired or needed. Equity is the dollar amount of assets owned by the business that is not offset by indebtedness. The rate of return on equity is the percentage of the owner's equity that is returned as net profit during the business year. Using the percent rate of return and the past year's price figures as guides, pricing for the current year can be developed. Alternative approaches to pricing include: direct cost pricing, profit margin pricing, time and materials pricing, and direct markup pricing.

Achievement Review

A. SHORT ANSWER

Answer each of the following questions as briefly as possible.

1. List four attributes of a good record system.
2. List the four basic types of records that are kept by a small business.
3. Indicate if the following information is provided by a profit-and-loss statement (A) or a balance sheet (B).
   a. shows the financial status of the business on a particular date
   b. lists the company's assets
   c. records the amount of money earned or lost over a period of time
   d. lists the company's liabilities
   e. itemizes expenses
   f. lists the net worth
   g. lists the net profit
4. Define the following terms:
   a. inventory shrinkage
   b. markup
   c. break-even point

5. Complete the following sentences
   a. Profits result from _____ sales volume without _____ fixed overhead costs.
   b. The dollar amount of assets owned by the business that is not offset by indebtedness is termed _____.
   c. The percentage of owner's equity that is returned as net profits during the business year is termed _____.
   d. Another term for equity is _____.
   e. The percentage rate of return on equity is calculated by dividing the net profit by the _____.
6. List four different approaches to pricing.
7. Of what value are financial ratios to a business person?

26

Computer Technology and the Ornamental Horticulture Industry

Objectives

Upon completion of this chapter, you will be able to

- describe the uses of computer technology in ornamental horticulture businesses.
- explain the major types of computer programs currently popular in the industry.
- discuss the benefits and limitations of computers to the industry.
- project future uses of computers in the industry.

THE TECHNOLOGY AND ITS USES

Throughout the history of our country each generation of Americans has witnessed some distinctive and unprecedented technological breakthrough thought impossible by earlier generations. The horse and buggy were retired by the automobile. The earth-bound passenger train was swept away by the airplane. The atmosphere-dependent plane has been surpassed by the gravity-defying rocket. Less dramatic but equally innovative has been the replacement of the stenographer, typewriter, and filing cabinet by the computer. Bordering on the edge of technological extinction are the T-square and drawing pencil, manually operated greenhouse controls, paper forms and reports, hand-mixed soils and growing media, traditional inventory control systems, and even some services of the U.S. Postal System.

Modern computers are able to process and store vast amounts of data in a variety of forms. They can accelerate calculations, project the effects of changes, allow instant recall of historical data, react to sensors and activate environmental systems, and turn drawings into electronic impulses and restructure them as drawings in far distant places. While doing all of these things computers never show the slightest sign of fatigue, bad temper, or stress.

Although most businesses across the nation and

the world have embraced computer technology and eagerly await each advancement, the green industries have been slow to accept the full array of computer services. Like so much of agriculture, the production of plants and related services can be imprecise. The image of the computer as a device demanding precision and consistency was perhaps the reason so many in the industry were hesitant to invest heavily in the technology.

With the increasing professionalism of ornamental horticulturists and the prominent roles of national organizations such as those listed in Appendix A, computer technology has found its way into the green industries' mainstream. As the number of computer programs for horticulture increases, the acceptance of the technology grows.

At present, horticulturists rely on computers for many of the same uses as do other businesses, including:

- correspondence
- customer accounts

- accounts payable
- payroll
- taxes
- mailing lists
- inventory records
- accounts receivable
- billing
- sales analysis

Industry-specific operations include:

- greenhouse environmental control
- soil and media mixing
- labeling of plants
- landscape specification preparation
- quantification of materials for building a landscape
- cost estimating
- landscape plant selection
- irrigation system design and graphics
- technical calculations, such as cut and fill
- landscape design graphics

FIGURE 26-1.
A LANDCADD Site planning and Landscape design module in use. (Courtesy of LAND-CADD International, Inc.)

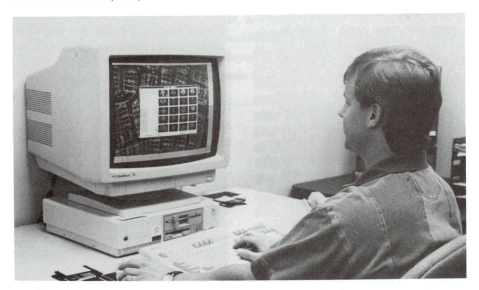

THE POPULAR PROGRAMS

Computer systems require machinery known as *hardware* and instructions to make the machinery operate, termed *software* or *programs*. The hardware sits, unable to do anything until given direction by the software in an electronic language that it can understand. Fortunately, the horticulturist using the computer need not understand the electronic language in order to make the program communicate with the hardware.

The hardware requirements include a means of entering data into the computer (input devices), a central processor of the data (the computer), a place to store the data once processed (hard or floppy disks), and a means of converting the data from the computer into printed form (output device). Input and output devices vary depending upon whether the data are written or graphic in form and what quality of output is desired.

The most common software programs are of three types:

1. word processing
2. numerical calculations
3. graphic visualization

Word processing is the successor to the typewriter. Letters, reports, memos, and other written documents are prepared and stored on word processors. Paragraphs can be moved or deleted, words can be changed or inserted, envelopes addressed, key phrases highlighted, lettering styles changed at will, and numerous other techniques applied instantaneously. Gone are smudged erasures and nameless form letters. Spelling errors and incorrect word choices can be minimized by using the proper software. Many good software programs are available and usable on the IBM personal computers, micro- and minicomputers, or IBM compatible computers that are now the standard for central processors.

Input for word processing is done on a keyboard that is similar in many ways to a typewriter. Output is usually from a printer that transfers the image on the computer's monitor to letterhead stationery or other paper being fed into the printer.

Numerical calculations enlist the capability of software programs that accomplish mathematical functions. Conventional business needs such as inventory control, accounts receivable and payable, payroll and taxes, billing, time records, income statements, and profit and loss statements are all commonly handled by computers.

Computer technology allows garden centers and nurseries to update their inventories when plants and other materials are purchased. One scan by a computerized register generates a printed receipt for the customer and simultaneously deducts the item from the inventory.

Landscapers are able to quantify materials from designs and convert the data into precise bids in a fraction of the time required by conventional measuring and counting methods. Last-minute changes to design specifications can be quickly input and the bid updated. Less time is required for calculations that in the past were time-consuming, tedious, and subject to human error. At the same time the accuracy of the calculations has probably improved, since the computer's insistence on certain data input has compelled companies to keep close track of labor hours, material costs, and equipment usage.

While a number of good software programs are being marketed to nursery and landscape companies, several enjoy especially widespread acceptance. One such system is the SLICE system developed by Thornton Computer Management Systems of Mainesville, Ohio. Unlike earlier programs that were directed to a generic business market, the SLICE system customized its software to the needs of the green industries. It is adaptive to a company's existing hardware or Thornton's own computer product. As part of a successful marketing campaign, backed up by a good product and customer service, the developers have made themselves highly visible to the garden center, nursery, and landscape industries and managed to implant their product's name in the minds of industry professionals as the number one computer system for their businesses. Other producers will surely continue to develop competing software products, and some landscape firms have even developed their own programs, usually as a modification of an existing generic package.

Input devices for numerical calculation programs include the keyboard, the scanner, and the digitizer, Figure 26-2. The *digitizer* permits direct measurement of lengths, areas, volumes, and perimeters from draw-

FIGURE 26-2.

A hand-held stylus is used to select commands and symbols on a digitizer or menu tablet. (Courtesy of GTCO Corporation)

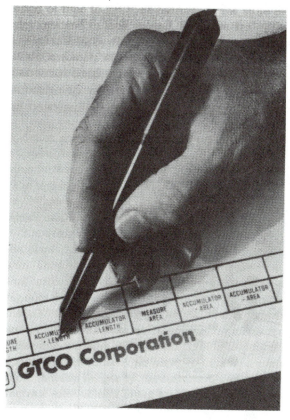

ings; so it is essential for the conversion of drawn data to numerical equivalents. The output device for numerical calculations is usually the printer.

Graphic visualization has gained acceptance within the landscape industry, but more slowly than word processing or numerical calculations. Two types of graphic visualization systems are used in the green industries, mostly by landscape designers and landscape architects: computer-aided design (CAD) and video imaging.

CAD Systems

CAD systems are accelerated drafting systems, designed to do what draftsmen have traditionally done with their T-square and pencils. Computer drafting has increased mainly because of a PC-based CAD software program named *AutoCAD* (a registered trademark of Autodesk, Inc.). Due to the necessity for hardware, software, and user training, AutoCAD is an expensive investment for a firm. It is not ready to use upon delivery to the office, moreover, because it is a general system, marketed to a variety of users. It is this generic nature of AutoCAD that makes it the dominant CAD system. However, to customize it for a specific type of drafting requires a third-party, discipline-specific software interface.

Landscape architects and landscape designers have most frequently selected LANDCADD as their discipline-specific software. Developed by a group of practicing professionals and released commercially in 1984, LANDCADD is marketed by LANDCADD, Inc., of Franktown, Colorado. Presently, LANDCADD offers modular programs that can be selected by firms to fit their needs thus avoiding the purchase of programs they will not use. At the heart of the software system is the Site Planning and Landscape Design Module. It permits the creation of two-dimensional plan views and elevations complete with contour lines, buildings, plant symbols, landscaping, and enrichment items. Surface textures, as well as lettering, can be added. As the design develops, the plant list is automatically created and tabulated. The module comes with its own menu of symbols to which users can add their own, further customizing the system. Figure 26-3 illustrates a landscape plan developed using a LANDCADD software package. Three-dimensional drawings of buildings, terrain, and grading plans are also possible using the Planning and Design Module.

An Irrigation Design Module is available as well as modules that permit material measurement, cost estimating, and plant selection, Figure 26-4. Another module provides a library of construction detail drawings that can be used directly or modified by the user before insertion into the landscape plans. Since success breeds competition, a number of competitive discipline-specific software systems have entered the market, each directed solely or in part to the landscape industry. Such competition assures a free market, innovative research and development, and reduced prices to the consumer.

FIGURE 26-3.

A LANDCADD design (Courtesy of Daniel International Corporation)

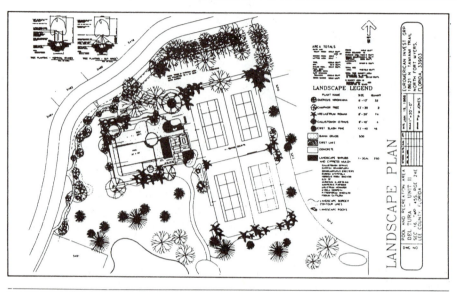

The input device for computer-aided design is usually a *mouse* or a *digitizing tablet and stylus*. The mouse is held in the designer's hand and rolled around the table surface while the cursor moves about on the screen. The tablet and stylus permit the user to place drawings onto the monitor's screen in a manner remotely similar to that of the traditional draftsman, Figure 26-5.

The output device for CAD system drawings may be the printers used for word processing, but the best quality of printout is obtained from a pen plotter, illustrated in Figure 26-6.

Video Imaging

Video imaging is a computerized technique that superimposes a landscape concept over a photographic image of the undeveloped landscape. The designer begins by taking several photographs of the building to be landscaped or filming it for several minutes using a camcorder. The photo or video image is then taken to the designer's studio and entered into the computer and retained by the system's memory. The designer then directs the computer to apply color photo images of plants, pavings, pools, lawn, and other landscape elements over the camcorder or photographic picture. The result is a full color suggestion of how a landscape will look before it is actually installed. It is an excellent sales tool to assist the client's visualization of the designer's ideas.

The leading video imaging system in America is Design Imaging Group, headquartered in Woodland Hills, California. The technology is so new that there is little competition in the marketplace. Figure 26-7 illustrates before and after images of landscapes as treated by the video imaging system. In the hands of a skilled operator, its effects are realistic and impressive. The system permits the presentation of a variety of design concepts to a client for selection, before time is spent on plan view designs that may never be constructed. At present, video imaging does not create scaled drawings from which the landscape can be built. However, the Design Imaging system is designed to interact (interface) with CAD systems

FIGURE 26-4.

A LANDCADD irrigation design (Courtesy of Kaden Landscape)

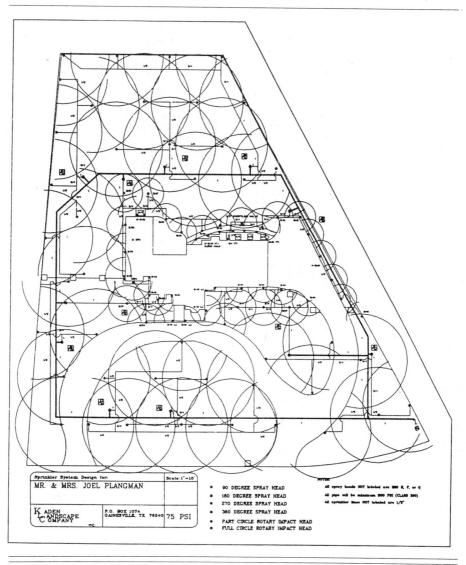

such as LANDCADD to apply color and realistic surface textures to the line drawings produced by the CAD system.

Input devices for the Design Imaging system are the video camcorder, tablet and stylus, and scanner. The graphic image is composed on the monitor and can then be reproduced as a color print or as a video-tape for presentation to a client audience of one or a hundred and one.

Greenhouse Environmental Controls

Enlisting the speed and alertness of computers to help grow plants in greenhouses has been made pos-

FIGURE 26-5.
A drawing pad and drawing boards with cordless transducer choices. (Courtesy of CalComp Digitizer Division)

FIGURE 26-6.
An A- through E-size dual-mode plotter (Courtesy of CalComp)

sible by the development of sensory input devices. They need not await human commands, but instead can sense and monitor temperatures, light, humidity, CO_2, wind velocity and direction, soluble salt levels, pH, and many other growth-related factors. The computerized environmental control system does three things:

1. *Sensing:* As just explained, the system objectively measures environmental factors using a series of independent sensors. The sensors are located both inside and outside the greenhouse range. They are not restricted to single locations within a house, which makes them a significant improvement over older, analog control systems.
2. *Decision making:* An independent computer monitors the sensors, analyzes the information they provide, and compares the existing environmental description against the grower's programmed instructions.
3. *Taking action:* The system activates equipment within the greenhouse to implement the decisions made.

Computerized environmental control systems are component systems whose parts are linked together by coaxial cables and computer chips, and whose data can be made available to the grower at any time, either in graphic displays or in digital or printed form via the computer monitor. Figure 26-8 illustrates a typical computerized environmental control system.

Unlike older greenhouse control systems, which are basically reactive systems (example: the temperature drops below a minimum and the thermostat

FIGURE 26-7.

Before and after example of computer imaging (Image by Gary Galpin, Design Imaging Group)

turns on the heat full force to warm the house), computerized systems can be programmed to respond almost continuously to its integrated sensory information. The result is a greenhouse environment that remains nearly constant, changing more slowly than the reactive systems. Temperature extremes, crop drying and watering, humidity levels, light conditions, and other growth factors adjust gradually. The assumption is that this stability results in less traumatic growing conditions and thus improved plant growth. To date, there is little research evidence to confirm or dispute this assumption. Logical though the probability is, scientific researchers are just be-

ginning to redirect their studies to evaluate the new technology. As computer technology continues to expand, it can be anticipated that such lags between technical claims and scientific confirmation will be common.

Representative greenhouse environmental control system manufacturers are:

- Oglevee Computer Systems in Connellsville, Pennsylvania
- Priva Computers, Inc., in Ontario, Canada
- Wadsworth Control Systems in Arvada, Colorado

THE BENEFITS AND LIMITATIONS OF COMPUTERS

Each technological advance carries with it advantages and disadvantages. If the advantages don't greatly outnumber the disadvantages, the technology is usually not accepted. Computer technology is no longer in its infancy, but neither has its development leveled off. Advancements that improve the computer's capabilities and/or offer manufacturers a competitive marketing advantage are announced regularly. The two edges of the technological sword are thus seen. The great benefits of time saving, reduced errors, rapid revisions, historical memory, and easy replication of tedious tasks are counterbalanced by the high costs of technological obsolescence, or the replacing or modifying of hardware and software and updating of employee training. There is also the danger of putting too much revenue-generating work into the hands of employee specialists with insufficient back-up replacements.

While secretaries, landscape estimators, accountants, specification writers, and others who deal primarily with words, numbers, and calculations have found computer technology easy to adjust to, many ornamental horticulturists have not. For workers whose input has always been accomplished largely through a keyboard, the computer has not asked them to change much. They still gather, copy, and report data in much the same manner that they have for years. The computer merely accelerates the amount of work they can accomplish and lessens their errors. It also replaces earlier forms of data

FIGURE 26-8.

Typical greenhouse computerized environmental control system (Courtesy of Timothy Toland)

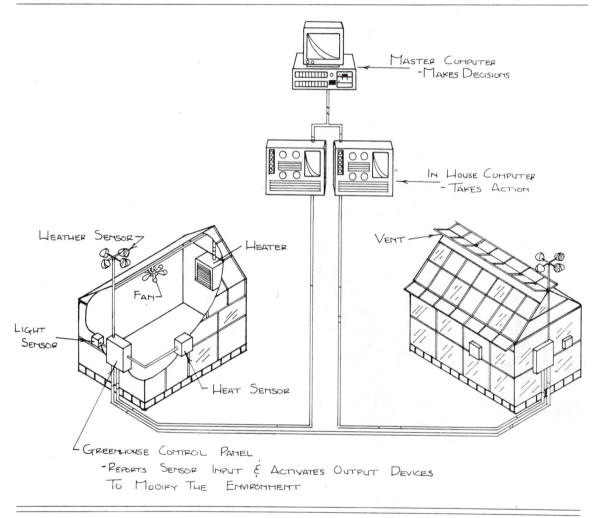

storage with newer forms, yet the end result is the same.

However, landscape designers, greenhouse and nursery growers, and other horticulturists have not had it as easy. To some, there is a need, real or imagined, to maintain a tactile relationship with the materials of their profession. Some greenhouse and nursery growers will believe that the plants are watered properly only after sticking their thumbs into the soil. They gain their professional satisfaction from

tending their greenhouse ranges and nursery fields personally, as men and women of the soil have done for centuries. The sophistication of the computer leaves them unimpressed, so it will take time to convince them that computer technology is an enhancement to their training and expertise, not a threat.

In similar fashion, some landscape architects and designers have held a psychological bias against the computerization of their workplace. They scoffed at the stiff imagery of the computer-generated drawings. Traditionally, the graphic appearance of the landscape plan, including its lettering and symbol styling, has been the hallmark of a designer or a firm. Early CAD designs *were* simplistic and mechanical. However, improved data banks of symbol and lettering styles plus programs that accept the designer's own stylings have removed most of the early objections. Recent software improvements have given CAD design systems the ability to produce perspective drawings and elevation views.

Still, to many landscape designers, the fun of designing, as well as the actual development of concepts, forms, and relationships, have been dependent upon the touch of pencil to paper. Like an electrical current, the flow of creativity is dependent upon contact among all of these elements. The computer does not allow that. It replaces free sketching with menu selections. It offers a data bank of symbols instead of spontaneous interpretations. It asks questions instead of offering answers, and requires exact responses from the designer if it is to serve. It is that mechanical quality of the input, not just the output, that may prove to be a limitation of the current technology, since many creative individuals are not proficient mechanics.

THE FUTURE

As a mechanical servant the computer's role in the industries of ornamental horticulture will continue to expand. One by one the objections and current limitations will be overcome. Systems that now use keyboards for data input may become voice driven in the future. Production systems that are now preprogrammed and allow little alteration can be expected to permit input by local growers, making them even more customized. Florists will be able to send wire orders worldwide over systems that simultaneously translate the order into the language of the receiving florist. Instant spelling and grammar checks will result from enlarged data bases. As more companies become linked nationally and internationally to all computers, communication and ordering may bypass not only the postal service and telephone, but send the FAX machine into the gadgeteering history book along with the eight-track musical tape.

Currently CAD systems have no understanding of what the lines and symbols of their graphics represent. They cannot distinguish between a shape that represents a pool of water and one that symbolizes a brick patio. The designer must tell the estimating software how to interpret the drawing before material calculations and cost estimates can be completed. Once the computer is able to recognize objects represented by the symbols and understand how they will be installed or constructed, then construction drawings, specifications, and cost estimates can be generated, modified, and published simultaneously with the designing.

Site analysis data may be entered into the CAD system directly from aerial photos or video images shot from elevated locations such as rooftops, planes, or even satellites. Optical and laser scanning systems may be the technology that permits this direct input of data. At some point in the future, voice commands may be used to direct or at least correct computer-assisted designing. Even the output products may change. Today's flat print taken from the monitor image may soon appear as a video walkthrough of the proposed design. Such a system will even be able to show how the selected plants will be sized at any age of interest to the client or the designer.

Computer advances are impacting all areas of ornamental horticulture. The technology is expanding to such an extent that computer systems will soon move from their present status as *aids* to the higher level of *assistant*. They will contribute positively to the further development and professionalism of the green industries.

SUMMARY

Computer technology is rapidly expanding its services throughout the industries of ornamental horticulture. Operations common to all businesses as well as those distinctive to the nation's flower shops, greenhouses, nurseries, and landscape and lawncare firms are now routinely aided by computers.

Programs such as SLICE, LANDCADD, Design Imaging, and several greenhouse environmental control systems are changing how plants are grown, inventories are monitored, plans are drawn, and estimates are prepared.

Ornamental horticulturists have been somewhat slow to accept the new technology compared to businesspeople in other disciplines. This is partly due to the difference in the way data must be entered into the systems, and partly due to conservatism and reluctance to accept change as improvement. Such cautious attitudes are changing though, and the acceptance of computers as valuable assistants is gaining momentum.

Computer technology is developing rapidly, so the future can be expected to bring new applications that will render obsolete many techniques that are now regarded as industry norms.

Achievement Review

1. List ten uses of computer technology that ornamental horticulturists and most other businesspeople use commonly.
2. List three uses of computers specific to each of the following horticulture professions:
   a. greenhouse grower
   b. nursery grower
   c. landscaper
3. Explain what the following computer software programs do.
   a. word processing
   b. CAD systems
   c. greenhouse environmental controls
4. Assume the role of a computer advocate. How would you respond to a landscape designer and a nursery grower who were contemptuous of computers and said there was no place for the technology in either of their businesses?
5. Without becoming totally impractical, offer a suggestion for a future technological advancement beyond the current capabilities of these computer programs:
   a. word processing
   b. CAD systems
   c. graphic imaging
   d. greenhouse environmental control
   e. automated irrigation
   f. estimating
   g. inventory control

27
Total Quality Management

Objectives

Upon completion of this chapter you will be able to

- explain how quality is defined and measured.
- define total quality management.
- explain the empowerment of employees.
- distinguish between internal and external customers.
- outline the steps in a Total Quality management program.

A NEW CONCERN FOR AN OLD IDEA

The end of World War II left the economy of the world tilted heavily in favor of the United States. Europe and most of Asia were left with only skeletal remains of their prewar industrial capability. American factories were intact and able to convert from the production of war goods to the manufacture of countless products and machines needed by the rest of the world. We had no competition on the world stage. As the rest of the world slowly rebuilt their industrial capabilities, comparison of their products to those made in the United States was the world-wide standard. Quality was defined as *made in America.*

Regardless of whether the business dealt in products or in services, most American companies operating in the postwar era focused their quality standard on the finished product or service and defined it to fit their personal image. Quality control was usually the final inspection step in the manufacturing or

service process, and products that failed to meet the prescribed quality standard were rejected. Similarly, services that were unsatisfactory were repeated in full or in part. The customer was not involved in the definition or the measurement of quality at all. Complaint or approval was the customer's only role.

Yet products and services that do not meet the customers' needs cannot be considered qualitative. A florist may fill an order for a flowering plant with the largest, most showy, healthiest specimen in the shop, believing he is giving the customer a top quality product. If the customer desires a plant for the center of her dining table, however, she will reject the plant as oversized and certain to inhibit conversation between guests seated around the table. If she is not satisfied with the plant, it is not a quality product. A nursery grower may set exacting standards for tree production. Every tree is shaped, sheared, and spaced to produce a block of plants that are exact duplicates. The nursery perceives and advertises its plants as being the highest quality. If the client wants matching trees to line a formal entry, then the trees possess the quality sought. If the client wants trees to create a naturalized woodland effect, the trees won't do the job. They lack quality as perceived by the customer, regardless of the grower's opinion or the price. A lawn that is installed to textbook standards may exhibit lush greenness and weed-free growth. Is it quality? Perhaps, if the client has water available to maintain it. If the local weather is droughty, then perhaps not.

In recent years, America's dominance of the world economy has lessened. Our former international customers are now our competitors. In matters of quality, the teacher has become the student and American businesspeople are having to observe and learn from other nations. The success of foreign manufacturers and service firms, most notably the Japanese, has been accomplished through hard work, clever management, and a different way of defining and measuring quality.

QUALITY DEFINED

While it is not easy to gain widespread acceptance of a new definition for a familiar term, the business world is beginning to understand that there is no quality standard for a product or service until the customers' satisfaction is registered. Giving the customers exactly what they want plus an extra measure of value without additional charge is or should be the goal of every business in America today.

Quality: Meeting the customer's requirements and exceeding those expectations.

This new definition means that the customer becomes the judge of quality and that the businessperson must find ways of measuring the customer's satisfaction. The new definition also means that the product or service must provide additional features that surpass the customer's expectations. If the business is a service firm, it may deliver the service at a level higher than the customer expected. If the business deals in products, those products may offer additional features without costing more than similar products without those features. Consider these examples:

Example 1. Two florists design equally well and charge the same for their service. Both deliver to customers' homes. For an order of cut flowers, both add extra greens and wrap them in green waxed tissue. At the time of delivery, the delivery person for one florist rings the customer's doorbell, hands the roses to the person answering, turns and leaves. The delivery person for the other florist rings the bell, smiles at the person who answers, says, "I've got a bouquet of fresh flowers for Mary Jones. Smith Florist really appreciates your business!" Tucked inside the wrapping is a packet of material that can be dissolved in water to prolong the life of the customer's flowers. Which florist has exceeded the customer's expectations?

Example 2: A landscape firm has a reputation as being high priced, yet continues to be successful. Why would customers regularly pay more for landscape services obtainable elsewhere at a lower price? Perhaps it is because they know that they will get exactly those materials and services they are expecting and on time. In addition, the company's crew will be dressed in clean uniforms each day, using clean equipment and vehicles. If the customer has a question or concern, there will always be someone on the site who can answer the question and resolve the concern. Other companies may be as competent technically, but allow their employees to work with their shirts off or report to work poorly groomed. The trucks may be dirty or rusted or of different colors. Customer questions may go unanswered or problems unresolved until someone with authority arrives to respond. The value added by the successful company explains its success.

Example 3: Two wholesale nurseries market a similar product assortment to local retail outlets and landscape contractors. One nursery digs the plants, delivers them, and bills the client. The other nursery does the same. When clients return unsatisfactory plants, one nursery replaces them without question. The other nursery replaces them, then follows up with a written or telephoned inquiry to determine why some of the plants were unsatisfactory to the customer. That nursery is measuring client satisfaction and at the same time learning something that may enable it to change and improve its production process.

TQM DEFINED

Putting the new definition of quality into the management methods of a company requires a departure from some deep-rooted traditions in American business. Now outdated are the beliefs that the company sets the standard of quality and that the time to measure standards is at the end of the process. If a shrub was unsatisfactory at the time of delivery, it was replaced. If a poinsettia had whiteflies when delivered, the grower apologized to the retailer and replaced it. If the patio wasn't graded correctly and surface water ran toward the house, it was torn out and done again. Heads may have rolled later back at the company. All such action awaited delivery of the product or service before applying standards or measuring client satisfaction. That was the old way, and it isn't working any more.

Total quality management (TQM) recognizes the client as the only real quality inspector of a company. It also regards the provision of a product or service as a *process,* a series of steps that begin with the first contact with the customer and proceed through to the delivery of the product or service, the billing, and the follow-up. The process may have many steps or only a few; but each step must be designed to fit efficiently with those that come before and after. Each step must also be totally understandable to the person responsible for accomplishing it. If each step in a process is understood by the person doing it, and every step of the process is done properly, then the result will be correctness. The product or service at the end of the process will meet the company's standards and satisfy the customer with accompanying cost efficiency and reduced waste. *As the quality goes up, the costs go down.*

Total quality management: Managing a company in a manner that allows continuous improvement of its services, products, processes, and organization to satisfy the requirements of its customers and exceed their expectations.

Horticultural businesses that want to apply TQM must begin by identifying all of the processes that make up the business. Then the steps within each process must be recognized and assessed to learn whether each step links efficiently to those on either side or whether there is duplication, omission, or an opportunity for confusion.

Here is an example of a sequence of processes in a flower shop.

1. Receive customer inquiry.
2. Make the sale.
3. Prepare the arrangement.
4. Deliver the arrangement.
5. Bill the customer.
6. Follow up to determine customer satisfaction.

Each of these processes can in turn be divided into several steps that flow in sequence. In our example, the process of preparing the arrangement might include this series of steps:

1. Order the flowers and greens.
2. Order the containers, picks, water-retention media, wire, tapes, and other construction materials.
3. Schedule the work.
4. Construct the arrangement.
5. Package the arrangement in preparation for delivery.

In a large shop and during a busy season several people may be involved in the process. Each employee must be trained to understand his or her particular process and how to do it efficiently and effectively. The employee must then pass the output of his or her process to the next person who will use that output in the accomplishment of the next process in the sequence. If each step is done correctly and each process proceeds with maximum efficiency, the job will be accomplished without waste and to the highest standard. Management attention can be focused on each step within the processes to isolate and correct problems before they affect the end product or service.

Here is another example of processes and steps in a landscape job:

Sequence of Processes

1. Sell the job.
2. Schedule the job.
3. Procure the materials.
4. Install the job.
5. Clean up.

6. Bill the client.
7. Honor all guarantees.
8. Follow up to determine client satisfaction.

Steps in the Procurement of Materials

1. Review the take-off list (quantities of materials needed).
2. Check sources for availability of materials.
3. Determine dates when materials are needed at the job site.
4. Order materials and specify delivery dates.
5. Receive delivery and verify the orders.

By monitoring each step in each process, managers are able to identify time-consuming repetitions, distracting sidetracks, or gaps that make the landscape service less than it could be. Correcting small problems within the process and training employees to do their jobs correctly and pass the output to the next employee in a form that permits the next step to get underway on schedule with all needed data or materials is the new role of managers.

INTERNAL AND EXTERNAL CUSTOMERS

Total quality management is the monitoring of processes and the people who perform those processes. It takes the spotlight off the finished product or service as the measurement of quality and instead scrutinizes the many little steps that lead to the finished product or service. While requiring a new definition of quality to recognize that it is client-driven, TQM also requires an expanded concept of who the customers are.

The traditional customer is the one who places the order and pays the bill. Every employee in the company must understand the obligation to satisfy that *external customer* and exceed his or her expectations. However, TQM recognizes other customers, *internal customers,* who are the employees of the company and who receive the output of the process that precedes theirs. Each internal customer-employee receives data or material from another employee, then does something with it or to it, and passes it along to his or her internal customer, the next employee working in the process. As an example, to get a bench of mums in flower by a particular date re-

quires that the cuttings be planted in the bench on an exact date weeks earlier. The propagator must take the cuttings, get them rooted, and deliver them to the grower exactly on time if the mature plants are to be in bloom on time. The propagator's internal customer is the grower awaiting the rooted cuttings. The grower's internal customer is the sales staff who have orders to fill from florists.

The larger a company is, the more internal customers it has. Employees need to know the exact form of the input they are to receive from the person whose work precedes theirs. They must also know what their role in the process is and in what form and standard their output must be passed to their internal customer. Each employee's work must be done completely and correctly before being passed along. Each employee within the company is simultaneously a provider of materials or services and an internal customer. As a provider the employee strives to satisfy his or her internal customer. In the role of internal customer, the employee is obligated to demand complete and correct input from the provider. Failure of any employee to reject faulty input will make the process defective and result in a substandard finished product or service, the ultimate disservice to the external customer.

Giving so much responsibility to employees *empowers* them in ways that are foreign to older, centralized management systems. Employee empowerment also permits workers to initiate the changes that they believe will improve the processes for which they are responsible.

Improvement Must Be Continuous. For companies seeking to implement a total quality management program, commitment and patience are essential. Improvements cannot be applied to the entire company by declaration or fiat. A company cannot close tonight and reopen tomorrow with a full TQM philosophy in place. Improvement is gradual, addressing one step and process at a time. The problem in a process is first identified. Then it must be determined if the problem results from improper employee performance or from some other reason. If employee malfunction is the reason for the problem, the person should be trained in the proper method and then required to perform properly thereafter. If the problem stems from another cause, it should be identified and a solution developed. Employees must then be trained in the new process. An evaluation period must be allowed to assure that the solution was a correct one and truly improves the process. If so, then the new process must be explained and taught to everyone in the company who is involved directly or indirectly with the process. The result of this improvement will be some type of saving in time or materials or other costs. Then the process begins again. One small positive change follows another in an ongoing program of incremental improvements. Total quality management is continuous. It never ends because improvements are never finished.

IMPLEMENTING TQM

The company that wants to make the transition to total quality management must do these things:

1. Recognize that quality can be defined and evaluated only by the external consumer client.
2. Analyze the processes that collectively comprise the products and services of the company.
3. Identify the steps within each process.
4. Scrutinize each step for problems and incrementally improve each step and each process.
5. Enlist the recognition by every employee that he or she is working to serve and satisfy the customer.
6. Identify to all employees their dual roles as a provider of materials or services within a process and as an internal customer with the right to expect the input from other providers to be complete and ready for use.
7. Empower all employees to respond directly to clients' questions and concerns.
8. Listen to employees as the most important source of suggestions for process improvements.
9. When processes are changed, evaluate them first, then train all involved employees and require their commitment to the new method.

10. Establish measurement tools, such as questionnaires or follow-up phone calls, to determine customer reaction to the company's products or services.
11. Recognize that TQM is a management method that is continuous and self-renewing.
12. Recognize that employees must accept and participate fully in TQM. If, after explanations and training are given, an employee is unable or unwilling to accept the management philosophy the employee should be separated from the company.

SUMMARY

The meaning of quality has changed from a standard measured by a company at the completion of its production or service to one defined by the customer. The role of every one of the company's employees is to perform his or her job in a way that contributes to the quality satisfaction of the consumer client. When company management and employees understand the total quality management philosophy, they accept that the company has both internal and external customers. The traditional external customer is the consumer who pays the bill. The internal customers are the employees themselves. Each worker is a provider of services or materials needed by another employee working within the company.

All processes of the company must be analyzed and the steps within each process identified. Then flaws, duplications, or omissions can be identified and corrected and training given to the employees. If, after evaluation, the new technique proves correct, all employees should be trained and expected to follow the new procedure.

Total quality management requires a company to incrementally improve its products, services, and organization continuously. The objective is to satisfy the requirements of its customers and to exceed their expectations.

Achievement Review

1. Define the following terms
   a. quality (the new meaning)
   b. total quality management
   c. internal customers
   d. external customers
   e. incremental improvements
2. What is meant by employee empowerment?
3. Outline the steps in a TQM program.

Glossary

abscission The fall of leaves or other plant parts from the plant.

accent plant A plant that has greater visual appeal than many other plants, yet is not as showy as a specimen plant.

accessory fruit Developed from one or more ovaries and includes the calyx and/or receptacle.

acclimatization The adjustment of an outdoor plant to interior conditions.

adobe A heavy, clay-like soil common to the U.S. Southwest.

adventitious A term usually applied to those roots or shoots that develop in unusual locations.

aeolian soil Soil that is transported and deposited by wind.

aeration The addition of air; for example, to the soil or to water.

aesthetic Attractive to the human senses.

after-ripening The changes that occur within a dormant seed that are necessary to permit its germination and natural growth.

aggregate fruit Developed from a single flower having a group of ovaries.

agronomy The science that deals with field crops and soil management.

alkaline Characterized by a high pH. Also called *basic*. The opposite of acid.

alluvial soil Soil that is carried in water such as rivers.

ammonification The conversion of the nitrogen in organic compounds to ammonia.

analysis The percentage of various nutrients in a fertilizer product. A minimum of three numbers on the package indicates the percentage of total nitrogen (N), available phosphoric acid (P_2O_5), and water soluble potash (K_2O), in that order.

annual A plant that completes its life cycle in one growing season.

anther In flowering plants, the part of the stamen that bears the pollen.

anthocyanin A common pigment in plants responsible for red coloring.

antitranspirant A liquid sprayed on plants to reduce water loss through transpiration. It helps prevent transplant shock, windburn, and sun scald.

apical meristem The tissue at the tip of a root or stem where growth occurs most rapidly.

apomixis The development of seed without the full sexual process occurring. The process of meiosis may be bypassed and the seed formed directly from a diploid megaspore.

arborist A tree maintenance specialist.

arid A term used in the description of landscapes where there is little usable water.

auxin A natural growth regulator occurring in plants. Auxins promote cell elongation.

balance sheet One part of a financial statement.

It shows the financial status of a business on the date of issue. The firm's assets, liabilities, and net worth as of that date are shown.

balled-and-burlapped A form of plant preparation in which a large part of the root system is retained in a soil ball. The ball is wrapped in burlap to facilitate handling during sale and transplanting.

bare root A form of plant preparation in which all soil is removed from the root system. The plant is lightweight and easier to handle during sale and transplanting.

bedding plant A herbaceous plant preseeded and growing in a peat or plastic pot or packet container.

blend A combination of the seeds of two or more cultivars of a single species.

bonsai The ancient Japanese craft of dwarfing trees.

break-even point The point in the operation of a business where total revenue equals total cost and there is neither a profit nor a loss.

breaker A greenhouse hose nozzle that diminishes the force of the water but not the rate of flow.

buccal spear The feeding apparatus of a nematode. It is used to puncture host cells and withdraw cellular fluids.

budding A type of grafting which uses a single bud as the scion material.

bulb A flowering perennial that survives the winter as a dormant fleshy storage structure.

calibration The adjustment of a piece of equipment so that it distributes a given material at the rate desired.

caliche A soil common to the American Southwest, characterized by high alkalinity and a calcareous hardpan deposit near the surface. The hardpan layer blocks drainage.

callus Wound tissue that develops from the parenchyma.

cambium Meristematic tissue that produces secondary phloem and xylem tissue (vascular cambium) or protective cork tissue (cork cambium).

canopy The collective term for the foliage of a tree.

carotene A common pigment in plants responsible for orange coloring.

caryogamy The fusion of two sexual nuclei. It occurs as part of plasmogamy in fungi.

cation exchange The capacity of colloidal soil particles to attract positively charged ions (cations) and to exchange one ion for another.

cell The basic structural unit of all plant and animal life.

chitin A nitrogenous polysaccharide compound that comprises most of the exoskeleton of insects.

chlorophyll Green pigment necessary for photosynthesis; located within the plastids of the plant cell.

chloroplast A specialized, subcellular structure within a plant cell that contains the chlorophyll.

chlorosis The yellowing of plant tissue for reasons other than a lack of light.

chromosomes Bodies within the nucleus of a cell that are composed of DNA and proteins. They determine heredity.

clone A group of new plants reproduced asexually from a single plant.

cold frame A low growing structure that uses the heat of sunlight passing through glass or plastic to provide the warmth needed to propagate, start, or harden-off plants.

colloidal A chemical state common to clay particles in which surface charges attract water and ions.

colluvial soil Soils that have moved in response to gravity, as after a landslide or mudslide.

color family One of the six major groupings of colors visible when white light is passed through a prism: red, orange, yellow, green, blue, and violet.

color scheme A grouping of colors.

compaction A condition of soil in which all air has

been driven from the pore spaces. Water is unable to move into and through the soil.

compatible graft A permanent union of the stock and scion in a grafted plant.

complete fertilizer A fertilizer containing nitrogen, phosphorus, and potassium, the three nutrients used in the largest quantities by plants.

complete flower A flower that contains all of the floral parts (petals, sepals, pistils, and stamens).

composite flower A grouping of many tiny flowers that gives the appearance of being a single blossom.

conditioning Preparation of soil to make it suitable for planting.

cone The reproductive structure of plants in the division Coniferopsida. Naked, unenclosed seeds are contained on the upper surface of each cone scale.

containerized A form of plant preparation for sale and transplanting. When purchased, the plant is growing with its root system intact in a container.

cool-season grass A turfgrass that is favored by daytime temperatures of 60° to 75° F.

cover crop A rapidly growing crop used to stabilize field soils and prevent erosion. It is planted after harvest of the major crop (as in a nursery) and before replanting of another major crop.

crop rotation Alternation of major crops and cover crops to permit rebuilding of the soil, soil stabilization, and a reduction of host-specific inoculum persisting in the soil.

crotch The point at which a branch meets the trunk of a tree or another larger branch.

crown The point at which the branches and root system of a shrub join.

cultivar An intentionally cultivated variety whose continuance is due primarily to propagation by horticulturists.

cuticle A waxy layer exterior to the epidermis of a plant. It is composed of cutin and is water resistant.

cytokinesis The division of the non-nuclear material in a cell during growth, including the formation of a new cell wall.

cytokinins Naturally-occurring hormones known to be active in the division of cells. They often interact with auxins to stimulate the initiation and growth of roots and shoots.

cytoplasm All of the living material in a cell other than the nucleus.

deciduous A type of plant that loses its leaves each autumn.

delayed incompatibility A graft union that appears permanent but is later found to be unsuccessful.

dicot An angiosperm (flowering plant) that has two seed leaves.

dihybrid cross A cross between parents that differ genetically in two or more independently inherited characteristics.

dioecious The male and female reproductive organs are borne on separate plants.

diploidal Having two sets of chromosomes.

dividend A profit paid to corporate shareholders.

dormancy A period of rest common to certain plants at certain stages of growth (for example, seeds) or at certain times of the year (for example, winter).

double dormancy A condition in seeds resulting from a hard, water-impermeable seed coat and an embryo that requires after-ripening.

drafting cloth A translucent plastic surface with a thin layer of linen bonded to it, used in landscape designing.

drainage The act of water passing through and beyond the root area of plants in a growing medium.

drum-lacing A technique for securing a balled-and-burlapped plant. Twine is wound around the ball and then laced in a zig-zag pattern.

ectoparasite A parasite whose life cycle occurs outside the host plant. The term is often applied to nematodes.

elevation view A method of illustration that provides two-dimensional views of the front, rear, or side of an object or area.

embryo The basic sporophytic plant that develops from a zygote inside an archegonium or an ovule.

endoparasite A parasite that completes most of its life cycle within a host plant. The term is often applied to nematodes.

epigynous A flower with an inferior ovary; that is, one attached to the stem below where the other flower parts are attached.

equity The dollar amount of assets owned by a business which is not offset by indebtedness. Equity = Total assets − Total liabilities

eradicant A pesticide that is applied to a plant when a pathogen or insect has arrived and must be killed immediately.

erosion The wearing away of the soil caused by water or wind.

espalier A pruning technique that allows trees or shrubs to develop only two-dimensionally. Espaliered plants have height and width, but little depth.

etiolation A condition resulting in a plant due to lack of light. The plant turns yellow, leaf size is reduced, and stems become weak and spindly.

evergreen A plant that retains its leaves all year.

exoskeleton The solid, outer portion of an insect's body. It serves as the point of attachment for muscles while protecting the softer tissues inside.

exotic plant A plant that has been introduced to an area by humans, not nature.

explant A piece of plant tissue used in tissue and organ culture.

fertilization The addition of nutrients to the growing medium through application of natural or synthesized products called fertilizers.

focal point The point of greatest visual attraction within a larger composition.

fogger A greenhouse hose nozzle that permits water to be applied as a fine mist to seedlings and tender plants. Also a means of delivering pesticides.

foot candle The amount of light produced by a standard candle at a distance of 1 foot. A foot candle (f.c.) is a measure of light intensity.

foundation planting The planting next to a building that helps it blend more comfortably into the surrounding landscape.

fungicide A chemical used for the control of fungi.

gamete A sex cell.

gametophyte The haploid phase of a plant's life when the sex cells (gametes) are produced.

ganglia Groups of specialized cells that are part of the nervous system of insects.

gene A part of the deoxyribonucleic acid (DNA) portion of a chromosome that determines heredity.

genotype The genetic composition of a plant.

geology The science that deals with natural history as recorded in rocks.

geotropism Movement in plants caused by gravity.

gibberellins A group of thirty or more closely related plant hormones that promote cell enlargement, often causing dramatic increases in plant height.

girdling The complete removal of a strip of bark around the main stem of a plant. After girdling, the ability of nutrients to pass from roots to leaves is lost, causing eventual death of the plant.

glacial till Soil deposited by glaciers.

glazing The covering for a greenhouse. Traditionally, the term refers to glass, but it can also be hard plastic.

grading Altering the existing slope or terrain of the land.

graft The union of two or more plant parts growing as a single plant.

graft union The place in the grafted plant where the stock and scion fuse.

green manure A crop planted as an alternate between the times of harvest and replanting of a major crop (as in nurseries). It serves as a cover crop but also improves soil structure, fixes nitrogen, and adds organic matter.

groundcover A low-growing, spreading plant, usually 18 inches or less in height.

guttation The slow exudation of liquid water from the hydathodes of a leaf.

haploidal Having one set of chromosomes.

hardening-off A reduction of the temperature and water given a plant in order to permit its survival under more stressful conditions.

hard paving Surfacing that is either poured in place, later hardening, or is set into place as modular units.

hardy Describes the ability of a plant to survive the winter in a cold climate.

heading back The shortening of a shrub branch. It is a pruning technique.

heaving Exposure of the root system of shallowly rooted plants caused by repeated freezing and thawing of the soil.

herbaceous Describes plants that are weaker and more succulent than woody plants. They lack a bark covering, and their twigs display little increase in diameter. Above-ground portions are often unable to survive the winter in cold climates.

herbicide A chemical that kills higher plants. The term usually means weed killers.

heterozygous A gene pair whose members are not alike.

high-analysis fertilizer A complete fertilizer with 30 percent or more of its weight in available nutrients.

holdfasts Appendages of certain vines that allow them to climb.

homozygous Both genes of a pair are alike.

hotbed A low growing structure that uses electric cables or heating pipes to provide the warmth needed to propagate, start, or harden-off plants.

hue A color in its most brilliant and unaltered state.

humus An organic colloidal component of soil resulting from the enzymatic breakdown of plant tissue.

hybrid vigor A phenomenon that occurs when the F_1 offspring of an inbred cross has qualities superior to that of either parent.

hydathode Openings in leaves, often at the tip, through which water can pass out in liquid form.

hydrology The science that deals with water, especially surface and ground water.

hyperplasia Growth, usually abnormal, resulting from an increase in the number of cells produced.

hypersensitivity The reaction of a plant cell to invasion by a pathogen or insect in which the cell dies so quickly that it fails to support further proliferation of the pest.

hypertrophy Growth, usually abnormal, resulting from an excessive increase in cell size.

hypha A thread-like filament that is part of a fungal thallus.

hypogynous A flower with a superior ovary; that is, one attached to the stem above the place where the other flower parts are attached.

hypoplasia Dwarfing resulting from a reduction in the number of cells produced.

hypotrophy Dwarfing resulting from a reduction in cell size.

imperfect flower A flower that has either pistils or stamens but not both.

incompatible graft The failure of a stock and scion to unite permanently in a grafted plant.

incomplete flower A flower lacking one or more of the floral parts.

incurve The center of a corner planting. It is the natural focus of attention and a logical place to use a specimen plant or other focal feature.

indoleacetic acid (IAA) The most commonly occurring natural auxin.

inert material The percentage, by weight, of material in a package that is not active. With packaged seed, it is the material that will not grow.

infected A condition in which a harmful agent is within a host plant.

infested A condition in which a harmful agent is on a host plant.

inflorescence Clusters of small flowers arranged on an axis.

inorganic fertilizer A fertilizer that is synthesized from chemicals that are not derived from living systems. Also known as a chemical fertilizer.

insecticide A chemical used for the control of insects.

intangible A quality denoting something that cannot be touched.

intensity The quality of visual strength or weakness that characterizes a color.

internode A region of the stem between two nodes.

interstock A piece of plant stem placed between the stock and scion, resulting in two graft unions. It is used to join two species that are incompatible with each other but not with the interstock.

inventory shrinkage Losses of perishable materials due to aging and death.

irrigation Application of water to a crop to maintain the proper balance of moisture and air in the soil.

jump-cutting A tree pruning technique that allows a scaffold limb to be removed without stripping off a long slice of bark as it falls. The technique requires three cuts.

juvenility A condition of vegetative growth during which a plant is incapable of flowering, even when the stimuli for flower initiation are present.

landscape design The arrangement of outdoor space in a way that serves the needs and desires of people without damage to natural ecological relationships.

lateral bud Any bud below the terminal bud on a twig.

layering A technique of asexual propagation that permits a severed stem to remain partially attached to the parent until roots are initiated at the cut. Afterwards the new rooted plant is separated from the parent.

leaching The passage of nutrient elements through the root region of the soil, making them unavailable to the plant.

lead branch The dominant branch of a tree.

lenticel A pore-like opening in stems and roots.

light quality The color of light emitted by a particular source.

lime A powdered material used to correct excess acidity in soil.

liner A rooted cutting ready for transplanting into a nursery field or container.

loam Soil which contains nearly equal amounts of sand, silt, and clay (a desirable condition).

low-analysis fertilizer A complete fertilizer with less than 30 percent of its weight in available nutrients.

luminosity The quality of certain colors that allows them to be seen under dim light.

lux The illumination received on a surface that is 1 meter from a standard light source known as unity. A lux is an international measurement of light intensity.

market The geographic area from which a business attracts most of its customers.

markup The difference between the wholesale cost of materials and their selling price.

maturity A condition of growth during which a plant is capable of flowering when the appropriate stimuli are provided.

megasporangium The structure in which megaspores are produced.

megaspore A haploid spore (N) that will develop into a female gametophyte.

meiosis A sequence of cell divisions that reduces the number of chromosomes in a cell by half.

meristematic tissue Plant tissue where the cells are actively dividing and rapid growth occurs. Regions of meristematic activity are found in the tips of stems and roots, in the cambium of the vascular system, and at the base of grass leaves.

metamorphosis The changes in insect form as they grow.

microsporangium The structure in which microspores are produced.

microspore A haploid spore (N) that will develop into a male gametophyte.

mitosis The normal division of cell nuclei that occurs as plants grow, involving no reduction and recombination of chromosomes.

mixture A combination of the seeds of two or more different species.

monocot An angiosperm (flowering plant) that has parallel veination and a single seed leaf.

monoculture The cultivation of a single species in abnormally large quantities.

monoecious Both male and female reproductive organs are borne on the same plant.

monohybrid cross A cross between parents that differ genetically in only one characteristic.

mulch A material placed on top of the soil to aid in water retention, prevent soil temperature fluctuations, and discourage weed growth.

multiple fruit Developed from multiple ovaries of multiple flowers borne on a single stalk.

mycelium A fungal thallus made of hyphae.

mylar A translucent plastic surface used in making landscape designs.

nanometer The unit of measurement for light wavelengths.

native plant A plant that evolved naturally within a given geographical region.

naturalized plant A plant introduced to an area as an exotic plant that adapts so well it may appear to be native.

necrosis A symptom of plant injury in which the tissue becomes desiccated and dies. Necrosis may be localized and limited or extensive.

nematicide A chemical used for the control of nematodes.

nitrification The conversion of ammonia to nitrite, then to nitrate.

nonselective herbicide A weed killer that kills all plant material with which it makes contact.

noxious weed A weed that is extremely undesirable and difficult to eradicate.

nucleus The subcellular structure within a cell that contains the chromosomes, nucleolus, and nucleoplasm.

nutrient ratio The proportion of each nutrient in a fertilizer in lowest terms. Example: A fertilizer with a 5-10-5 analysis has a 1-2-1 nutrient ratio.

organ A grouping of related tissues to perform a complex function.

organ culturing The reproduction of plants from tiny pieces of plant organs.

organic fertilizer A fertilizer composed of compounds derived from living organisms. The nutrients are made available to the plants more slowly than with inorganic fertilizers. The cost of the nutrients is greater than with inorganics.

outcurve The outer regions of a corner planting.

overhead costs Fixed costs in the operation of a business. They include administrative salaries, advertising costs, rent or mortgage payments, telephone and utility costs. They are not directly related to the cost of doing business on any single day or for any one client.

oviparous A classification for eggs that hatch inside the female. The term is applied to insects.

ovipositor A special organ for laying eggs, found in female insects.

ovoviviparous A term applied to insects that do not lay eggs but whose young develop inside the female.

ovule A megasporangium. In seed plants it will develop into the seed.

parasite An injurious agent that is biological and

infectious or infestious. It is incapable of manufacturing its own food and derives its sustenance from the cells of other organisms.

parthenocarpy The initiation of fruit-set without pollination. It is accomplished by the application of either auxins or gibberellins to the pistils of flowers.

parthenogenesis Reproduction from eggs that have not been fertilized; common in certain insects.

pasteurization The elimination of undesirable weeds, microbes, and insects from soil. It is accomplished with steam or with chemical fumigants. It does not kill all life in the soil, only the harmful organisms.

pathogen The causal agent in plant diseases.

perennial A plant that lives more than two growing seasons. It often is dormant during the winter.

perfect flower A flower that has both pistil and stamens.

perigynous A flower with a superior ovary and the petals and sepals fused to form a tube-like structure around but separate from the ovary.

permanent wilting Wilting that results from severe water deficiency in the soil. Transpiration reduction alone will not alleviate the condition. Water must be added to the soil or the plant will die.

personnel management The direction of workers in a manner that brings out their best efforts and attitudes on behalf of the business.

perspective view A method of illustration that provides a three-dimensional view of an object or area.

pesticide A chemical used for the control of insects, plant pathogens, or weeds.

petiole scar The scar left on a twig by the leaf petiole of the previous year's growth.

pH A measure of the acidity or alkalinity of soil. A pH of 7.0 is considered neutral. Ratings below 7.0 are acidic; above 7.0 basic (alkaline).

phenotype The external physical appearance of a plant.

phloem Part of the vascular tissue of a plant. It carries organic materials from the leaves to other parts of the plant.

photoperiodism The effect of varying durations of light exposure on plant growth and development.

photosynthesis A process unique to green plants in which sugar is manufactured from water and carbon dioxide in the presence of chlorophyll. Light energy drives the chemical reaction, and oxygen is released in the process.

phototropism A growth movement in plants in response to light.

pinching Removal of the terminal shoot on a flowering branch to permit development of lateral shoots for a fuller plant. The technique is applied to herbaceous more than to woody plants.

pistil The female reproductive organ of a flower. It is located in the center of the flower.

pith The parenchymatous tissue that is located in the center of a dicotyledonous plant stem. A plant's pith may be described as solid, chambered, or hollow depending upon the species.

plasma membrane Thin, continuous, semipermeable material surrounding the protoplast and controlling the substances that can move into or out of the cell.

plasmogamy The fusion of two compatible gametes.

plastid A specialized unit within the cytoplasm of a cell. It is usually involved in food manufacture and storage.

plug A small piece of sod used to start a new lawn or repair an established one. Plugging is a time-consuming means of installing a lawn, common to the southern United States.

plugs Rooted seedlings ready for transplanting with their root systems intact, thereby eliminating transplant shock.

plug sheet A shallow, chambered flat used to produce plugs.

polarity In plant propagation, the term refers to the formation of new shoots at the end of a cutting nearest the tip and the formation of roots at the end nearest the crown.

pollination A transfer process necessary for sexual reproduction. In flowering plants, pollen is transferred from the anther to the stigma. In nonflowering plants, pollen is transferred from a microsporangium to an ovule.

profit-and-loss-statement One part of a financial statement. It records how much money a business earned or lost over a period of time.

propagation Reproduction that is deliberately controlled and manipulated.

protectant A pesticide applied to a plant before a pathogen or insect arrives to kill the injurious agent when it does arrive.

protoplast The living matter of a plant cell.

proximal/distal balance A form of balance in which the on-site landscape counterbalances the off-site landscape.

pruning The removal of a portion of a plant for better appearance, improved health, controlled growth, or attainment of a desired shape.

purity The percentage, by weight, of pure grass seed in a mixture.

quality Meeting the customers' requirements and exceeding their expectations.

ramet An individual plant within a clone.

respiration A breaking-down process that uses oxygen and enzyme catalysts to oxidize sugar to carbon dioxide and water. In the process, energy is produced.

rodenticide A chemical used for the control of rodents.

root cap A mass of cells that covers the apical meristem (growing region) in a root to protect it as the root grows through the soil.

rosetted The condition of a plant in which leaf development is good but internodal growth is retarded.

runner A stem that grows along the ground and forms new plants at one or more nodes.

scaffold branches The branches of a tree that create the canopy. They arise from and are secondary to the lead branch.

scarification The breaking of a seed coat otherwise impervious to water to permit water uptake by the embryo. It is a pregermination treatment.

scion The shoot portion of a grafted plant.

sedentary soil A soil that weathers from bedrock and remains in place.

selective herbicide A weed killer formulated to be effective only against particular plants, such as annual grasses or broadleaved species.

senescence The aging of a plant or plant part.

shade A color hue darkened by the addition of black.

shrub A multistemmed woody plant, smaller than a tree.

side dress The application of fertilizer along the side of a row of crop plants at a time most satisfactory to growth of the plants.

simple fruit Developed from a single ovary.

site An area of land having potential for development.

slow-release fertilizer A slow-action fertilizer in which nitrogen content is in a form not soluble in water. The nitrogen is released more slowly into the soil for more efficient uptake by plants.

sod Established turf which is moved from one location to another.

soft paving Loose aggregate materials used as surfacings.

soil The thin outer layer of the earth's crust, made up of weathered minerals, living and nonliving organisms, water, and air.

soil separates Groups of soil particles formed as bedrock weathers. Depending upon the particle size, a separate may be clay, silt, sand, or gravel.

soil structure The arrangement of soil particles into aggregates.

soil texture The composition of a soil as determined

by the proportion of sand, silt, and clay that it contains.

soluble salts Fertilizer elements that dissolve in water in the soil. In excess quantities, they can limit the availability of water to the plants or may reach excessive levels in the plant tissue, thereby causing harm, even death, to the plants.

species A category of plant classification distinguishing a plant from all others.

specimen plant A plant distinguished by some unusual visual quality such as its shape or colors.

spiracles Small openings along the side of the thorax and abdomen through which insects breathe.

sprig A piece of grass shoot. It may be a piece of stolon or rhizome or lateral shoot. Sprigging is a vegetative method of establishing a lawn.

stamen The male reproductive organ of a flower.

stigma In flowering plants, the part of the pistil that receives pollen during pollination.

stipule A leaflike appendage located where the petiole of a leaf joins the stem.

stipule scar A mark left on a stem after the stipules fall away. They adjoin the petiole scar.

stock The root portion of a grafted plant.

stolon An aerial shoot that takes root after coming into contact with the soil.

stratification The chilling period required by some dormant seeds to accomplish the after-ripening necessary for germination.

sucker A shoot that originates from a plant's underground root system.

sun scald A temperature-induced form of winter injury. The winter sun thaws the above-ground plant tissue, causing it to lose water. The roots remain frozen and unable to replace the water. The result is drying of the tissue.

tangible A quality denoting something that can be physically touched.

taxonomy The systematic classification of plants.

temporary wilting Wilting from which the plant will recover as soon as the rate of transpiration falls below the rate of water uptake by the plant.

tender Describes the inability of a plant to survive the winter in cold climates. It is also used to describe the vulnerability of a young plant to weather extremes.

tendrils Appendages of certain vines that allow them to climb.

terminal bud The bud at the end of a twig. It exercises apical dominance over the lateral buds below it.

thallus The vegetative body of a fungus.

thatch A layer of organic residue located on the soil surface. It can be in varying stages of decomposition.

thinning out The removal of a shrub branch at or near the crown. It is a pruning technique.

tiller The lateral shoot of a bunch-type grass. Tillers are produced from axillary buds located within the leaf sheath.

tint A color hue lightened by the addition of white. In flowers, a tint appears as a pastel color.

tissue A grouping of similar cells to perform a common function.

tissue culturing The reproduction of plants from tiny pieces of undifferentiated plant tissue.

tone A color hue grayed by the addition of both white and black.

topiary The pruning of plants into unnatural shapes such as animals, architectural features, or geometric forms. The technique is associated with formal gardens.

topography A record of an area's terrain.

top working The grafting of scions onto the existing framework and root system of a large, established tree. It is done with fruit and nut trees to produce a crop faster than natural growth permits.

total quality management Managing a company in a manner that allows continuous improvement of its services, products, processes, and organization to satisfy the requirements of its customers and exceed their expectation.

trace element A nutrient essential to the growth of plants but needed in far less amounts than the major elements.

transition Used in floral design, the term refers to the use of flowers in a size sequence similar to the way they unfold in nature; that is, largest at the center, smallest at the edges.

transpiration The loss of water vapor from a plant.

transported soil Soils that are moved by the forces of nature. *See* colluvial, alluvial, aeolian, and glacial soils.

twining A manner by which certain vines are able to climb.

vacuole A cavity within the cytoplasm containing the cell sap.

variety A classification of a plant that recognizes some characteristic distinguishing it from others of the same species.

vascular bundle The conducting tissue of a plant. It is composed of the xylem and the phloem.

vellum A translucent paper used in landscape designing.

venation The pattern of the veins in a leaf.

vendor A supplier of materials needed in the operation of a business.

vine A plant with a vigorous central lead shoot and a long, linear growth habit.

warm-season grass A turfgrass that is favored by daytime temperatures of 80° to 95° F.

water sprout A small shoot of a tree or shrub that develops along the trunk.

weed A plant growing where it is not wanted and having no economic value.

windburn The drying out of plant tissue (especially evergreens) by the winter wind.

winter injury Any damage done to elements of the landscape during the cold weather season.

woody Describes plants that have a corky outer surface of bark covering their older stems. Woody plants usually survive the winter, and woody stems normally increase in diameter each year.

working drawing A copy of a landscape design done on heavy paper or plastic film. It is used repeatedly during the construction of a landscape and must be very durable.

wound paint A sealing paint used over plant wounds of 1 inch or more in diameter after pruning.

xanthophyll A common pigment in plants responsible for yellow coloring.

xylem Part of the vascular tissue of a plant. It carries water and minerals from the roots to the above-ground plant parts.

zygote A diploidal plant cell.

Appendix A

PROFESSIONAL AND TRADE ORGANIZATIONS FOR HORTICULTURE AND RELATED AREAS

AGRONOMY, SOIL

AMERICAN SOCIETY OF AGRONOMY
677 S. Segoe Road
Madison, WI   53711

CROP SCIENCE SOCIETY OF AMERICA
677 S. Segoe Road
Madison, WI   53711

PROFESSIONAL GROUNDS MANAGEMENT
SOCIETY
7 Church Lane
Suite 13
Pikesville, MD   21208

SOIL CONSERVATION SOCIETY OF AMER-
ICA
7515 N.E. Ankeny Road
Ankeny, IA 50021

SOIL SCIENCE SOCIETY OF AMERICA
677 S. Segoe Road
Madison, WI   53711

WEED SCIENCE SOCIETY OF AMERICA
309 W. Clark Street
Champaign, IL   61820

BOTANY

AMERICAN ASSN. OF BOTANICAL GAR-
DENS AND ARBORETA
P.O. Box 206
Swarthmore, PA   19081

BOTANICAL SOCIETY OF AMERICA
University of Kentucky
Lexington, KY   40506

BULBS

NETHERLANDS FLOWER-BULB INSTITUTE
Corporate Plaza II
51 Cragwood
S. Plainfield, NJ   07080

BUSINESS MANAGEMENT

GARDEN INDUSTRY OF AMERICA, INC.
2501 Wayzata Boulevard
Minneapolis, MN   55440

NATIONAL LAWN AND GARDEN DISTRIB-
UTORS ASSN.
1900 Arch Street
Philadelphia, PA   19103

NATIONAL SMALL BUSINESS ASSN.
1604 K Street N.W.
Washington, DC   20006

FERTILIZERS, CHEMICALS

FERTILIZER INSTITUTE
1015 18th Street N.W.
Washington, DC   20036

NATIONAL FERTILIZER SOLUTIONS ASSN.
8823 N. Industrial Road
Peoria, IL   61615

POTASH/PHOSPHATE INSTITUTE
2801 Buford Highway N.E.
Suite 401
Atlanta, GA   30329

FLORIST

FLORAFAX INTERNATIONAL, INC.
4175 S. Memorial Drive
Tulsa, OK   74145

FLORISTS' TRANSWORLD DELIVERY ASSN.
29200 Northwestern Highway
Southfield, MI   48037

SOCIETY OF AMERICAN FLORISTS
901 N. Washington Street
Alexandria, VA   22314

TELEFLORA
12233 W. Olympic Boulevard
Suite 140
Los Angeles, CA   90064

FLOWERS, PLANTS

ALL-AMERICA ROSE SELECTIONS, INC.
P.O. Box 218
Shenandoah, IA   51601

AMERICAN CAMELLIA SOCIETY
P.O. Box 1217
Fort Valley, GA   31030

AMERICAN DAHLIA SOCIETY, INC.
345 Merritt Avenue
Bergenfield, NJ   07621

AMERICAN HIBISCUS SOCIETY
206 N.E. 40th Street
Pompano Beach, FL   33064

AMERICAN IRIS SOCIETY
7414 E. 60th Street
Tulsa, OK   74145

AMERICAN PRIMROSE SOCIETY
2568 Jackson Highway
Chehalis, WA   98532

AMERICAN RHODODENDRON SOCIETY
14635 S.W. Bull Mountain Road
Tigard, OR   97223

AMERICAN ROSE SOCIETY
P.O. Box 30,000
Shreveport, LA   71130

BEDDING PLANTS, INC.
P.O. Box 286
Okemos, MI   48864

BROMELIAD SOCIETY
647 S. Saltair Avenue
Los Angeles, CA   90049

HOLLY SOCIETY OF AMERICA, INC.
407 Fountain Green Road
Bel Air, MD   21014

INTERIOR PLANTSCAPE ASSN.
2000 L Street N.W.
Suite 200
Washington, DC   20036

LIVING PLANT GROWERS ASSN.
1419 21st Street
Sacramento, CA   95814

NORTH AMERICAN LILY SOCIETY, INC.
P.O. Box 476
Waukee, IA   50263

GOVERNMENT AGENCIES

ANIMAL AND PLANT HEALTH INSPEC-
    TION SERVICE
6505 Belcrest Road
Hyattsville, MD   20782

SOIL CONSERVATION SERVICE
P.O. Box 2890
Washington, DC   20013

GRASS

AMERICAN SOD PRODUCERS ASSN.
4415 W. Harrison
Hillside, IL   60162

BETTER LAWN AND TURF INSTITUTE
AND THE LAWN INSTITUTE
991 W. Fifth Street
Marysville, OH   43040

PROFESSIONAL LAWN CARE ASSN.
1225 Johnson Ferry Road
Suite B
Marietta, GA   30067

HORTICULTURE

AMERICAN HORTICULTURAL SOCIETY
P.O. Box 0105
Mt. Vernon, VA   22121
and
7931 E. Boulevard Drive
Alexandria, VA   22308

HORTICULTURAL RESEARCH INSTITUTE,
INC.
2000 L Street N.W.
Suite 200
Washington, DC   20036

INTERNATIONAL FEDERATION OF LAND-
SCAPE ARCHITECTS
Lisbon, Portugal

LANDSCAPING

AMERICAN INSTITUTE OF LANDSCAPE
ARCHITECTS
602 E San Juan Ave.
Phoenix, AR   85012

AMERICAN SOCIETY OF LANDSCAPE AR-
CHITECTS
1750 Old Meadow Road
McLean, VA   22101

ASSOCIATED LANDSCAPE CONTRACTORS
OF AMERICA
12200 Sunrise Valley Drive
Suite 150
Reston, VA 32091

COUNCIL OF TREE AND LANDSCAPE AP-
PRAISERS
1250 I Street N.W.
Suite 504
Washington, DC   20005

NATIONAL BARK PRODUCERS ASSN.
301 Maple Avenue W.
Tower Suite 504
Vienna, VA   22180

NATIONAL LANDSCAPE ASSN.
2000 L Street N.W.
Suite 200
Washington, DC   20036

PROFESSIONAL GROUNDS MANAGEMENT
SOCIETY
7 Church Lane
Suite 13
Pikesville, MD   21208

NURSERIES

AMERICAN ASSOCIATION OF NURSERY-
MEN
230 Southern Building
Washington, DC   20005

MAILORDER ASSN. OF NURSERYMEN, INC.
210 Cartwright Boulevard
Massepequa Park, NY   11762

WHOLESALE NURSERY GROWERS OF
AMERICA, INC.
2000 L Street N.W.
Suite 200
Washington, DC   20036

OUTDOOR LIVING

CALIFORNIA REDWOOD ASSN.
591 Redwood Highway
Suite 3100
Mill Valley, CA   94941

GARDEN INDUSTRY OF AMERICA, INC.
2501 Wayzata Boulevard
Minneapolis, MN   55440

GOLF COURSE SUPERINTENDENTS ASSN.
OF AMERICA
1617 St. Andrews Drive
Lawrence, KS   66044

KEEP AMERICA BEAUTIFUL, INC.
99 Park Avenue
New York, NY   10016

NATIONAL SWIMMING POOL INSTITUTE
2111 Eisenhower
Alexandria, VA   22314

PEST CONTROL

ENTOMOLOGICAL SOCIETY OF AMERICA
4603 Calvert Road
College Park, MD   20740

INTERNATIONAL   PESTICIDE   APPLICA-
   TORS, INC.
a.k.a. Washington Tree Service
20057 Ballinger Road N.E.
Seattle, WA   98155

NATIONAL PEST CONTROL ASSN.
8100 Oak Street
Dunn Loring, VA   22027

WEED SCIENCE SOCIETY OF AMERICA
309 W. Clark Street
Champaign, IL   61820

POWER EQUIPMENT, PARTS

OUTDOOR   POWER   EQUIPMENT   INSTI-
   TUTE, INC.
1901 L Street N.W.
Washington, DC   20036

SMALL   ENGINE   SERVICING   DEALERS
   ASSN., INC.
P.O. Box 6312
St. Petersburg, FL   33736

SEEDS

AMERICAN SEED TRADE ASSN.
1030 15th Street N.W.
Suite 964
Washington, DC   20005

SOIL CONDITIONERS

CANADIAN   SPHAGNUM   PEAT   MOSS   IN-
   FORMATION BUREAU
928 Broadway
New York, NY   10010

PERLITE INSTITUTION, INC.
6268 Jericho Turnpike
Commack, NY   11725

TREES

AMERICAN FOREST INSTITUTION
1619 Massachusetts Avenue N.W.
Washington, DC   20036

AMERICAN FORESTRY ASSN.
1319 18th Street N.W.
Washington, DC   20036

AMERICAN SOCIETY OF CONSULTING AR-
   BORISTS
315 Franklin Road
N. Brunswick, NJ   08902

NATIONAL ARBORISTS ASSN., INC.
1400 Wantagh Avenue
Suite 207
Wantagh, NY   11793

NATIONAL CHRISTMAS TREE ASSN.
611 E. Wells Street
Milwaukee, WI   53202

SOCIETY OF MUNICIPAL ARBORISTS
7447 Old Dayton Road
Dayton, OH   45427

Appendix B

SELECTED READINGS FOR THE STUDY OF ORNAMENTAL HORTICULTURE

The Science of Ornamental Horticulture

Agrios, George N. *Plant Pathology.* New York: Academic Press Inc., 1978.

Bailey, Liberty Hyde. *Manual of Cultivated Plants.* New York: Macmillan Publishing Co. Inc., 1949.

Bailey, Liberty Hyde, et. al. *Hortus Third: A Concise Dictionary of Plants Cultivated in the United States and Canada.* New York: Macmillan Publishing Co. Inc., 1976.

Baker, Kenneth, and R. James Cook. *Biological Control of Plant Pathogens.* San Francisco: W. H. Freeman & Co. Publishers, 1974.

Bannister, Peter. *An Introduction to Physiological Plant Ecology.* New York: Halsted Press. 1976.

Barden, John, et al. *Plant Science.* New York: McGraw-Hill, 1987.

Barth, Friedrich. *Insects and Flowers: The Biology of a Partnership.* Princeton, New Jersey: Princeton University Press, 1985.

Beevers, Harry. *Respiratory Metabolism in Plants.* Evanston, Illinois: Row, Peterson, 1961.

Berrie, Alex. *An Introduction to the Botany of the Major Crop Plants.* London: Bellmawr, 1977.

Billings, William. *Plants and the Ecosystem.* Belmont, California: Wadsworth Publishing, 1978.

Bleasdale, J. K. A. *Plant Physiology in Relation to Horticulture.* London: Macmillan Publishing Co. Inc., 1974.

Bohn, Hinrich, Brian McNeal, and George O'Connor. *Soil Chemistry.* New York: John Wiley & Sons Inc., 1979.

Bray, C. M. *Nitrogen Metabolism in Plants.* New York: Longman Inc., 1983.

Bridwell, Ferrell. *Landscape Plants: Their Identification, Culture & Use.* New York: Delmar Publishers, Inc., 1994.

Brooks, Audrey, and Andrew Halstead. *Garden Pests and Diseases.* New York: Simon & Schuster, 1980.

Buol, S. W., F. D. Hole, and R. J. McCracken. *Soil Genesis and Classification.* Ames, Iowa: Iowa State University Press, 1980.

Burns, George W. *The Plant Kingdom.* New York: Macmillan Publishing Co. Inc., 1974.

Carter, Walter. *Insects in Relation to Plant Disease.* New York: John Wiley & Sons Inc., 1973.

Cravens, Richard. *Pests and Diseases.* Alexandria, Virginia: Time-Life Books, 1977.

Cronquist, Arthur. *How to Know the Seed Plants.* Dubuque, Iowa: W. C. Brown Co., 1979.

Cunningham, John. *Common Plants: Botanical and Colloquial Nomenclature.* New York: Garland Publishing Inc., 1977.

Dirr, Michael. *Manual of Woody Plants.* Champaign, Illinois: Stipes Publishing Co., 1990.

Dodge, Bertha. *It Started in Eden: How the Plant Hunters and the Plants They Found Changed the Course of History.* New York: McGraw-Hill Inc., 1979.

Donahue, Roy L., and Raymond Miller. *Soils: An Introduction to Soils and Plant Growth.* Englewood Cliffs, New Jersey: Prentice-Hall Inc., 1989.

Dropkin, Victor. *Introduction to Plant Nematology.* New York: John Wiley & Sons Inc., 1980.

Flegmann, A. W., and Raymond George. *Soils and Other Growth Media.* London: Macmillan Publishing Co. Inc., 1975.

Fletcher, J. T. *Diseases of Greenhouse Plants.* New York: Longman, 1984.

Fry, William. *Principles of Plant Disease Management.* New York: Academic Press Inc., 1982.

Galston, Arthur, Peter Davies, and Ruth Satter. *The Life of the Green Plant.* Englewood Cliffs, New Jersey: Prentice-Hall Inc., 1980.

Graf, Alfred. *Exotica, Series 4 International: Pictorial Cyclopedia of Exotic Plants from Tropical and Near–tropic Regions.* East Rutherford, New Jersey: Roehrs Company Publishing, 1982.

Greulach, Victor. *Plant Function and Structure.* New York: Macmillan Publishing Co. Inc., 1973.

Greulach, Victor, and J. Edison Adams. *Plants: An Introduction to Modern Botany.* New York: John Wiley & Sons Inc., 1976.

Hammett, K. *Plant Propagation.* New York: Drake Publishing, 1973.

Hartmann, Hudson, Wm. J. Flocker, and Anton Kofranek. *Plant Science: Growth, Development and Utilization of Cultivated Plants.* Englewood Cliffs, New Jersey: Prentice-Hall Inc., 1988.

Headstrom, Birger. *The Families of Flowering Plants.* South Brunswick, New Jersey: A. S. Barnes & Co. Inc., 1978.

Healey, B. J. *The Plant Hunters.* New York: Scribner Book Companies Inc., 1975.

Heywood, Vernon. *Plant Taxonomy.* London: Edward Arnold, 1976.

Hudson, Norman. *Soil Conservation.* Ithaca, New York: Cornell University Press, 1981.

Janick, Jules. *Horticulture Science.* San Francisco: W. H. Freeman & Co. Publishers, 1986.

Janick, Jules, et. al. *Plant Science: An Introduction to World Crops.* San Francisco: W. H. Freeman & Co. Publishers, 1981.

Jeffrey, Charles. *An Introduction to Plant Taxonomy.* New York: Cambridge University Press, 1982.

Johnson, Warren, and Howard Lyon. *Insects That Feed on Trees and Shrubs: An Illustrated Practical Guide.* Ithaca, New York: Cornell University Press, 1976.

Koopowitz, Harold. *Plant Extinction: A Global Crisis.* Piscataway, New Jersey: Stone Wall Press, 1983.

Krishnamoorthy, H. N, ed. *Gibberellins and Plant Growth.* New York: John Wiley & Sons Inc., 1975.

LaCroix, J. Donald, ed. *Plants, People and Environment.* New York: Macmillan Publishing Co. Inc., 1979.

Leopold, Aldo Carl. *Plant Growth and Development.* New York: McGraw-Hill Inc., 1964.

Levitt, Jacob. *Responses of Plants to Environmental Stresses.* New York: Academic Press Inc., 1980.

Liu, Cheng, and Jack B. Evett. *Soils and Foundations.* Englewood Cliffs, New Jersey: Prentice-Hall Inc., 1981.

McDaniel, Gary. *Ornamental Horticulture.* Reston, Virginia: Reston Publishing Co., 1979.

Matthews, Richard E. *Plant Virology.* New York: Academic Press Inc., 1970.

Maxwell, Fowden, and Peter Jennings, eds. *Breeding Plants Resistant to Insects.* New York: John Wiley & Sons Inc., 1980.

Moore, David, ed. *Green Planet: The Story of Plant Life on Earth.* New York: Cambridge University Press, 1982.

Nelson, Richard, *Breeding Plants for Disease Resistance: Concepts and Applications.* University Park, Pennsylvania: Pennsylvania State University Press, 1973.

Nickell, Louis. *Plant Growth Regulators: Agricultural Uses.* New York: Springer-Verlag New York Inc., 1982.

Noggle, Glen, and George Fritz. *Introductory Plant Physiology.* Englewood Cliffs, New Jersey: Prentice-Hall Inc., 1976.

North, C. *Plant Breeding and Genetics in Horticulture.* New York: John Wiley & Sons Inc., 1979.

Pirone, Pascal. *Diseases and Pests of Ornamental Plants.* New York: John Wiley & Sons Inc., 1978.

Postlethwait, Samuel. *The Audio-tutorial Approach to Learning, Through Independent Study and Integrated Experiences.* Minneapolis: Burgess Publishing Co., 1972.

Pyenson, Louis. *Fundamentals of Entomology and Plant Pathology.* Westport, Connecticut: AVI Publishing Co., 1980.

Raven, Peter H., Ray Evert, and Helena Curtis. *Biology of Plants.* New York: Worth Publishing, 1976.

Roberts, Daniel, and Carl Boothroyd. *Fundamentals of Plant Pathology.* San Francisco: W. H. Freeman & Co. Publishers, 1972.

Roberts, Daniel, et. al. *Fundamentals of Plant Pest Control.* San Francisco: W. H. Freeman & Co. Publishers, 1978.

Sanchez, Pedro. *Properties and Management of Soils in the Tropics.* New York: John Wiley & Sons Inc., 1976.

Schmidt, Clifford et al. *Hands-on Botany: An Audio-tutorial Approach.* New York: John Wiley & Sons Inc., 1976.

Sharvelle, Eric. *Plant Disease Control,* Westport, Connecticut: AVI Publishing Co., 1979.

Sill, Webster. *Plant Protection: An Integrated, Interdisciplinary Approach.* Ames, Iowa: Iowa State University Press, 1982.

Smith, Kenneth. *Plant Viruses.* New York: John Wiley & Sons Inc., 1977.

Smith, Kenneth. *A Textbook of Plant Virus Diseases.* New York: Academic Press Inc., 1972.

Sopher, Charles, and Jack Baird. *Soils and Soil Management.* Reston, Virginia: Reston Publishing Company, 1982.

Stevens, Russell. *Plant Disease.* New York: Ronald Press, 1974.

Stevenson, Forrest Frederick, and Thomas Mertens. *Plant Anatomy.* New York: John Wiley & Sons Inc., 1976.

Stone, Doris. *The Lives of Plants.* New York: Scribner Book Companies Inc., 1983.

Streets, Rubert. *The Diagnosis of Plant Diseases: A Field and Laboratory Manual Emphasizing the Most Practical Methods for Rapid Identification.* Tucson, Arizona: University of Arizona Press, 1978.

Swift, Lloyd. *Botanical Bibliographies.* Minneapolis: Burgess Publishing Co., 1970.

Thomas, Emrys, and M. R. Davey. *From Single Cells to Plants.* New York: Springer-Verlag New York Inc., 1975.

Thompson, Louis, and Frederick Troeh. *Soils and Soil Fertility.* New York: McGraw-Hill Inc., 1978.

Ting, Irwin. *Plant Physiology.* Reading, Massachusetts: Addison-Wesley Publishing Co. Inc., 1982.

Tippo, Oswald, and William Louis Stern. *Humanistic Botany.* New York: W. W. Norton & Co. Inc., 1977.

Tisdale, Samuel, Werner Nelson, and James Beaton. *Soil Fertility and Fertilizers.* New York: Macmillan Publishing Co. Inc., 1985.

Treshow, Michael. *Environment and Plant Response.* New York: McGraw-Hill Inc., 1970.

Troeh, Frederick, J. Arthur Hobbs, and Roy Donahue. *Soil and Water Conservation for Productivity and Environmental Protection.* Englewood Cliffs, New Jersey: Prentice-Hall, 1980.

Vankat, John. *The Natural Vegetation of North America: An Introduction.* New York: John Wiley & Sons Inc., 1979.

Villiers, Trever. *Dormancy and the Survival of Plants.* London: E. Arnold, 1975.

Walsh, Leo, and James Beaton, eds. *Soil Testing and Plant Analysis.* Madison, Wisconsin: Soil Science Society of America, 1973.

Weaver, Robert J. *Plant Growth Substances in Agriculture.* San Francisco: W. H. Freeman & Co. Publishers, 1972.

Welsh, James. *Fundamentals of Plant Genetics and Breeding.* New York: John Wiley & Sons Inc., 1981.

Westcott, Cynthia, and R. Kenneth Horst. *Westcott's Plant Disease Handbook.* New York: Van Nostrand Reinhold, 1990.

Willson, Mary. *Plant Reproductive Ecology.* New York: John Wiley & Sons Inc., 1983.

Wilson, M. Curtis, Donald Schuder, and Arwin Provonsha. *Insects of Ornamental Plants.* Prospect Heights, Illinois: Waveland Press, 1982.

Zadoks, Jan, and Richard Schein. *Epidemiology and Plant Disease Management.* New York: Oxford University Press, 1979.

The Craft of Ornamental Horticulture

Allen, Oliver, and the editors of Time-Life Books. *Decorating with Plants.* Alexandria, Virginia: Time-Life Books, 1978.

Ball, Victor. *The Ball Bedding Plant Book.* West Chicago, Illinois: George J. Ball, Inc., 1977.

Beard, James. *Turfgrass: Science and Culture.* Englewood Cliffs, New Jersey: Prentice-Hall Inc., 1983.

Benz, Morris. *Flowers: Abstract Form II.* Houston, Texas: San Jacinto Publishing Company, 1979.

Benz, Morris. *Flowers: Geometric Form.* College Station, Texas: San Jacinto Publishing Co., 1980.

Berninger, Miriam, and Lou Berninger. *Flower Arranging: 150 Illustrated Concepts.* Madison, Wisconsin: Berninger, 1972.

Bickford, Elwood, and Stuart Dunn. *Lighting for Plant Growth.* Kent, Ohio: Kent State University Press, 1972.

Booth, Norman, and James Hiss. *Residential Landscape Architecture.* Englewood Cliffs, New Jersey: Prentice-Hall, 1991.

Brooklyn Botanic Garden. *Origins of American Horticulture: A Handbook.* Brooklyn, New York: Brooklyn Botanic Garden, 1968.

Bunt, A. C. *Modern Potting Composts: A Manual on the Preparation and Use of Growing Media for Pot Plants.* University Park, Pennsylvania: Pennsylvania State University Press, 1976.

Carpenter, Phillip, Theodore Walker, and Frederick Lanphear. *Plants in the Landscape.* San Francisco: W. H. Freeman & Co. Publishers, 1975.

Clements, Julia. *Flowers in Praise: Church Festivals and Decorations.* London: B. T. Batsford, 1977.

Crockett, James, and the editors of Time-Life Books. *Flowering Shrubs.* New York: Time-Life Books, 1972.

Cutler, Katherine. *Flower Arranging for All Occasions.* Garden City, New York: Doubleday Publishing Co., 1981.

Daniel, William, and R. P. Freeborg. *Turf Manager's Handbook.* Cleveland: Harvest Publishing Co., 1979.

Dietz, Marjorie. *Landscaping and the Small Garden.* New York: Doubleday Publishing Co., 1973.

Edmonds, John. *Container Plant Manual.* London: Grower Books, 1980.

Emmons, Robert. *Turfgrass Science and Management.* Albany, New York: Delmar Publishers Inc., 1984.

Evans, L. T. *Daylength and the Flowering of Plants.* Menlo Park, California: W. A. Benjamin, 1975.

Free, Montague, and Marjorie Dietz, revisor and editor. *Plant Propagation in Pictures.* Garden City, New York: Doubleday Publishing Co., 1979.

Fretz, Thomas, Paul Read, and Mary Peele. *Plant Propagation Lab Manual.* Minneapolis: Burgess Publishing Co., 1979.

Furuta, Tokiyi. *Environmental Plant Production and Marketing.* Arcadia, California: Cox Publishing Co., 1976.

Hackett, Brian. *Planting Design.* New York: McGraw-Hill Inc., 1979.

Hennebaum, Leroy. *Landscape Design: A Practical Approach.* Reston, Virginia: Reston Publishing Co., 1981.

Hartmann, Hudson, Dale Kester, and Fred Davis. *Plant Propagation: Principles and Practices.* Englewood Cliffs, New Jersey: Prentice-Hall Inc., 1990.

Hixson, Bill. *Flower Arranging.* Tokyo: Keisen Engei Center, 1971.

Hope, Frank. *Turf Culture: A Complete Manual for the Groundsman.* Poole, Dorset: Blanford Press, 1978.

Hutchinson, William. *Plant Propagation and Cultivation,* Westport, Connecticut: AVI Publishing Co., 1980.

Ingels, Jack. *Landscaping: Principles & Practices.* New York: Delmar Publishers, Inc., 1992.

MacQueen, Sheila. *Sheila MacQueen's Complete Flower Arranging.* New York: Times Books, 1980.

Mastalerz, John, ed. *Bedding Plants: A Manual on the Culture of Bedding Plants as a Greenhouse Crop.* University Park, Pennsylvania: Pennsylvania Flower Growers, 1985.

McDaniel, Gary. *Floral Design and Arrangement.* Reston, Virginia: Reston Publishing Co., 1981.

Melby, Pete. *Simplified Irrigation Design*. Mesa, Arizona: PDA Publishers, 1988.

Menninger, Edwin. *Color in the Sky: Flowering Trees in Our Landscapes*. Stuart, Florida: Horticulture Books, 1975.

Menninger, Edwin, et. al. *Flowering Vines of the World*. New York: Hearthside Press, 1970.

Mierhof, Annette. *The Dried Flower Book: Growing, Picking, Drying, Arranging*. New York: E. P. Dutton Inc., 1981.

Motloch, John. *Introduction to Landscape Design*. New York: Van Nostrand Reinhold, 1990.

Nelson, William. *Landscaping Your Home*. Urbana, Illinois: Cooperative Extension Service, University of Illinois, 1975.

Niemczyk, Harry. *Destructive Turf Insects*. Fostoria, Ohio: Gray Printing Company, 1981.

Orcutt, Georgia. *Successful Planters*. Farmington, Michigan: Structures Publishing Company, 1977.

Penn, Cordelia. *Landscaping with Native Plants*. Winston-Salem, North Carolina: John F. Blair, Publisher, 1982.

Pool, Mary, ed. *20th Century Decorating, Architecture and Gardens: 80 Years of Ideas And Pleasure from House and Garden*. New York: Holt, Rinehart & Winston, 1980.

Poole, Hugh, and Dennis Pittenger. *Care of Foliage and Flowering Plants for Retail Outlets*. Columbus, Ohio: Ohio State University Press, 1980.

Prockter, Noel. *Simple Propagation: A Book of Instructions for Propagation by Seed, Division, Layering, Cuttings, Budding and Grafting*. New York: Scribner Book Companies Inc., 1977.

Robinette, Gary, ed. *Landscape Planning for Energy Conservation*. Reston, Virginia: Environmental Design Press, 1977.

Robinette, Gary. *Plants, People and Environmental Quality*. Washington, D.C.: U.S. Department of the Interior, 1972.

Sprague, Howard. *Turf Management Handbook: Good Turf for Lawns, Playing Fields, and Parks*. Danville, Illinois: Interstate Printers and Publishers, 1982.

Still, Steven. *Herbaceous Ornamental Plants*. Champaign, Illinois: Stipes Publishing Company, 1988.

Taloumis, George. *Winterize Your Yard and Garden*. Philadelphia: J. B. Lippincott Co., 1976.

Toogood, Alan. *Propagation*. New York: Stein and Day, 1981.

Turgeon, Alfred J. *Turfgrass Management*. Englewood Cliffs, New Jersey: Prentice-Hall, 1991.

Vargas, J. M. *Management of Turfgrass Diseases*. Minneapolis: Burgess Publishing Co., 1981.

Viertel, Arthur. *Trees, Shrubs and Vines: A Pictorial Guide to the Ornamental Woody Plants of the Northern United States Exclusive of Conifers*. Syracuse, New York: Syracuse University Press, 1970.

Walker, Theodore. *Planting Design*. New York: Van Nostrand Reinhold, 1991.

Watkins, James. *Turf Irrigation Manual: The Complete Guide to Turf and Landscape Sprinkler Systems*. Dallas: Telsco Industries, 1977.

Wetherell, D. F. *Introduction to In Vitro Propagation*. Wayne, New Jersey: Avery Publishing Group Inc., 1982.

Wyman, Donald. *Wyman's Gardening Encyclopedia*. New York: Macmillan Publishing Co. Inc., 1977.

The Professions of Ornamental Horticulture

American Association of Nurserymen. *American Standards for Nursery Stock*. Washington, D.C.: American Association of Nurserymen, 1973.

Davidson, Harold, and Roy Mecklenburg. *Nursery Management: Administration and Culture*. Englewood Cliffs, New Jersey: Prentice-Hall Inc., 1981.

Hannebaum, Leroy. *Landscape Operations: Management, Methods and Materials*. Englewood Cliffs, New Jersey: Prentice-Hall, 19XX.

Marshall, Lane. *Landscape Architecture: Guidelines to Professional Practice*. Washington, D.C.: American Society of Landscape Architects, 1981.

McDonald, Elvin. *Plants as Therapy*. New York: Praeger Publishers, 1976.

Pinney, John. *Beginning in the Nursery Business*. Chicago: American Nurseryman, 1985.

Pinney, John. *Your Future in the Nursery Industry*. New York: R. Rosen Press, 1982.

Simonds, John. *Landscape Architecture: A Manual of Site Planning and Design.* New York: McGraw-Hill Inc., 1983.

Stanley, John, and Alan Toogood. *The Modern Nurseryman.* Boston: Farber and Farber, 1981.

The Business of Ornamental Horticulture

Briggaman, Joan. *Small Business Record Keeping.* Albany, New York: Delmar Publishers Inc., 1983.

Cassell, Dana. *How to Advertise and Promote Your Retail Store.* New York: American Management Associates, 1983.

Chapman, Elwood. *Supervisor's Survival Kit.* Chicago: Science Research Associates, 1986.

Clegg, Peter, and Derry Watkins. *The Complete Greenhouse Book: Building and Using Greenhouses from Coldframes to Solar Structures.* Charlotte, Vermont: Garden Way Publishing Co., 1978.

Correll, Philip, and Jane Pepper, eds. *Energy Conservation in Greenhouses.* Newark, Delaware: University of Delaware Press, 1977.

Cypert, Samuel. *Writing Effective Business Letters, Memos, Proposals and Reports.* Chicago: Contemporary Books, 1983.

Diamond, Jay, and Gerald Pintel. *Introduction to Contemporary Business.* Englewood Cliffs, New Jersey: Prentice-Hall Inc., 1975.

Dobyns, Lloyd, and C. Crawford-Mason. *Quality or Else.* Boston: Houghton-Mifflin, 1991.

Evans, L. T. *Daylength and the Flowering of Plants.* Menlo Park, California: The Benjamin-Cummings Publishing Co., 1975.

Evered, James. *Shirt Sleeves Management.* New York: AMACOM, 1981.

Feucht, James, and Jack Butler. *Landscape Management.* New York: Van Nostrand Reinhold, 1988.

Haimann, Theodore, and Raymond Hilgert. *Supervision: Concepts and Practices of Management.* Cincinnati: South-Western Publishing Company, 1972.

Harris, O. Jeff. *Managing People at Work: Concepts and Cases in Interpersonal Behavior.* Santa Barbara: Wiley, 1976.

Hurd, R. G., and G. F. Sheard. *Fuel Saving in Greenhouses: The Biological Aspects.* London: Grower Books, 1981.

Jucius, Michael. *Personnel Management.* Homewood, Illinois: Richard D. Irwin Inc., 1975.

Loen, Ernest. *Personnel Management Guides for Small Business.* Washington, D.C.: Small Business Administration, 1974.

Lorentzen, John. *The Manager's Personnel Problem Solver: A Handbook of Creative Solutions to Human Relations Problems in Your Organization.* Englewood Cliffs, New Jersey: Prentice-Hall Inc., 1980.

Maedke, Wilmer, Mary Robek, and Gerald Brown. *Information and Records Management.* Encino, California: Glencoe Publishing Co., 1981.

Mastalerz, John. *The Greenhouse Environment: The Effect of Environmental Factors On the Growth and Development of Flower Crops.* New York: John Wiley & Sons Inc., 1977.

Mills, Kenneth, and Judith Paul. *Successful Retail Sales.* Englewood Cliffs, New Jersey: Prentice-Hall Inc., 1979.

Myers, Marvin. *Managing Without Unions.* Reading, Massachusetts: Addison-Wesley Publishing Co., 1976.

National Greenhouse Manufacturers Association. *Standards: Design Loads in Greenhouse Structures; Ventilating and Cooling Greenhouses; Greenhouse Heat Loss.* The Association, 1981.

Nelson, Kennard. *Flower and Plant Production in the Greenhouse.* Danville, Illinois: Interstate Printers and Publishers, 1978.

Nelson, Paul. *Greenhouse Operation and Management.* Englewood Cliffs, New Jersey: Prentice-Hall, 1990.

Ocko, Judy, and M. L. Rosenblum. *Advertising Handbook for Retail Merchants: You're the Secret Ingredient in Good Ads.* New York: National Retail Merchants Association, 1981.

Place, Irene, and David Hyslop. *Records Management: Controlling Business Information.* Reston, Virginia: Reston Publishing Co., 1982.

Reynolds, Robert, and W. R. Luckham. *Business Management Techniques for Nurserymen: A Practical Guide to Decision Making.* Reston, Virginia:

Environmental Design Press, 1979.

Steinbrunner, Marion. *Greenhouses: From Design to Harvest.* Blue Ridge Summit, Pennsylvania: Tab Books, 1982.

Sullivan, Glenn, Jerry Robertson, and George Staby. *Management for Retail Florists: With Applications to Nurseries and Garden Centers.* San Francisco: W. H. Freeman & Co. Publishers, 1980.

Tenner, Arthur, and Irving DeToro. *Total Quality Management.* Reading, Massachusetts: Addison-Wesley, 1992.

Wells, Walter. *Communications in Business.* Boston: Kent Publishing Co., 1981.

Whitcomb, Carl. *Production of Landscape Plants.* Stillwater, Oklahoma: Lacebark Publications, 1987.

Index